RADIOSTERILIZATION OF MEDICAL PRODUCTS 1974

The following States are Members of the International Atomic Energy Agency:

AFGHANISTAN
ALBANIA
ALGERIA
ARGENTINA
AUSTRALIA
AUSTRIA
BANGLADESH
BELGIUM
BOLIVIA
BRAZIL
BULGARIA
BURMA
BYELORUSSIAN SOVIET SOCIALIST REPUBLIC
CANADA
CHILE
COLOMBIA
COSTA RICA
CUBA
CYPRUS
CZECHOSLOVAKIA
DEMOCRATIC PEOPLE'S REPUBLIC OF KOREA
DENMARK
DOMINICAN REPUBLIC
ECUADOR
EGYPT
EL SALVADOR
ETHIOPIA
FINLAND
FRANCE
GABON
GERMAN DEMOCRATIC REPUBLIC
GERMANY, FEDERAL REPUBLIC OF
GHANA
GREECE
GUATEMALA
HAITI
HOLY SEE
HUNGARY
ICELAND
INDIA
INDONESIA
IRAN
IRAQ
IRELAND
ISRAEL
ITALY
IVORY COAST
JAMAICA
JAPAN
JORDAN
KENYA
KHMER REPUBLIC
KOREA, REPUBLIC OF
KUWAIT
LEBANON
LIBERIA
LIBYAN ARAB REPUBLIC
LIECHTENSTEIN
LUXEMBOURG
MADAGASCAR
MALAYSIA
MALI
MAURITIUS
MEXICO
MONACO
MONGOLIA
MOROCCO
NETHERLANDS
NEW ZEALAND
NIGER
NIGERIA
NORWAY
PAKISTAN
PANAMA
PARAGUAY
PERU
PHILIPPINES
POLAND
PORTUGAL
ROMANIA
SAUDI ARABIA
SENEGAL
SIERRA LEONE
SINGAPORE
SOUTH AFRICA
SPAIN
SRI LANKA
SUDAN
SWEDEN
SWITZERLAND
SYRIAN ARAB REPUBLIC
THAILAND
TUNISIA
TURKEY
UGANDA
UKRAINIAN SOVIET SOCIALIST REPUBLIC
UNION OF SOVIET SOCIALIST REPUBLICS
UNITED KINGDOM OF GREAT BRITAIN AND NORTHERN IRELAND
UNITED REPUBLIC OF CAMEROON
UNITED STATES OF AMERICA
URUGUAY
VENEZUELA
VIET-NAM
YUGOSLAVIA
ZAIRE
ZAMBIA

The Agency's Statute was approved on 23 October 1956 by the Conference on the Statute of the IAEA held at United Nations Headquarters, New York; it entered into force on 29 July 1957. The Headquarters of the Agency are situated in Vienna. Its principal objective is "to accelerate and enlarge the contribution of atomic energy to peace, health and prosperity throughout the world".

Printed by the IAEA in Austria
June 1975

PROCEEDINGS SERIES

RADIOSTERILIZATION OF MEDICAL PRODUCTS 1974

PROCEEDINGS OF THE SYMPOSIUM ON
IONIZING RADIATION FOR STERILIZATION OF
MEDICAL PRODUCTS AND BIOLOGICAL TISSUES
HELD BY THE
INTERNATIONAL ATOMIC ENERGY AGENCY
AT BOMBAY, 9-13 DECEMBER 1974

INTERNATIONAL ATOMIC ENERGY AGENCY
VIENNA, 1975

RADIOSTERILIZATION OF MEDICAL PRODUCTS 1974
IAEA, VIENNA, 1975
STI/PUB/383
ISBN 92–0–010475–4

FOREWORD

Ionizing radiation is increasingly being employed for the sterilization of hermetically sealed pre-packed ready-for-use medical supplies, ranging from complicated instruments to sutures and from pharmaceutical starting materials to biological tissue preparations. So far, a total of about fifty commercial-scale gamma irradiation facilities and a lesser number of electron accelerators have been commissioned, and the radiation sterilization of medical products has already become a well established industrial process in some of the technologically advanced countries. In recent years several developing countries have also introduced radiation sterilization practices while a growing number are in the planning stage for installing this technology in the near future.

The many advantages of these practices over the conventional methods of sterilization by autoclaving, dry heat or ethylene oxide gas are an important reason for this rapid development. At present, next to power production, the radiation sterilization of public health-care supplies is one of the most important industrial applications of large radiation sources.

During the last ten years the IAEA has strongly encouraged this development, actively fostering the accumulation and dissemination of relevant information through its international meetings and co-ordinated research activities. The Agency's first symposium on this topic was held in Budapest in 1967. Since then a great deal of new technical information and practical experience has accumulated from the operation of radiation sterilization plants in the Member States as well as from relevant research activities on the applicability of this technique on a wider range of new heat-sensitive medical products.

The Agency's second symposium on Ionizing Radiation for Sterilization of Medical Products and Biological Tissues was held in Bombay in December 1974, to summarize and review critically all the current information and experience on the practices including the regulatory aspects of the processes involved and of the final products. One hundred scientists attended from 26 countries, including 18 developing countries, and two international organizations. The proceedings include papers on the microbiological control of the process, monitoring of the sterilizing efficiency of irradiation facilities, practical dosimetry systems, the establishment of sterilizing radiation dose, and the radiation sterilization of sutures, devices, pharmaceuticals and biological tissue preparations. Studies on the nature and the mechanisms of chemical, physical and immunological alterations of the medical products induced by the sterilizing radiation doses and the possible means of preventing them received particular attention.

With a reasonable balance of scientists from developing and developed countries it was possible to have an effective exchange of views on problems in different developing countries. Some of the discussions centred on the identification and analysis of the local factors, an essential starting-point when introducing this technology into developing countries having significantly different conditions of general hygiene, climate, manpower and economy. The Agency's Recommendations for Radiation Sterilization of Medical Products, a central point of discussion, was revised and up-dated in the light of new developments and experience since its original formulation in 1967.

It is hoped that the proceedings will serve as a guide and valuable source book for scientific and technical personnel engaged in radiation sterilization of medical products, as well as for radiation biologists, microbiologists, pharmaceutical quality control experts in the national pharmacopoeias, and experts in radiation dosimetry.

Sincere thanks and appreciation are due to the authorities of the Government of India for hosting the Agency symposium and providing the excellent facilities that helped greatly towards the success of the meeting.

EDITORIAL NOTE

The papers and discussions have been edited by the editorial staff of the International Atomic Energy Agency to the extent considered necessary for the reader's assistance. The views expressed and the general style adopted remain, however, the responsibility of the named authors or participants. In addition, the views are not necessarily those of the governments of the nominating Member States or of the nominating organizations.

Where papers have been incorporated into these Proceedings without resetting by the Agency, this has been done with the knowledge of the authors and their government authorities, and their cooperation is gratefully acknowledged. The Proceedings have been printed by composition typing and photo-offset lithography. Within the limitations imposed by this method, every effort has been made to maintain a high editorial standard, in particular to achieve, wherever practicable, consistency of units and symbols and conformity to the standards recommended by competent international bodies.

The use in these Proceedings of particular designations of countries or territories does not imply any judgement by the publisher, the IAEA, as to the legal status of such countries or territories, of their authorities and institutions or of the delimitation of their boundaries.

The mention of specific companies or of their products or brand names does not imply any endorsement or recommendation on the part of the IAEA.

Authors are themselves responsible for obtaining the necessary permission to reproduce copyright material from other sources.

CONTENTS

EFFECTS OF STERILIZING RADIATION DOSE ON THE CONSTITUENTS OF MEDICAL PRODUCTS (Session III)

APPLICATION OF RADIATION STERILIZATION TO MEDICAL PRODUCTS OF BIOLOGICAL ORIGIN (Session IV)

TECHNOLOGICAL ASPECTS OF RADIATION-STERILIZATION FACILITIES (Session V)

RADIATION STERILIZATION OF PHARMACEUTICAL SUBSTANCES (Session VI)

REPORTS ON CURRENT STATE OF RADIATION STERILIZATION OF MEDICAL PRODUCTS IN MEMBER STATES (Sessions VII and VIII)

REPORT OF WORKING GROUP ON THE REVISION OF THE IAEA RECOMMENDED CODE OF PRACTICE FOR RADIATION STERILIZATION OF MEDICAL PRODUCTS (Session IX)

MICROBIOLOGICAL CONTROL ASPECTS OF RADIATION STERILIZATION

(Session I)

Chairman: B.B. GAITONDE (India)

TESTS OF THE VALIDITY OF A MODEL RELATING FREQUENCY OF CONTAMINATED ITEMS AND INCREASING RADIATION DOSE

A. TALLENTIRE, A.A. KHAN
University of Manchester,
Manchester,
United Kingdom

Abstract

TESTS OF THE VALIDITY OF A MODEL RELATING FREQUENCY OF CONTAMINATED ITEMS AND INCREASING RADIATION DOSE.

The ^{60}Co radiation response of Bacillus pumilus E601 spores has been characterized when present in a laboratory test system. The suitability of test vessels to act as both containers for irradiation and culture vessels in sterility testing has been checked. Tests have been done with these spores to verify assumptions basic to the general model described in a previous paper. First measurements indicate that the model holds with this laboratory test system.

INTRODUCTION

A previous paper (1) set out the derivation of a theoretical model describing the dependence of the proportion of items contaminated (P) in a population of items on radiation dose (D). Employing certain assumptions, the model was evaluated for a range of doses, and curves relating P and D were constructed. The effect on curve shape of changing the initial average number of micro-organisms on items (a(0)) and of varying the parameters of microbial sensitivity to radiation was determined. From a consideration of these determinations came the concept that continuous monitoring of the microbiological quality of radiation sterilized items and the process of their manufacture might be achieved by observing the behaviour of P as a function of dose at levels below those used for sterilization (2). Applied to a practical situation this concept became known as 'the sub-process dose technique'.

Just how valuable this concept is in practice is difficult to assess without performing a large number of measurements of P on actual production items. Such an exercise is laborious and hardly justified at the present time when the concept is still being developed. Even so it is worth noting that the feasibility of applying sub-process dose techniques for monitoring the production of radiation sterilized disposable plastic syringes (3) and surgical dressings (4) has already been demonstrated.

In our view a next obvious step in the development of the concept is to test the validity of the theoretical model in circumstances where relevant variables are under control. This necessitates the use of a laboratory test system, which can be employed to simulate production items, consisting of specific micro-organisms, which respond in a predictable way to radiation, distributed in a known manner with a chosen frequency. In addition, the test system should preferably be of a form that these variables can be changed at will and that tests for the absence of surviving organisms can be readily done with a high degree of confidence. Using such a model comparisons can be made between expected probabilities of contaminated items (P) derived from evaluations of the theoretical model and observed proportions of contaminated items (P measured). The latter will be obtained by giving doses of radiation which are fractions of the sterilization dose to the laboratory test system possessing its precisely defined and known microbial loads.

The present paper describes experimental work done to test the assumptions basic to the model, together with our first attempts at verifying it using a laboratory contrived test system.

MATERIALS AND METHODS

The test organism

Spores of *Bacillus pumilus* E601 were used as the test organism. This organism was chosen because of its relatively low sensitivity to high energy radiation and its non-fastidious growth requirements (5,6). Additionally, from inspection of published data (7) it appeared that the response of *pumilus* spores to graded doses of radiation can be wholly described by a simple mathematical expression. Preliminary experiments showed that successive aliquots drawn from a single suspension of *B.pumilus* spores contained, within the limits of normal random sampling, the same number of viable spores per unit volume; this property held for suspensions ranging in concentration from 2×10^1 to 7×10^9 spores per aliquot. Furthermore, these spores are known to exhibit a highly reproducible response to γ-radiation, as measured by dose/survival data, when irradiated under standard conditions of test (8). Initially then it seemed that aliquots of a suspension of *B.pumilus* E601 spores could constitute the microbial load in a laboratory test system.

Spores were produced as a surface growth (9 days at 37°C) on 'Oxoid' Potato Dextrose Agar (code CM139). They were harvested in distilled water, washed 3x by centrifugation, resuspended in distilled water and held at 4°C. The viability of the stock suspensions, as measured by microscopic and plate counts, was about 90%, and over a period of one year the response of spores to γ-radiation under defined conditions of test was invariant.

The culture media for spores surviving irradiation

Dose-ln survival curves were constructed from estimates of numbers of spores surviving radiation doses made using a surface plating technique (9). 'Oxoid' Blood Agar Base (code CM55) was the plating medium.

Sterility tests were done with 'Oxoid' Nutrient Broth No.2 (code CM67) as the culture medium. It was prepared double strength.

According to the maker's specifications, the composition of Nutrient Broth No.2 is, with the addition of agar, identical with Blood Agar Base. Routinely 500ml lots of both media were prepared and sterilized by autoclaving at 121°C for 30min.

The irradiation facility

Irradiation was carried out in a ^{60}Co source of the 'HOTSPOT' type. Up to nine 2ml aliquots of spore suspension, contained in test vessels (q.v.) located in a specially constructed jig, could be irradiated simultaneously with a given radiation dose to a high degree of precision. The dose-rate was determined using ferrous sulphate dosimetry done in the same geometry as that used for irradiation of spore aliquots. At the beginning of this series of experiments the dose-rate was 6.42 krads/min. The degree of constancy of dose absorbed in aliquots using the jig is shown by the fact that with three successive replicate determinations, the variation within optical density values derived from 9 samples of dosimetric solution does not differ significantly from the variation between the 3 replicates. Replicate variation amounted to less than 1%, a value that is within the limits of error of this dosimetric method (10).

The test vessels

Two main requirements determined our choice of test vessels. These were (a) the need, on occasions, to submit the entire irradiated aliquot of spore suspension to a test for sterility, and (b) the need to minimise the frequency of occurrence of 'false positives' in such tests. Test vessels were 5ml clear glass vials (51mm long, 15mm diameter) possessing flanged necks. Each was closed with a latex rubber stopper, held firmly in position by means of a crimped aluminium collar. These vessels served both as containers for irradiation of spores in suspension and, when required, as culture vessels in tests conducted after irradiation with the object of revealing whether or not viable spores remained.

The irradiation of spores and the methods of scoring survivors

In generating data for construction of dose-ln survival curves, nine 2ml aliquots, drawn from a common suspension of spores in equilibrium with air, were sealed under air in vials. Each aliquot was then exposed, without stirring during irradiation, to a different dose of γ-radiation. The fraction of surviving spores was estimated in the usual way from plate counts performed on a suitable dilution of each irradiated aliquot and on a control aliquot that received no irradiation.

In tests designed to measure P (the proportion of contaminated items in a population of items given sub-process irradiation), aliquots of a suspension representing items of known microbial load were treated in a manner identical to that described above, except that all were exposed to the same dose of radiation. After irradiation, 2ml of double strength Nutrient Broth was injected into each vial using a repetitive dispensing syringe. Vials and contents were then incubated at 37°C for 7 days with periodic checks for microbial growth as indicated by the appearance of turbidity in the culture medium.

RESULTS AND DISCUSSION

Dose-ln survival curves

A prerequisite of evaluating P using the theoretical model is a knowledge of the expression relating the reduction in numbers of surviving organisms to dose. Initially then an examination was made of the survival characteristics of *B.pumilus* spores irradiated in the conditions in which they occur in the laboratory test system.

Dose/survival data collected over a period of 2 weeks from 74 aliquots drawn from a common spore suspension, are plotted in Fig.1. The response

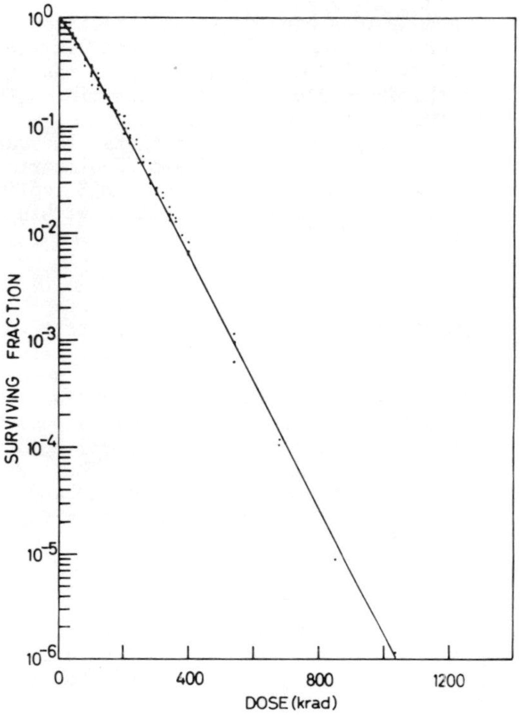

FIG.1. Semi-logarithmic plot of dose/survival data derived from ^{60}Co irradiation of B. pumilus spores in aqueous suspension at an initial concentration of 5×10^7 viable spores/ml. The curve is the fit of the expression $N/N(0) = 1-(1-e^{-kD})^n$ to the data and gives values of k and n of 0.0136 $krad^{-1}$ and 1.44 respectively.

traced by the points is typical of that seen with many species of spores; on a semi-logarithmic plot it takes the form of a small convex shoulder at high survival levels becoming essentially linear at low levels. This shape resembles closely the 'multiple hit' curve, usually expressed as

$$N/N(0) = 1-(1-\exp(-k\ D))^{n} \qquad (1)$$

where $N/N(0)$ is the fraction of micro-organisms surviving a radiation dose D, and k and n are constants. These constants can be used to characterise wholly the response of cell populations to irradiation, the value of n indicating the extent of the shoulder and the value of k the slope of the linear portion of the dose response curve.

Assuming that the survival data, depicted in Fig.1, satisfy expression (1) except for random variations, they were fitted to the 'multiple hit' curve by the method of Tyler and Dipert (11), employing a statistical weighting factor of 1. The fitted curve appears in Fig.1 as a continuous line. It is apparent that this curve describes closely the γ-radiation response of B.pumilus spores under present test conditions. The derived values of k and n are 0.0136 ± 0.0001 krad and 1.44 ± 0.06 respectively.

TABLE I. VALUES OF k AND n DERIVED FROM SURVIVAL DATA ORIGINATING FROM SEVEN REPLICATE EXPERIMENTS DONE OVER A PERIOD OF SIX MONTHS.

No.spores/ 2ml aliquot ($x10^{-7}$)	k ($x 10^3$) $krad^{-1}$	s.e.[a] of k ($x 10^3$)	$\log_e n$	s.e. of $\log_e n$	n
4.8	13.9	0.2	0.405	0.186	1.50
7.0	13.6	0.3	0.345	0.194	1.41
8.0	13.9	0.2	0.490	0.168	1.63
8.0	14.0	0.2	0.422	0.144	1.52
8.0	13.6	0.5	0.570	0.293	1.77
8.0	13.7	0.3	0.790	0.177	2.20
10.0	13.8	0.2	0.712	0.121	2.04

[a] s.e. = standard error.

Given in Table I are values of constants, k and n which have been calculated using data from experiments similar in design to that depicted in Fig.1. Each experiment was done employing between 7 and 9 aliquots drawn from one of four different spore suspensions. The good agreement amongst the tabulated values, and between these and values derived from Fig.1 provides evidence of the reproducibility and the stability of the response of the test organism.

One effect under examination in tests of the validity of the theoretical model is consequent upon variations in the average number of micro-organisms present prior to irradiation. In such tests, the initial average number/2ml aliquot should be fixed over as wide a range of values as practicable; this requirement determined our collection of dose/survival data for B.pumilus spores over the experimental limits of spore concentration. Fig.2 shows responses (k and n values) plotted against spore concentration. Clearly, response to radiation, measured in this manner, is constant for spore concentrations between 10^3 and 10^9/aliquot, but at concentrations above about 10^9, both constants take consistently low values. It is recalled that spores are irradiated in aqueous suspension that is initially in equilibrium with air and unstirred during irradiation. With these conditions, there is a real likelihood of oxygen depletion occurring on irradiation at high spore concentrations; this would result in a reduction in response. Experiments in which suspensions at 5×10^{10} spores/aliquot were gassed continuously with air during irradiation support this explanation, since k and n values are similar to those seen at low spore concentrations unstirred.

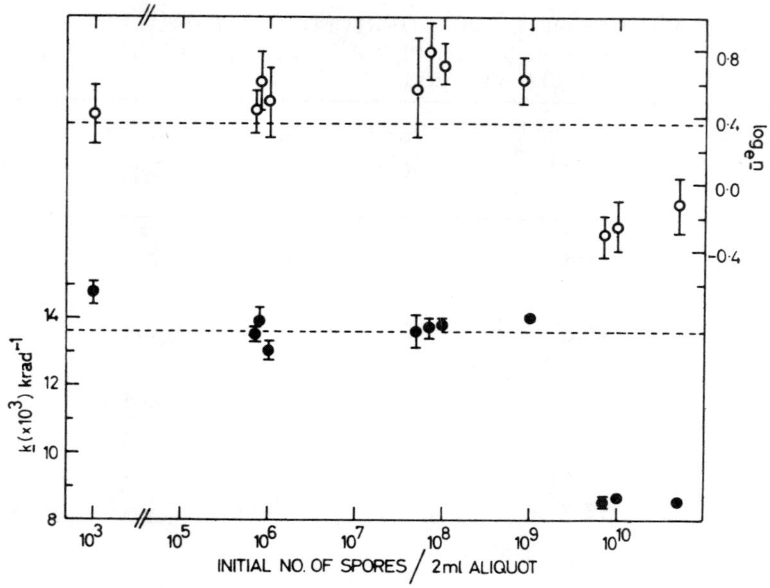

FIG.2. Radiation response, as measured by values of k and n, of B. pumilus spores as a function of initial spore concentration - suspensions unstirred during ^{60}Co irradiation. The dashed lines denote values of these constants derived from data given in Fig.1.

A consideration of the collected dose/survival curves has led to adoption of expression (1) for relating numbers of surviving B.pumilus spores in the laboratory test system and dose. For present predictions of $\underline{P}$, the proportion of contaminated aliquots after exposure to a radiation dose $\underline{D}$, expression (1) was included in the model with $\underline{k}$ and $\underline{n}$ taking values of 0.0136 krad^{-1} and 1.44 respectively, the only restriction being that $\underline{a}(0)$, the initial average number of spores/aliquot, is set below 10^9.

Assumption inherent in sterility testing

In the development of the model it was assumed that one or more micro-organisms, unirradiated or surviving a radiation treatment, will give rise to visible turbidity after incubation in liquid culture media. Our test of this assumption is based upon the fact that numbers of organisms present in successive samples drawn from a single suspension of low mean number/sample volume are distributed according to a Poisson series (12).

Suspensions were prepared by serial dilution to contain suitably low mean numbers of viable B.pumilus spores/0.1ml suspension. Estimates of means, obtained from plate counts performed on appropriate dilutions in the series showed that means fell between 0.1 and 2.7. Thirty 0.1ml samples were drawn in lots from suspensions and inoculated into separate 4ml quantities of sterile Nutrient Broth No.2. The frequencies of cultures showing growth after incubation at 37°C for 7 days were noted. Experiments similar in

TABLE II. OBSERVED AND EXPECTED FREQUENCIES OF 'POSITIVES' FOR SAMPLES DRAWN IN LOTS OF 30 FROM 8 DIFFERENT UNIRRADIATED SPORE SUSPENSIONS AND TESTED FOR STERILITY

Mean No.spores/0.1ml sample of suspension	Observed frequency (O)	Expected frequency (E)
0.10	2	2.856
"	4	"
"	4	"
"	3	"
"	1	"
"	2	"
"	4	"
"	3	"
"	2	"
"	1	"
0.20	9	5.439
"	6	"
"	3	"
"	8	"
"	9	"
"	6	"
"	6	"
"	5	"
"	2	"
"	4	"
0.71	17	15.252
0.78	20	16.248
0.90	17	17.805
2.12	27	26.400
2.34	28	27.111
2.70	28	27.984

design were also done with spores containing low means of viable irradiated spores. Doses of radiation, given to suspensions prior to diluting to mean values between 0.47 and 2.24/0.1ml, were sufficient to reduce the original population of spores by a factor of 10^{-1}, 10^{-4} or 10^{-5}. Tables II and III list the observed frequencies of growth (O) obtained with unirradiated and irradiated spore suspensions respectively, together with corresponding expected frequencies (E). Calculations of the values of E were based upon

$$E = (1 - p(0)) \qquad (2)$$

where p(0), the first term of the Poisson series (e^{-m}), is the probability that 0.1ml of suspension contains no viable spores; it was derived using the appropriate estimate of the mean number of spores/0.1ml suspension (m). Obviously, E represents the frequencies of growth originating from one or more viable spores and provided that O does not differ significantly from it, the original assumption is verified.

TABLE III. OBSERVED AND EXPECTED FREQUENCIES OF 'POSITIVES' FOR SAMPLES DRAWN IN LOTS OF 30 FROM 5 DIFFERENT IRRADIATED SPORE SUSPENSIONS AND TESTED FOR STERILITY

Mean No. spores/0.1ml sample of suspension	Observed frequency (O)	Expected frequency (E)
0.47[a]	11	11.250
"	12	"
"	13	"
"	7	"
"	10	"
"	12	"
"	12	"
"	11	"
"	9	"
"	9	"
0.67[a]	16	14.649
0.75[a]	15	15.831
2.02[b]	22	26.022
2.24[c]	28	26.808

[a] survivors of a dose sufficient for 5-log cycles of inactivation.
[b] survivors of a dose sufficient for 4-log cycles of inactivation.
[c] survivors of a dose sufficient for 1-log cycle of inactivation.

Two kinds of analyses have been done on the separate sets of data derived from sampling unirradiated or irradiated suspensions. The residuals (O-E) when plotted against expected values (E) are randomly scattered about a line with slope not significantly different from zero (i.e. there is no evidence of correlation between residuals and expected values). Where replicate estimates of observed frequencies (O) were made by withdrawal of 10 lots of samples from suspensions containing a mean number of viable spores of 0.1, 0.2 or 0.47/0.1ml, values of χ^2, derived from the statistic

$$\sum_{i=1}^{10} \frac{(O-E)^2}{E},$$

are not significant at $p = 0.05$. Thus all present evidence supports the original assumption. These findings also provide a clear demonstration that a survivor, as recognised by the formation of a visible colony following incubation of irradiated spores on the solidified agar medium, possesses an equal chance of recognition by appearance of a visible turbidity in the broth medium. This is a requirement of the present tests of the model, since

TABLE IV. VALUES OF PROPORTIONS OF CONTAMINATED ALIQUOTS (P measured) OBTAINED AFTER IRRADIATION IN LOTS OF 27, AND VALUES OF PROBABILITIES (P) PREDICTED BY THE MODEL

Initial average spores/aliquot a(0)	Dose (krad)	P (measured)	P
60	312.5	0.59, 0.78[a]	0.7
	370.0	0.37, 0.55	0.4
	425.0	0.26, 0.26	0.2
6000	650.0	0.59, 0.55	0.7
	720.0	0.37, 0.44	0.4
	775.0	0.29, 0.26	0.2

[a] Measurements derived from duplicate lots.

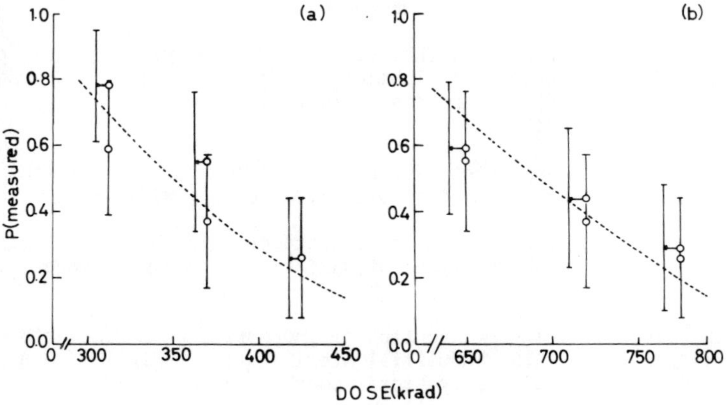

FIG.3. The fit of values of P(measured), taken from Table IV, to the curves predicted by the model relating probabilities of contaminated items (P) and radiation dose for initial average numbers of viable spores (a(0)) of (a) 60 and (b) 6000. The curves (dashed lines) were derived from the expression:

$$0 = \exp\left| -a(0)\{ 1-\exp(-0.0136D))^{1.44}\} \right|$$

where P = 1 - 0

The error bars about the values of P(measured), open circles, are 95% confidence limits.

measurements of P employ turbidity as the criterion of survival, while predictions of values of P generated from the model are based on survival data where the criterion is the formation of a colony.

Tests of validity of the model

Our first tests aimed at verifying the general theoretical model have been done with numbers of spores distributed in a normal manner in aliquots of suspensions containing on average 60 and 6000 spores/aliquot. Lots of 27 aliquots of a given average spore number, in the form of the laboratory test system, have each been given one of three graded doses of radiation. After irradiation, the contents of each vial were tested for sterility. Values of P (measured), the proportion of each lot of 27 vials that was contaminated, appear in Table IV, along with the evaluations of expected probabilities (P) of contaminated vials obtained from the model employing expression (i) and constants relevant to the system.

A visual comparison of the measured proportions of contaminated aliquots, P (measured), with probabilities (P) predicted by the model is given in Fig.3. Confidence limits about P (measured) were estimated from a normal approximation to the binomial sampling problem with a correction for discontinuity. The good agreement between measured and predicted values is taken as a clear indication that the general model, as conceived initially (1), holds for a laboratory contrived test system and is worthy of further development.

ACKNOWLEDGEMENT

This work was supported in part by a research award of the U.K. Panel on Gamma and Electron Irradiation.

REFERENCES

(1) TALLENTIRE, A., DWYER, J., LEY, F.J., Microbiological quality control of sterilized products: Evaluation of a model relating frequency of contaminated items with increasing radiation treatment, J.Appl. Bacteriol. 34 (1971) 521.

(2) TALLENTIRE, A., Aspects of microbiological control of radiation sterilization, Int. J. Radiat. Sterilization 1 (1973) 85.

(3) LEY, F.J., et al., Radiation sterilization: Microbiological findings from subprocess dose treatment of disposable plastic syringes, J.Appl. Bacteriol. 35 (1972) 53.

(4) WHITE, J.D.M.,"Biological control of industrial gamma radiation sterilization ", Ch.6, Industrial Sterilization (PHILLIPS, G.B., MILLER, W.S., Eds), Duke University Press, North Carolina (1973).

(5) PEPPER, R.E., BUFFA, N.T., CHANDLER, V.L., Relative resistances of micro-organisms to cathode rays, Appl.Microbiol. 4 (1956) 149.

(6) DARMADY, E.M., et al., Radiation sterilization, J.Clin.Path. 14 (1961) 55.

(7) BURT, M.M., LEY, F.J., Studies on the dose requirement for the radiation sterilization of medical equipment. I. Influence of suspending media, J.Appl.Bacteriol. 26 (1963) 484.

(8) SHAICKLY, M.A.S., M.Sc. Thesis, University of Manchester, Manchester (1971).

(9) TALLENTIRE, A., JACOBS, G.P., Radiosensitization of bacterial spores by ketonic agents of differing electron affinities, Int.J.Radiat.Biol. 21 (1972) 205.

(10) SPINKS, J.W.T., WOODS, R.J., An Introduction to Radiation Chemistry, Wiley, New York (1964).

(11) TYLER, S.A., DIPERT, M.H., On estimating the constants of the 'multi-hit' curve using a medium speed digital computer, Phys.Med.Biol. 7 (1962) 201.

(12) COWELL, N.D., Morisetti, M.D., Microbiological techniques - some statistical aspects, J.Sci.Fd.Agric. 20 (1969) 573.

DISCUSSION

K. H. CHADWICK: You have shown that with a laboratory system you can obtain a very reproducible radiobiological effect. Can you say how reproducible you expect the bacteriological test strips to be when used to test the efficiency of the radiation treatment in industrial sterilization plants in different parts of the world?

A. TALLENTIRE: In the present work we are dealing with a model bacteriological test system and the variability between responses to radiation from one occasion to another is very small. I refer you to Table I of my paper. In using bacteriological test pieces to measure the relative effectiveness of different sterilization facilities, I would expect a much greater variation than we see in our experiments. One significant source of variation in plant testing would be the dose itself and the precision with which the effective dose can be measured.

K. H. CHADWICK: Does that mean that the variation in the test strips used in plants could be related to imprecise dosimetry? I would agree that the dosimetry might not be precise, but my limited experience in radiobiology tells me that radiobiological sensitivity is affected by almost anything. Could this result from both variation in dosimetry and radiobiological sensitivity?

A. TALLENTIRE: Yes, I am sure there would be variations originating from both dosimetric and microbiological sources, and that these would contribute to the overall variability. However, I believe that the variation associated with certain of the bacteriological test strips now in routine use could be reduced if they were differently designed.

K. H. CHADWICK: In the Table showing k and n versus concentration, there is a drop in k and n at about 10^{10} cm^3. Is this purely a change in the shape of the curve or also a change in sensitivity?

A. TALLENTIRE: The drop in response seen at high spore concentrations is a change in sensitivity, which is reflected in changes in both k and n values. The lower responses are probably brought about by oxygen depletion in suspensions of high cell concentrations, as indicated by the fact that they assume regular values if the suspensions are continuously aerated during γ-irradiation.

K. SUNDARAM: Perhaps you could clarify two points. First, what was the medium in which the test organisms were irradiated, and, second, is your model system capable of accommodating the radiation dose variability which is encountered in actual practice? I mean by this that in your paper you describe only conditions whose k is constant, and n variable, or vice-versa.

A. TALLENTIRE: The washed microorganisms were irradiated while suspended in triple-distilled water. As regards your second point, the model could readily be modified to take account of variations in dose, and it is our intention to look into these effects at a later stage.

PRE-STERILIZATION CONTAMINATION OF DISPOSABLE MEDICAL PRODUCTS AND THE CHOICE OF MINIMUM STERILIZATION DOSE

V. HORÁKOVÁ, E. BURIÁNKOVÁ
SVUT Centre for Research and
Application of Ionizing Radiation, Brno,
Czechoslovakia

Abstract

PRE-STERILIZATION CONTAMINATION OF DISPOSABLE MEDICAL PRODUCTS AND THE CHOICE OF MINIMUM STERILIZATION DOSE.

The bacterial contamination was assessed on randomly taken samples of blood-transfusion devices, donor sets, intra-uterine contraceptive devices and inserters, surgical gloves and dressing material prior to sterilization. The quantitative and qualitative efficiency of six nutrient media was compared. The best results were obtained with the enriched "Universal" medium. It was confirmed that the contamination of plastic products was low compared with dressing material. Most frequently, Gram-positive aerobic spore-forming rods and Gram-positive cocci were found on non-sterile medical disposable products. A method was tested to obtain a general informative picture of the resistance of bacteria on products. The methods used for choosing the dose for radiation sterilization of medical products are discussed.

1. INTRODUCTION

The possibility of using ionizing radiation for sterilization of medical materials, especially disposable medical products, has particularly aroused the interest of the medical profession in Czechoslovakia. The first radiation plant, based on the principle of a ^{60}Co source in technological equipment manufactured by the Canadian firm Atomic Energy Ltd., has been in use since 1972, but it has not been possible to meet all the demands of manufacturers of disposable medical products, even though the costs of this sterilization means are higher compared with other available methods.

As sterilized products are used in human medicine, the management of such plants requires great responsibility, and the most objective criteria for evaluating the results of the sterilization process, the effectiveness of which depends on the correct choice of radiation dose, must be introduced. Many authors have discussed the choice and extent of minimum dose to be used [1-7]. Although views regarding the proper choice of sterilization dose vary, all emphasize the evaluation of pre-sterilization contamination, which can give a clear picture of the production hygiene and makes it possible to estimate the probable efficiency of the given sterilization process.

This paper contributes some facts about methods and criteria for evaluating radiation sterilization of medical products, especially with regard to determining pre-sterilization contamination and choice of sterilization dose.

2. MATERIALS AND METHODS

2.1. Determination of the pre-sterilization contamination

Pre-sterilization contamination was detected on packed non-sterilized samples of the following medical products: surgical gloves, blood-transfusion devices, VOR 02 donor sets, DANA-SUPER intra-uterine contraceptive devices (IUD) and inserters, DANA-COR IUDs, and SVUTIN dressing material. Two hundred test samples of these products were selected at random from the production line at the end of work-shifts during the last week of each annual quarter. The inner and outer surfaces of the items, including the inner side of the package, were tested for contamination.

Samples, depending on size, or irregular complicated shape, were divided into smaller parts and rinsed for 30 min in saline solution with 0.1% Tween 80. With the exception of the solutions from the flushed surgical gloves, the rinses were further filtrated through membrane filters, a Czech product SYNPOR (50 mm size, average pore size 0.3 μm). Microorganisms concentrated on filters were incubated on plates with blood agar for 4 d at a temperature of 35°C and a further 2 d at a temperature of 22°C.

The rinses from the surgical gloves were pipeted in 10-ml amounts each to sterile Petri dishes and mixed with 20 ml 45°C warm "Universal" medium enriched with 1% glucose [8]. The colonies were counted after 5 d cultivation at 35°C.

Gross colony-morphology and conventional microscopy of Gram-stained preparations were performed.

2.2. Comparison of quantitative and qualitative efficiency of various media

When the initial contamination of SVUTIN dressing material and VOR 02 donor sets was estimated, the quantitative and qualitative efficiencies for the following media were compared:

Blood agar (beef extract 0.3%, peptone 1%, NaCl 0.5%, agar 2%, defibrinated sheep blood 5%, pH 7.2);

"Universal" medium (beef heart extract 1000 ml, peptone 0.2%, NaCl 1%, gelatin 0.4%, yeast extract 0.6%, glucose 1%, agar 2%, pH 7.2);

Enriched "Universal" medium (above described medium with 5% defibrinated sheep blood added);

Dextrose yeast agar (yeast extract 0.5%, peptone 0.5%, dextrose 1%, agar 2%, pH 7.2);

Pads with nutrient agar, glucose and Chinese blue (product of Czechoslovak firm Imuna, n.p.);

Pads with malt extract agar (product of Czechoslovak firm Imuna, n.p.).

The membrane filters after filtration of rinses were placed on the above media in groups of the same number of samples. They were incubated for 6 d at a temperature of 35°C, with the exception of selective media for the isolation of yeasts and moulds that were kept at room temperature. Macroscopic and microscopic countings of results followed.

2.3. General informative estimation of radioresistance of germs on disposable products

After refiltering solutions from rinsed medical products to observe the pre-sterilization contamination, 50% of membrane filters were dried for

18 h at 35°C and 65-70% relative humidity. The dried filters were sealed in sterile polyethylene bags and irradiated in a ^{60}Co irradiator, Type J-6000 of the Atomic Energy of Canada, Ltd. at a dose of 2.5 Mrad. After irradiation the filters were transferred to plates with blood agar and incubated for 6 d at a temperature of 35°C. If any growth of colonies was observed, these were isolated, biochemically identified and their inactivation curves were determined.

3. RESULTS AND DISCUSSION

3.1. Choice of optimal method for determining pre-sterilization contamination, with special attention to nutrient media used

The frequently used nutrient media were compared to find the optimal environment, which would enable the widest spectrum of microorganisms present on disposable medical products to be isolated.

As can be seen from the initial contamination level of SVUTIN dressing materials (see Table I), the highest count was found on blood agar, where an average of 93 microorganisms was cultivated per product. The other media yielded substantially less favourable results: the average count per product determined on Dextrose yeast agar was 25.5, and on "Universal" medium 38.8. But when the "Universal" medium, proposed by Czerniawski and Kotelko [8], was supplemented with glucose and defibrinated sheep blood, it proved to be most effective. The results can be seen in Table II, where a comparison of three nutrient media is made for an assessment of pre-sterilization contamination of VOR 02 donor sets. The average count determined per item increased to 70.9 microorganisms, when enriched "Universal" medium was employed, whereas an average of only 27 was estimated with blood agar.

The evaluation of microscopic differences in the isolated types of germs showed that similar results could be obtained with blood agar and "Universal" medium. In the case of SVUTIN dressing material, aerobic spore-forming

TABLE I. PRE-STERILIZATION CONTAMINATION OF 'SVUTIN' DRESSING MATERIALS. COMPARISON OF DIFFERENT NUTRIENT MEDIA

No. of germs per item	% of tested products				
	BA[a]	UM[b]	DYA[c]	MA[d]	NGA[e]
0	0	0	0	0	0
1-10	0	0	10	50	45
11-100	67	96.7	86.7	50	55
101-500	33	3.3	3.3	0	0

[a] Blood agar. [b] "Universal" medium. [c] Dextrose yeast agar. [d] Malt extract agar. [e] Nutrient agar with glucose and Chinese blue.

TABLE II. PRE-STERILIZATION CONTAMINATION OF 'VOR 02' DONOR SETS. COMPARISON OF DIFFERENT NUTRIENT MEDIA

No. of germs per item	% of tested products		
	EUM[a]	BA[b]	SA[c]
0	0	0	8.9
1-10	26.9	31.3	74.7
11-100	62.7	64.2	8.9
101-500	8.9	4.5	7.5
501-1000	1.5	0	0

[a] Enriched "Universal" medium. [b] Blood agar. [c] Sabouraud agar.

TABLE III. MORPHOLOGICAL TYPES OF MICROORGANISMS ON 'SVUTIN' DRESSING MATERIALS BEFORE STERILIZATION

Morphological types of germs	% of isolated strains				
	BA[a]	UM[b]	DYA[c]	MA[d]	NGA[e]
Aerobic spore-forming rods	68.0	70.0	70.6	37.2	94.9
Gram-positive cocci	15.7	16.1	10.0	0	0
Gram-negative aerobic rods	0.8	0.7	0	0	0
Gram-positive aerobic rods	8.6	6.6	3.6	1.2	0
Moulds	2.6	1.3	3.6	50.0	5.1
Yeasts	4.3	5.3	12.2	11.6	0

[a] Blood agar. [b] "Universal" medium. [c] Dextrose yeast agar. [d] Malt extract agar. [e] Nutrient agar with glucose and Chinese blue.

rods were the most frequently isolated germs, representing 68% with blood agar and 70% with "Universal" medium. Further, Gram-positive cocci were found in 15.7, and 16.1% respectively, from the total count of isolated strains (see Table III). Other types of germs (Gram-negative and positive aerobic rods and fungi) were determined in a minority of cases. The results obtained from VOR 02 donor sets were also very similar with blood agar and enriched "Universal" medium. As can be seen from Table IV, the majority

TABLE IV. MORPHOLOGICAL TYPES OF MICROORGANISMS ON 'VOR 02' DONOR SETS BEFORE STERILIZATION

Morphological types of germs	% of isolated strains		
	EUM[a]	BA[b]	SA[c]
Aerobic spore-forming rods	15.6	18.5	5.6
Gram-positive cocci	54.0	54.6	4.0
Gram-negative aerobic rods	1.4	1.1	0
Gram-positive aerobic rods	23.7	22.8	13.4
Yeasts	1.0	0	22.2
Moulds	3.8	2.0	53.2
Actinomycetes	0.5	1.0	1.6

[a] Enriched "Universal" medium. [b] Blood agar. [c] Sabouraud agar.

of microorganisms were Gram-positive cocci (blood agar 54.6%, "Universal" medium 54%), Gram-positive rods (blood agar 22.8%, "Universal" medium 23.7%), and Gram-positive spore-forming strains (blood agar 18.5%, "Universal" medium 15.6%).

The other investigated media, mainly Sabouraud agar and malt extract agar, are selective and favour growth of the moulds and yeasts. They may be used as complementary media. On pads with nutrient agar, glucose and Chinese blue, spore-forming rods predominated, but even those were found in significantly smaller amounts than on other nutrient media. For this reason, this medium was judged unsuitable.

The choice of nutrient medium is one of the factors which greatly influence the routine microbiological evaluation of the pre-sterilization count levels. As stated by Christensen and co-workers [5], a suitable method for counting microorganisms on disposable equipment prior to sterilization should provide for the findings of even small numbers of staphylococci, aerobic and anaerobic spore-forming rods, micrococci, coryneform rods, streptococci and fungi. From this point of view the enriched "Universal" medium [8] seems to be suitable because it gives better results than media consisting only of tryptone, meat extract and dextrose. The "Universal" medium was recommended in the Collection of Regulations Recommending the Radiation Sterilization of Disposable Medical Equipment [9].

Practical experience has confirmed that the laboratory technique for counting germs on different medical products has to be chosen individually according to shape and consistency of examined items. The simple method of concentrating germs by means of membrane filters [3] cannot always be used. Our experiences with the surgical gloves are an example. We had to mix the immediate solution with the nutrient medium because the powder from the gloves blocked the pores of the membrane filters.

TABLE V. INITIAL COUNTS OF MICROORGANISMS ON SOME DISPOSABLE MEDICAL PRODUCTS

No. of germs per item	% of tested products						
	Surgical gloves	Blood transfusion devices	VOR 02 donor sets	DANA-SUPER IUDs	DANA-SUPER Inserters	DANA-COR IUDs	SVUTIN dressing material
0	76.4	12	0	10.8	8.1	0	0
1-10	21.6	56	31.3	78.4	35.1	54.5	0
11-100	2	32	64.2	10.8	47.3	41.8	67
101-500	0	0	4.5	0	5.4	3.7	33
501-2000	0	0	0	0	4.1	0	0

3.2. Results of pre-sterilization contamination on some medical disposable products

As reported in several publications [10-14], usually not many viable germs are found on disposable products, especially if production is carried out under good hygienic conditions. Our results, shown in Table V, prove this to be true, particularly when plastic instruments are considered. In the case of blood-transfusion devices, DANA-SUPER IUDs and inserters there were no microorganisms found in 8.1 to 12% of tested samples. In a significantly larger fraction of examined products, a contamination between 1 to 100 microbes was assessed. An average of 14 microorganisms per blood-transfusion device and the maximum 111 germs per single test-piece were found. Donor sets VOR 02 were contaminated on an average with 27 germs and the maximum found was 208. DANA-SUPER IUDs had an average count of 5.6 germs and a maximum of 66. DANA-COR IUDs were contaminated on an average by 28.6 microorganisms, and the maximum found on one item was 446. The average count determined for DANA-SUPER inserters was high compared with other disposable products mentioned, even though it is a simple device. The average of 84.6 germs per item was caused by a high number of microorganisms which were found only on three tested samples, where 950, 1420 and 2000 viable cells were noticed. Further investigation showed that the contamination occurred because there had been a change in the production technique and accidently a highly contaminated rinsing preparation had been used to decrease the inner surface friction. This experience is just another proof that it is necessary to control microbiological conditions of production, which can be competently checked by a bacteriologist [13, 15].

Extremely good results were obtained when the initial contamination of surgical gloves was estimated — 76.4% were found to be free of contamination, while of the remaining 23.6%, only 1 to 28 colonies were cultivated. One sample had an average of 1.4 germs. The favourable results were probably due not only to good hygienic production conditions, but to the partly anti-bacterial properties of rubber.

TABLE VI. MORPHOLOGICAL TYPES OF MICROORGANISMS ON SOME DISPOSABLE MEDICAL PRODUCTS

Morphological types of germs	% of isolated strains						
	Surgical gloves	Blood transfusion devices	VOR 02 donor sets	DANA-SUPER IUDs	DANA-SUPER Inserters	DANA-COR IUDs	SVUTIN dressing material
Gram-positive cocci	31.4	47.4	54.6	43.8	44.7	37.4	15.7
Aerobic spore-forming rods	28.6	28.1	18.5	15.0	17.5	37.4	68.0
Gram-negative aerobic rods	0	3.5	1.1	14.5	6.2	5.4	0.8
Gram-positive aerobic rods	0	14.0	22.8	10.0	12.9	3.5	8.6
Yeasts	0	0	0	5.3	7.2	7.0	4.3
Moulds and actinomycetes	40.0	7.0	3.0	11.4	11.5	9.3	2.6

On the other hand, the most unfavourable picture was observed in SVUTIN dressing material. In 67% of the examined samples, 1 to 100 bacteria were found per item, and in the remaining 33%, 100 to 460 bacteria were assessed. The average number of germs found per item was 93. This may be due to the fact that the basic raw materials used were already very contaminated, and further treatment does not have such an inactivation effect, as for instance the heat fabrication of plastic materials.

The analysis of the total picture of investigated types of medical disposable products clearly proves that the initial contamination increases with the number of manual operations as compared with semi-automation or even complete automation of production. An irregular distribution of singularly highly contaminated pieces can also be expected, as shown by our example of the DANA-SUPER inserters.

The morphological differences in the microorganisms observed on the medical disposable products have revealed some interesting data, as seen from Table VI. Plastic products were mostly contaminated with cocci (37.4 to 54.6%), whereas surgical gloves and dressings were contaminated with spore-forming germs (68.6% in surgical gloves and 70.6% in SVUTIN dressings). It is necessary to realize that particularly among these kinds of microorganisms were found the forms most resistant to radiation [16-18].

3.3. Estimate of microorganism resistance on medical products selected for radiation sterilization

To evaluate the efficiency of the sterilization process, it is not sufficient just to have a relatively wide knowledge about the pre-sterilization contamination of products to be sterilized, since another basic question arises which limits the sterilization process – the resistance of the contaminating microorganisms to the sterilizing medium. A general informative evaluation of resistance, which was tested on five different medical products, showed no occurrence of germs which would survive a dose of 2.5 Mrad. Igali [14], whose recommended method was used in this work, also published similar conclusions of experiments with syringe needles.

However, a weakness of this method lies in the fact that, to recover germs by rinsing and membrane filtration, the natural micro-environment of the products is disturbed. The microorganisms are inactivated naturally, influenced also by other factors, and this changes the picture of the actual inactivation effect of radiation. To avoid this problem it seems to be more important to focus attention on observing the influence of subnormal process doses, as suggested by some authors [10, 14, 17, 19].

3.4. Choice of dose

It is evident that, before the sterilization dose can be established for different kinds of medical products, it is necessary to know the contamination level prior to sterilization [20]. However, very often radiation plants are not directly connected to manufacturers but form independent economic units offering a sterilization service to different producers, which makes a regular daily control of production impossible. If the production hygiene is controlled by the manufacturer, the inactivation effect of radiation is not considered. For this reason we based the choice of the radiation dose on an examination of at least 200 samples once every three months. On these randomly selected samples, the pre-sterilization contamination and the general informative resistance to radiation were tested according to methods described in sections 2.1 and 2.3. The possibility of seasonal and periodical variations was considered, and the test-pieces were withdrawn from normal production at the end of work-shifts [6, 11].

If the average initial count did not exceed 100 germs per item, with a possible maximum up to 500, a minimum radiation dose of 2.5 Mrad was used [21]. The microbiological control of the sterilization process by means of biological indicators, as well as tests of efficiency of the radiation facility, were performed throughout [5, 20]. Up to now we have received no complaints of non-sterility of disposable medical products sterilized by a minimum 2.5-Mrad dose.

A sudden increase of initial counts or occurrence of resistant microorganisms cannot be excluded when testing random samples once every three months. It does provide a certain potential hazard. The number of samples to be examined in order to obtain an adequate picture also remains a problem limiting the reproducibility of results.

Czerniawski and Stolarczik [7] suggested establishing the sterilization dose on the basis of estimating resistance of at least 1000 strains isolated from disposable products, production environment, and people directly involved in the manufacture. To calculate the minimum dose, the initial counts of microorganisms on products, the resistance of isolated strains expressed in D-10 values, and the required margin of safety, were considered. The practical tests were carried out in a factory producing syringe needles, where 8314 strains were isolated and tested for resistance. It was found that 7.81% of the strains survived in a fraction of 10^{-8} of doses 2 to 4 Mrad. The authors calculated that products contaminated on an average of

100 germs per item had to be sterilized by a 3-Mrad dose if a safety margin of 10^{-6} was required. This method [7] evaluates the pre-sterilization situation by a complex means. But it is laborious regarding the quantity of microbiological tests. From this aspect it is more promising to examine the model developed by Tallentire et al. [19] and tested in practice by Ley et al. [17].

An evaluation of sterilization efficiency must focus on the microbiological situation before the sterilizing process. However, the chosen methods should be realistic for routine practice without needlessly increasing the costs of radiation sterilization.

REFERENCES

[1] LEY, F.J., TALLENTIRE, A., Pharm. J. 195 (1965) 216.
[2] CHRISTENSEN, E.A., HOLM, N.W., JUUL, F., The Basis of the Danish Choice of Dose for Radiation Sterilization of Disposable Equipment, Risø Rep. No.140 (1966).
[3] CHRISTENSEN, E.A., MUKHERJI, S., HOLM, N.W., Microbiological Control of Radiation Sterilization of Medical Supplies. I. Total Count on Medical Products (Disposable Syringes and Donor Sets) Prior to Radiation Sterilization, Risø Rep. No.122 (1968).
[4] CHRISTENSEN, E.A., KALLINGS, L.O., FYSTRO, D., Microbiological control of sterilization procedures and standards for sterilization of medical equipment, Ugeskr. Laeger. 131 (1969) 2123.
[5] CHRISTENSEN, E.A., in Manual on Radiation Sterilization of Medical and Biological Materials, Tech. Rep. Ser. No.149, IAEA, Vienna (1973) 131.
[6] TUMANYAN, M.A., RAKITSKAJA, G., NIKOLAJEVA, L., ANDREJEVA, G., PAVLOV, E., "Microbiological fundaments of the radiation sterilization", CMEA Conf. on Radiation Sterilization, Baku (1972).
[7] CZERNIAWSKI, E., STOLARCZIK, L., "Microbiological questions of radiation sterilization. The choice of sterilization dose", (in Russian) presented at CMEA Working Group Meeting, Problems of Radiation Sterilization, Suchumi (1973).
[8] CZERNIAWSKI, E., KOTELKO, K., "Suitability of the 'Universal' nutrient medium for the microbiological control", (in Russian), CMEA Conf. Radiation Sterilization, Baku (1972).
[9] CMEA, "Collection of regulations recommending the radiation sterilization of disposable medical equipment" (in Russian), CMEA Standing Commission for Peaceful Uses of Atomic Energy, Moscow (1973).
[10] TATTERSALL, K., "Problems of microbial contamination in prepacked preparations", Ionizing Radiation and the Sterilization of Medical Products (Proc. Symp. Risø 1964), Taylor and Francis, London (1965) 15.
[11] COOK, A.M., BERRY, R.J., "Pre-sterilization contamination on disposable hypodermic syringes: Necessary information for the rational choice of dose for radiation sterilization", Radiosterilization of Medical Products (Proc. Symp. Budapest, 1967), IAEA, Vienna (1967) 295.
[12] LEY, F.J., CRAWFORD, C.G., KELSEY, J.C., "Report from the United Kingdom", Radiosterilization of Medical Products, Pharmaceuticals and Bioproducts, Techn. Rep. Ser. No.72, IAEA, Vienna (1967) 40.
[13] CHRISTENSEN, E.A., HOLM, N.W., JUUL, F.A., "Radiosterilization of medical devices and supplies", Radiosterilization of Medical Products (Proc. Symp. Budapest, 1967), IAEA, Vienna (1967) 265.
[14] IGALI, S., " The assessment of the initial contamination of medical disposable products, and the connection of the radiation resistance and the sterilization dose" (in Russian), CMEA Conf. Radiation Sterilization, Baku (1972).
[15] WIGERT, R., UNGER, G., "Proposals for methodical control of the hygienic state of the production of disposable supplies sterilized by radiation" (in Russian), presented at CMEA Working Group Meeting on the Problems of Radiation Sterilization, Suchumi (1973).
[16] CHRISTENSEN, E.A., "Radiation Resistance of Bacteria and the Microbiological Control of Irradiated Medical Products", Sterilization and Preservation of Biological Tissues by Ionizing Radiation (Proc. Panel, Budapest, 1969), IAEA, Vienna (1970) 1.
[17] LEY, F.J., WINSLEY, B., HARBORD, P., KEALL, A., SUMMERS, T., J. Appl. Bacteriol. 35 (1972) 53.
[18] LEY, F.J., in Manual on Radiation Sterilization of Medical and Biological Materials, Techn. Rep. Ser. No.149, IAEA, Vienna (1973) 37.
[19] TALLENTIRE, A., DWYER, J., LEY, F.J., J. Appl. Bacteriol. 34 (1971) 521.

[20] INTERNATIONAL ATOMIC ENERGY AGENCY, "Recommended code of practice for radiosterilization of medical products", Radiosterilization of Medical Products (Proc. Symp. Budapest 1967), IAEA, Vienna (1967) 423.

[21] HORÁKOVÁ, V., "The choice of the sterilization dose for disposable medical supplies" (in Russian), CMEA Conf. Radiation Sterilization, Baku (1972).

DISCUSSION

F. FERNANDES: In your attempt to establish the pre-sterilization count you limited your study to the determination of aerobic bacteria. Why have the anaerobes been excluded?

Vlasta HORÁKOVÁ: We performed pre-sterilization bacterial counts of samples made mainly of plastic materials. The conditions applied, i.e. shape and type of packaging, were unfavourable for the survival of anaerobes. In view of this we disregarded the presence of anaerobic microorganisms.

P.V. IYER: Do you submit reports to the manufacturers, either periodically or on a routine basis, on the results of your tests with pre-sterilized samples of their products, so as to enable them to improve manufacturing conditions and procedures?

Vlasta HORÁKOVÁ: Yes, we do. In addition, we try to induce manufacturers to set up new laboratories and to organize routine control.

SYNERGISTIC EFFECTS OF HEAT AND IRRADIATION TREATMENT (THERMORADIATION) IN THE STERILIZATION OF MEDICAL PRODUCTS

C.A. TRAUTH, Jr.
Biosystems Studies Division

H.D. SIVINSKI
Biosystems Research Department,
Sandia Laboratories,
Albuquerque, N.Mex.,
United States of America

Abstract

SYNERGISTIC EFFECTS OF HEAT AND IRRADIATION TREATMENT (THERMORADIATION) IN THE STERILIZATION OF MEDICAL PRODUCTS.

This paper describes a generic class of sterilization processes in which properly chosen combinations of radiation and heat synergistically inactivate many bacteria and viruses. Treatments with optimal combinations are shown to offer the possibility of using lower total doses and lower temperatures than would be required separately for sterilization. This results from easier elimination of heat-labile, radioresistant organisms and radiolabile, heat-resistant organisms, and from synergistic inactivation of organisms which are both radioresistant and heat resistant. These processes depend upon temperature, dose-rate, and time in fairly complex ways; therefore, an analytical framework in which they can be defined is also presented.

1. INTRODUCTION

In the sterilization of medical products, populations of biological systems can be inactivated by sufficient exposure to either heat or ionizing radiation; indeed, these agents are probably involved in the majority of commercial pasteurization and sterilization processes in use today. What may be less well known is that an appropriately selected combination of heat and ionizing radiation may offer appreciable advantages as a pasteurizing or sterilizing treatment over either agent used separately. This occurs for two reasons. First, a properly chosen combination treatment sterilizes both heat-labile and radiolabile organisms rapidly. Because there are radioresistant organisms that are relatively heat-labile, and heat-resistant organisms that are relatively radiolabile, this in itself can offer considerable advantage. The organisms of greatest concern, then, are those that are both heat-resistant and radioresistant. The second reason for the potential advantage of combinations is that they may be chosen to inactivate organisms in this last class in a synergistic way. In this paper, (1) it is demonstrated that such a synergism exists when dose-rate and temperature are properly chosen, (2) some of the properties of synergistic inactivation are described, and (3) a means for designing sterilization cycles in which radiation and heat combinations are used is presented.

By way of historical perspective, prior to the mid-1960s, a number of investigators had observed that some combinations of heat and ionizing radiation inactivate biological systems at a rate greater than the sum of the independent thermal and radiation inactivation rates. Such a synergism had been observed in a variety of biological systems, including active proteins [1-8], viruses [9], spores [10-12], bacteria [13, 14], yeast [15], mammalian cells [16, 17], and human cancers [18-21]. During this same period, other investigators were unable to find evidence of temperature effects in radiation sterilization. For example, in Ref. [22] it is reported that the radiation target sensitivity of øX174 phage is independent of temperature; in a later study [23], a temperature dependence was observed. As will be seen below, both reports are correct. However, much better understanding of the way in which the existence of a synergism between heat and radiation depends on process parameters has recently been achieved.

Section II summarizes much of the data, gathered since 1968 at Sandia Laboratories, supporting the conclusion that, when properly chosen, combinations of heat and radiation act synergistically to inactivate biological systems. In addition, data relating to some of the properties of such combinations are presented. Section III presents an analytical framework which generally describes the effect of temperature on radiation-sterilization cycles. As will become apparent, such a framework is an absolute necessity if the application of heat and radiation combinations in the sterilization of medical devices is to be considered.

It should be remarked that the measure of the thermal influence on radiation effects will generally be the synergism evidenced in combined treatments involving both. This is illustrated in a hypothetical way in Fig. 1. Here

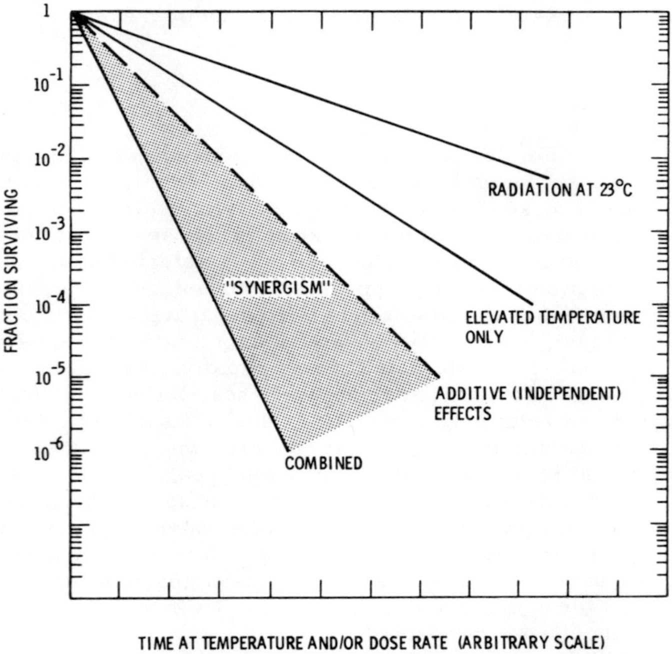

FIG. 1. Illustration of synergistic response to simultaneous application of heat and ionizing radiation.

it is imagined that radiation at some fixed dose-rate and ambient temperature yields a logarithmic survivor curve, as indicated. Heat at a fixed elevated temperature is assumed to do likewise. If heat and radiation effects are independent phenomena, simultaneous application of these treatments should be additive, so that the dashed survivor curve in Fig. 1 results. When the actual survivor curve differs significantly from the dashed curve in a combined environment, this behaviour is ascribed to thermal influence on the radiation effects. Because, in all the examples, the combined effects are greater than the hypothetical additive effects, the discussion of thermal influence in radiosterilization is often stated in terms of this synergistic response as it is illustrated in Fig. 1.

2. EXISTENCE OF HEAT-ACCELERATED RADIATION EFFECTS

Most of the early work done at Sandia Laboratories was performed with dry Bacillus subtilis var. niger spores by M. C. Reynolds and various co-workers. To them must go considerable credit for a thoroughly comprehensive experimental programme in which temperature, dose-rate, and various environment parameters were investigated over rather broad ranges.

Figure 2 presents some early data [24] on the inactivation of dry B. subtilis spores at 125°C and a dose-rate of about 50 krad/h. As can be seen, this temperature alone provides a reasonable sterilization regime, and

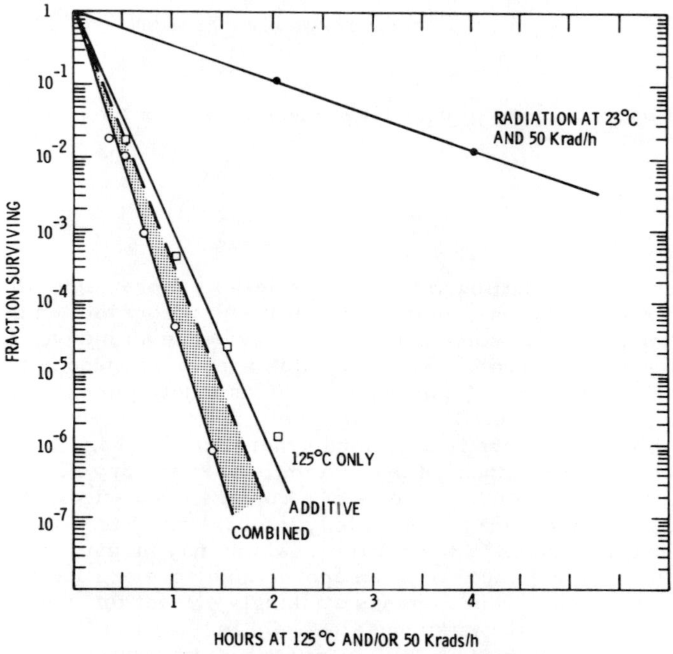

FIG. 2. Inactivation of dry Bacillus subtilis var. niger spores at a dose-rate of 50 krad/h and 125°C separately and simultaneously.

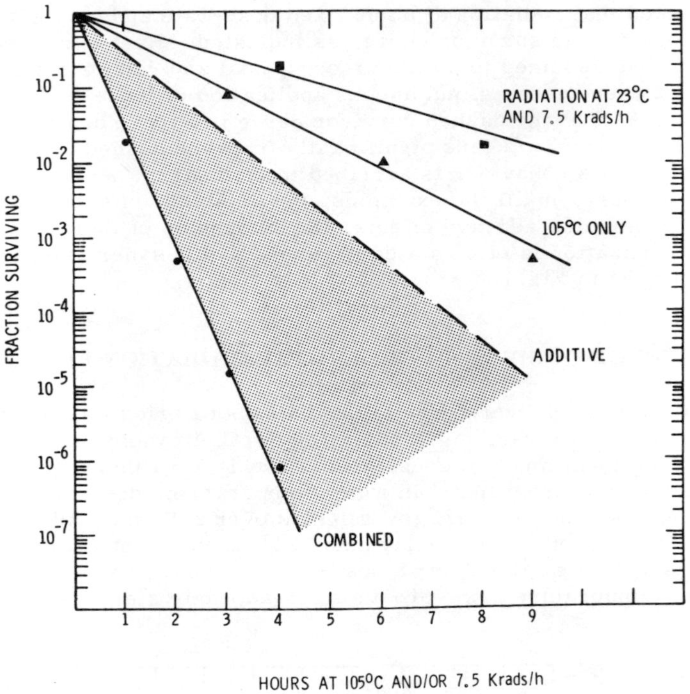

FIG.3. Inactivation of dry Bacillus subtilis var. niger spores at a dose-rate of 7.5 krad/h and 105°C separately and simultaneously.

its addition to the radiation yields a considerably more rapid sterilization cycle than radiation alone, most of which results from the temperature itself. Nevertheless, there is some synergism in evidence as indicated in the figure. By lowering the temperature to 105°C, this synergism becomes more evident [25], as is shown in Fig. 3 for dry B. subtilis inactivation at about 7.5 krad/h. Here again, the heat sterilization, by itself, is more effective than the radiation; however, there is a decided synergism. In Fig. 4, at 95°C and about 50 krad/h, a pronounced synergism is seen for dry B. subtilis spores in which the heat, itself, plays less of a role than radiation [26]. While the data in Figs 2, 3 and 4 may be similarly interpreted, Fig. 4 most strongly indicates that increases in temperature (which may or may not have much effect by themselves) can lead to an appreciable acceleration of radiation-sterilization cycles. This becomes strikingly evident for unidentified soil spores with a high heat resistance at 125°C, as shown in Fig. 5 [27, 28]. In this case, there is essentially no direct heat effect; however, the heat in combination with ionizing radiation yields a treatment exceedingly more effective than the radiation alone (at ambient temperature).

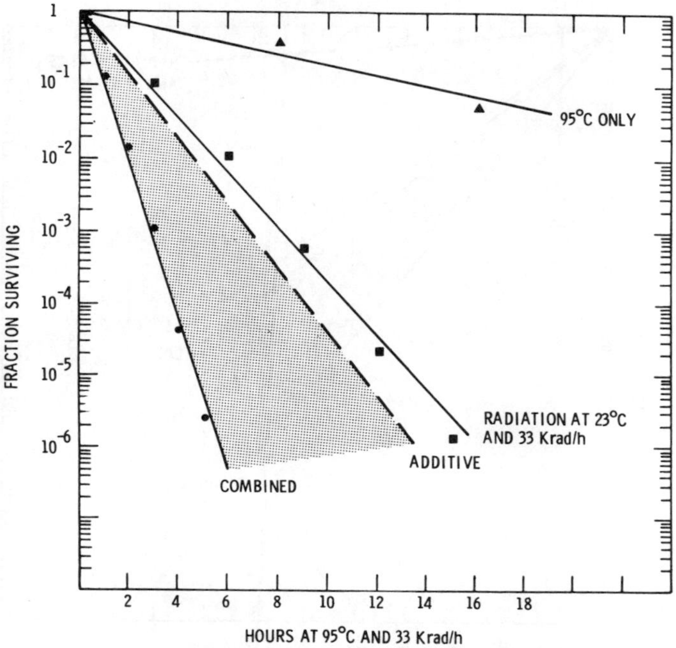

FIG.4. Inactivation of dry Bacillus subtilis var. niger spores at a dose-rate of 33 krad/h and 95°C separately and simultaneously.

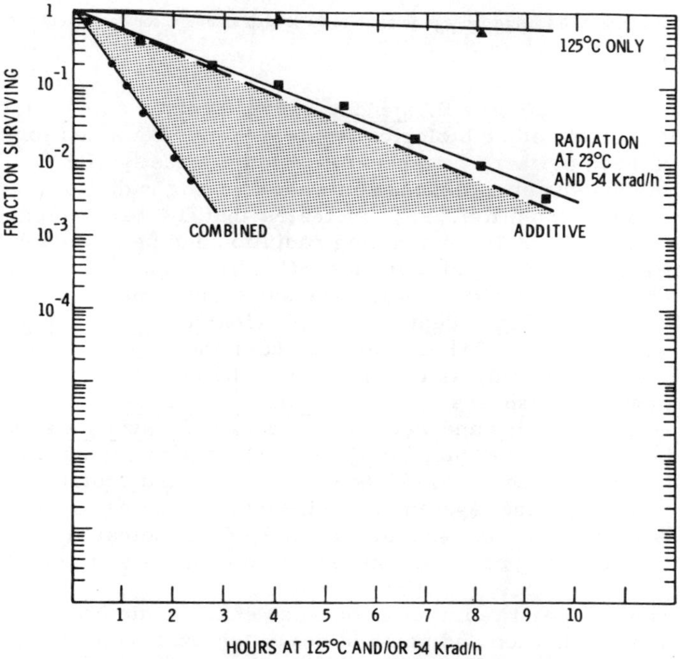

FIG.5. Inactivation of dry, unidentified heat-resistant soil spores at a dose-rate of 54 krad/h and 125°C separately and simultaneously.

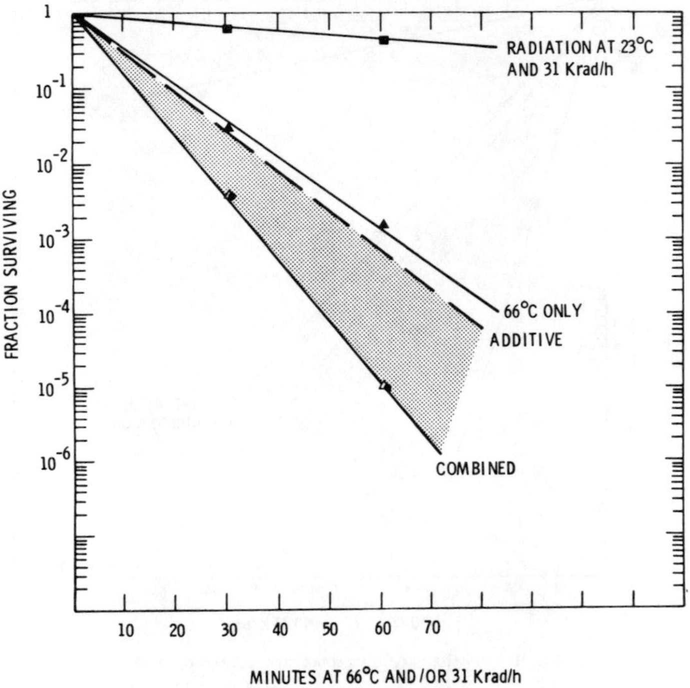

FIG.6. Inactivation of T4 bacteriophage in broth at 31 krad/h and 66°C separately and simultaneously.

As synergism with dry B. subtilis spores became apparent and understanding of it grew, other biological systems were examined for this behaviour. Trujillo has been primarily responsible for the investigation into the "universality" of the synergistic response between ionizing radiation and heat. Figure 6, adapted from Ref. [29], indicates that T4 bacteriophage is inactivated synergistically by ionizing radiation and heat when in solution. The single-stranded RNA polio virus (LSC) also is inactivated synergistically in solution as shown in Fig. 7 [30]. The same phenomenon is observed in vegetative bacteria: Fig. 8 depicts the behaviour of Escherichia coli at 50°C and 25 krad/h in broth [27]. In this instance, the organisms are undergoing division during treatment, as can be seen in the control. David [31] has shown that anaerobic spores of Clostridium sporogenes exhibit synergistic inactivation in both a dry and wet state as do wet B. subtilis spores.

This is a rather brief picture of the work done at Sandia Laboratories to demonstrate that a synergism exists between heat and radiation in the inactivation of various biological systems. While much remains to be done, it nonetheless seems fair to conclude that these data indicate that the possibility exists for accelerating radiation sterilization with the addition of heat (and conversely).

Combined heat and radiation processes share some characteristics of radiation-sterilization and some of heat-sterilization processes, and they have some characteristics peculiarly important only to themselves.

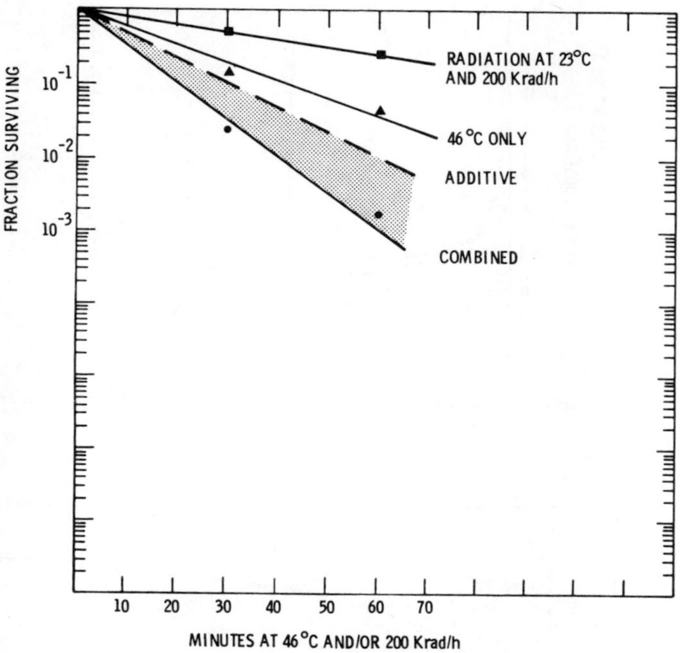

FIG.7. Inactivation of poliovirus (LSC) in MEM complete at 200 krad/h and 46°C separately and simultaneously.

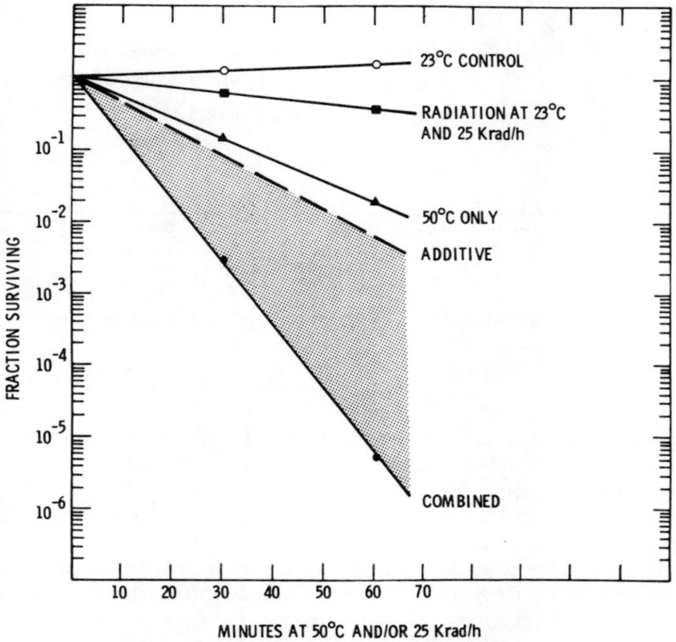

FIG.8. Inactivation of Escherichia coli in broth at 25 krad/h and 50°C separately and simultaneously.

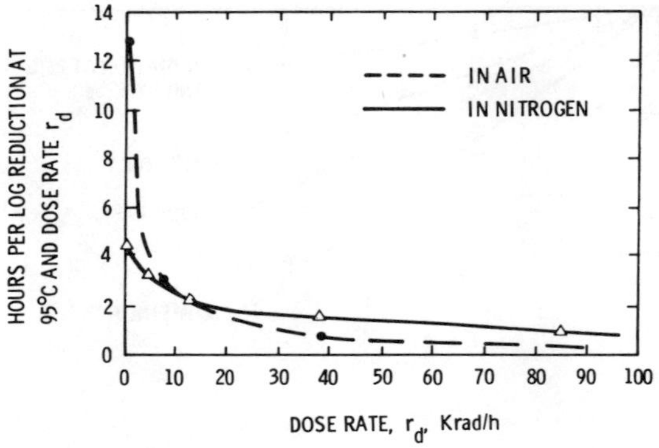

FIG.9. Dry Bacillus subtilis var. niger spore inactivation at 95°C at various dose-rates in air and nitrogen.

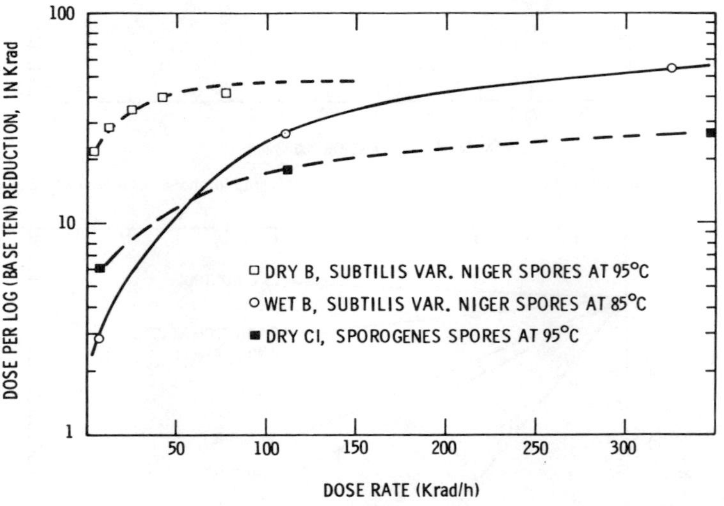

FIG.10. Dose-rate effects in wet and dry Bacillus subtilis var. niger spores and dry Clostridium sporogenes spores.

For example, depending upon dose-rate, heat and radiation combinations may exhibit an oxygen effect approximately the same as that encountered in radiation sterilization. This is shown in Fig. 9 for dry B. subtilis var. niger spores at 95°C [24]. Significantly, there are dose-rates at which little or no oxygen effect is encountered.

For a fixed dose-rate, the temperature at which one encounters a given rate of inactivation by dry heat and radiation combinations is generally greatest for bacterial spores. Thus, the organisms that are most resistant to heat and radiation combinations tend to be bacterial spores, as is the case with heat sterilization. But, unlike dry-heat sterilization, combined heat and radiation processes exhibit essentially no dependency upon relative humidity (RH) over a large region (20 to 60%) of ambient RH values [24].

Finally, Fig. 10 shows that combined processes exhibit a distinct dose-rate effect. Here the dose required to reduce a population of B. subtilis var. niger spores one log (base 10) in both a dry state and in water is shown as a function of dose-rate at elevated temperature (95°C dry and 85°C in water) [24, 31]. In the biological systems investigated, a general dose-rate dependence has been observed when the systems are synergistically inactivated by heat and radiation combinations. Spores tend to exhibit a less pronounced dose-rate dependency even at ambient temperature.

In designing combined heat and radiation-sterilization cycles, this dose-rate dependency may play a large role because relatively low dose-rates may be desired in order to take advantage of the lower dose per log reduction required by some biological systems in this dose-rate region.

This raises an obvious practical question. How does one design sterilization cycles involving combinations of heat and ionizing radiation?

3. AN ANALYTICAL FRAMEWORK FOR THE DESCRIPTION OF HEAT-ACCELERATED RADIATION EFFECTS

To answer the question just posed, a description of the rate of inactivation, k, of biological systems, as a function of temperature, T, and dose-rate, r_d, is required:

$$k = f(T, r_d)$$

much as is now done in the form [27]

$$E[n(t)] = n(0)\, e^{-D/D_{37}} = n(0)\, e^{-\frac{r_d}{D_{37}}\cdot t} \tag{1}$$

for an exponentially reduced population after a radiation dose D at ambient temperature. Here,

$$k_R = \frac{r_d}{D_{37}} \tag{2}$$

is the rate at which inactivation occurs at dose-rate r_d, n(t) is the number of survivors at time t, and E denotes expectation (or mean). The parameter, D_{37}, varies with species and environmental conditions [32]. For organisms which are inactivated in the logarithmic manner assumed in Eq. (1), the major concern in defining sterilization cycles is in determining the magnitude of D_{37} for the organisms that are being sterilized. By definition, this parameter is independent of dose-rate; therefore, the experimental determination of D_{37} is relatively straightforward for any particular organisms and set of environmental conditions.

A similar situation exists in describing heat inactivation. This is commonly done in the form [27]

$$E[n(t)] = n(0)\, 10^{-t/D} \tag{3}$$

where D is the decimal reduction time (base 10). Here D depends not only on the organisms and environmental conditions but also upon temperature, T.

To best describe the temperature dependence of D upon T, Eq. (3) will be rewritten

$$E[n(t)] = n(0)\, e^{-k_T t} \tag{4}$$

where $k_T = \log_e 10/D$ is the rate of thermal inactivation. J.P. Brannen, of Sandia Laboratories has extensively studied models of thermal inactivation [33-36], and found that an Eyring kinetic representation for k_T leads to models which have a high compatibility with survival data at different temperatures. In this approach, let

$$k_T = \frac{\kappa T}{h}\, e^{\Delta S^{\neq}/R}\, e^{-\Delta H^{\neq}/RT} \tag{5}$$

where κ is Boltzman's constant, h is Planck's constant, and R is the gas constant; $\Delta S^{\neq}$ and $\Delta H^{\neq}$ are the (sum of the) activation entropy and enthalphy of the underlying reaction (s); and T is expressed in degrees Kelvin.

In the application of Eqs (4) and (5), slightly more work is required than in applying Eq. (1) for radiation inactivation. In the thermal case, a minimum of a k_T value (or equivalently, D-value) at <u>two</u> different temperatures is required. This permits the solution of Eq. (5) for the values of the parameters $\Delta S^{\neq}$ and $\Delta H^{\neq}$ and yields a description of k_T as a function of temperature. This, in turn, may be substituted into Eq. (4) to describe survival in cases where inactivation is exponential.

If the inactivating effect of heat and radiation combinations were simply the sum of the separate effects of heat and radiation, a model which would adequately represent these effects would be

$$E[n(t)] = n(0)\, e^{-k \cdot t} \tag{6}$$

where

$$k = k_T + k_R \tag{7}$$

with k_R and k_T given by Eqs (2) and (5), respectively. Such a model describes the (theoretical) additive effects of heat and radiation in earlier figures. Clearly, as a representation of synergistic effects, such a model is inadequate. The rate k must be greater than that given by Eq. (7). Dugan developed a model, based on Eqs (6) and (7), which generally adequately represents the synergistic behaviour [37, 29]. This was done by letting

$$k = k_T + k_R + k_{TR} \tag{8}$$

where the rate k_{TR} is that needed to increase $k_T + k_R$ to the value of k which actually describes the survival curve. This is shown in Fig. 11. The parameter k_{TR} provides a description of the "rate of synergism". Since k_{TR} cannot be measured directly, its value is determined from measurable rates k, k_R, k_T according to $k_{TR} = k - (k_T + k_R)$. In spite of this, Dugan was able to deduce a generic form for k_{TR}:

$$k_{TR} = C' \, r_d^{\alpha/T}\, e^{-\beta/T} \tag{9}$$

in which C', α, and β are constants. This form is motivated by the possible involvement of free radicals as described in Refs [27] or [29].

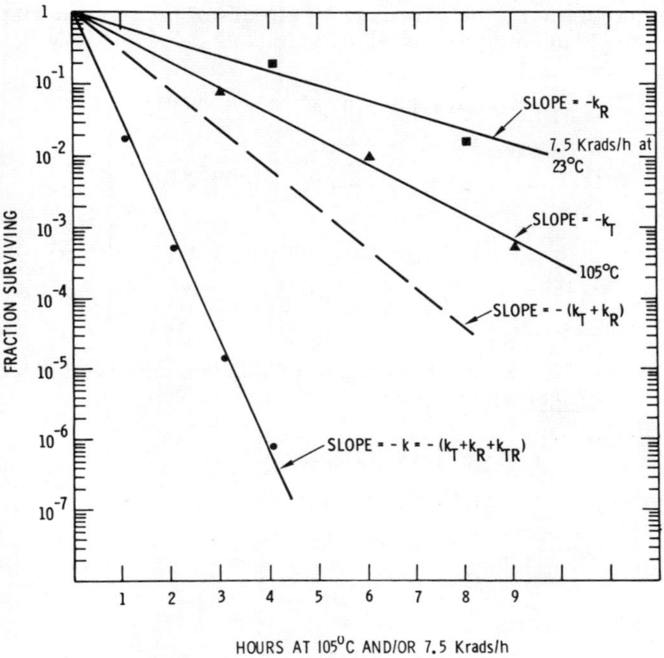

FIG.11. Illustration of rate terms in model describing combined heat and radiation treatments.

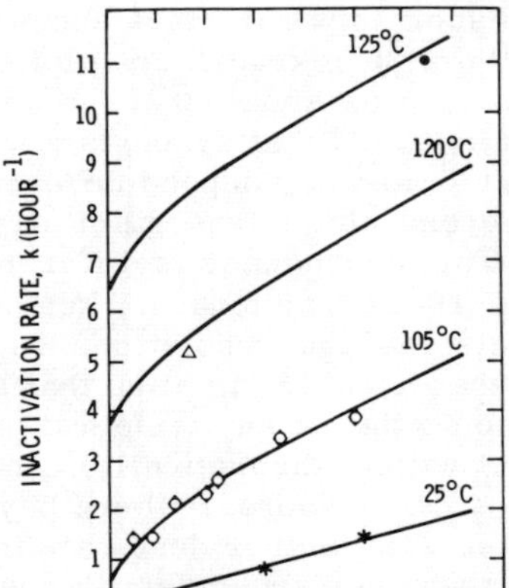

FIG.12. Comparison of model prediction (solid lines) with inactivation rates obtained from survivor data for B. subtilis var. niger spores in a dry state.

In summary, then, the inactivating effects of heat and radiation combinations may be represented by the expressions

$$
\left.
\begin{aligned}
E[n(t)] &= n(0)\, e^{-(k_T + k_R + k_{TR})\cdot t} \\
\text{with} \qquad k_T &= \frac{\kappa T}{h}\, e^{\Delta S^{\neq}/R}\, e^{-\Delta H^{\neq}/RT} \\
k_R &= C r_d \\
k_{TR} &= C' r_d^{\alpha/T} e^{-\beta/T}
\end{aligned}
\right\} \qquad (10)
$$

Figure 12 shows the general agreement of this model with data for dry B. subtilis var. niger spores when

$$
\begin{aligned}
\Delta S^{\neq} &= 12.63 \text{ entropy units} \\
\Delta H^{\neq} &= 33\,590 \text{ calories/mole} \\
C &= 0.0234 \text{ krad}^{-1} \\
C' &= e^{16.27} \\
\alpha &= 159.12 \\
\beta &= 12\,944.1.
\end{aligned}
$$

Similar compatibility exists with T4 bacteriophage [29] and LSC polio virus [30].

At this point, such statements as "synergism exists for 'properly' chosen combinations of heat and radiation", which have been used liberally above can be made more precisely. The behaviour of k_R, k_T, and k_{TR} and their sum, k, for the B. subtilis parameters above as a function of temperature at the fixed dose-rate of 30.6 krad/h are plotted in Fig. 13. From this figure, it can be seen that, below about 60°C and above about 135°C, $k_R + k_T$ will be much larger than k_{TR}, and that synergism will not be readily apparent at this dose-rate. In that k_R and k_{TR} depend differently on dose-rate, these limits of temperature, beyond which synergism may not be evident, will also depend upon dose-rate. This dependence is different for dry B. subtilis spores (where α/T of Eq. (10) is fractional for temperatures of concern) than it is for, say T4 bacteriophage in solution. In the latter case, $\alpha = 481$ [29] and α/T is greater than 1, which implies that synergism increases indefinitely with dose-rate so that at any temperature a dose-rate high enough to obtain synergistic inactivation (theoretically) can be found. This would not be true of the dry B. subtilis spores, where beyond some dose-rate synergism should decrease with further dose-rate increases.

It is evident then that whether synergism is observed depends on both temperature and dose-rate, and in ways which are fairly complex compared with the dependence of heat-sterilization cycles on temperature or radiation-sterilization cycles upon dose-rate. Without a model, such as that given by Eq. (10), a strictly experimental determination of the two-dimensional (temperature and dose-rate) region of synergism for any test organism could

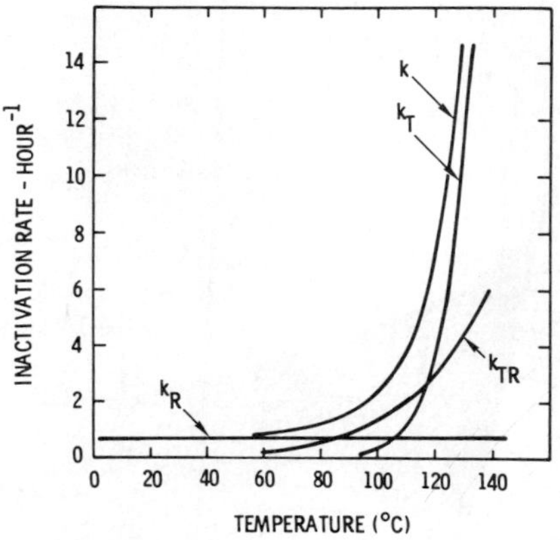

FIG.13. The behaviour of the individual inactivation rate parameters for B. subtilis var. niger spores in a combined heat and radiation environment utilizing gamma radiation at 31 krad/h.

be exceedingly time-consuming and expensive. To obtain results at 10 temperatures and 10 dose-rates (in all combinations) would require at least 200 experiments. To do a comparable job of obtaining values for the model parameters would require only about 15 experiments; about 20 more experiments would be required to check randomly the model's predictive capability within the two-dimensional region of interest. With a smaller data base of this sort, the amount of synergism to be expected as a function of dose-rate and temperature can be predicted as shown in Fig. 14 [26] for dry B. subtilis spores; and these predictions can be compared with experimental results at 10 points within such a region that were not used in making the predictions. From a figure such as Fig. 14, dose-rate required for maximum synergism at any given temperature can be determined or, conversely, at any given dose-rate, the temperature required for optimum synergism can be determined. What is not apparent from Fig. 14 is the process time required at various dose-rate/temperature combinations, but this information may also be obtained from a model such as that given in Eq. (10). An example of how this might be done is given in Fig. 15 in which the relationship between temperature, dose-rate, total dose, and sterilization time for an 8-log (base 10) decrease in dry B. subtilis spores is shown [27]. If an 8-log decrease in population is desired within the constraints of no greater than 110°C temperature, no greater total dose than 200 krad and no greater sterilization time than 8 h, the shaded region shown in Fig. 15 must be used. This implies that the dose-rate must lie between about 10 and 50 krad/h and that the temperature be between about 100 and 110°C. If a process temperature of 105°C is chosen, the dose-rate within the region 10 to 50 krad/h that yields the greatest synergism may be selected by referring to Fig. 14 or by direct computation as shown in Fig. 16 [24]. The actual value is about 20 krad/h as seen in Fig. 16. Thus, within the constraint imposed and a choice of 105°C, the optimal dose-rate is 20 krad/h and the process time is about 8 h.

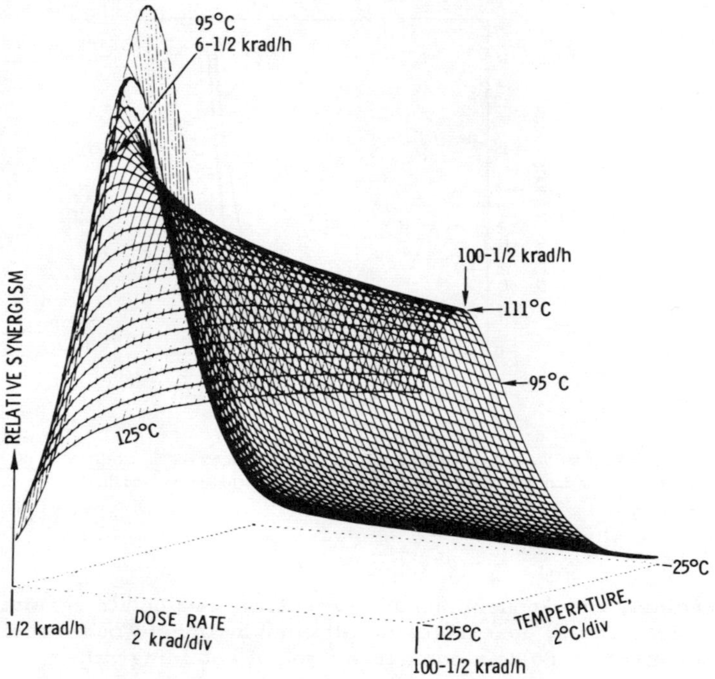

FIG.14. The relative synergism $k/(k_T + k_R)$ of dry Bacillus subtilis var. niger spores as a function of temperature and dose-rate (from Ref.[26]).

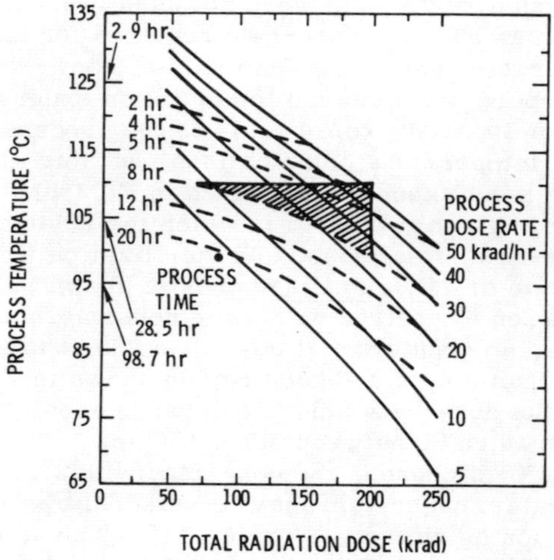

FIG.15. Time, temperature, dose and dose-rate interactions yielding an 8-log reduction in a population of dry Bacillus subtilis var. niger spores.

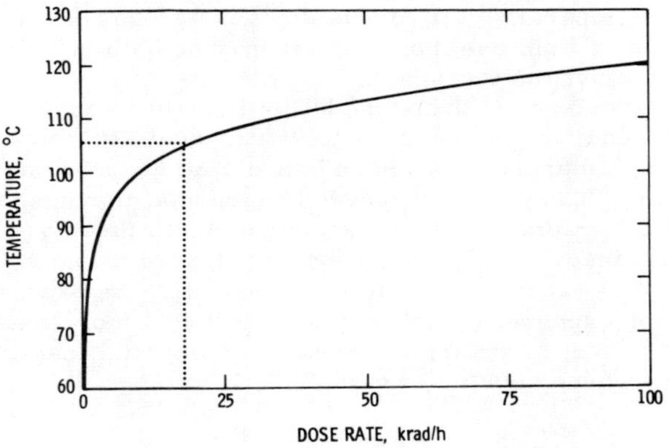

FIG.16. The combination of dose-rate and temperature resulting in the greatest synergistic inactivation of dry Bacillus subtilis var. niger spores. (From Ref.[26]).

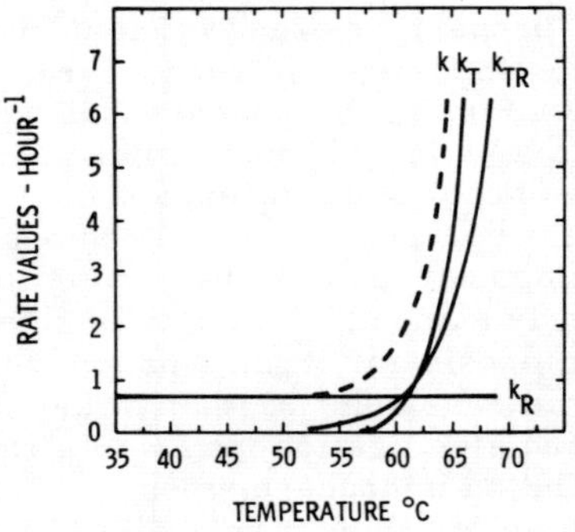

FIG.17. Behaviour of model parameters for T4 bacteriophage in broth at 31 krad/h. The temperature region in which synergism may be observed is narrow at this dose-rate.

It becomes fairly evident that a quantitative description of the combined effects of heat and radiation processes is highly desirable because it presents the only practical way in which to optimize sterilization cycles. An experimental program is still necessary, of course, to determine the model parameter values appropriate to specific test organisms and sterilization conditions. In this respect, some observations seem appropriate. First, the data should lie, at least partially, in the temperature/dose-rate domain where synergism is to be found. Experience indicates that this occurs most noticeably at temperatures below (but "near") those which rapidly sterilize by themselves; this is a good starting point experimentally. Several dose-rates (high, intermediate, and low, for example) need to be investigated at these temperatures. For a fixed dose-rate, the temperature range for synergistic inactivation may be much broader for dry systems than for those in solution. Evidence of this may be obtained by comparing Fig. 13 for dry B. subtilis spores with Fig. 17 for T4 bacteriophage in solution.

Experience in comparing "wet versus dry" systems is limited, and the conclusion is not a firm one; however, it may be helpful in selecting an experimental temperature range.

Finally, experience with the model by Eq. (10) indicates that it is a good one as long, of course, as inactivation is exponential in character. Any model is merely an approximation and should be regarded as no more than such. Brannen [38, 39], has recently developed a more general model of thermal effects in radiation sterilization which, in theory, will represent non-exponential inactivation and antisynergistic behaviour. There is some indication that this latter may be of concern [40]. As experience with it grows, this more general model may well be found more desirable than the one described above; however, in cases of exponential inactivation, the model presented here should serve well.

4. SUMMARY AND DISCUSSION

A synergistic inactivation of many biological systems occurs when properly chosen combinations of heat and ionizing radiation are used. There is no firm evidence that this statement is not applicable to all biological systems. The phrase "properly chosen" is most important, and a quantitative means exists for defining the temperature/dose-rate region in which synergism is to be found, based on much less data than would be required to do this (less confidently) in a strictly experimental way. At this time, reports of no synergism between heat and radiation appear to occur when investigators are working outside these "properly chosen" regions.

The potential of using heat to accelerate radiation sterilization cycles for medical devices is clearly evident, in fact it is potentially beneficial in two ways. First, many radioresistant organisms are relatively heat-labile (for example, viruses and radioresistant vegetative cells). Thus, sufficient heat as an adjunct to radiation sterilization may well eliminate many radioresistant organisms by itself. The main concern, then, is with only those organisms which are both radioresistant and heat resistant, which are generally to be found among bacterial spores. The best approach, which takes advantage of the second area of potential benefit, is to design combination heat and radiation cycles which inactivate the most resistant of these organisms synergistically. Precisely what combinations should be used will depend upon what organisms in this class are apt to be present on the items being sterilized or on the organism used as a test standard and upon the nature of the items themselves (e.g., solutions or devices). In this paper, a theoretical framework by means of which such processes can be defined has been presented. Also, it has been observed that, for some combinations of temperature and dose-rate, the combined heat and radiation processes are less sensitive to some environmental parameters than either radiation or heat processes separately.

Much work is still needed with bacterial spores that are both heat resistant and radioresistant in combined heat and radiation environments. Sandia Laboratories' experience in this area is limited to naturally occurring dry spores (which were only modestly radioresistant, see Fig. 5) and viruses in solution (which are heat-labile). It appears that this is the first step in assessing the practicality of using heat to accelerate radiation sterilization of medical devices. The potential for important gains is certainly implied

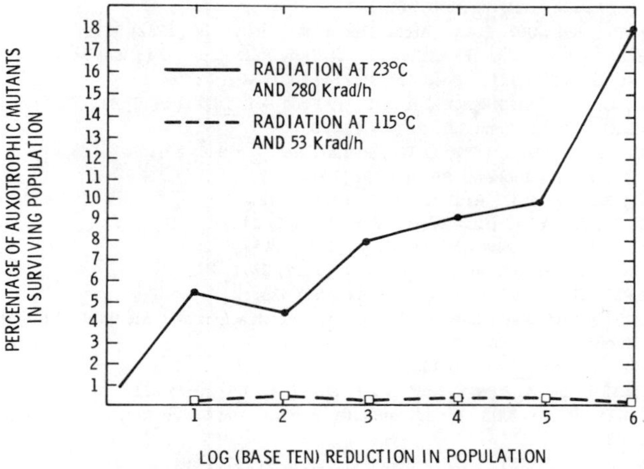

FIG.18. Auxotrophic mutants induced in populations of dry Bacillus subtilis var. niger spores by radiation at ambient and elevated temperature.

by the results obtained with, say, naturally occurring spores, but precise behaviour is unknown. Emborg [41] has recently reported data in this area, but there is still insufficient information available to characterize the behaviour of heat-resistant, radioresistant organisms in the framework presented above.

Also, very preliminary data [42] suggest that heat and radiation combinations may produce fewer mutations among survivors than radiation at room temperature. R.T. Dillon of Sandia Laboratories is attempting to assess the effect of heat, radiation, and combinations thereof at producing auxotrophic mutations in dry B. subtilis var. niger spores. This is a very large class of mutants, and conclusions drawn from its study may be applicable to other classes; however, this is unknown. Data shown in Fig.18 indicate that the per-cent mutations among survivors induced by severe thermoradiation cycles is near that which is a natural background level for the spore stock being investigated, while those induced by radiation may be as high as 15%. It is too early to know how generally valid such behaviour might be, but these results would seem to offer yet another potential way in which combinations of heat and radiation may be of benefit in the sterilization of medical devices.

REFERENCES

[1] SETLOW, R., Proc. Natl. Acad. Sci. U.S. 38 (1952) 166.
[2] POLLARD, E.C., POWELL, W., REAUME, S., Ibid., p.173.
[3] SETLOW, R., DOYLE, B., Arch. Biochem. Biophys. 46 (1953) 31.
[4] SETLOW, R., DOYLE, B., Arch. Biochem. Biophys. 48 (1954) 441.
[5] BRUSTAD, T., (1961), Radiat. Res. 15 (1961) 139.
[6] BRUSTAD, T., in Biological Effects of Neutron and Proton Irradiations, (Proc. Symp. Upton, 1963) 2, IAEA, Vienna (1964) 404.
[7] AUGENSTEIN, L.G., BRUSTAD, T., MASON, R., in Advances in Radiation Biology 1 (AUGENSTEIN, L.G., MASON, R., QUASTLER, H., Eds), Academic Press, N.Y. (1964) 227.

[8] FLUKE, D.J., Radiat. Res. 28 (1966) 677.
[9] ADAMS, W.R., POLLARD, E.C., Arch. Biochem. Biophys. 36 (1952) 311.
[10] POWERS, E.L., WEBB, R.B., EHRET, C.F., Radiat. Res. Suppl. 2 (1960) 94.
[11] KAN, B., GOLDBLITH, S.A., Food Res. 22 (5) (1957) 509.
[12] LICCIARDELLO, J.J., NICKERSON, J.T.R., J. Food Sci. 27(3) (1962) 211.
[13] LICCIARDELLO, J.J., J. Food Sci. 29 (1964) 469.
[14] STAPLETON, G.E., EDINGTON, C.W., Radiat. Res. 5 (1956) 39.
[15] WOOD, T.H., Arch. Biochem. Biophys. 52 (1954) 157.
[16] BELLI, J.A., BONTE, F.J., Radiat. Res. 18 (1963) 272.
[17] DESCHNER, E., GRAY, C.H., Radiat. Res. 11 (1959) 115.
[18] MÜLLER, C., München, Med. Wernschr. 57 (1910) 1490.
[19] WESTERMARK, N., Skand. Arch. Physiol. 52 (1927) 257.
[20] LANGENDORFF, H.M., Strahlentherapie 64 (1939) 512.
[21] WOEBER, K., in Ultrasonic Energy: Biological Investigations and Medical Applications (KELLY, E.,Ed.) Univ. of Illinois Press (1965) 137.
[22] GINOZA, W., Nature 199 (1963) 453.
[23] HOTZ, G., MÜLLER, A., Proc. Nat. Acad. Sci. U.S. 60 (1968) 251.
[24] SIVINSKI, H.D., REYNOLDS, M.C., in Life Sciences and Space Research X, Akademie-Verlag, Berlin (1972) 33.
[25] REYNOLDS, M.C., GARST, D.M., Space Life Sci. 2 (1970) 394.
[26] REYNOLDS, M.C., BRANNEN, J.P., in Radiation Preservation of Food (Proc. Symp. Bombay, 1972), IAEA, Vienna (1973).
[27] SIVINSKI, H.D., GARST, D.M., REYNOLDS, M.C., TRAUTH, C.A. Jr., TRUJILLO, R.E., WHITFIELD, W.J., in "Industrial Sterilization" (BRIGGS PHILLIPS, G., and MILLER, W.S., Eds), Duke Univ. Press (1973) 305.
[28] REYNOLDS, M.C., LINDELL, K.F., DAVID, T.J., FAVERO, M.S., BOND, W.W., Appl. Microbiol. 28 (3) (1974) 406.
[29] TRUJILLO, R., DUGAN, V.L., Biophys. J. 12 (1972) 92.
[30] TRUJILLO, R., DUGAN, V.L., Heat Accelerated Radioinactivation of Attenuated Poliovirus, to be published.
[31] DAVID, T.J., Masters Thesis, Univ. of New Mexico, Albuquerque, N.M., (1973).
[32] DUGAN, V.L., TRUJILLO, R., J. Theo. Biol. 44 (1974) 397.
[33] BRANNEN, J.P., Math. Biosci. 2 (1968) 165.
[34] BRANNEN, J.P., Biophys. 7 (1970) 55.
[35] BRANNEN, J.P., J. Theo. Biol. 32(2) (1971) 331.
[36] BRANNEN, J.P., GARST, D.M., Appl. Microbiol. 23(6) (1972) 1125.
[37] DUGAN, V.L., Space Life Sci. 2 (1971) 498.
[38] BRANNEN, J.P., Sandia Laboratories Research Rep. SAND74-0269 (1974).
[39] BRANNEN, J.P., Radiat. Res., to be published.
[40] GRAIKOSKI, J., Dissertation Abstr. 22(2) (1961) 394.
[41] EMBORG, Claus, Appl. Microbiol. 27(5) (1974) 830.
[42] DILLON, R.T., Abstr. 6th Int. Spore Conf., East Lansing, Mich. (1974) No.1.12,8.

DISCUSSION

K. SUNDARAM: I must congratulate you on a very fine paper on the synergistic effects of temperature and radiation. I'd like to ask if you have considered radiation damage to products that may be unstable at high temperature.

C.A. TRAUTH: With regard to damage at high temperature, our experience relates primarily to electronic potting compounds (and associated solid-state electronics) and similar polymers. With these materials we have observed that radiation-induced point defects actually anneal more rapidly at elevated temperatures, which results in less damage. Beyond this I have no information on materials. I would suggest, however, that combination treatments require considerably smaller total doses than conventional radiation treatments, since they work much faster. In view of this, one might anticipate less radiation damage.

B.L. GUPTA: Could you say something about the mechanism underlying the synergistic effect which you observed?

C.A. TRAUTH: I can only speculate as to its nature. Some of those working in this area, including myself, are inclined to believe that temperature accelerates indirect chemical damage mechanisms, most probably from free radicals, both through influence on the reaction rate and through effects on the configuration and stability of the substrate being attacked by the radicals. However, our evidence at this juncture is purely circumstantial. I'm afraid that is all I can say about it.

A. TALLENTIRE: Could you elaborate on the synergistic effects seen in wet systems? Are they as extensive as in dry systems? Further, you mentioned the need for "properly chosen" process conditions to achieve maximum effectiveness. Could it not also be that the organisms must be present under "properly chosen" conditions when combined treatments are applied? What, for example, is the influence on synergistic effects of other variables (apart from oxygen and water, which you mentioned) contributing to the cells' microenvironment?

C.A. TRAUTH: Synergistic effects can be as great in wet systems as in dry ones in terms, for example, of the magnitude of K_{TR}. The biggest difference, as I see it, is that once the thermal effects set in, they become dominant much faster in wet-heat inactivation than in dry-heat sterilization. This tends to limit the temperature region over which synergism is observed, as illustrated in my paper.

As for environmental factors other than oxygen and water, I'm sure that some will affect the nature of the synergistic response in certain systems. At the same time, the virus data which I presented were obtained in MEM complete and the phage data in dilute broth, so that, at least for these systems, organic materials do not eliminate synergism. We have not yet investigated this with spores.

A. TALLENTIRE: I feel myself that for the moment we ought to exercise a great deal of caution in our approach to the application of combined heat and radiation treatments in practice.

C.A. TRAUTH: Yes, I agree completely. The organisms of interest, the environmental conditions and the nature of the items to be sterilized will doubtless all affect the potential usefulness of this treatment. Every situation should be carefully assessed on its merits. On the other hand, I would not like to see an over-cautious approach that might preclude active investigation of the value of heat and radiation combinations.

H.M. ROUSHDY: Can you offer any radiobiological explanation for the increased evidence of microbiological mutations you described in connection with the synergistic effect of radiation and heat? I should also be interested in knowing whether the increased mutation rate is confined mainly to specific groups and strains of microorganisms or are your findings more general?

C.A. TRAUTH: Some very preliminary work by Dr. R.T. Dillon has shown that, when inactivated synergistically, dry B. subtilis spores show a decreased incidence of auxotrophic mutation among survivors. In our investigations, we have so far examined only auxotrophic mutants in these spores in the dry state and we cannot with confidence make generalizations for any other system or set of conditions.

K.H. MORGANSTERN: Since the temperature range is limited, one cannot cover a wide range of values. But the radiation dose-rate can be varied over many orders of magnitude. Have you observed the same synergistic effects at much higher dose-rates, for example, Mrad/h or Mrad/s

C.A. TRAUTH: We have investigated synergism at much higher dose-rates than is evident from my paper. Synergism has been observed in dry B. subtilis spores at dose-rates up to several Mrad/h, using ^{60}Co, and almost up to this level with viruses in solution. In addition, a synergistic response has been observed in dry spores, using pulsed radiation at a dose-rate of about 20 krad/5 ns.

B.B. GAITONDE (Chairman): Have you studied the growth characteristics of the surviving organisms, for example, those in the Clostridida group, such as Cl. botulinum or Cl. tetani?

C.A. TRAUTH: We have looked only briefly at the detailed nutritional requirements of B. subtilis survivors that are auxotrophic mutants. We have not yet worked with any anaerobic organisms.

A. KEALL: Can you tell us anything about the economics of the thermoradiation treatment you describe? I would agree that the technique is of theoretical interest but I have some doubt as to its industrial and/or commercial value, even when applied in specially constructed plants.

C.A. TRAUTH: Clearly, a complete answer can be given only for a specific product under specific conditions. Since I do not work in the medical sterilization industry, I cannot give you such specific information. I feel, however, that heat and radiation combinations should not be viewed only in the narrower context of those items which are being radiosterilized today, but rather as a potentially useful technique, first, in current problem areas; second, for new, perhaps more complex, products, e.g. implanted electronics; and third, as an adjunct to current and future heat-sterilization processes.

Furthermore, the question is not strictly one of economics. Benefits, too, must be considered. For example, there are many complex, potted electronic devices which cannot reasonably be sterilized with heat or radiation separately but which can withstand a combination treatment. I think this makes the treatment cost-effective for these devices. More generally, the addition of radiation to heat-sterilization cycles (especially dry heat) may permit considerable reduction in heat, hence in this context there will be considerable economic benefit.

As an adjunct to radiation sterilization, heat offers potential benefit, for example, when one is dealing with viruses. This is particularly true at a time when the aim is to decrease total doses for certain classes of products, notably solutions and tissues. While I cannot quote any figures to support my argument, I consider that this is another area of potential economic benefit from combination treatments.

A further advantage that should be taken into account in any cost-benefit analysis is throughput. Combined treatments can be considerably more rapid. Simply comparing the cost of heat with that of radiation is not enough.

The final point I would like to make is that heat and radiation combinations are not a panacea for all ills but rather a technology of which one should be aware. I am confident that there will come a time and a context within which these treatments will prove cost-effective.

INACTIVATION OF B. Pumilus SPORES BY COMBINATION HYDROSTATIC PRESSURE-RADIATION TREATMENT OF PARENTERAL SOLUTIONS

Pamela A. WILLS
Australian Atomic Energy Commission
Research Establishment,
Sutherland, N.S.W.,
Australia

Abstract

INACTIVATION OF B. Pumilus SPORES BY COMBINATION HYDROSTATIC PRESSURE-RADIATION TREATMENT OF PARENTERAL SOLUTIONS.

Bacterial spores are inactivated by moderate hydrostatic pressures. The radiation dose required to sterilize radiation sensitive pharmaceuticals can be considerably reduced using a combination hydrostatic pressure-radiation treatment. This paper describes a combination pressure-radiation sterilization process using Bacillus pumilus spores suspended in water, 0.9% saline, and 5% dextrose solutions. The optimum temperatures for spore inactivation at 35 MPa and the degree of inactivation at 35, 70 and 105 MPa applied for times up to 100 min have been determined. Inactivation was greatest in saline and least in dextrose. Spores in dextrose were only slightly less radiation resistant than in saline or water. It was calculated that the radiation dose required for sterilization could be halved with appropriate compression treatment. Examples of combinations of pressure-radiation suitable for sterilization are given. One combination is compression at 105 MPa for 18 min for a dose of 1.25 Mrad.

INTRODUCTION

Conventional sterilization techniques using heat or radiation may be unsuitable for liquids sensitive to degradation by these agents. The amount of heat and/or radiation required to sterilize such materials can be considerably reduced if they are treated with hydrostatic pressure either before or during sterilization. This has been shown for milk [1, 2] and other foods [3].

Bacterial spores are inactivated by moderate hydrostatic pressures [4, 5]. Inactivation is preceded by initiation of germination and the degree of inactivation is dependent on spore species, pressure, compression time, temperature, pH, ionic strength and other factors [1, 6]. Because some spores always survive compression, as with other germination methods [7], pressure cannot be used alone as a sterilization process. A combination hydrostatic pressure-radiation treatment for sterilization of pharmaceutical solutions is described in this paper.

MATERIALS AND METHODS

Test organism and production of spores

Nutrient broth, inoculated from a single colony of Bacillus pumilus, was incubated at 37°C for 24 hours. The culture was used to seed Roux flasks containing soybean agar [8]. After incubation at 37°C for 3 days and 32°C for 4 days (at least 95% sporulation), spores were harvested in chilled sterile water. The suspension was centrifuged in an HT International Centrifuge at 3,000 x g for 15 minutes, the supernatant discarded, and the residue left standing overnight at 1°C. After resuspension in water, the spores were washed three times. Centrifugation between washes was at 10,000 x g for 5 minutes. The suspension was then heated at 80°C to inactivate any remaining vegetative cells, rapidly cooled, washed three more times and finally resuspended in sterile water at a concentration of about 2 x 10^9 spores/ml. More than 98% of spores were fully refractile.

Test solutions

Three test solutions were used: sterile water for injection, 0.9% w/v sodium chloride (Merck) and 5% w/v D-glucose (B.D.H.). Sterilization was at 121°C for 20 minutes for water and saline; the dextrose solution was membrane (0.45 µm) filtered. Working suspension of B. pumilus spores were prepared at a concentration of about 8 x 10^7 spores/ml test solution. The pH values of the suspensions were 7.0 (water), 6.7 (saline) and 6.6 (dextrose).

Pressurization

For each experiment, three replicate 1 ml plastic syringes containing about 0.5 ml spore suspension were sealed with Parafilm and inserted in a stainless steel pressure vessel. Pressure was generated either by an SKF or an Enerpac stainless steel hand-pump. For both pumps the compression fluid was water, but because the SKF pump operated from a silicone oil reservoir, a stainless steel membrane was inserted into the pressure line to separate the oil from the water. Compression was measured by a National Instrument Company gauge with divisions every 0.35 MPa. The pressure vessels were immersed in a thermostatically controlled water bath. Temperature inside the pressure vessel was monitored using a thermistor probe.

Determination of optimum temperature for compression effect

Spore suspensions of each test solution were compressed to 35 MPa for 15 minutes at temperatures between 30°C to 70°C. The degree of change with temperature was measured by refractility counts on air dried smears. At least 500 spores were counted. The fraction of phase bright spores surviving compression was calculated.

Determination of inactivation and survival after compression at 35, 70 or 105 MPa

Viable counts were performed before and after compression at 35, 70 or 105 MPa for times up to 100 minutes, at the optimum temperature determined for each suspension. Dilutions prepared in 0.065 M potassium phosphate buffer, pH 6.8, were heated at 75°C for 20 minutes to inactivate spores which had been germinated but not inactivated by the pressure. Pour-plates were prepared from the dilutions using nutrient agar (Oxoid nutrient broth and 1.2% Oxoid Agar No.3) and incubated at 32°C. The plates were counted after 48 hours. Further incubation did not increase the number of colonies. Results were calculated from counts of at least two dilutions [9].

Irradiation

Specially cleaned glass tubes 8 mm (I.D.) x 60 mm containing 1 ml spore suspension were sealed so that irradiation was essentially anoxic. Using a polystyrene foam stand 12 tubes could be irradiated simultaneously in a ^{60}Co source with cylindrical geometry. Dose rate was measured by ceric-cerous dosimeters [10]. The mean dose rate was 385 krad/h and variation in dose between the tube positions was less than 4%. Viable counts were carried out essentially as described in the preceding section.

RESULTS AND DISCUSSION

Dormant spores are fully refractile when viewed by phase contrast microscopy. Germinated spores lose refractility and darken. The change in refractility was used to determine the optimum temperature for pressure-initiated germination of B. pumilus spores at a constant pressure of 35 MPa applied for 15 minutes.

The results for the three spore suspensions are shown in a semi-logarithmic plot of the fraction of spores remaining fully refractile against temperature (Fig.1). At 35 MPa the optimum temperatures for germination, and hence inactivation of B. pumilus spores, were 51°C, 52°C and 54°C for spores suspended in 0.9% saline, 5% dextrose, and water respectively.

Temperature is one of the most critical parameters influencing the sensitivity of spores to compression. The optimum temperature varies with species and generally increases with increasing heat resistance [1]. The optimum temperature determined for each solution at 35 MPa was also used to estimate spore inactivation at pressures of 70 and 105 MPa. Only slight shifts in optimum temperature occur between these pressures [2].

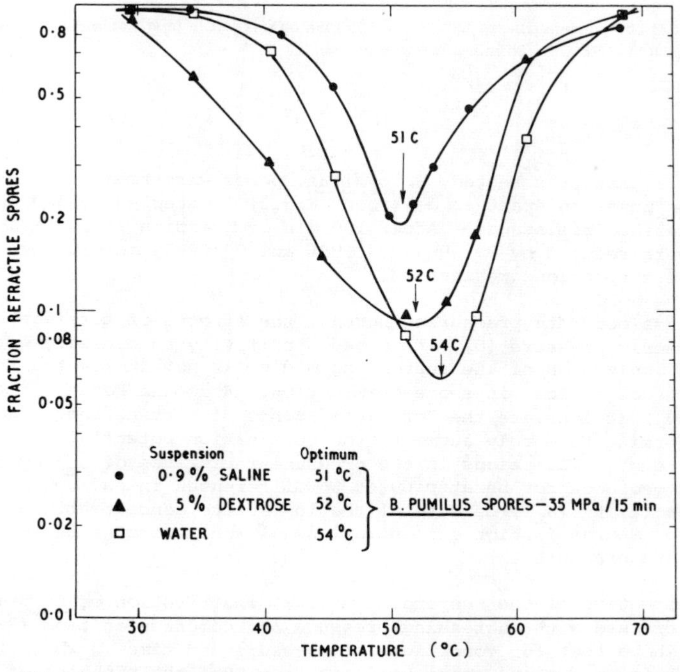

FIG.1. Effect of 15-min compression at 35 MPa applied at different temperatures on refractility of B. pumilus spores.

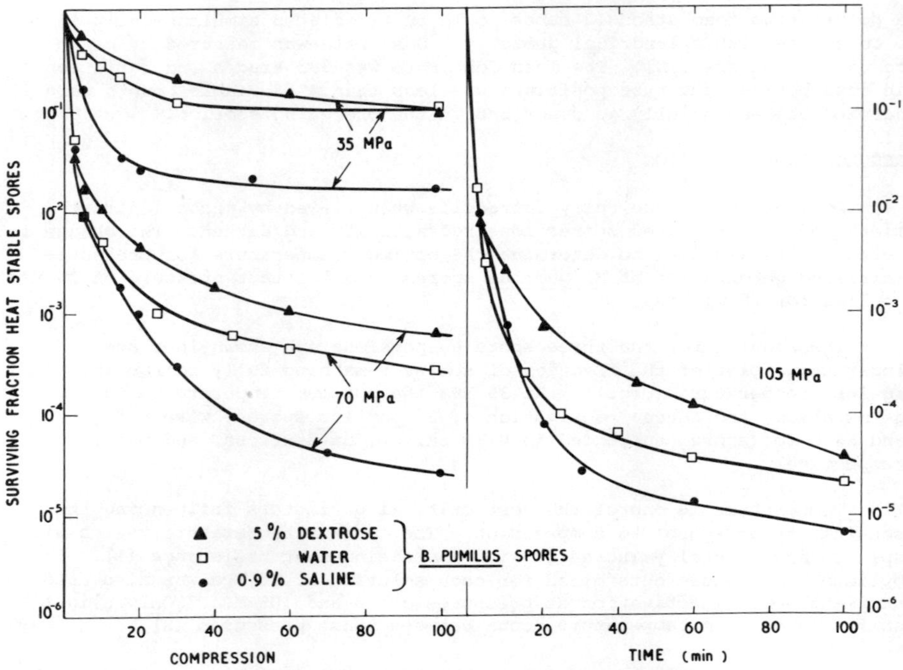

FIG.2. Effect of compression time at 35 MPa, or 105 MPa on survival of heat stable B. pumilus spores. Spores were heated at 75°C for 20 min after decompression.

At all pressures tested, B. pumilus spores were most sensitive in 0.9% saline and most resistant in dextrose (Fig.2). Water suspended spores were of intermediate resistance. After 100 min compression at 105 MPa, spore numbers were reduced by 99.996%, 99.998% and 99.9993% in glucose, water, and saline suspensions respectively.

Because moderate pressures increase the effects of germinants acting at atmospheric pressure [6], increased sensitivity to pressure with increased ionization of the suspending medium is not unexpected in view of the importance of ions in spore germination. Rode and Foster [11] have postulated that ions are the "critical agents of germination", and that organic germinants merely augment "the germination potential of particular ionic systems". Variations in the pressure resistance of the spores in the different media cannot be attributed to differences in pH, or to differences in water activity (a_w), because of the low solute concentrations used in the solutions. Spores in high concentrations of ionic solutes become more resistant to pressure [5].

Irrespective of the suspending medium, inactivation of B. pumilus spores increased with increasing pressure and compression time (Fig.2). It has been shown that for conditions of pressure and time in which the overall change in B. pumilus spores is less than 90%, the kinetics of initiation of germination and inactivation of the spores is a first-order

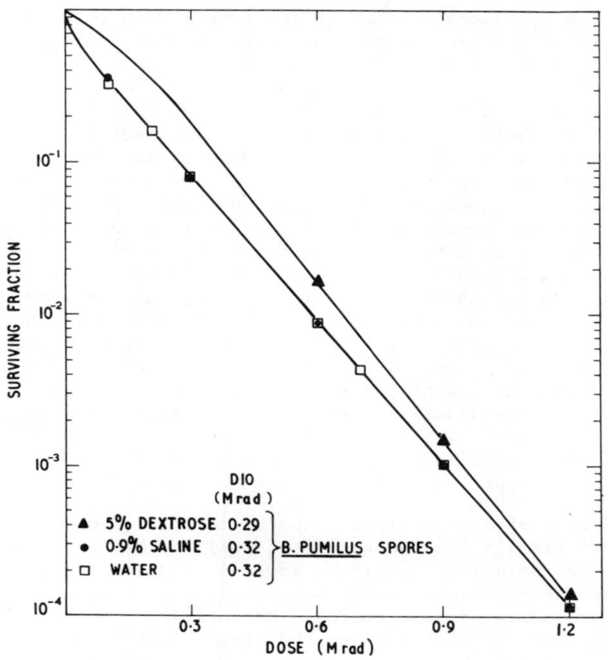

FIG.3. Radiation resistance of B. pumilus spores.

consecutive reaction [12]. When these conditions are exceeded, increased inactivation of the spores is not directly proportional to increases in compression time, and the inactivation time curves flatten out. At 105 MPa, tenfold reductions in the numbers of spores suspended in saline occurred after 1.5, 3, 9, 18 and 73 minute compression. Spores suspended in dextrose and water reacted similarly. This effect has also been noted for a variety of Bacillus sp. suspended in milk [1]. The phenomenon of 'tailing' is common to all germination methods [7].

In Fig.3, the logarithm of the fraction of spores surviving irradiation has been plotted against dose. The exponential portions of the curves have been fitted by linear regression analysis. Analysis of variance showed that the lines for B. pumilus spores suspended in water and saline can be superimposed. The radiation decimal reduction dose (D_{10}) for spores in these solutions was 0.32 Mrad. Briggs and Yazdany [13] also obtained this D_{10} value for anoxic B. pumilus spores and showed that survival was not influenced by the presence of up to 8% NcCl. For B. pumilus spores irradiated in dextrose, the dose survival curve has a slight shoulder, and the D_{10} value was 0.29 Mrad. Both curves converge at a dose of 1.25 Mrad, giving an inactivation factor of 10^4.

Previous work has shown that a combination treatment of hydrostatic pressure with radiation has an additive effect on spore inactivation and is not synergistic. Provided that the degree of inactivation achieved by separate treatments is known, the overall reduction in spore numbers for simultaneous compression-radiation or for compression followed by irradiation will be the sum of the two treatments. Experimental evidence of this effect has been reported elsewhere [1, 2].

It is generally assumed that an efficient sterilization process acceptable to the pharmaceutical industry is one which results in not more than one unsterile item in a million items processed. The inactivation factor required for sterilization will therefore be determined by the average initial contamination level of the untreated items. Extrapolation from Fig.3 shows that the anoxic inactivation factor for B. pumilus spores in typical parenteral solutions is 10^8 at the normal radiation sterilizing dose of 2.5 Mrad. Thus this dose could be safely used to sterilize 1×10^6 items with an average contamination level of 100 organisms per item. However, the same sterilizing efficiency can be achieved when the radiation dose is halved to 1.25 Mrad, if appropriate pressure treatment is applied either simultaneously or before irradiation. Inactivation factors of 10^4 for B. pumilus spores are easily achieved at pressures of 105 MPa. An example of a suitable pressure-radiation sterilization treatment to inactivate B. pumilus spore contaminants of saline solutions would be a pressure of 105 MPa applied at 51°C for 18 minutes combined with a radiation dose of 1.25 Mrad. Other combinations of pressure-radiation are given in Table 1.

The principle of spore inactivation by means of a combination pressure-radiation sterilization treatment, illustrated in this paper by one bacterial species and three solutions, could be applied to most pharmaceutical solutions. All spore species are inactivated by pressures up to about 300 MPa. Higher pressures protect spores. Because different species have different temperature optima for pressure inactivation, in practice compression would probably be carried out over a temperature range determined after identification of the natural contaminants.

The main advantage of a pressure-radiation combination treatment is the comparatively low radiation dose required to effect sterility. It is conceivable that pharmaceuticals which undergo radiolytic breakdown at the normal sterilizing dose of 2.5 Mrad may be unaffected at the reduced radiation doses applied in the combination treatment. The proposed method could have limitations for solutions at extreme pH, for solutions with high ionic contents, or those containing germination inhibitors [6]. These conditions would not be expected to occur in pharmaceutical solutions.

TABLE I PRESSURE-RADIATION LEVELS REQUIRED FOR AN EFFICIENT STERILIZATION PROCESS (NOT MORE THAN ONE UNSTERILE ITEM PER MILLION)

Average contamination level per item	Radiation dose (Mrad)	Pressure (MPa)	Compression time at optimum temperature (minutes)		
			0.9% Saline	Water	5% Dextrose
10	0.95	105	18	25	67
	1.25	70	19	27	62
100	0.95	105	75	> 100	> 100
	1.25	105	18	25	67

(Pressure conditions based on inactivation of B. pumilus spores)

REFERENCES

[1] WILLS, P.A., CLOUSTON, J.G., GERRATY, N.L., "Microbiological and entomological aspects of the food irradiation program in Australia", Radiation Preservation of Food (Proc. Symp. Bombay, 1972), IAEA Vienna (1973) 231.

[2] WILLS, P.A., Effects of hydrostatic pressure and ionising radiation on bacterial spores, At. Energy Aust. 17 (1974) 2.

[3] GOULD, G.W., Inactivation of spores in food by combined heat and hydrostatic pressure, Acta Aliment. 2 (1973) 377.

[4] CLOUSTON, J.G., WILLS, P.A., Initiation of germination and inactivation of Bacillus pumilus spores by hydrostatic pressure, J. Bacteriol. 97 (1969) 684.

[5] SALE, A.J.H., GOULD, G.W., HAMILTON, W.A., Inactivation of bacterial spores by hydrostatic pressure, J. Gen. Microbiol. 60 (1970) 323.

[6] GOULD, G.W., SALE, A.J.H., Initiation of germination of bacterial spores by hydrostatic pressure, J. Gen. Microbiol. 60 (1970) 335.

[7] GOULD, G.W., JONES, A., WRIGHTON, C., Limitations of the initiation of germination of bacterial spores as a spore control procedure, J. Appl. Bacteriol. 31 (1968) 357.

[8] SMITH, N.R., GORDON, R.E., CLARK, F.E., Aerobic spore forming bacteria. USDA Monograph No.16, Wash., (1952).

[9] FARMILOE, F.J., CORNFORD, S.J., COPPOCK, J.B.M., INGRAM, M., The survival of B. subtilis spores in the baking of bread, J. Sci. Food Agric. 5 (1954) 292.

[10] MATTHEWS, R.W., Potentiometric estimation of megarad dose with the ceric-cerous system, Int. J. Appl. Radiat. Isot. 23 (1972) 179.

[11] RODE, L.J., FOSTER, J.W., Ionic germination of spores of Bacillus megaterium QM B1551, Arch. Mikrobiol. 43 (1962) 183.

[12] CLOUSTON, J.G., WILLS, P.A., Kinetics of initiation of germination of Bacillus pumilus spores by hydrostatic pressure, J. Bacteriol. 103 (1970) 140.

[13] BRIGGS, A., YAZDANY, S., Effect of sodium chloride on the heat and radiation resistance and on the recovery of heated or irradiated spores of the genus Bacillus, J. Appl. Bacteriol. 33 (1970) 621.

MICROBIOLOGICAL PROBLEMS OF RADIATION STERILIZATION CONTROL OF DISPOSABLE MEDICAL PRODUCTS

V. HORÁKOVÁ
SVUT Centre for Research and Application of Ionizing Radiation, Brno, Czechoslovakia

Abstract

MICROBIOLOGICAL PROBLEMS OF RADIATION STERILIZATION CONTROL OF DISPOSABLE MEDICAL PRODUCTS.

Dose-response curves were determined for three strains of cocci and seven strains of aerobic spore-forming rods after irradiation by two different ^{60}Co sources and Van de Graaff electron accelerator. Besides the test strains Streptococcus faecium A_2l, Bacillus sphaericus C_IA and Bacillus pumilus E 601, some strains isolated from irradiated vaccines and animal diets, or found among common air-contaminating bacteria and pathogenic cocci were examined. The efficiency of the used radiation sources was compared. The control of the microbiological efficiency of radiation sterilization is discussed regarding routine practice.

1. INTRODUCTION

In recent years, discussion has dealt with the definition of the term "sterility" and, above all, how to control it. In the last Pharmacopoea Bohemoslovenica III [1] it is stated that those materials and subjects may be considered sterile that have been deprived of all viable germs, and their sterility proved by an appropriate examination. The required examination is mainly direct testing of products randomly taken from the sterilized batch. However, the criticism of conventional sterility testing [2-5] has caused attention to now be paid to the biological control of the efficiency of the sterilization process.

2. MATERIAL AND METHODS

The microbiological efficiency of the investigated radiation sources was tested by determining the inactivation curves of the following strains:

Micrococcus radiodurans CCM 1700 (A.W. Anderson, strain R_1) [6], Streptococcus faecium A_2l [7], Staphylococcus aureus CCM 1770; Bacillus sphaericus C_IA [7]; Bacillus pumilus E 601 [8], Bacillus subtilis var. niger NCTC 10073; Bacillus subtilis CCM 1817; Bacillus subtilis M, Bacillus sp. 29 B and Bacillus sp. 2b 1.

Preparation of test pieces for irradiation:

Non-sporulating strains were incubated on plates with blood agar or Dextrose yeasts agar for 48 h and spore-forming strains on nutrient agar for 5 days at a temperature of 35°C. After incubation, a suspension in saline solution was prepared and pipetted in 0.05-ml amounts on sterile polyethylene foils.

The test pieces were air-dried at 35°C and a relative humidity of 65-70% for 24 h.

After being sealed into polyethylene bags, the test pieces were exposed to six graded doses of ionizing radiation in a range from 0.1 to 3 Mrad. The experimental details described in Procedures for Counting of Surviving Organisms [7] were further used in order to estimate the number of surviving cells. An average of six test pieces was calculated.

The irradiation of the tested strains was carried out parallelly by three sources:

Van de Graaff electron accelerator, TUR Dresden, GDR (intensity 50 and 100 μA, voltage 1.5 MeV, dose-rate 2.15, and 4.4×10^7 rad/min),

^{60}Co source – ^{60}Co Irradiator Type J-6000, Atomic Energy of Canada, Ltd. (dose-rate 23.01 krad/h),

^{60}Co Gamma-cell Irradiation Unit, Atomic Energy of Canada, Ltd. (dose-rate 336.15 krad/h).

3. RESULTS AND DISCUSSION

In determining the biological efficiency of irradiation facilities, attention was paid to three test strains, Bac. pumilus E 601, Bac. sphaericus C_IA and Str. faecium A_2l, recommended and obtained through the IAEA [8].

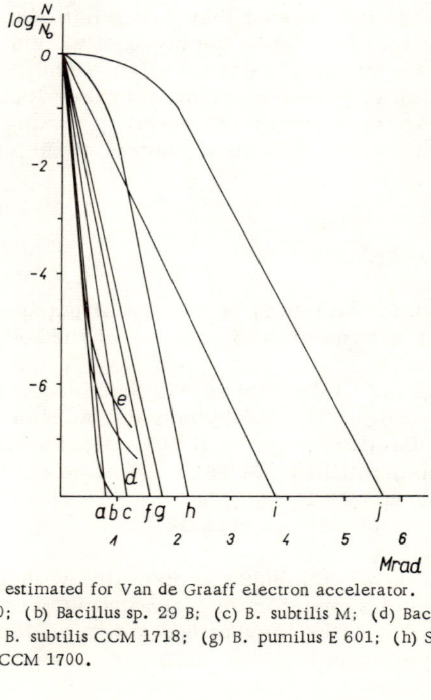

FIG.1. Inactivation curves estimated for Van de Graaff electron accelerator.
(a) Staph. aureus CCM 1770; (b) Bacillus sp. 29 B; (c) B. subtilis M; (d) Bacillus sp. 2b 1; (e) B. subtilis var niger NCTC 10073; (f) B. subtilis CCM 1718; (g) B. pumilus E 601; (h) Str. faecium A_2l; (i) B. sphaericus C_IA; (j) Mic. radiodurans CCM 1700.

To have a comparison, the experiments were extended on an estimate of dose-response curves for several other strains isolated from irradiated animal diets and vaccines (Bac. subtilis M, Bacillus sp. 29 B, and Bacillus sp. 2b 1); further, for common air spore-forming rods Bac. subtilis CCM 1718, which often cause accidental complications by sterility testing; and finally for pathogenic cocci Staph. aureus CCM 1770.

In Fig. 1 are semi-logarithmic plots of dose-response curves obtained for an electron accelerator, showing that the strains Mic. radiodurans CCM 1700, Bac. sphaericus C_IA and Str. faecium A_2l, were found to be the most resistant. A population of Mic. radiodurans germs was reduced by 8 log cycles after absorbing a 5.7-Mrad dose. A surviving fraction of 10^{-8} was assessed for the strain B. sphaericus when irradiated by a 3.8-Mrad dose, and for Str. faecium after exposure to a 2.25-Mrad dose. All other investigated strains were more sensitive. Initial counts were reduced by the inactivation factor 10^8 after irradiation by doses lower than 2 Mrad, including the strain B. pumilus E 601. The steepest inactivation curve was observed by Staph. aureus CCM 1770, and the inactivation factor 10^8 was assessed at a dose as low as 0.8 Mrad.

Similar relations were stated when bacterial cultures were irradiated by ^{60}Co isotopes, as shown in Fig. 2. The initial count of Mic. radiodurans germs decreased to a surviving fraction (10^{-8}) after receiving a 4.3-Mrad dose. The test pieces of B. sphaericus had to be exposed to a 3.2-Mrad dose, and those of Str. Faecium to 2.25 Mrad in order to obtain similar results. The strains B. subtilis CCM 1718 and Staph. aureus CCM 1770 survived 1.3- and 0.73-Mrad doses, respectively, in the surviving fraction (10^{-8}), and proved to have the lowest resistance.

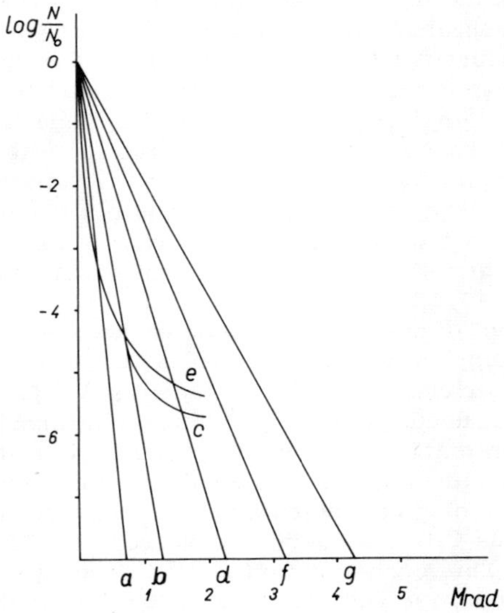

FIG. 2. Inactivation curves estimated for ^{60}Co irradiator.
(a) Staph. aureus CCM 1770; (b) B. subtilis CCM 1718; (c) B. pumilus E 601; (d) Str. faecium A_2l; (e) B. subtilis var. niger NCTC 10073; (f) B. sphaericus C_IA; (g) Mic. radiodurans CCM 1700.

The dose-response curves transposed to the semi-logarithmic plot showed almost straight lines, corresponding to a simple exponentional relation between survival and radiation dose. A clear shoulder was observed on inactivation curves of Str. faecium $A_2$1 after irradiation up to 1 Mrad and, in the case of Mic. radiodurans CCM 1700, up to a 2-Mrad dose, but only when beta rays were used. The presence of a steep initial slope and a pronounced tail were found by four spore-forming strains: B. subtilis var. niger NCTC 10073, Bacillus sp. 29 B, Bacillus sp. 2b 1 and B. pumilus E 601. The concave shape of an inactivation curve characterizes a microorganism population that is not homogeneous regarding resistance [3, 10]. We noticed a spontaneous dissociation of colonies in R and S forms in the case of the strain B. pumilus E 601. However, attempts to isolate the morphologically different colonies and to define their resistance separately remained without success. It cannot be excluded that these two morphologically different types may also possess a different resistance to radiation.

We expected to find differences in estimated counts of germs after irradiation in the industrial facility, ^{60}Co Irradiator J-6000, or in the Gammacell Irradiation Unit. In the latter case the dose-rate was more than 10 times higher, and also the average temperature exceeded the conditions in the industrial irradiator by 20°C. But the differences were not significant with the exception of the strain Str. faecium $A_2$1, where a surviving fraction of 10^{-8} was assessed at a dose of 1. 8 Mrad after exposure in the Gamma-cell chamber, whereas in the industrial irradiation facility it was necessary to use a higher dose of 2.25 Mrad.

Evident differences were, however, demonstrated when exposures to gamma or beta rays were compared. As follows from the referred data, it was always necessary to use higher beta-ray doses to inactivate similar counts of germs, compared with gamma radiation. The difference in the value of the doses of gamma or beta radiation increased with increasing resistance of the tested microorganisms. Therefore, in the case of the most resistant strain, Mic. radiodurans CCM 1770, the dose used in the accelerator exceeded by 1.4 Mrad the dose employed in ^{60}Co source so as to ensure the same inactivation factor. Anyway, the fact cannot be generalized, since only small number of strains was tested. An exception was again observed with the strain Str. faecium $A_2$1, but only when the irradiation was carried out in the industrial facility; then the dose reducing a given population of germs by a factor of 10^8 was the same as when it was irradiated by the electron accelerator. A higher efficiency of gamma radiation is also documented by other authors [7, 9, 11, 12].

From the aspect of resistance, shape of dose-response curves, and the reproducibility of experiments, the most reliable results were obtained with the strains B. sphaericus C_IA and Str. faecium $A_2$1 (see Fig. 3). This is why they are, in addition, used exclusively to test the microbiological efficiency of the industrial irradiator whenever the process is altered. Besides, biological indicators are applied in the daily routine control of radiation sterilization. The biological indicators are prepared from spores of the strain B. sphaericus C_IA and contain an average of 10^6 viable germs per single test piece. This concentration is still acceptable under our conditions and does not yield positive results after absorption of the minimum 2. 5-Mrad dose. Compared with the data of other authors [3, 7] the D-10 values of the strains B. sphaericus C_IA and Str. faecium $A_2$1 estimated by us are lower.

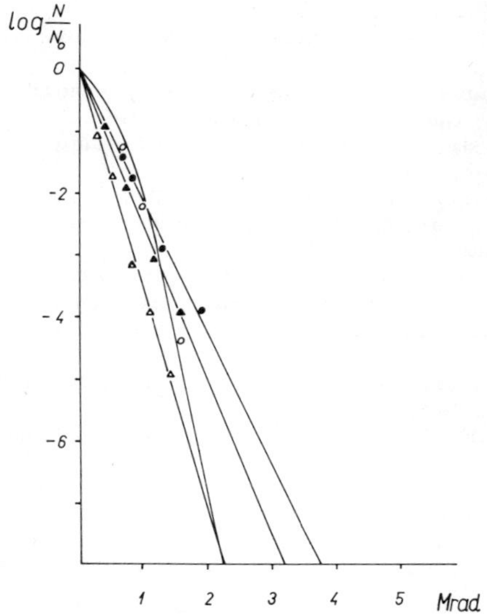

FIG.3. Inactivation curves of test strains Streptococcus faecium $A_2$1 and Bacillus sphaericus C_IA.
▲ ● B. sphaericus C_IA.
△ ○ Str. faecium $A_2$1.

○ ● Van de Graaff electron accelerator.
△ ▲ ^{60}Co irradiator.

The reason may lie in the fact that the test pieces for the estimation of dose-response curves were not prepared from suspensions in serum broth [8], which might be a protection against radiation, but the microorganisms were suspended in saline solution.

In Czechoslovakia radiation sterilization is permitted and introduced by Pharmacopoea Bohemoslovenica III [1], which suggests materials to be sterilized, basic radiation sources to be used in industrial practice, and a minimum recommended dose of 2.5 Mrad. The authorities require microbiological control of the radiation facility and the sterilization process by means of standard strains. When the control is performed regularly together with physico-chemical dosimetry, it is permitted for the conventional sterility testing on items taken at random from final products to be avoided. Testing by biological efficiency monitors does not prove that the sterilization dose was chosen correctly, taking into consideration the count and resistance of germs contaminating the product prior to sterilization. But, as it is based on radiation resistant strains, this control may give good information about the course of the sterilization treatment.

REFERENCES

[1] PHARMACOPOEA BOHEMOSLOVENICA III, Avicenum, Prague (1970).
[2] ARTANDI, C., "Aims and standards of sterilization treatments", Ionizing Radiation and The Sterilization of Medical Products (Proc. Symp. Risø, 1964), Taylor & Francis, London (1965) 85.

[3] CHRISTENSEN, E.A., "Radiation resistance of bacteria and the microbiological control of irradiated medical products", Sterilization and Preservation of Biological Tissues by Ionizing Radiation (Proc. Panel, Budapest, 1969), IAEA, Vienna (1970) 1.
[4] IGALI, S., "Microbiological control of products sterilized by radiation" (in Russian), CMEA Conf. Microbiological and Hygienic Questions of Radiation Sterilization, Bad Brambach (1970).
[5] CRAWFORD, C.G., in Manual on Radiation Sterilization of Medical and Biological Materials, Techn. Rep. Ser. No. 149, IAEA, Vienna (1973) 153.
[6] ANDERSON, A.W., NORDAN, H.C., CAIN, R.F., PARRISH, G., DUGGAN, D., Studies on a radio-resistant micrococcus, I. The isolation, morphology, cultural characteristics and resistance to gamma radiation, Fd. Technology 10 (1956) 575.
[7] Procedures for Counting of Surviving Organisms, Research Establishment Risø, CE-10-1968 (1968).
[8] INTERNATIONAL ATOMIC ENERGY AGENCY, "Recommended Code of Practice for Radiosterilization of Medical Products", Radiosterilization of Medical Products (Proc. Symp. Budapest 1967), IAEA, Vienna (1967) 423.
[9] LEY, F.J., in Manual on Radiation Sterilization of Medical and Biological Materials, Techn. Rep. Ser. No. 149, IAEA, Vienna (1973) 37.
[10] LAWRENCE, C.A., BLOCK, S.S., Disinfection, Sterilization and Preservation, Lea and Febiger, Philadelphia (1968).
[11] TATTERSALL, K., "Problems of microbial contamination in prepacked preparations", Ionizing Radiation and the Sterilization of Medical Products (Proc. Symp. Risø 1964), Taylor & Francis, London (1965) 15.
[12] CHRISTENSEN, E.A., RINGERTZ, O., HOLM, N.W., in Manual on Radiation Sterilization of Medical and Biological Materials, Techn. Rep. Ser. No. 149, IAEA, Vienna (1973) 131.

DISCUSSION

Kikiben PATEL: What is the situation with regard to the possible presence of pyrogens in the medical product?

Vlasta HORÁKOVÁ: Pyrogenicity is very carefully tested for whenever a new product is to be put on the market. If the results of the tests are favourable, industrial production is permitted. There is no further requirement for routine control of pyrogenicity, but attention to the level of bacterial contamination before the radiation sterilization is recommended. Regular pyrogenicity testing is carried out only when ethylene-oxide sterilization is applied.

H. M. ROUSHDY: Your curves show the inactivation effect of ^{60}Co gamma radiation on certain strains of microorganisms to be significantly greater than that of an accelerated electron beam. Can you explain the mechanism underlying this phenomenon of differential radiosensitivity in terms of the RBE.

Vlasta HORÁKOVÁ: The differences have also been observed by other authors, as I stated. We did not attempt to find an explanation for them. I assume the different RBEs may be due to the low energy of the Van de Graaff electron accelerator used (1.5 MeV). The differences might be less apparent when using higher energy.

A. TALLENTIRE: I feel that extreme caution should be exercised in reporting dose/ln survival curves that are concave in shape, i.e. curves that exhibit a "tail". There have been several occasions in the past when it was originally thought that this form of curve described the true response to radiation, but it was later found to be due to a technical fault!

Vlasta HORÁKOVÁ: I agree. The methods used, cultivation conditions and so forth, play a major role in determining the dose/response curves. We plan to make further studies of the conditions under which we obtained results for a number of the spore-forming strains described.

DOSIMETRY ASPECTS OF RADIATION-STERILIZATION PRACTICES

(Session II)

Chairman: K. H. CHADWICK (Netherlands)

ВЫБОР ДОЗЫ ОБЛУЧЕНИЯ ДЛЯ СТЕРИЛИЗАЦИИ МЕДИЦИНСКОЙ ПРОДУКЦИИ

В.В.БОЧКАРЕВ, Е.П.ПАВЛОВ, В.В.СЕДОВ,
В.Г.ХРУЩЕВ, Э.Г.ТУШОВ, Г.А.КОНЯЕВ
Институт биофизики Министерства здравоохранения СССР,
Москва,
Союз Советских Социалистических Республик

Abstract—Аннотация

CHOICE OF THE IRRADIATION DOSE FOR THE STERILIZATION OF MEDICAL PRODUCTS.

The principles for selecting the appropriate dose for the radiation sterilization of medical products are set forth, taking into account the initial contamination of the product, the radiation sensitivity of contaminants and the required level of reliability of sterilization. The initial contamination level of certain medical preparations (glucose and radiopharmaceuticals) is established and the radiation sensitivity of the isolated contaminants is determined in terms of Dio indices. Of the microorganisms isolated the most common were staphylococci, streptococci, Gram-negative bacteria, Aspergilli, Penicillia and yeast fungi. Spore-forming types of microorganisms were isolated with a frequency of the order of 10^{-2}. The radiation sensitivity in terms of D_{10} was established for more than 3000 strains of microorganisms. For 75 - 85% of the strains the D_{10} indices gave 10 - 40 krad and for 0.2 - 1% more than 100 krad. The experimental data were subjected to computer analysis which confirmed the adequacy of the techniques used to determine radiosensitivity. The authors then calculated by computer the radiation sterilizing doses for different degrees of initial contamination of the products with reliability coefficients of 10^6 and 10^8. For sterilizing radiopharmaceuticals and glucose solutions these doses are 0.8 - 1 Mrad. Verification experiments show that these doses give reliable sterilization of the medical products concerned.

ВЫБОР ДОЗЫ ОБЛУЧЕНИЯ ДЛЯ СТЕРИЛИЗАЦИИ МЕДИЦИНСКОЙ ПРОДУКЦИИ.

Излагаются принципы выбора величины дозы облучения для стерилизации медицинской продукции, основанные на учете инициальной контаминации стерилизуемой продукции, радиационной чувствительности контаминантов и требований к надежности стерилизации. Приводятся данные определения инициальной контаминации некоторых медицинских препаратов (глюкоза, радиофармацевтические препараты), а также результаты изучения радиационной чувствительности выделенных контаминантов по показателям $Д_{10}$. Среди выделенных микроорганизмов преобладали стафилококки, стрептококки, грамотрицательные бактерии, аспергиллы, пенициллы и дрожжеподобные грибы. Спорообразующие виды микроорганизмов выделялись с частотой порядка 10^{-2}. Радиационная чувствительность по показателям $Д_{10}$ определена у свыше 3000 штаммов микроорганизмов. Для 75-85% штаммов показатели $Д_{10}$ составляли 10-40 крад, а для 0,2-1% они превышали 100 крад. Путем обработки экспериментальных данных на компьютере получена информация о достаточности проведенных исследований и рассчитаны величины стерилизующих доз радиации при различной инициальной контаминации продукции и коэффициентах надежности 10^6 и 10^8. Для стерилизации радиофармацевтических препаратов и растворов глюкозы эти дозы составляли 0,8-1 Мрад. Приведены результаты экспериментальной проверки эффективности установленных доз облучения, показавшей надежность стерилизации данной медицинской продукции в указанных дозах.

Известно, что стерилизующая доза облучения зависит от инициальной контаминации продукции, радиочувствительности контаминантов и требуемой степени надежности стерилизации. Однако, несмотря на то, что такой подход известен давно, вопрос об установлении обоснованной стерилизующей дозы на производстве медицинской продукции пока еще не решен. В чем же состоят трудности и каковы проблемы?

ТАБЛИЦА I. РАДИАЦИОННАЯ ЧУВСТВИТЕЛЬНОСТЬ МИКРООРГАНИЗМОВ, ОБСЕМЕНЯЮЩИХ РАДИОФАРМАЦЕВТИЧЕСКИЕ ПРЕПАРАТЫ И ПРОИЗВОДСТВЕННЫЕ ПОМЕЩЕНИЯ ПРЕДПРИЯТИЯ, ВЫПУСКАЮЩЕГО ЭТИ ПРЕПАРАТЫ

Микроорганизмы	$Д_{10}$ (крад)						
	10-20	21-40	41-60	61-80	81-100	100	Всего
Грамположительные палочки	34 (1,65%)	70 (3,4%)	60 (2,9%)	58 (2,81%)	38 (1,84%)	22 (1,07%)	282 (13,68%)
Staph. albus	239 (11,6%)	420 (20,38%)	5 (0,24%)	-	-	-	664 (32,23%)
Staph. aureus	53 (2,57%)	152 (7,37%)	-	-	-	-	205 (9,93%)
Streptococcus spp.	21 (1,01%)	36 (1,74%)	20 (0,97%)	9 (0,43%)	-	-	86 (4,21%)
Bac. subtilis	-	-	-	-	28 (1,35%)	2 (0,09%)	30 (1,44%)
Грамотрицательные кокки	33 (1,6%)	78 (3,78%)	13 (0,63%)	4 (0,19%)	-	-	128 (6,2%)
Грамотрицательные палочки	51 (2,47%)	201 (9,76%)	49 (2,38%)	17 (0,82%)	-	-	318 (16,43%)
Pseudomonus aeruginosa	-	-	1 (0,04%)	-	-	-	1 (0,04%)
Aspergillus spp.	26 (1,26%)	69 (3,34%)	27 (1,31%)	2 (0,09%)	9 (0,43%)	-	133 (6,44%)
Penicillium spp.	17 (0,82%)	44 (2,13%)	31 (1,05%)	19 (0,99%)	10 (0,40%)	-	121 (5,4%)
Дрожжевые грибы	19 (0,92%)	42 (2,03%)	24 (1,16%)	6 (0,29%)	1 (0,04%)	-	92 (4,44%)
Всего:	493 (23,89%)	1112 (53,92%)	230 (10,68%)	115 (5,52%)	86 (4,13%)	24 (1,15%)	2060 (100%)

Прежде всего следует отметить, что проведением соответствующих санитарно-гигиенических мероприятий можно добиться поддержания инициальной контаминации радиационно стерилизуемой продукции в каких-то подходящих пределах. Точно также разработаны достаточно простые методы определения инициальной контаминации продукции и известны критерии надежности стерилизации. Проблемы возникают, во-первых, в области изучения радиационной чувствительности микроорганизмов, контаминирующих продукцию и производство, и, во-вторых, в том, каким образом учитывать результаты изучения радиочувствительности микрофлоры при установлении стерилизующей дозы облучения.

Что касается изучения радиационной чувствительности микроорганизмов, определяемой по показателям $Д_{10}$ и кривым выживаемости, то основным сдерживающим фактором в проведении этих исследований является, очевидно, их значительная трудоемкость. Однако, другим немаловажным отрицательным фактором является и то обстоятельство, что весьма нелегко в конкретном случае определить нужный объем исследований по изучению радиочувствительности контаминантов.

Положение значительно улучшилось после того, как в исследовательской работе стало возможным применить современную вычислительную технику. Это позволило использовать математический аппарат для обработки экспериментальных данных с целью установления величины стерилизующей дозы облучения.

В настоящем докладе приведены результаты изучения радиационной чувствительности микроорганизмов на двух крупных предприятиях медицинской промышленности СССР, данные, полученные при обработке этих результатов на компьютере с целью определения достаточности объема исследований и установления величин стерилизующих доз облучения, а также результаты проверки эффективности установленных режимов стерилизации.

Выделение микроорганизмов и контроль стерильности продукции после облучения проводили с помощью обычных микробиологических методов исследования. В качестве питательных сред использовали агар Хоттингера (75 мг% аминного азота, pH 7,2-7,4) с добавлением 2% дрожжевого гидролизата и 10% инактивированной бычьей сыворотки, а также тиогликолевую среду Сабуро. О радиочувствительности микроорганизмов судили по характеру кривых "доза-эффект" и по величине $Д_{10}$. Последнюю определяли по известной формуле:

$$Д_{10} = \frac{Д}{\log_{10} S_0 - \log_{10} S_1}$$

где Д – доза облучения, S_0 – исходная популяция, S_1 – количество выживших клеток после облучения популяции в дозе Д. Исходя из экспериментально найденных величин $Д_{10}$ и инициальной контаминации, а также требуемого коэффициента безопасности, рассчитывали с помощью компьютера величину стерилизующей дозы облучения, используя известное уравнение:

$$\frac{1}{KN} = \sum_i \delta_i \cdot 10^{-\frac{Д}{Д_{i_{10}}}} \qquad [1]$$

где K – коэффициент безопасности, N – инициальная контаминация продукции, δ_i – доля микроорганизмов в микрофлоре, для которых $Д_{10}^{i-1} \le Д_{10}^{i} \le Д_{10}^{i+1}$.

При изучении радиационной чувствительности микроорганизмов на производстве радиофармацевтических препаратов (РФП) получены следующие результаты (табл.I). Среди свыше 2000 штаммов микроорганизмов, выделенных на производстве РФП методом случайной выборки, преобладали радиочувствительные варианты. Показатели $Д_{10}$ для 77,8% штаммов не превышали 40 крад, а микроорганизмы, для которых показатели $Д_{10}$ превышали 100 крад, встречались с частотой всего 1,15%. Микрофлора изученного производства была представлена стафилококками, стрептококками, грамположительными и грамотрицательными палочками и грибами. Спорообразующие виды (преимущественно Bac. subtilis) встречались с частотой менее 2%. Это обстоятельство имеет практическое значение, поскольку спорообразующие микроорганизмы обычно отличаются радиорезистентностью.

Сходные результаты были получены при изучении радиочувствительности 1000 штаммов микроорганизмов, выделенных методом случайной выборки на производстве глюкозы (табл.II).

Экспериментальные данные изучения радиационной чувствительности микроорганизмов, выделенных на производстве медицинской продукции, были подвергнуты анализу на компьютере по принципу случайной выборки. Оказалось, что характер распределения вариантов в экспериментальной выборке терял зависимость от ее величины, начиная с выборки, состоящей из 300 штаммов. Из этого следует, что случайная выборка, состоящая из 300-500 выделенных на производстве штаммов, достаточна для того, чтобы охарактеризовать радиочувствительность микрофлоры данного производства.

На основании проведенных исследований были рассчитаны на компьютере по приведенному выше уравнению стерилизующие дозы облучения для каждого из изученных производств. Коэффициент надежности принят равным 10^6 и 10^8, что соответствует требованиям, обычно предъявляемым к надежности стерилизации медицинской продукции [2]. Как видно из табл.III, стерилизующие дозы для соответствующих уровней инициальной контаминации на обоих производствах оказались близкими. Это закономерно, поскольку, как отмечалось выше, существенных различий в спектре радиационной чувствительности микроорганизмов, выделенных на каждом из производств, не было.

Нами была изучена инициальная контаминация продукции обоих производств. На производстве РФП она составляла 10^1 клеток на единицу продукции, а на производстве глюкозы – 10^2. Как можно видеть из результатов исследований, приведенных в табл.III, стерилизующие дозы облучения для продукции с инициальной контаминацией такого уровня и данной микрофлоры не должны превышать 1 Мрад.

Для экспериментальной проверки эффективности установленных стерилизующих доз облучения были проведены исследования с естественно и искусственно контаминированными образцами продукции. В качестве тест-микроорганизмов использовали неспоровые (Staph. aureus штамм 209P, E. coli штамм №675, Proteusvulgasis штамм №1, Streptococcus falcium штамм $A_2$1) и споровые формы микробов (Bac. subtilis, Bac. sphaericus C_1A). Штаммы Strept. faecium и Bac. sphaericus получены нами от доктора Christensen (Дания).

ТАБЛИЦА II. СПЕКТР РАДИАЦИОННОЙ ЧУВСТВИТЕЛЬНОСТИ КОНТАМИНАНТОВ, ВЫДЕЛЕННЫХ НА ПРОИЗВОДСТВЕ АМПУЛИРОВАННЫХ РАСТВОРОВ ГЛЮКОЗЫ

Радиационная чувствительность микроорганизмов ($Д_{10}$ в крад)	Количество выделенных штаммов	Относительное содержание (%) штаммов среди изученных культур
1-20	477	47,7
21-40	391	39,1
41-60	89	8,9
61-90	31	3,1
81-99	10	1,0
100-120	2	0,2

ТАБЛИЦА III. ДОЗЫ ОБЛУЧЕНИЯ ДЛЯ СТЕРИЛИЗАЦИИ РАСТВОРОВ ГЛЮКОЗЫ И РАДИОФАРМАЦЕВТИЧЕСКИХ ПРЕПАРАТОВ, УСТАНОВЛЕННЫЕ НА ОСНОВАНИИ ИЗУЧЕНИЯ РАДИОЧУВСТВИТЕЛЬНОСТИ МИКРОФЛОРЫ ПРОИЗВОДСТВА ЭТОЙ ПРОДУКЦИИ

Вид производства	Коэффициент безопасности	Инициальная контаминация продукции		
		10^1	10^2	10^3
		Стерилизующие дозы облучения (крад)		
Раствор глюкозы (ампулы)	10^6	530	640	760
	10^8	760	880	1000
Радиофармацевтические препараты	10^6	520	640	750
	10^8	750	870	990

Поглощенные дозы облучения 0,2-0,4 Мрад обеспечивали фактор инактивации неспоровых форм микробов порядка 10^6-10^8. Для споровых форм микробов эти факторы инактивации достигались при дозах облучения около 1 Мрад. Ввиду низкой естественной инициальной контаминации продукции все 200 образцов, облученных в дозах 0,4-0,5 Мрад, оказались, как и следовало ожидать, стерильными. Таким образом, мы считаем, что облучение в дозе 1 Мрад обеспечивает достаточный фактор инактивации для микроорганизмов, контаминирующих продукцию изученных производств, а достигаемая надежность стерилизации отвечает требуемому уровню.

Величины стерилизующих доз облучения, по данным наших исследований, являются значительно более низкими, чем применяемые в настоящее время в промышленности в других странах. Так, в США, Великобритании, Франции и ФРГ обычно используют дозу облучения

2,5 Мрад. Однако, надо подчеркнуть, что такой обработке подвергаются изделия и материалы, содержащие микроорганизмы в сухом состоянии. Наличие в среде воды (и некоторые другие факторы), как в нашем случае, значительно повышает радиационную чувствительность микроорганизмов, что позволяет намного снизить дозы облучения, необходимые для стерилизации. Здесь уместно также отметить, что доза 2,5 Мрад применяется на производствах с весьма различными "микробиологическими" условиями. И поскольку этот режим стерилизации оказался эффективен в этих различных условиях, то, очевидно, для производств с хорошими санитарно-гигиеническими условиями доза 2,5 Мрад может оказаться завышенной.

В настоящее время существуют различные подходы к разработке режимов стерилизации радиационным методом. Например, скандинавские специалисты при выборе дозы облучения исходят из радиационной чувствительности высокорадиорезистентных микроорганизмов, выделенных в производственных условиях, без должного учета их относительного содержания в микрофлоре, обсеменяющей производство. Поэтому в Дании и Швеции приняты более высокие дозы облучения (3,5-4,5 Мрад), чем в других странах.

Однако, как мы уже отмечали [3], такой подход не отражает реальной ситуации на производстве. Так, из опубликованных исследований известно, что высокорадиорезистентные варианты микробов встречаются в производственной пыли и на изделиях с частотой, не превышающей одной клетки на 10^3-10^4 микробов смешанной популяции. Обсемененность изделий (шприцы, изделия службы крови) в среднем не превышает 10^2 микробов на одно изделие [4]. Таким образом, на 1 млн. изделий содержится не более 10^8 микробов. Это означает, что в указанной популяции может быть не более 10^4-10^5 микробов, обладающих высокой радиорезистентностью. Как показывает анализ кривых "доза-эффект", для стерилизации такого количества радиорезистентных микробов достаточна доза облучения, равная 2,5 Мрад. Радиочувствительная часть этой популяции, а она составляет свыше 99% всех микробов, стерилизуется при облучении дозами до 1,5 Мрад [5]. Основной и существенный недостаток указанного выше методического подхода состоит в том, что он приводит к завышению стерилизующих доз облучения. Это может отрицательно отразиться на качестве изделий и на технико-экономической эффективности радиационного метода стерилизации в целом, а также значительно сузить область его применения.

Для разработки оптимальных режимов радиационной стерилизации Ley и Tallentire [6] предложили изучать и отрабатывать режимы стерилизации на изделиях, "естественно" загрязненных в условиях конкретного производства, искусственно повысив инициальную контаминацию. При этом имеется в виду следующее. В течение определенного времени на предприятии выпускаются изделия без соблюдения гигиенических правил (обработка изделий персоналом без предварительного мытья рук, повышенная запыленность помещений, прекращение дезинфекции помещений и т.д.), что приводит к резкому (на 2-3 порядка) увеличению обсемененности изделий микробами, представляющими местную естественную микрофлору. После упаковки и стерилизации проверяют стерильность изделий. Величина контрольной выборки устанавливается в зависимости от требуемой степени надежности. Таким способом можно решить вопрос об эффективности различных режимов стерилизации. Смысл этой методики заключается в том, что, во-первых, выбор дозы основывается на реальной радиочувствительности "естественной" для данного производства микрофлоры (и в этом основная положительная сторона метода) и, во-вторых, количество отбираемых для изучения эффективности стерилизации изделий сокращается в силу их более высокой инициальной контаминации. Но при такой постановке исследований происходит значительное микробное загрязнение производства, а выпускаемая продукция сильно обсеменяется. Поэтому такую методику, по-видимому, целесообразно применять в опытно-промышленных условиях, а не на серийном производстве.

Особого рассмотрения заслуживает вопрос о том, каким образом устанавливать стерилизующие дозы облучения на вновь организуемых производствах. В этих условиях мы не исключаем возможности применения эталонных радиорезистентных штаммов, таких как Bac. pumilus E601, Streptococcus faecium $A_2$1, Bac. sphaericus C_1A и др. Однако, при этом необходимо снимать кривые "доза-эффект" в подходящих условиях (в изделиях, растворах препаратов и т.д.). Мы не считаем правильным применять специальную обработку микробов (например, лиофилизация в растворе сывороточного альбумина) с целью повышения их радиорезистентности. По нашему мнению, наиболее подходящими тест-микроорганизмами являются споры микробов.

Другой важный вопрос, который возникает при изучении эффективности стерилизации с применением тест-микроорганизмов, связан с определением требуемого фактора инактивации. Здесь необходимо принять во внимание то обстоятельство, что радиорезистентные варианты микробов встречаются на производстве с частотой порядка 10^{-2}. Таким образом, для обеспечения нужной надежности стерилизации (коэффициент безопасности 10^6) продукции, имеющей инициальную контаминацию 10^2 клеток, требуется инактивировать не больше 10^6 радиорезистентных микробов. По этому фактору инактивации и следует выбирать величину стерилизующей дозы облучения на кривой "дозы-эффект", снятой в условиях, соответствующих тем, которые будут на производстве. Установленные таким способом режимы стерилизации должны рассматриваться в качестве временных.

В целом же мы полагаем, что действительно научно обоснованный подход в выборе величины дозы облучения должен базироваться на изучении естественной микробной контаминации с учетом реального спектра радиационной чувствительности контаминантов. Этот спектр показывает относительную долю микроорганизмов с определенным уровнем радиочувствительности (величины $Д_{10}$). Зная инициальную контаминацию продукции, а также учитывая требуемый коэффициент надежности, можно установить стерилизующую дозу облучения.

Затронутые в настоящем докладе вопросы отнюдь не исчерпывают круг проблем, связанных с выбором оптимальных режимов стерилизации медицинской продукции. Мы не рассматривали таких важных вопросов, как влияние различных факторов на радиационную чувствительность контаминантов, значение возможной контаминации продукции вирусами и т.д. В отношении последних известно, что они отличаются обычно значительной радиорезистентностью. Но, по-видимому, проблема вирусной контаминации возникает только при стерилизации продукции биологического происхождения, особенно при стерилизации различных гомотрансплантатов, и, что важно, дозы облучения в 1 Мрад и выше обеспечивают значительный фактор инактивации вирусов.

ЛИТЕРАТУРА

[1] БОЧКАРЕВ, В.В., ТУШОВ, Э.Г. и др., Медицинская Радиология, 12 (1973) 34.
[2] KALLINGS, L.O., Radiosterilization of Medical Products, IAEA, Vienna (1967) 433.
[3] БОЧКАРЕВ, В.В., ПАВЛОВ, Е.П. и др., Журнал Микробиологии, Эпидемиологии и Иммунологии, 8(1973) 56.
[4] CHRISTENSEN, E., MUKHERJI, S., HOLM, N., Arch. Pharm. Chemi 75 (1968).
[5] CHRISTENSEN, E., Sterilization and Preservation of Biological Tissues by Ionizing Radiation, IAEA, Vienna (1970) 1.
[6] LEY, F., TALLENTIRE, A., Pharm. J. 193 (1964) 216.

DISCUSSION

P.V. IYER: In your country, have you found any seasonal differences in the radiosensitivity of the organisms that might have an effect on the determination of the radiation dose?

E.P. PAVLOV: No, over the two years that we have been conducting the investigations, we have not found any significant differences from season to season in the spectra for radiosensitivity of the contaminating microflora.

N.G.S. GOPAL: Do not dissolved compounds in aqueous solutions affect the inactivation factors of microorganisms in them, perhaps as a result of competition between the solutes and microorganisms for the free radicals?

E.P. PAVLOV: For the most part, we determined the D_{10} values in salt solutions and in distilled water (i.e. in the case of microbial spores). But in some instances the radiosensitivity determinations were made in solutions of different drugs (not containing the SH-group). In such cases the radiosensitivity level did not usually fall below — and was often even higher than — the values for the former.

FACILITY CALIBRATION, THE COMMISSIONING OF A PROCESS, AND ROUTINE MONITORING PRACTICES

K.H. CHADWICK
Association Euratom-ITAL,
Wageningen,
The Netherlands

Abstract

FACILITY CALIBRATION, THE COMMISSIONING OF A PROCESS, AND ROUTINE MONITORING PRACTICES.
Radiation dosimetry provides an independent control of the radiation sterilization process. The various dosimetric measurements which should be carried out in calibrating a facility and commissioning a process are discussed. Some attention is given to batch-operated plants, the calibration of a dosimetry system and accuracy and precision in use. One or two dosimetry systems which are being developed for megarad doses are briefly discussed.

DOSIMETER CALIBRATION

The best advice which can be given to the processor is: take a dosimeter system which is well developed, has a standard procedure for its use (4), is roughly water equivalent (plastics, liquid chemicals) and is best suited to your own situation. Do not try to develop your own dosimeter system. In spite of the many varied radiation effects measurable at megarad doses (5) surprisingly few dosimetry systems have been developed to a routine use in many laboratories throughout the world. Many liquid chemical dosimeters are very sensitive to impurities and one of the more promising liquid systems seems to be the ethanol-chlorobenzene dosimeter (6), several plastic film dosimeters have been developed and the three perspex systems appear to have achieved the widest use in this field till now (7).

The perspex systems can all be purchased together with a calibration curve of optical density change versus dose but this assumes that the read-out instrument, in this case a spectrophotometer, is in proper working order. Figure 1 illustrates the calibration of clear perspex using four different spectrophotometers and illustrates the errors which may be made by assuming a calibration curve and using a measuring system which is not working properly. It is in fact always advisable to make ones own calibration using ones own read-out unit, this has the added advantage that the processor gains experience in the use of the system and an idea of the reproducibility or precision he can expect to achieve. Many processors, however, do not have the possibility of making their own calibration and, in order to overcome the possible errors arising from faulty spectrophotometer use, a series of perspex dosimeters (clear, red or amber), which have been irradiated to a number of known doses, can now be purchased. Using these, the processor can set up his own calibration curve using his own read-out unit with the same type of dosimeter he will use in practice. It is important to realise that the induced optical density is not stable in any of the perspex types and that any time period, between the calibrating irradiation at the supplier and the read-out by the processor, which exceeds 2 or 3 days, will lead to fading corrections which cause a decrease in the accuracy of the calibration. (Figure 2). An alternative solution to this problem is a complete check of

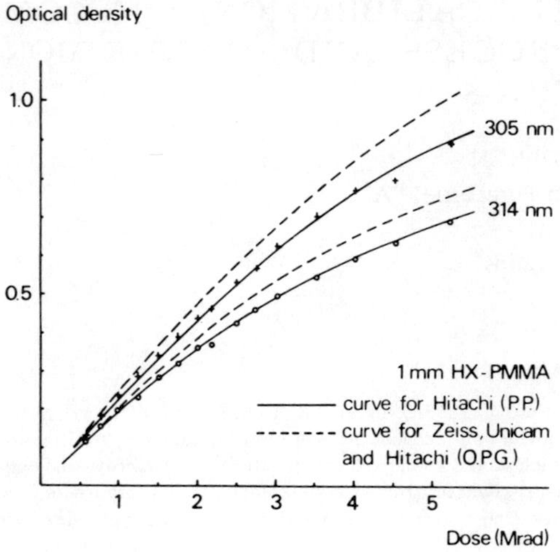

FIG.1. The calibration curves for 1-mm HX clear perspex measured on four different spectrophotometers using the same dosimeter samples. One spectrophotometer gives a different curve of OD versus dose.

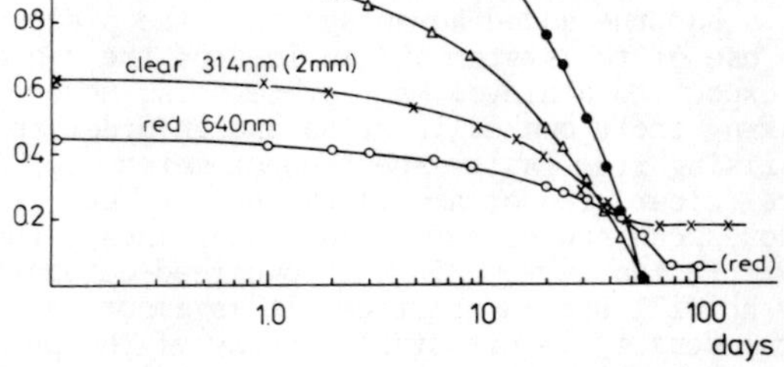

FIG.2. Air or oxygen induced fading of the different PMMA types after irradiation illustrating the stable OD remaining in clear HX perspex and the complete fading in the coloured perspex. The fading rate is exaggerated as the samples have been stored under 1 atm pressure of oxygen.

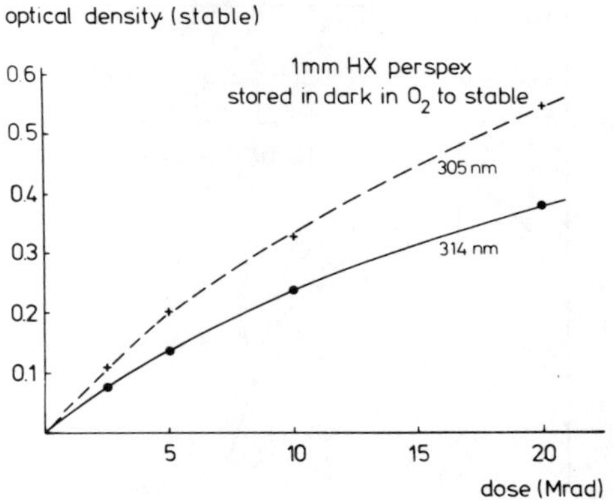

FIG.3. The stable OD-dose relationship for 1-mm clear HX perspex stored in air or oxygen in the dark.

the read-out unit, wavelength, OD scale and straylight, for example, plus a set of samples, similar to the dosimeter, having a known stable OD value in the wavelength region of interest, which cover the range of OD values used in the dosimetry. In the case of clear perspex it may well be possible to make a series of such samples as the radiation-induced OD contains a very stable portion which remains after oxygen fading and which also exhibits a dose response. (Fig. 3). We have irradiated several samples to high doses and have followed the stability of this remaining species for some six months now and the results can be considered promising.

If all precautions are taken and the standard procedure for the dosimeter system chosen is followed it should be possible to achieve a calibration of the complete dosimeter system to an accuracy of 3-5%, a value which I consider to be quite satisfactory for radiation sterilization at the processing stage. Using the perspex systems a precision of better than 2% (1 σ) can be achieved and this seems to be a good target level for other systems.

INTERCOMPARISON

In view of the increasing use of radiation for sterilization throughout the world and the concommitent increase in international trade in products sterilized by radiation it seems to me that it is time that an independent organisation such as the IAEA set up an international procedure for the intercomparison of radiation dose in the megarad range. This would make it easier for public health authorities to accept and approve products which were radiation sterilized in another country. It would also give the processor added confidence in his dosimetry technique and a reliable source of standardization. Such an intercomparison is at present being considered within the European Common Market under the auspices of the European Commission.

DOSIMETRY ACCURACY IN PRACTICE

Having a dosimeter system which is calibrated to 5% accuracy, has a precision of 2% and agrees with international norms is not sufficient. A dosimeter in practice, where it may be irradiated at different dose-rates

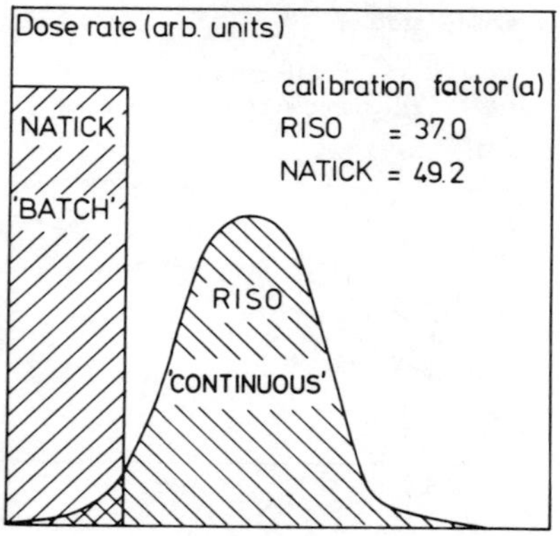

FIG.4. The dependence of the calibration factor (a) of the oxalic acid dosimeter on the exposure pattern for gamma rays (1).

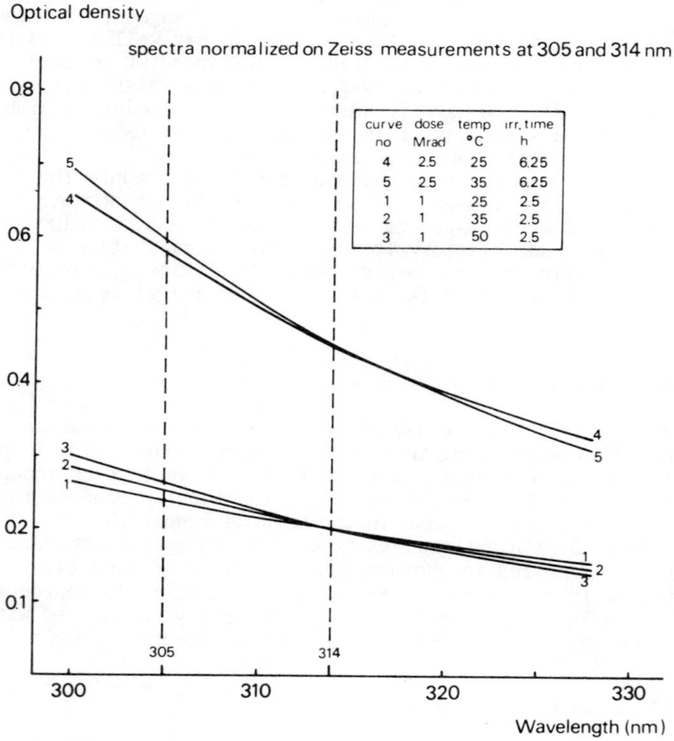

FIG.5. The change in the OD spectrum in the clear HX perspex dosimeter caused by irradiation at different temperatures.

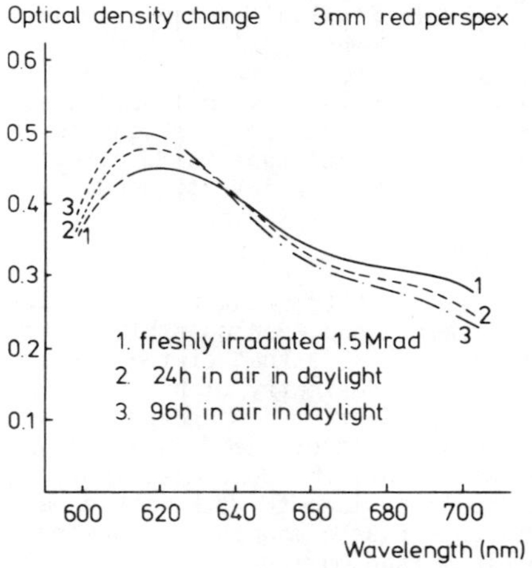

FIG.6. The change in the OD spectrum in the 3-mm red 4034 perspex dosimeter with time at room temperature.

Table I.

The maximum and minimum dose given by different dosimeters relative to the dose in water for measurements in thick unit density targets irradiated in a four pass ^{60}Co plaque bulk irradiator and the apparent dose uniformity ratio (D_{max}/D_{min}).
Data taken from reference 11 which are based on the theoretical calculations of Brynjoflsson (10).

Target thickness	20 cm			25 cm		
Dosimeter	D_{max}	D_{min}	D_{max}/D_{min}	D_{max}	D_{min}	D_{max}/D_{min}
Water	1.0	1.0	1.27	1.0	1.0	1.45
Cerric Sulphate (0.1 M Ce)	1.10	1.20	1.17	1.075	1.21	1.29
Perspex (PMMA)	0.95	0.95	1.28	0.95	0.95	1.46
LiF	1.01	1.015	1.26	1.005	1.015	1.44
PVC	1.16	1.21	1.22	1.13	1.30	1.26
Quartz, SiO_2-glass	1.07	1.11	1.22	1.05	1.13	1.35

of scattered radiation, possibly at differing temperatures, over a period of several hours, may behave somewhat differently than in calibration.
Figures 4, 5 and 6 and table I illustrate differences which may arise: the oxalic acid dosimeter is very dose-rate dependent (1) the clear perspex dosimeter exhibits a change in OD spectrum which depends on time and temperature and means that best stability is obtained using 314-316 nm as the measurement wavelength (8) the red perspex dosimeter also exhibits a change in OD spectrum and 640 nm is the best measurement wavelength (9), and some

dosimeters which are not water equivalent can give an erroneous estimate of dose when a considerable component of the dose arises from scattered radiation, as is the case in a bulk irradiator (10, 11).

These problems are not always easy to find, investigate and eliminate, which is one of the reasons why the processor should choose a well-developed system where hopefully all the important "side-effect" problems which may affect the accuracy in use have been investigated.

FACILITY CALIBRATION

In some cases the firms constructing the irradiation facility will make some initial dose measurements and when possible the processor should use his own dosimetry system along side that of the firm, in other cases he is left alone to investigate the dosimetry of his plant. Initial measurements usually make use of dummy product of a standard bulk density and the first important step is the measurement of the dose distribution in a typical product package in a full plant. The dose distribution itself is of little interest, no attempt should be made to work with a mean dose in the product as the mean dose is in this case a meaningless number, what is important is the position(s) of minimum (and maximum) dose in the product, and the Uniformity Ratio (D_{max}/D_{min}) which should be compared with the calculated design value.

In a continuously operating automatic plant the product will usually be shuffled so that the dose varies only in the depth in the product, and the positions of maximum and minimum dose will be reasonably predictable. In a batch-operated facility, variations in dose may occur in the other two dimensions of the product, as well as in the depth, so that the positions of maximum and minimum dose may be more difficult to predict and greater care should be taken in making dose-distribution measurements.

Using the position of minimum dose the facility calibration curves can be determined. In a batch-operated facility the relation of minimum dose to time is linear but will not pass directly through the origin as a small, but possibly non-negligible, dose will be accumulated during the raising and lowering of the source.

In an automatic facility where the packages remain in each position for a fixed "dwell" time and are shuffled quickly from position to position the relation of the minimum dose to the total cycle time (dwell time + shuffle time) will be linear but will not necessarily pass directly through the origin. Normally the shuffle time will be constant and during this time the first and last boxes are also transported in and out of the facility so that a constant entrance and exit dose is accumulated. These "transit" doses may be very small compared to the total dose but the purpose of the measurement of the facility calibration is to determine this. (Fig. 7). In an automatic facility with a continuously moving conveyor where the entrance and exit conveyor speed is directly related to the speed past the source the relation of minimum dose to the reciprocal of the conveyor speed will be linear and will pass through the origin.

When the facility calibration curve has been set up using dummy product at one bulk density and it has been determined whether the curve passes through the real origin or an "imaginary" origin (point X in Fig. 7), curves can be generated for different bulk densities by making a single measurement at each density and joining these points with the experimentally determined origin.

In many practical situations the source in use is not filled with uniform product and in this case the facility is adjusted so that the minimum dose in the heaviest product in the source is the effective dose approved for the process. This means that lighter products are to some extent overdosed, and, although this may be considered as a safety factor, I wonder why irradiation plants are not equipped with a series of product boxes of

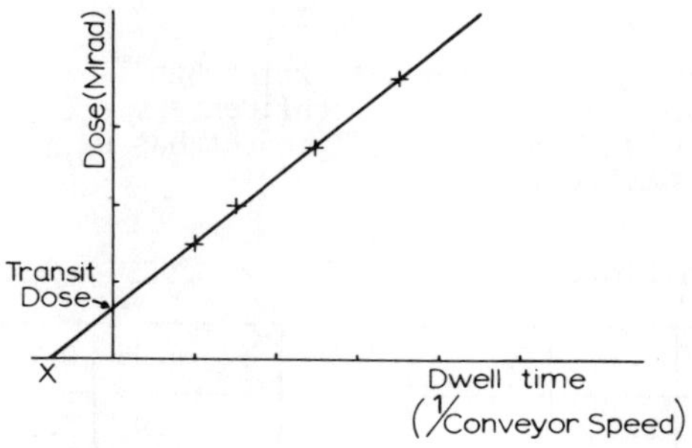

FIG.7. A typical facility calibration curve showing the determination of the "transit dose".

differing depths, each chosen to accept a small defined bulk density range, so that the facility could be run at a fixed rate, regardless of filling, and so that the uniformity ratio would be approximately constant for all the differing products and so that the minimum dose in all the product was of approximately the same value. This would reduce the amount of dosimetry needed in the facility although it might introduce some mechanical and packaging problems.

COMMISSIONING THE PROCESS

Even when the source is filled with a uniform product the bulk density of the product will vary from box to box and this means that the process is statistical in nature and that it is impossible to state absolutely that all the product receives more than the approved effective dose.

In order to examine the statistics of the process the minimum dose is determined in a number (n) of randomly selected product boxes and the mean minimum dose $\overline{D}_{min}$ and the standard error S_{BD} are calculated. Because $\overline{D}_{min}$ and S_{BD} are only estimates of the real values and because they are random variables the tolerance limit statement, which defines the statistical expectation for the process, can only be made with a given probability (or confidence) attached.

If the authorized minimum dose for a certain process is D_{eff}, then the processor can guarantee with a degree of confidence $\gamma = (1-\alpha)$, that there is only a probability $\beta = (1-P)$, that the minimum dose in any one box of product is less than D_{eff} by setting the plant parameter such that

$$D_{eff} = \overline{D}_{min} - kS_{BD} \qquad \text{(see Ref.12)}$$

This is a lower tolerance limit and k is a factor which is dependent upon n, γ and β and which can be obtained from statistical tables for a one sided frequency distribution (4, 13)
e.g. if γ =.95 (95% confidence level), β =.05 (5% chance of underdosing) and n = 20 then k = 2.4.
Table II indicates the dependence of k on n, γ and β.

The product kS_{BD} can be considered as a safety factor which is used to guarantee that the process is carried out correctly. Although the value of S_{BD} is dependent upon the bulk density variations it also contains implicitly a term which is related to the precision of the dosimeter system used. If the precision is good S_{BD} will depend mainly on bulk density variations and from the dosimetric point of view the process is optimalized. If the precision is poor and contributes to a larger S_{BD} then the safety factor will be too large

Table II.

The dependence of k, the factor with which SBD is multiplied, on the choice of the level of statistical confidence $\gamma = (1-\alpha)$, the probability $\beta = (1-P)$ that a box of product has $D_{min} < D_{eff}$ and the number of boxes sampled (n).

1. $\gamma = 0.95$ (95% Confidence level)

n \ β	0.05	0.1	0.15	0.2
8	3.4	2.75	2.2	1.9
10	3.0	2.4	2.0	1.7
15	2.6	2.1	1.8	1.5
20	2.4	1.9	1.6	1.4
30	2.2	1.8	1.5	1.3

2. $\gamma = 0.975$ (97.5% Confidence level)

n \ β	0.05	0.1	0.15	0.2
8	3.9	3.2	2.5	2.2
10	3.4	2.8	2.2	1.9
15	2.8	2.3	1.9	1.7
20	2.6	2.1	1.7	1.5
30	2.4	1.9	1.5	1.4

and the process is less efficient and less economic than it could be. This is an additional reason why the processor should choose a well-developed dosimeter system with a standard procedure.

ROUTINE MONITORING

In a well-designed continuously operating automatic facility it should be simple to monitor continuously the various plant parameters such as source position, package position, cycle time or conveyor speed etc. Thus, once a process has been commissioned using dosimetry, and the plant parameters are known, a continuous automatic control of these parameters should be sufficient to ensure that the plant is operating correctly. It is probably advisable to make random checks using dosimeters, more to check the source than to check the process.

In a batch-operated plant the position is different, handling and control will be, in general, manual and only semiautomatic, and there will be more chance for human error. In this case, I see no alternative to the use of dosimeters in each product package, either in the position of minimum dose, or when the dose distribution is well known, in a standard position where the dose can be directly related to the minimum dose in the product.

All product packages should have a Go-No-Go Radiation Indicator which changes colour in the dose region near to the approved effective dose for the process. This helps inventory control of what has or has not been irradiated and prevents a double irradiation.

DOSIMETER SYSTEMS

Three different perspex dosimeter systems are at present in use in sterilization plants; all three systems, clear HX, red 4034 and amber 3042 can be purchased together with calibration curves at the measurement wavelengths plus samples which have been irradiated to defined radiation doses from the UK Panel for Electron and Gamma Irradiation 35-37a, Finsbury Square, London EC 2.

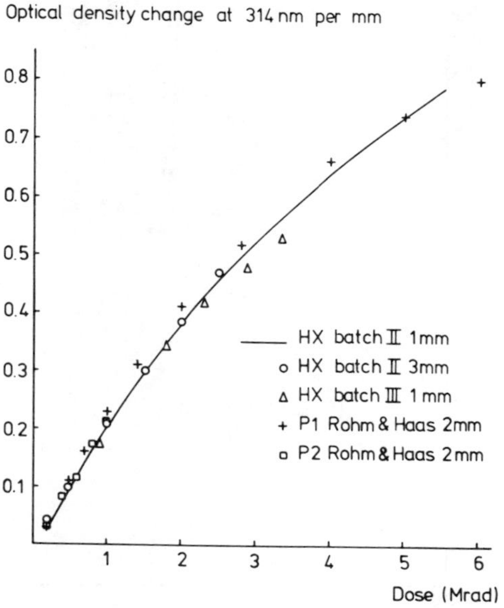

FIG.8. The OD-dose relationship of five clear perspex sheets of differing manufacture and thickness plotted as change in OD at 314 nm/mm thickness (ITAL measurements).

Clear HX perspex - Fig. 8 shows the OD versus dose relationships for a variety of different perspex types of differing thickness normalised to 1 mm thickness. The constancy of the response is an important point in favour of the use of this material as a dosimeter. Because of temperature/time instabilities 314-316 nm is the most stable measurement wavelength. The dosimeters can be obtained in thicknesses of 1 and 3 mm. At 314 nm the 3-mm sample can just be used to read out 2.5 Mrad and this could be advantageous when the irradiation times are long and oxygen diffusion during irradiation will be relatively more important in the 1-mm samples than in the 3 mm. Ideally a 2-mm dosimeter would improve on the fading problem associated with the 1-mm samples whilst giving a useful OD measurement up to ± 5 Mrad. After irradiation the OD fades on oxygen diffusion and eventually reaches a very stable value which could be useful as a secondary check on the radiation dose.

Red 4034 perspex - obtainable in 3 mm thickness, "pre-aged" packed in aluminium/plastic foil, this means that the unirradiated OD cannot be measured which will slightly reduce the precision. Measurement is at 640 nm to avoid temperature-time instability, this perspex is less sensitive and gives an OD reading of ± 0.6 at 2.5 Mrad. The induced OD is instable and fades on oxygen diffusion almost completely (Fig. 2). The dose relationship saturates more obviously than that of clear perspex at about 5-6 Mrad. Temperatures above 60-70 °C cause a considerable development of colour round 600 nm.

Amber 3042 perspex - also obtainable in 3 mm thickness, prepacked in the aluminium/plastic foil. The irradiated OD spectrum has two peaks for measurement, 603 nm and 651 nm. Treatment up to 55°C does not affect the

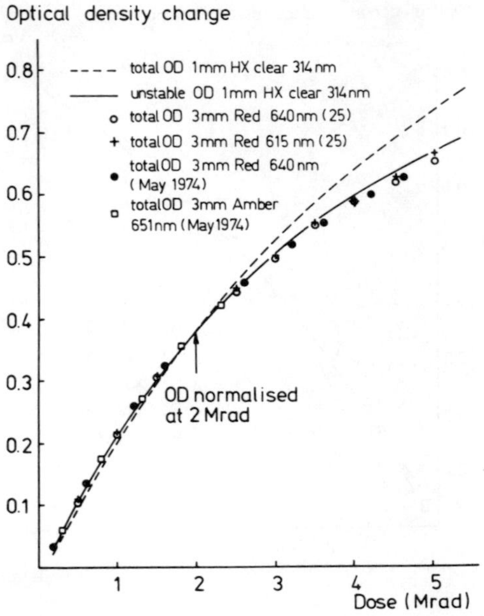

FIG.9. Comparison of the OD-dose relationships for clear, red and amber perspex when normalized at 2 Mrad.

603-nm peak but decreases the 651-nm peak, above 80°C the complete spectrum fades. In oxygen or air the OD fades completely (fig. 2). This perspex is sensitive and 603 nm cannot be used above 1.5 Mrad (OD = 1.6, 3 mm thick), the 651 nm is useful up to ∿ 2 Mrad (OD = 1.4) but I do not believe that it can be used in 3-mm thickness for doses above 2.5 Mrad unless a very special spectrophotometer is available. This system measuring at 603 nm could be a most useful system if it were made available in 2-mm and 1-mm thicknesses.

Figure 9 shows the dose relationships for red and amber perspex normalised at 2 Mrad to the total OD and unstable OD curve of the clear HX perspex. It can be seen that all three perspex types exhibit a threshold dose at ± 100 krad, and that the red and amber follow most closely the unstable OD curve for the clear perspex. The red and amber exhibit no stable OD response, saturate more obviously than the clear, making high dose measurements less sensitive, but, as the radiation-induced signal fades in time more or less completely, they have a doubtful advantage in that they could be re-used several times.

Radiochromic Dye Films - these thin film dosimeter systems developed by McLaughlin (14, 15) and by Harrah and Bishop (16,17) deserve special mention because they are tissue equivalent, stable before and after irradiation and are dose-rate independent up to 10^{15} rad/s. They are ideal for electron dosimetry, can be read out on a spectrophotometer at wavelengths in the 600-nm region, cover a dose range from 100 krad - 20 Mrad and can be obtained commercially from Far West Technology Inc. 330 South Kellogg, California 93017.

Ethanol-chlorobenzene - this dosimeter system has been developed by Dvornik *et al.* (18, 19) and extended to routine measurements for sterilization doses by Foldiak *et al.* (6) by using a high-frequency oscillometric measurement of conductivity. Although this system uses a liquid, which may be considered by some to be a disadvantage in an industrial process, it is reported to be insensitive to contaminants and to give a stable readout for at least two years after irradiation. The measurement technique is sensitive to local stray capacitance changes and also places high requirements on the dimensions of the plastic container of the dosimeter solution. Even so, this is one of the few techniques which is being developed to a routine use. I would personally like to see a standard procedure for this dosimeter system plus more information on dose-rate limitations, other possible side effects and the availability of the read-out instrument.

CONCLUSION

Dosimetry can provide an independent control of the radiation sterilization process; therefore, it is important that the processor choose a well-developed system and become thoroughly acquainted with it. The choice, calibration and intercomparison of the dosimeter system are important steps in setting up the radiation process. The statistical nature of the process should be understood and taken into account when the process is commissioned. Although there are several well-developed dosimetry systems available each one has some shortcomings and there does not appear to be one which is completely ideal.

ACKNOWLEDGEMENT

I would like to acknowledge the discussions I have had with L. Chandler, D. Ehlermann, W.L. McLaughlin, F.X. Rizzo and Y. Takashima during the preparation of the IAEA Dosimetry Manual for Food Irradiation which have certainly influenced a great part of this paper. This publication is contribution number 1152 of the Biology Division of the European Commission. The work is supported by the Euratom contract 094 BIAN and by the Dutch Ministry of Agriculture and Fisheries.

REFERENCES

[1] HOLM, N.W., BJERGBAKKE, E., Experiences in Radiosterilization of Medical Products, IAEA-159, Unpubl. Doc. (1974) 125.

[2] PHILIPS, G.O., Experiences in Radiosterilization of Medical Products, Ibid., p. 193.

[3] JUUL, F., CHRISTENSEN, E.A., HOLM, N.W., Experiences in Radiosterilization of Medical Products, Ibid., p. 153.

[4] CHANDLER, L., et al., Dosimetry for the Radiation Processing of Food, IAEA, Vienna, 1974, in Press.

[5] McLAUGHLIN, W.L., in Technical Developments and Prospects for Sterilization by Ionizing Radiation, Proc. Johnson-Johnson Symp., Multiscience Publications (1974) In Press.

[6] FOLDIAK, G., HORVATH, Z., STENGER, V., in Dosimetry in Agriculture, Industry, Biology and Medicine (Proc. Symp. Vienna, 1972), IAEA, Vienna (1973) 367.

[7] CHADWICK, K.H., International course on the metrology of ionizing radiation (CASNATI, Ed.), held in Varenna, 1974. In Press.

[8] CHADWICK, K.H., in Dosimetry in Agriculture, Industry, Biology and Medicine (Proc. Symp. Vienna, 1972), IAEA, Vienna (1973) 563.

[9] WHITTAKER, B., Radiation dose and dose-rate measurements in the Megarad range, UK Panel, Gamma and Electron Irradiation Symp. (1970) 11.

[10] BRYNJOLFSSON, A., "A significant correction factor in gamma ray dosimetry", Radiation Chemistry (HART, E.J., Ed.) 1, Advanc. Chem. Ser. 81 Amer. Chem. Soc. (1968) 550.

[11] WEISS, J., RIZZO, F.X., Manual on Radiation Dosimetry (HOLM and BERRY, Eds), Marcel Dekker, Inc. N.Y. (1970) 231.

[12] CHADWICK, K.H., in Technical Developments and Prospects for Sterilization by Ionizing Radiation, Proc. Johnson-Johnson Symp., Multiscience Publications (1974). In Press.

[13] OWEN, D.B., Factors for one-sided tolerance limits and for variable sampling plans, S.C.R.-607 (1963) distributed by N.T.I.S., US Dept. of Commerce, 5285 Port Royal Rd., Springfield, Va. 22151.

[14] McLAUGHLIN, W.L., Manual on Radiation Dosimetry (HOLM and BERRY, Eds), Marcel Dekker, Inc. N.Y. (1970) 377.
[15] HUMPHERYS, J.C., CHAPPELL, S.E., McLAUGHLIN, W.L., JARRETT, R.D., N.B.S. Interim Rep. 73-413 (1973).
[16] HARRAH, L.A., Radiat. Res. 41 (1970) 229.
[17] BISHOP, W.P., HUMPHERYS, K.C., RANDTKE, P.T., Rev. Sci. Inst. 44 (1973) 443.
[18] DVORNIK, I., Manual on Radiation Dosimetry (HOLM and BERRY, Eds), Marcel Dekker, Inc. N.Y. (1970) 345.
[19] RAZEM, D., DVORNIK, I., in Dosimetry in Agriculture, Industry, Biology and Medicine (Proc. Symp. Vienna, 1972), IAEA, Vienna (1973) 405.

DISCUSSION

B.L. GUPTA: You showed two intercomparisons of the HX perspex dosimeters. The second intercomparison was an improvement over the first. Was this due to the material used or to instrumental or human error?

A second point is that you have suggested that it is sufficient to calibrate an irradiation facility once, after which other monitoring devices can then be used. In that case is it not sufficient to use standard dosimeters like the Super Fricke, and give a lower dose?

K.H. CHADWICK: The intercomparison I showed actually related to radiobiology experiments performed at different institutes in Europe at a dose of 200 rad and involving the use of LiF TL dosimeters. The intercomparison was very useful in that after the first round a report was written up advising on the dosimetry procedures; the institutes getting poorer results were able to improve their work considerably.

In reply to your second question, I do not see any possibility of sending a dosimeter system through part of an industrial plant. If you are really going to monitor the process, you have to use a dosimeter reading up to 2.5-3 Mrad throughout the plant as a whole.

K.S. AGGARWAL: I don't think that at the present stage we have any really reliable dosimeters for industrial use. We have encountered difficulties with all the systems tried out — perspex, caesium, and the oxalic acid system you mentioned. I should therefore like to ask whether you think the perspex is accurate enough to be used in plant calibration and routine operations?

Secondly, you mentioned that the calibration of perspex may vary with the spectrophotometer. Does this mean that the calibration curve recommended by the UK Panel on Gamma and Electron Irradiation is of little use to the user of the dosimeter?

K.H. CHADWICK: I agree that there does not appear to be an absolutely perfect dosimeter system available for this dose range at the moment. Most systems have snags. The perspex system is the one with which I am most familiar and I personally favour it.

In reply to your second question, there certainly may be some variation in the response of spectrophotometers, which are usually not serious but still affect the measured response of some dosimeters. The UK calibration curve is of considerable value. If you make your own calibration and it doesn't tally with the UK curve, you should be very wary, first of all, of your measuring equipment. But when this is thoroughly checked and found to be satisfactory, the UK curve should be made use of. The cause of any further differences should be sought in the estimation of dose in the calibration facility.

K.H. MORGANSTERN: I don't think one should be too worried about the 10% variation in intercalibration which you mentioned, when there appears to be as much as 300% variation in what our colleagues feel is a proper sterilization dose. Wouldn't you say that what is really important industrially is the internal consistency — day to day or hour to hour?

K.H. CHADWICK: We do not know that the variation in industrial plants is less than 10%, although we hope it is. To speak of 300% variation in the sterilization dose is somewhat overstating the case. In general, 2.5 Mrad is the accepted sterilization dose. But what is more important is that although some products may be authorized for sterilization at lower doses, the authorization is based on the achievement of a minimum dose — and for this reason we need good dosimetry.

I think that what is important industrially is to start the process going, using dosimetry, and to follow it up with checks on internal consistency from day to day, which is essential if routine dosimetry monitoring is not going to be required.

K. KRISHNAMURTHY: I think you said that there was no need for routine process control dosimetry once the process parameters had been monitored and kept within the limits prescribed by the commissioning tests. In a service plant like ISOMED, we get a variety of products ranging from cotton to saline sets and with varied density. How is it possible to have process control without dosimetry?

K.H. CHADWICK: When an industrial plant is engaged in irradiating a mixed product of varying bulk density, it should be commissioned to give the required minimum dose in the heaviest or densest material processed; this dose will then be exceeded in the less dense products.

J. LAIZIER: In applying electron beam treatments we find that we have to use "continuous" as opposed to "point" measurements of the delivered dose, as a result of the instabilities that may occur in the instruments, and also due to the low penetration. PVC film seems to be utilized widely for this purpose. Have you any comment to make on the characteristics of the PVC dosimeter?

K.H. CHADWICK: For many reasons I do not think that PVC film is a real radiation dosimeter. For one thing, it is dose-rate dependent, and also dependent on the plasticizer used. It is also very batch dependent. Furthermore, the film needs thermal development treatment after irradiation and this can be inconvenient. However, I agree that under more or less fixed irradiation conditions PVC film could be used to monitor the radiation process.

USE OF CERIC SULPHATE AND PERSPEX DOSIMETERS FOR THE CALIBRATION OF IRRADIATION FACILITIES

R.D.H. CHU, M.T. ANTONIADES
Atomic Energy of Canada Limited,
Commercial Products,
Ottawa, Canada

Abstract

USE OF CERIC SULPHATE AND PERSPEX DOSIMETERS FOR THE CALIBRATION OF IRRADIATION FACILITIES.

The conditions encountered during the calibration of a ^{60}Co radiation sterilization plant may differ considerably from the conditions under which the dosimetry systems are calibrated. Variations in the dose-rate, temperature, and energy spectrum may affect the dosimeter response. The effects of these parameters on ceric sulphate and Perspex dosimeters were studied and it was concluded that the ceric sulphate and Perspex-type Red 4034 dosimetry systems are both suitable for irradiator calibration provided that adequate precautions are taken in their use.

INTRODUCTION

Testing by means of microbiological reference standards of six research facilities and ten industrial radiation sterilization plants by the Accelerator Department at Risö in co-operation with the IAEA indicated that significant errors could exist in the dosimetry calibration at the commissioning of medical products sterilization irradiators. Emborg et al. [1] reported that the six research radiation facilities showed the same microbiological efficiency as the reference sources at Risö, whereas six of the eight industrial ^{60}Co radiation sterilization plants and one of the two industrial linear accelerator plants yielded microbiological countings which indicated differences in microbiological efficiency corresponding to differences in absorbed dose of 10-40%.

Holm and Bjergbakke [2] suggested that industrial ^{60}Co radiation sterilization plants may present difficulties in dose calibration greater than those for research facilities or linear accelerators. Mechanical limitations on the conveyor speed usually preclude the direct measurement of the absorbed dose under the normal processing conditions by a primary dosimetry system such as the Fricke dosimeter. Heat-transfer problems due to the long product dwell time make calorimetry measurements impractical.

For the calibration of ^{60}Co radiation sterilization plants, dosimetry systems capable of measuring absorbed doses in the range from about 0.5 to 5.0 Mrad are usually used. These dosimetry systems are calibrated at positions of known dose-rates which have been measured by a primary dosimetry system such as the Fricke dosimeter. Calibrations may be performed in such differing conditions as in a lead-shielded self-contained irradiator such as the AECL Gammacell 220, at a position within an array of source pencils near the bottom of a water-filled pool, or at a reference position adjacent to the source in the industrial ^{60}Co radiation sterilization plant.

The conditions encountered during the normal processing in the ^{60}Co radiation sterilization plant usually differ considerably from the conditions under which the dosimetry systems were calibrated. Wide variations in the instantaneous dose-rate occur during the transport of the dosimeter in a product package around the ^{60}Co source while the dosimeter calibration is usually performed at a constant dose-rate. Some increase in temperature occurs during the irradiation process in a radiation sterilization plant. Calibration of the dosimeters may be performed at ambient room temperatures or, if the calibration is performed at a high dose-rate in a lead-shielded self-contained irradiator, the temperature may be as high as 25°C above ambient if a cooling system is not provided.

The energy spectrum of the gamma radiation reaching the product in a ^{60}Co radiation sterilization plant is degraded by Compton scattering during the passage through source and product material. Difference in the energy spectrum may give a dosimeter response different from the dose absorbed by the product, especially if the dosimeter contains significant quantities of high atomic number materials. In lead-shielded self-contained irradiators and in underwater positions used for dosimeter calibrations the energy spectrum may also contain significant amounts of degraded radiation. Calibration positions adjacent to a source in air would receive smaller amounts of degraded radiation.

Dosimetry systems based on the radiation-induced reduction of ceric sulphate or the darkening of perspex are commonly used for the calibration of industrial ^{60}Co radiation sterilization plants. This paper examines the conditions in a typical industrial ^{60}Co radiation plant and shows how these irradiation conditions may affect the response of ceric sulphate or perspex dosimeters. Other sources of error in the use of these dosimetry systems are also discussed in order to establish an accurate dosimetry method for the calibration of industrial ^{60}Co radiation sterilization plants.

INDUSTRIAL ^{60}Co RADIATION STERILIZATION PLANTS

The JS-6500 Irradiation Plant shown in Fig. 1 is a typical ^{60}Co radiation sterilization plant designed and manufactured by Atomic Energy of Canada Limited (AECL), Commercial Products. In this irradiator design products in corrugated fiberboard boxes are moved on roller conveyors past a vertical source plaque consisting of a number of cylindrical ^{60}Co source elements. There are two levels and two rows of product containers on each side of the source. Movement from one row to another and from the upper level to the lower level is achieved by means of a series of pneumatic pushers and transfer devices. Movement through the source pass mechanism is not continuous but intermittent. The time spent by the product box in each of the irradiation positions on the conveyors is determined by a master timer. The product flow is shown in Fig. 2.

A range of box sizes can be processed by the JS-6500 Irradiation Plant. For the maximum box size (58.2 cm long by 48.2 cm wide by 91.2 cm high) 49 product boxes are in the irradiator at any instant. With a source content of 200 000 Ci of ^{60}Co the required timer setting for a minimum dose of 2.5 Mrad is about 22 min for a product density of 0.20 g/cm^3. The total time that the product spends in the irradiator is about 18 h. The minimum

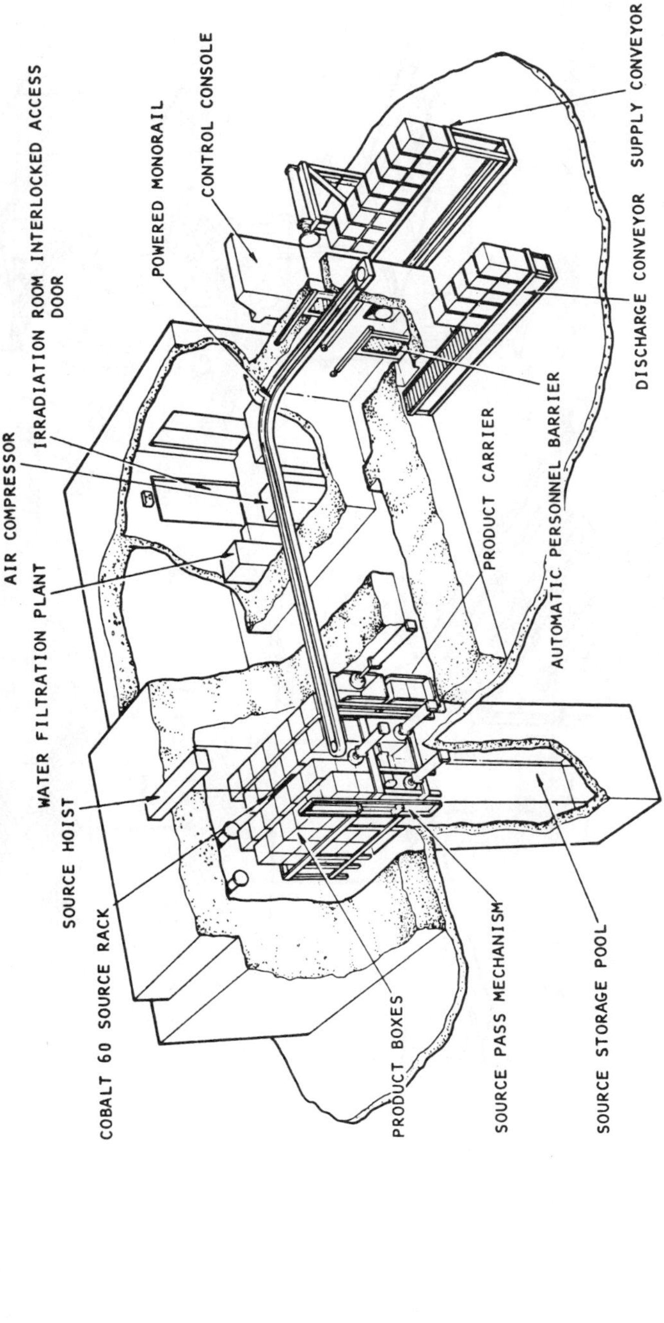

FIG.1. JS-6500 irradiation plant.

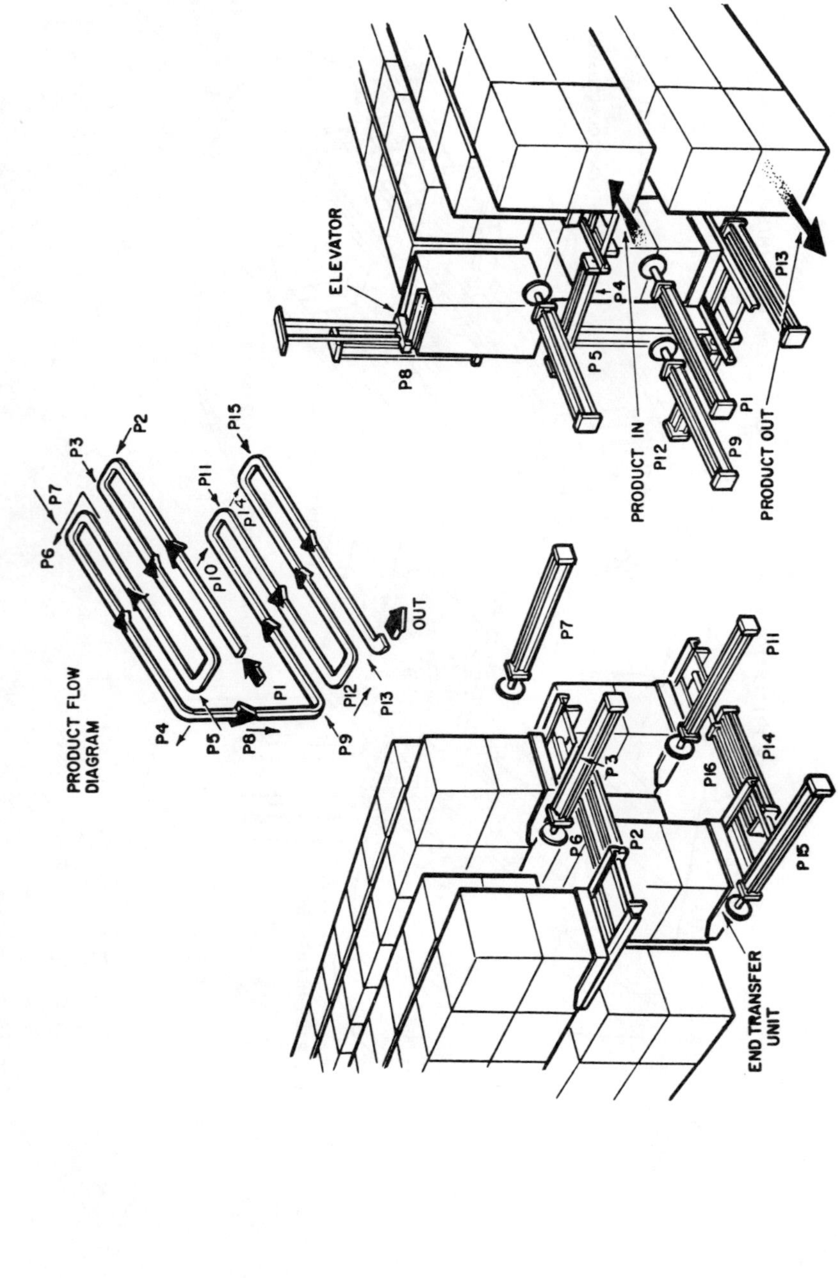

FIG.2. Product flow path in JS-6500 irradiation plant.

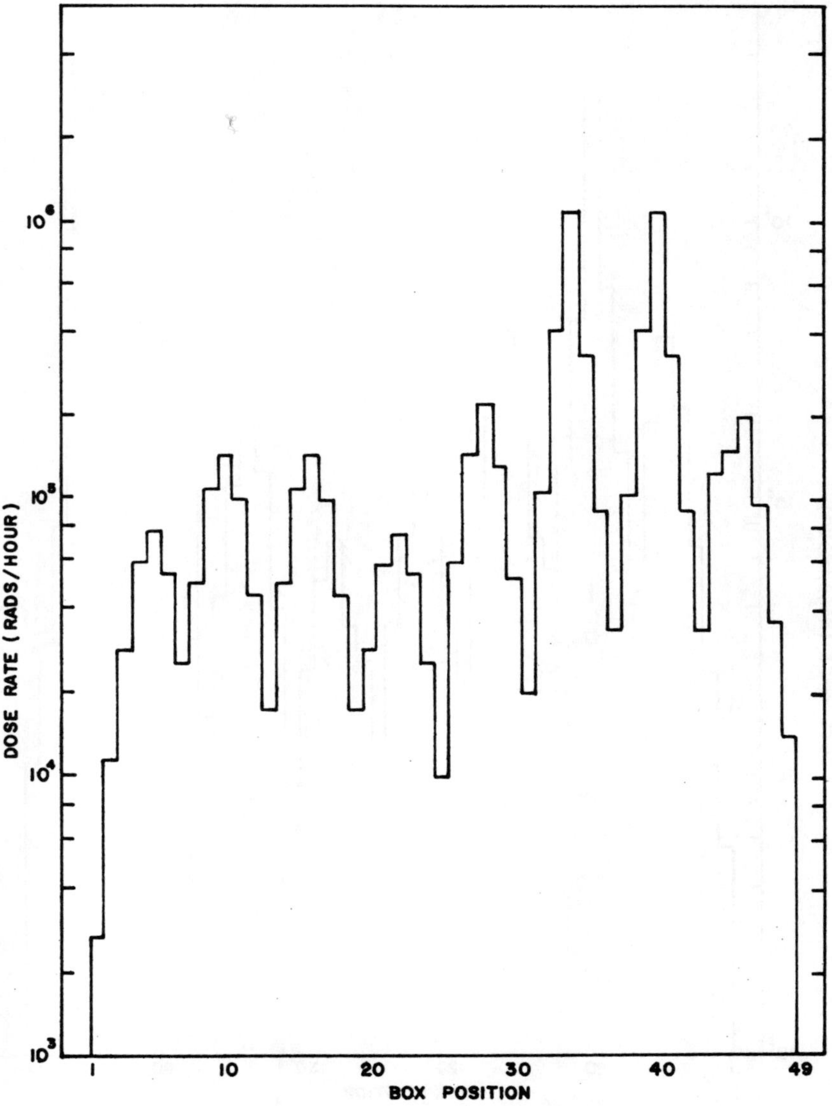

FIG.3. Dose-rates at top of box at each of the 49 box positions in the JS-6500 irradiation plant. Source activity = 200 000 Ci ^{60}Co. Product density = 0.2 g/cm^3.

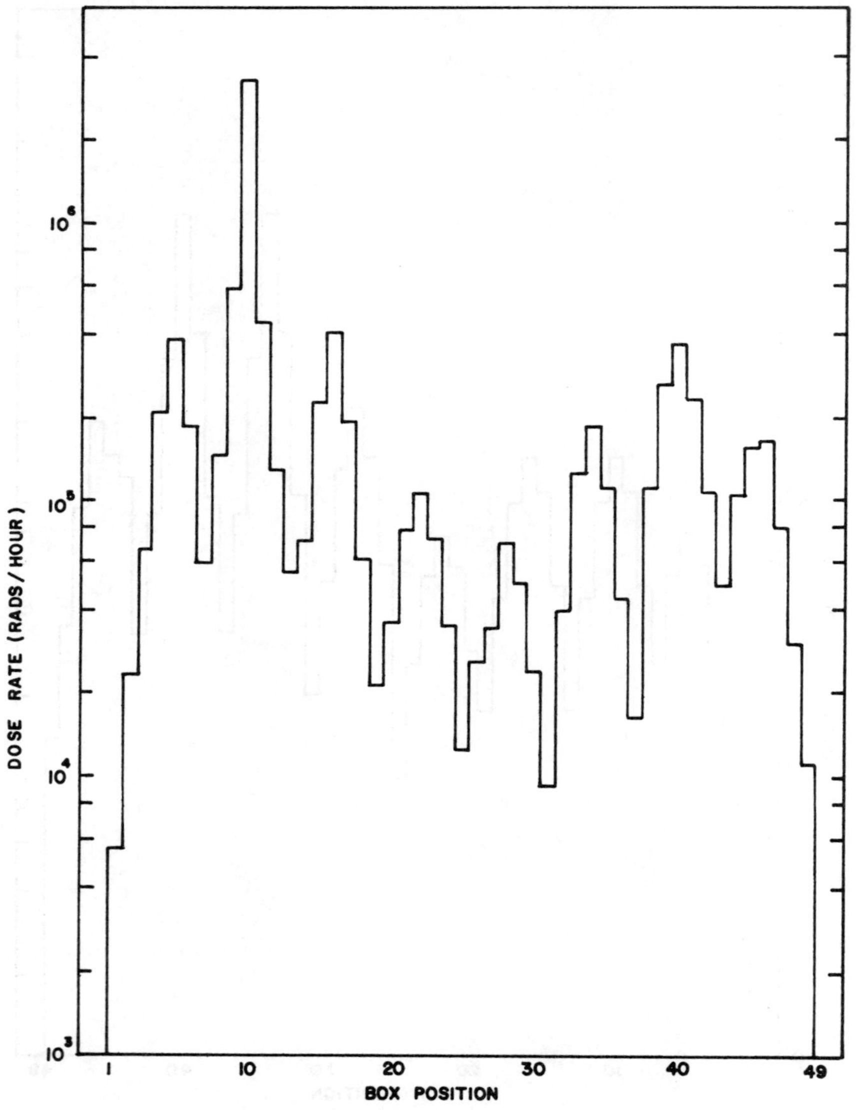

FIG.4. Dose-rates at side of box at each of the 49 box positions in the JS-6500 irradiation plant. Source activity = 200 000 Ci ^{60}Co. Product density = 0.2 g/cm^3.

dose received by the product occurs near the top and the bottom of the product box. The maximum dose occurs at the side of the box and is higher than the minimum dose by a factor of about 1.37. For a minimum dose of 2.5 Mrad the maximum dose is about 3.43 Mrad.

The calculated dose-rates at the top of the product box at each of the 49 positions in the irradiator are shown in Fig. 3. Figure 4 shows the dose-rates at the side of the product box. For the 200 000-Ci ^{60}Co source the instantaneous dose-rate varies from 3000 rad/h to 3.3 Mrad/h. The average dose-rate during the passage of the product through the irradiator is 0.14 to 0.19 Mrad/h.

The temperature increase during irradiation was measured by placing a bi-metallic recording thermometer in product boxes passing through the normal process cycle. It was found that the temperature could remain at the ambient room temperature or could rise as much as 20°C during the irradiation. The higher increases in temperature were attributed mainly to increases in the temperature in the irradiation chamber owing to absorption of the gamma radiation in the source, conveyors and the concrete walls. The temperature rise depended on the source activity, the room ventilation, the length of time the source plaque was raised, and whether or not a heat exchanger was installed in the source storage pool.

Brynjolfsson [3] has calculated how the degradation of the gamma rays penetrating various distances in water can affect the radiation doses absorbed in various dosimeters. To determine how the energy degradation may affect dosimeters placed in product boxes passing through a JS-6500 Irradiation Plant a calculation was made of the water-equivalent attenuation by 0.2-g/cm^3 product at the top of a product box located at the various positions in the irradiator.

CERIC SULPHATE DOSIMETRY

Ceric sulphate solution can be used to measure absorbed doses in the range from 0.1 to 100 Mrad. The range of the dose to be measured by the dosimeter determines the ceric ion concentration to be used. For measurements in the dose range from 1 to 4 Mrad for the calibration of ^{60}Co radiation sterilization plants 0.015M, ceric ammonium sulphate solution prepared according to a procedure similar to ASTM Standard D3001-71 [4] is used by AECL.

An advantage in the use of ceric sulphate is that the response, like that of the Fricke dosimeter, can be described in terms of the primary radiolytic yields from the irradiation of water. Experience with the use of ceric sulphate dosimetry in the calibration of ^{60}Co radiation sterilization plants has shown this dosimetry system to be well suited for this use.

Stock solutions of 0.1M ceric sulphate stored in the dark are stable for over a year. Diluted dosimeter solutions in flame-sealed glass ampoules are stable for at least several months, both before and after irradiation. The glass ampoules are conditioned by filling with dosimeter solution, irradiating to about 1.5 Mrad, and setting aside for 24 h. The irradiated solution is then decanted and the ampoules are rinsed well with dosimeter solution. The ampoules are then filled and flame-sealed.

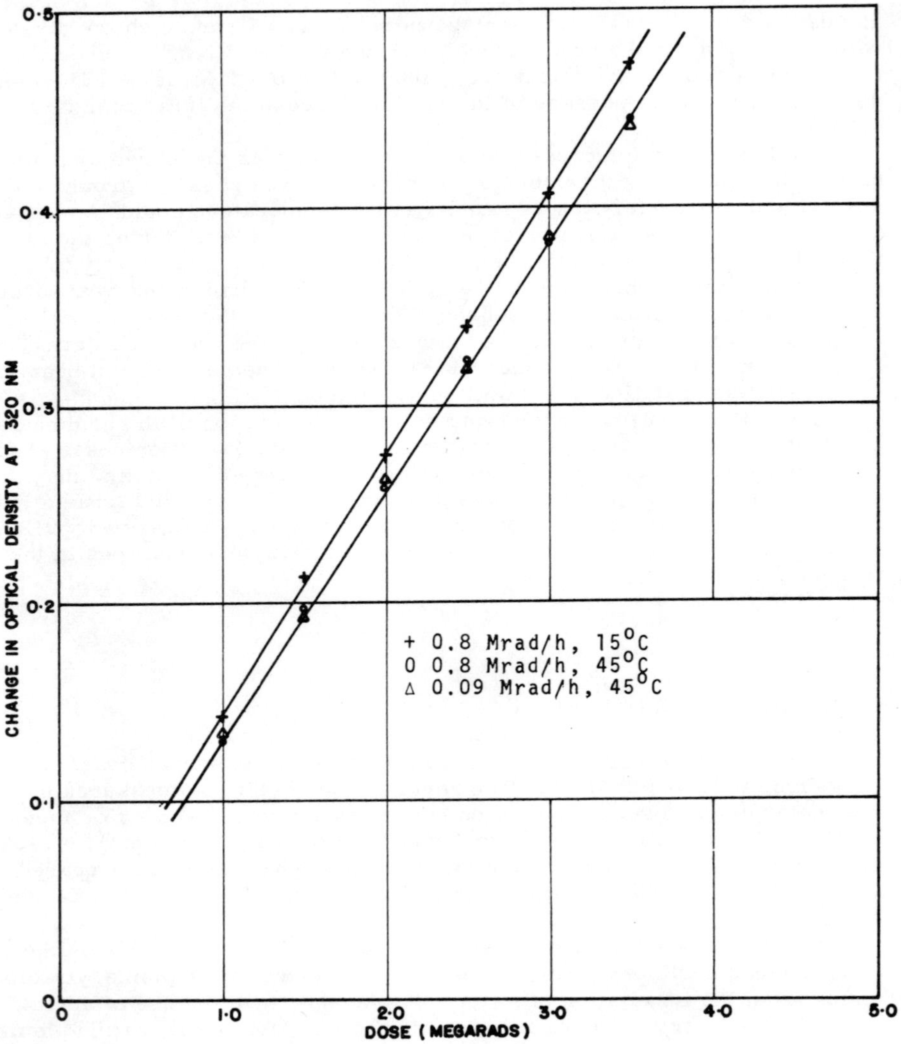

FIG.5. Ceric sulphate calibration curves for irradiation at 15 and 45°C. Change in optical density read at 320 nm on Beckman DB.

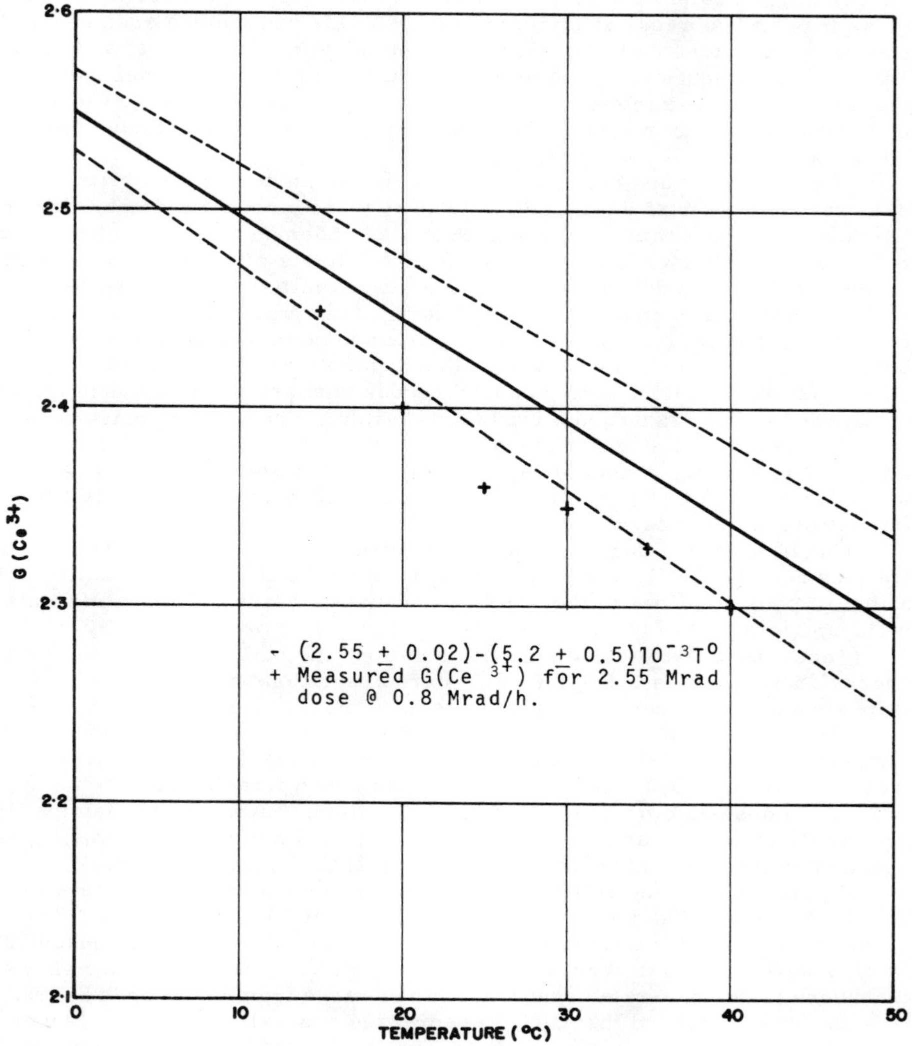

FIG. 6. Variation of $G(Ce^{3+})$ with irradiation temperature.

For the calibration of a ^{60}Co radiation sterilization plant, ceric sulphate dosimeters are prepared prior to the commencement of the irradiator installation, usually about five weeks prior to the dosimetry. Following the dosimetry, the irradiated ceric sulphate dosimeters are returned to AECL for reading. Dosimeters may be read within several days after irradiation or as long as two weeks after irradiation. As the dosimeters are prepared and read under identical conditions, variations in measurements and analytical techniques are minimized. To ascertain that no change has occurred in the dosimeters owing to the storage and handling conditions a calibration check is made using unirradiated dosimeters kept under the same conditions.

The two major disadvantages of the ceric sulphate system are the extreme sensitivity of the reduction yield to trace amounts of impurities and the deviation from water-equivalent response under degraded spectral energy conditions. Matthews [5] has shown that the effects of trace impurities can be inhibited by the addition of cerous ions to the initial solution and that satisfactory results could be obtained using mains water and no special cleaning of glassware. Both ceric sulphate and combined ceric-cerous solutions have been used by AECL with acceptable results. At present, ceric sulphate solutions with no added cerous ions are normally used for the calibration of ^{60}Co radiation sterilization plants. The batch-to-batch consistency of the $G(Ce^{3+})$, the number of ions of ceric reduced per 100 eV of absorbed radiation, is confirmed by routine calibration checks. Some variation in the $G(Ce^{3+})$ is found between different batches but this is taken into account in the dosimeter calibration.

The effect of the degraded energy spectrum for the ceric sulphate concentrations used by AECL was estimated from the calculated transmission factors at the various positions in the JS-6500 Irradiation Plant and Brynjolfsson's [3] build-up factor data for transmission in water. For product density of 0.2 g/cm^3 and 0.015$\underline{M}$ ceric sulphate solution, the build-up effect due to the energy degradation is 2.9%. If ceric-cerous solutions of 0.01$\underline{M}$ ceric sulphate and 0.01$\underline{M}$ cerous sulphate are used the build-up effect is 3.7%.

The variation in the response of ceric sulphate dosimeters with temperature and dose-rate was studied by irradiating dosimeters in two AECL Gammacell 220 lead-shielded self-contained irradiators which differed in dose-rate by a factor of 8.5. The temperature in the Gammacell 220 irradiators was varied by surrounding the dosimeters with a small glass coil containing water circulated from a constant temperature bath.

Figure 5 shows the calibration curves for ceric sulphate dosimeters irradiated in the high dose-rate Gammacell 220 at 15°C and at 45°C. The calibration curve for irradiation at 45°C in the low dose-rate Gammacell 220 is also shown. This differs little from the high dose-rate curve and shows that there is no dose-rate dependence in the range from 0.093 to 0.8 Mrad/h.

Matthews [6] found that $G(Ce^{3+})$ decreased approximately linearly with increase in irradiation temperature. For aerated ceric sulphate solutions $G(Ce^{3+})$ was given by the empirical equation:

$$G(Ce^{3+}) = (2.55 \pm 0.02) - (5.2 \pm 0.5)\, 10^{-3}\, T^0$$

Figure 6 shows the $G(Ce^{3+})$ calculated from this equation and the $G(Ce^{3+})$ measured in irradiations to 2.55 Mrad in the high dose-rate Gammacell 220.

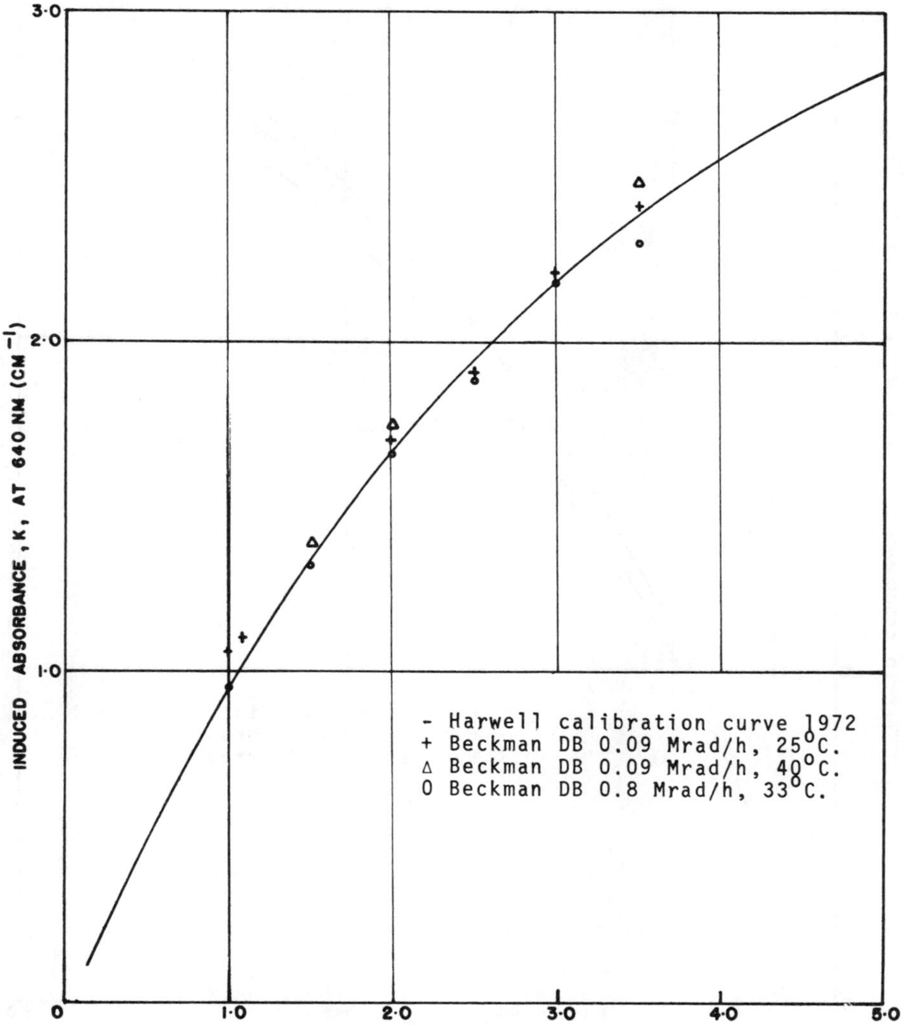

FIG.7. Harwell calibration curve for Perspex-type Red 4034 and Beckman DB readings for dosimeters irradiated in AECL Gammacell 220 irradiators. Calibration curve for batch of dosimeters drawn by Harwell in 1972.

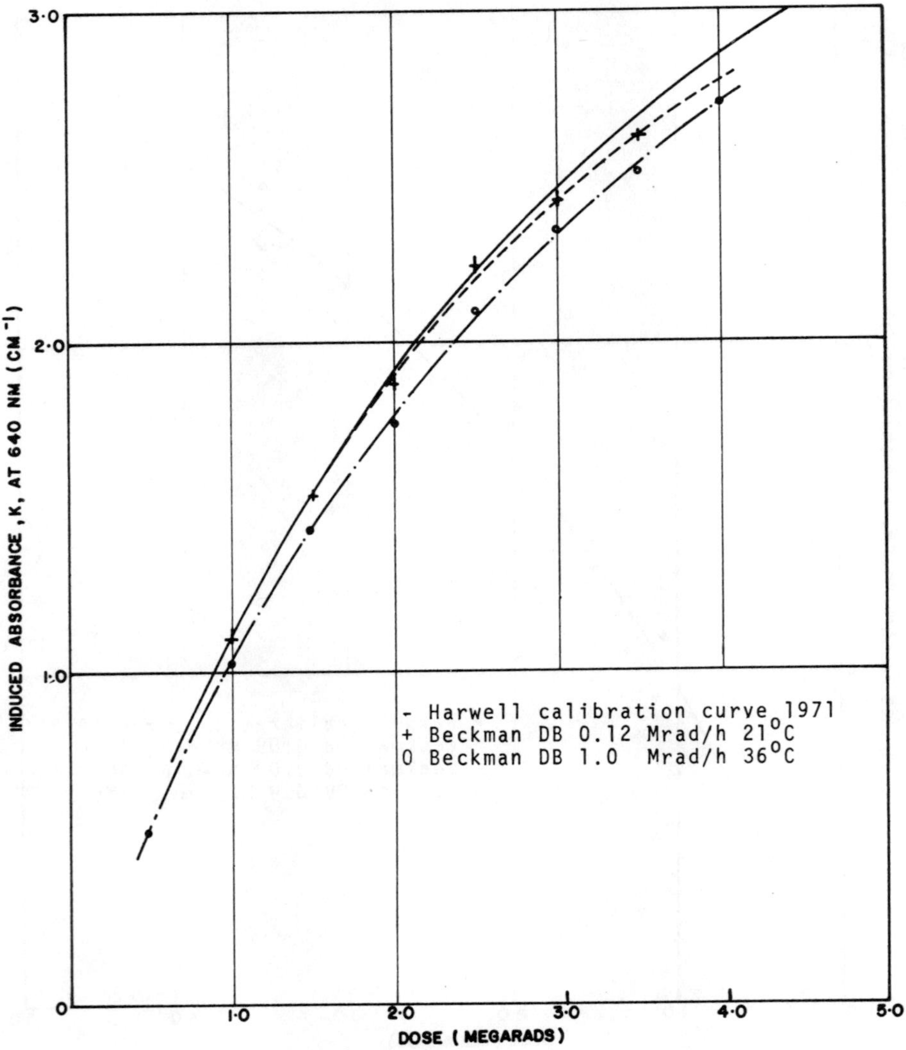

FIG. 8. Harwell calibration curve for Perspex-type Red 4034 and Beckman DB readings for dosimeters irradiated in AECL Gammacell 220 irradiators. Calibration curve for batch of dosimeters drawn by Harwell in 1971.

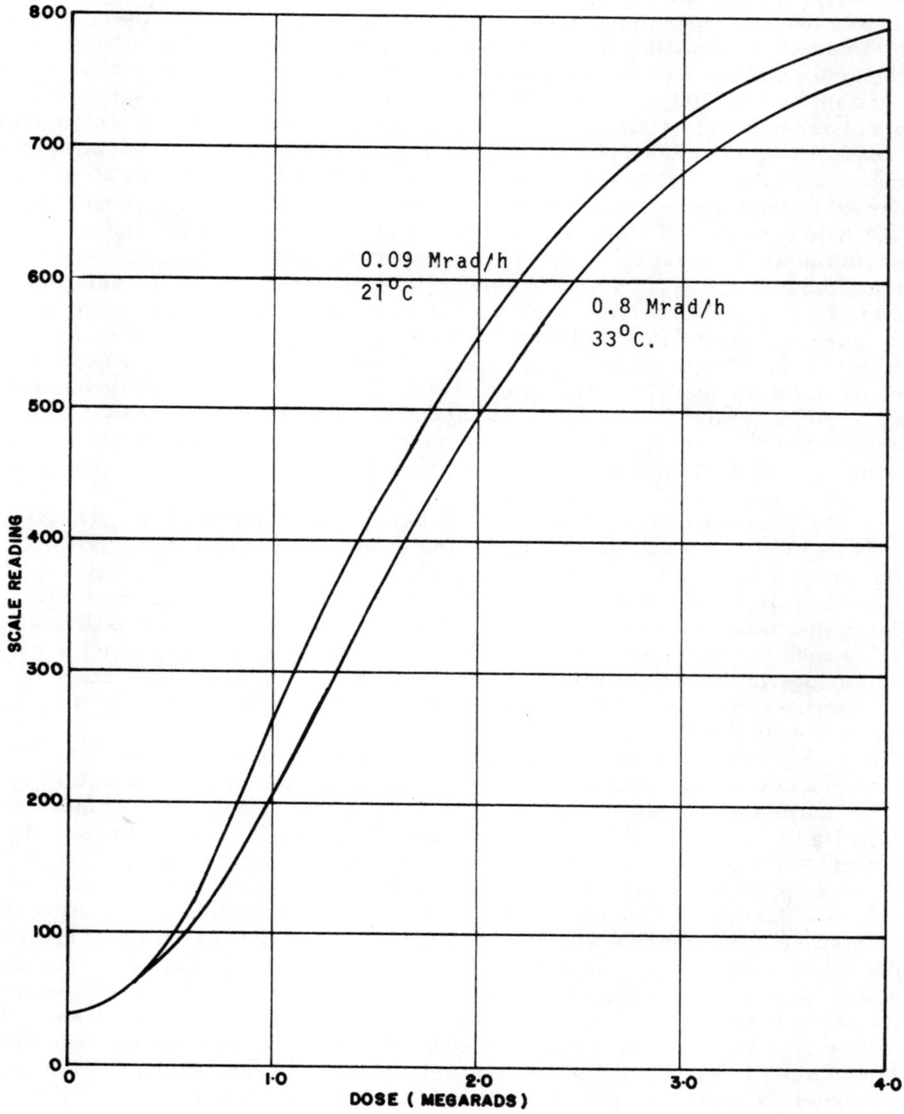

FIG. 9. Calibration curves for AECL Red Perspex dosimeters irradiated in AECL Gammacell 220 irradiators and read on AECL Reader type BC-2.

PERSPEX DOSIMETRY

The measurement of the radiation-induced optical density change in some forms of dyed or clear perspex provides a suitable method of dose determination. Commercial red-dyed Perspex type Red 4034 manufactured by Imperial Chemical Industries Ltd. (ICI), in sheets of 3-mm nominal thickness darkens on irradiation owing to the formation of a new absorption band in the 600-700 nm region of the spectrum [7]. Perspex-type Red 4034 sheets are cut into strips for use in a spectrophotometer and given an ageing treatment to accelerate the normal air and water-vapour absorption in fresh material. Packaged dosimeters are supplied, together with a calibration curve, by the Atomic Energy Research Establishment, Harwell, Berkshire, U.K.

The response of Perspex-type Red 4034 dosimeters to ^{60}Co gamma radiation at different dose-rates and temperatures was studied to determine the effects of these parameters during the calibration of a ^{60}Co radiation sterilization plant. Figure 7 shows the Harwell calibration curve for a recently obtained batch of Red 4034 dosimeters together with readings obtained on a Beckman DB spectrophotometer for irradiations at two dose-rates in Gammacell 220 irradiators. The readings are in good agreement with the Harwell curve and indicate about 4% difference in the calibration curves at 0.093 and 0.8 Mrad/h. Irradiations performed at temperatures from 15°C to 40°C indicated that there was little change in the response with temperature.

The dose-rate dependence for this batch of Red 4034 dosimeters was less than that found for an older batch of Red 4034 dosimeters obtained from Harwell. Figure 8 shows the Harwell calibration curve together with the Beckman DB readings for dose-rates of 0.12 and 1.0 Mrad/h. For this batch the combined temperature and dose-rate effect was found to give a difference in the response of about 8%. This difference in the results obtained with the two batches indicated that dose-rate and temperature effects may vary from batch to batch or may vary with the storage time of the Red 4034 dosimeters.

From Brynjolfsson's build-up factor data [3] the build-up effect for Perspex dosimeters owing to the energy degradation in a ^{60}Co radiation sterilization plant was estimated to be about 1.0%. However, the mechanism causing the colour change may not be directly dependent on the absorbed energy and may be a function of the gamma energy spectrum.

Red 4034 dosimeters are at present being used concurrently with ceric sulphate dosimeters during the calibration of ^{60}Co radiation sterilization plants. The dosimeters are returned to AECL for reading on the Beckman DB spectrophotometer and may be read up to two weeks following irradiation with little correction for post-irradiation fading. In some instances the Red 4034 dosimeters were also read directly after irradiation on a spectrophotometer available on site. The small post-irradiation changes when read at 640 nm allows calibration checks to be made between different spectrophotometers using irradiated dosimeters. Ellis [8] has pointed out how problems such as stray light in spectrophotometry can influence the dosimeter readings. In intercomparisons we have performed using different spectrophotometers we have found differences in the calibration curves, particularly between grating and prism spectrophotometers. Prior to the use of a spectrophotometer for Red 4034 dosimetry the calibration curve should be checked by reading dosimeters irradiated to known doses.

Perspex HX is a special grade of ICI clear Perspex manufactured to specifications supplied by the UK Panel for Electron and Gamma Irradiation. Studies performed using 1-mm-thick dosimeters read at 305 nm as recommended by the UK. Panel showed that Perspex HX was not suitable for irradiation during the calibration of the ^{60}Co radiation plant and return of the dosimeters to AECL for reading. Non-reproducible results when compared with ceric sulphate dosimeters were attributed to variations in the storage conditions following irradiation affecting the post-irradiation fading. Chadwick [9] showed that storage time, storage temperature, irradiation time, and irradiation temperature affect the Perspex HX measurements at 305 nm and that these factors have a minimal effect at 314 nm. He found that the dose measured by Perspex HX at an industrial ^{60}Co radiation sterilization plant at 305 nm indicated a dose of 3.3 Mrad while the measurement at 314 nm indicated the correct dose of 2.7 Mrad. This difference in the measurement at 305 nm is consistent with our results. We have not tried reading Perspex HX at 314 nm.

AECL supplies injection-moulded red Perspex cylinders of ICI Diakon Moulding Compound type 4037 for use for routine dosimetry in ^{60}Co radiation sterilization plants. The transmission of light in the range 600-700 nm is read on a special read-out instrument, the AECL Reader type BC-2, and the dose received is obtained from a calibration curve (Fig. 9). By controlling the thickness of the cylinders during the injection moulding, the need to measure the thickness of each individual dosimeter and to correct for the variation in thickness as required for Red 4034 and Perspex HX dosimeters is eliminated. However, this dosimetry system is found to be dose-rate dependent as shown in Fig. 8 and is not suitable for direct use in an irradiator calibration. As a large number of dosimeters can be read in a short time and the read-out instrument is sufficiently robust for travel, this system is useful for determination of the dose distribution in the product box and for approximate measurements to ascertain correct operation of the irradiator. The dose-rate dependence means that there will be some difference in the response in a ^{60}Co radiation sterilization plant from that obtained under static irradiation conditions. The calibration must be checked against ceric sulphate and modifications made to the calibration curve for the actual processing conditions in which the dosimetry system is to be used.

CONCLUSIONS

Ceric sulphate and Perspex-type Red 4034 dosimeters have both been shown to be suitable for use in the calibration of ^{60}Co radiation sterilization plants. The good post-irradiation stability of both of these dosimetry systems enables readings of these dosimeters to be taken at AECL, thus minimizing errors due to instrumentation or to analytical techniques.

For ceric sulphate dosimetry the temperature during irradiation should be taken and the readings corrected for the temperature. The dose-rate variations in ^{60}Co radiation sterilization plants will have negligible effects on the dosimeter response. The effect of variations in the energy spectrum is minimal provided that low cerium concentrations are used. Extreme care in the preparation of ceric sulphate dosimeters must be taken to eliminate effects of trace impurities. Purity requirements for water and chemicals are less stringent if cerous ions are added to the ceric solution. The $G(Ce^{3+})$ for each batch of dosimeters should be determined.

The present batch of Perspex-type Red 4034 dosimeters from Harwell showed no temperature dependence and a dose-rate dependence of about 4% for dose-rates from 0.093 to 0.8 Mrad/h. However, a previous batch of Red 4034 dosimeters had shown a larger dose-rate dependence. Each new batch of Red 4034 dosimeters should be checked for temperature and dose-rate dependence before use and the calibration curve used for determining the absorbed dose should be obtained at approximately the same temperature and average dose-rate as in the ^{60}Co radiation sterilization plant. Effects of energy degradation on the relative amounts of energy absorbed in perspex dosimeters and water-equivalent product will be small but some uncertainty exists concerning whether or not the colour change in Perspex is directly proportional to the energy absorbed.

REFERENCES

[1] EMBORG, C., CHRISTENSEN, E.A., KALLINGS, L.O., ERIKSEN, W.H., BJERGBAKKE, E., HOLM, N.W., "Control of the microbiological efficiency of radiation sterilization plants by means of B. sphaericus, strain C_IA, Str. faecium, strain A_21 and Coli Phage T_1", Experiences in Radiation Sterilization of Medical Products, IAEA-159, Unpubl. Doc. (1974) 53.

[2] HOLM, N.W., BJERGBAKKE, E., "Dosimetry procedures for radiation sterilization of medical and bio-medical products", Ibid., p.125.

[3] BRYNJOLFSSON, A., "A Significant Correction Factor in Gamma Ray Dosimetry, Radiation Chemistry, (HART, E.J., Ed.) 1, Advances Chem. Ser. 81 Amer. Chem. Soc. (1968) 550.

[4] ASTM STANDARD D3001, "Standard method of test for absorbed gamma and electron radiation dose with the ceric sulphate dosimeter", Annual Book of ASTM Standards (1971) Part 35.

[5] MATTHEWS, R.W., An evaluation of the ceric-cerous system as an impurity-insensitive megarad dosimeter, Int. J. Appl. Radiat. Isotopes 22 (1971) 199.

[6] MATTHEWS, R.W., Effect of solute concentration and temperature on the ceric-cerous dosimeter, Radiat. Res. 55 (1973) 242.

[7] WHITTAKER, B., "Red perspex dosimetry", Manual on Radiation Dosimetry (HOLM, N.W., BERRY, R.J., Eds), Marcel Dekker Inc., New York (1970) 363.

[8] ELLIS, S.C., "Problems in spectrophotometry and their influence in radiation measurements", Radiation Dose and Dose Rate Measurements in the Megarad Range, Proc. Symp. National Physical Laboratory and U.K. Panel on Gamma and Electron Irradiation, U.K. (1970) 18.

[9] CHADWICK, K.H., "The choice of measurement wavelength for clear HX perspex dosimetry", Dosimetry in Agriculture, Industry, Biology and Medicine (Proc. Symp. Vienna, 1972), IAEA, Vienna (1973) 563.

DISCUSSION

B.L. GUPTA: What kind of containers do you use for the ceric sulphate dosimeter?

R.D.H. CHU: We use standard 2-ml neutraglass ampoules.

B.L. GUPTA: Are they stoppered or flame-sealed?

R.D.H. CHU: The ampoules are flame-sealed.

K. KRISHNAMURTHY: You mentioned that you have found variations in the G-value of the ceric sulphate from batch to batch. Could you please indicate the range of variation found?

R.D.H. CHU: At an ambient temperature of about 25°C, the batch-to-batch variation in the G-value ranges from approximately 2.35 to 2.45.

K. KRISHNAMURTHY: I should like to comment in this connection that we have found that the ceric sulphate system gives reproducible values (± 3%) under industrial conditions, i.e. when products of the same type are being processed, although great care must be taken in this work. There is, in fact, no other dosimeter readily available today that gives this level of reproducibility (which can be counterchecked by radiolytic reduction in a standard dose field).

K.S. AGGARWAL: Mr. Chu, you mentioned that Matthews' method of using the Ce^{4+}-Ce^{3+} system was better than that based on Ce^{4+} alone. That being the case do you have any particular reason for still using the Ce^{4+} system? I ask this because my experience with the cerium system is that, first, it is temperature-dependent, hence it cannot be used in a commercial plant where the temperature varies from position to position; and, second, in service irradiation plants, which process products of varying densities simultaneously, the degraded γ-spectrum keeps changing and so it is difficult to correct the error due to the energy degradation.

R.D.H. CHU: We find no problem with the ceric system, provided that the dosimeters are prepared and read under controlled laboratory conditions. We feel that Matthews' potentiometric method using a ceric-cerous system has advantages for on-site reading as it reduces errors due to purity of materials during the dilution of solutions, as well as errors due to the analytical techniques. We shall be using both methods concurrently for a while, but intend to base our recommended timer settings on the ceric dosimeters read at AECL.

At present we measure the temperature to which the product is exposed during the irradiation cycle and use the appropriate G-value for the average irradiation temperature. We do not find much temperature variation from position to position within the plant.

Our calculations have shown that the correction required for energy degradation is small in the case of the cerium concentrations employed. Variation in the energy response with variation in product density will be negligible.

INDUSTRIAL AND DOSIMETRIC ASPECTS OF RADIATION STERILIZATION IN AUSTRALIA

Pamela A. WILLS, J.G. CLOUSTON,
R.W. MATTHEWS
Australian Atomic Energy Commission
Research Establishment,
Sutherland, N.S.W.,
Australia

Abstract

INDUSTRIAL AND DOSIMETRIC ASPECTS OF RADIATION STERILIZATION IN AUSTRALIA.

Radiation sterilization has been practised in Australia since 1960. Medical goods, pharmaceuticals, lyophilized reagents used for pharmaceuticals, and biological tissues (fascia lata in particular) are sterilized at three ^{60}Co plants and at a small facility at the Australian Atomic Energy Commission's Research Establishment. All processing is carried out at 2.5 Mrad in accordance with the Australian Code of Good Manufacturing Practice for Therapeutic Goods. However, the Code does not preclude the use of lower doses in special circumstances. A ceric-cerous megarad dosimeter has been developed. It is used for calibration of plant and routine dosimetry. Dosimeters are measured potentiometrically using a direct reading meter.

INTRODUCTION

Ionizing radiation was accepted in Australia as an effective sterilization method fourteen years ago. This paper describes the current situation with regard to irradiation facilities and gives details of medical products being radiation sterilized commercially. Some legal aspects of radiation processing are discussed. A direct reading ceric-cerous dosimeter system developed for megarad dosimetry is described.

INDUSTRIAL ASPECTS

Irradiation facilities

The first and largest ^{60}Co radiation plant in the world designed for industrial sterilization was built in 1959 at Dandenong, Victoria, Australia for Gamma Sterilization Pty. Limited [1]. Basic design information for the plant (capacity 2 x 10^6 curies ^{60}Co) was supplied by the United Kingdom Atomic Energy Authority. At the time of its construction, the primary purpose of the plant was the sterilization of imported goat hair bales used in manufacturing carpets. Because of new technology in the carpet industry, the plant is no longer used for this purpose. It is now operated for the manufacture of wood-plastic composites and, on a contract basis, for the sterilization of material, including medical goods and biological tissues.

In 1971 a second commercial ^{60}Co radiation plant was built, also at Dandenong, Victoria, for Tasman Vaccine Laboratory (Aust) Pty. Limited, whose parent company has operated a radiation plant in New Zealand since 1966. TVL's Dandenong plant, designed and installed by Atomic Energy of Canada Limited, (JS6500, capacity 1 x 10^6 curies ^{60}Co), is used for the company's own products and for contract work. The basic contract charge of \$A1.50/ft^3 (0.028 m^3)/2.5 Mrad, decreases with increasing assured volumes. At the end of 1974, the plant was loaded with approximately 4 x 10^5 curies ^{60}Co. Dwell time for low density material is about 10 hours with one carton (0.22 m^3) leaving the cell every 10 minutes.

A third radiation plant, also an AECL JS6500 unit, became operational at Botany, New South Wales, late in 1972. It is owned by Johnson & Johnson Pty. Limited and is used exclusively for its own products and those of its subsidiaries. It contains about 2.5 x 10^5 curies ^{60}Co.

In addition to these commercial irradiation plants, a limited amount of space for contract irradiations is available at the Australian Atomic Energy Commission's Research Establishment, Lucas Heights, New South Wales. These irradiations are carried out in a storage pond using both spent fuel elements and ^{60}Co as radiation sources [2]. Dose rates vary between several thousand rad and 1 x 10^6 rad/h. Two special 'hot' ^{60}Co rigs with dose rates of 2 Mrad/h and 3.9 Mrad/h have been installed in the pond for radiation sterilization of short-lived radiopharmaceuticals. These cylindrical rigs consist of 12 equispaced tubes (2.2 cm x 35.4 cm x 11.3 cm). Each tube contains two ^{60}Co rods. Overdose ratio is 1.33.

Materials to be irradiated are usually packed in cardboard or metal cylinders 24 cm dia. x 60 cm height, volume 0.028 m^3. Cylinders 30 cm in height are occasionally used for small quantities. Two large cylinders or one small cylinder, are loaded into a stainless-steel watertight can and located in the pond either between the fuel elements or in the ^{60}Co rigs. The two cylinders are inverted half way through the irradiation period to ensure even dosing. The maximum-minimum dose ratio is 1.2.

Although radiation efficiency in the storage pond at the AAEC facility near Sydney is very low (about 5%), the operation is labour intensive and charges are comparatively high ($A1.50/$ft^3$ (0.028 m^3)/Mrad), it provides an alternative to transporting goods 1,200 miles to the Victorian plants. It also gives local companies an opportunity to develop techniques for application to their products. Sutures form the major part of medical goods treated but the facility is being increasingly used for other items as industries become aware of the application to their products (Table I). In 1970, only 13 containers of medical goods (excluding sutures) were irradiated for commercial companies. By 1973, the number rose to 222 containers.

All the plants, except Johnson & Johnson, carry out radiation treatment of specialized hospital equipment unsuited for sterilization by any other method. Hospitals are charged at a special reduced rate. The minimum handling charge is also reduced considerably.

TABLE I. QUANTITY OF MEDICAL GOODS (EXCLUDING SUTURES) RADIATION STERILIZED AT AAEC

Year	Number of 24 cm x 60 cm cylinders		
	Hospitals	Industry	Total
1970	55	13	68
1971	101	15	116
1972	88	75	163
1973	139	222	361

AAEC - Industry liaison

After the commissioning of Johnson & Johnson's radiation plant, the AAEC conducted a training course, with particular emphasis on health and safety aspects, for personnel involved in the operation of the plant. Tasman Vaccine Laboratories (Aust) Pty. Limited, have similarly requested the AAEC to provide some training for their personnel. Lectures in radiation technology and radiation sterilization are included in a graduate radioisotope course conducted by the Australian School of Nuclear Technology, Lucas Heights, New South Wales.

Advisory and consultative services on radiation sterilization are provided by the AAEC. Advice is free, but customers pay for consultation. By means of research contracts, industrial organisations can engage the AAEC to work on their specific problems.

Medical goods sterilized by radiation

Large volumes of sutures, syringes and syringe needles, and disposable dressing packs are treated. The Australian market for syringes is 80,000,000 per annum. Most of them are fabricated from polystyrene, although recently a newly established company commenced manufacturing syringes using a radiation-resistant grade of polypropylene. Dressing packs contain such items as dressings, swabs, trays, forceps, bandages, trolley covers, nasal packing, eye and foam pads, paraffin or glycol nets, and wooden spatulas. These packs are widely used in New Zealand but they have not been readily accepted by Australian hospitals, partly because many of the large hospitals have considerable capital and staff investments in central sterilizing departments. However, hospital resistance is gradually being overcome, and radiation sterilization of these packs is now regarded as a growth area.

Other medical products treated include: catheters made from rubber, PVC, polyethylene, stainless steel (embolectomy, Bard Foley, dwell, packing, thoracic); drains (Yeates, Penrose, thoracic); tubes (connecting, tracheotomy, endotracheal, gastric, Forregar, sump); colostomy rods and cannulae; blood transfusion sets, bags (polyethylene, nylon, PVC, cellophane); vials, bottles, test tubes (polystyrene or polyethylene); surgical gloves (rubber); swabs (cotton, cotton and rayon); dressings, bandages, drapes (paper or polyethylene); and petri dishes.

Larger items or specialized assemblies which are gamma sterilized as required are: Surgivac and Haemivac units (pump, catheters, tubes); kidney perfusion sets (nylon bags, cannulae, plastic tubing); parts for technetium generators (polyethylene tubing, PVC adaptors, needles, clamps, terylene and glass fibre discs); leads and catheters for pacemakers; tubing used on by-pass machines for open-heart surgery; Dermatome and Humby stainless-steel blades.

Approximately 230 Keratomes and Graefe knives, used in eye surgery, have been gamma sterilized by one hospital in the past year. Radiation is the only suitable sterilization method for these knives because of the extreme delicacy of their cutting edges and the difficulty in obtaining suitable racks for their storage. After use the instruments are cleaned, dried, and replaced in plastic containers which are then sealed in Steripeel packets for re-sterilization.

Sterilization of pharmaceuticals

In Australia, commercial radiation sterilization has only been applied to a few pharmaceuticals all of which are for topical use. The antibiotics neomycin, polymyxin and bacitracin are radiation sterilized either separately or combined as a dusting powder. Radiation sterilization has been used for ophthalmic preparations containing physostigmine salicylate (in an oil base), sodium sulphacetamide, or mercuric oxide. Radiation-sterilized talc has been marketed by one company; however, another company will not use radiation for talc, as they have found that at 2.5 Mrad the characteristic perfume of their product is lost. A lubricating cream (Biosorb) is sterilized by radiation. One hospital uses radiation to sterilize normal saline required for perfusing kidney transplants.

There seems little likelihood that there will be much demand for radiation sterilization of internally administered drugs, particularly in view of recently amended regulations which demand extensive testing of new drugs before licensing. The question of radiation sterilization of some ophthalmic preparations is also being re-examined by the Australian Department of Health.

Radiation sterilization of lyophilized reagents for preparing technetium-99m radiopharmaceuticals

The radionuclide ^{99m}Tc, widely used in scintigraphy, can be complexed with various chemicals which localize in specific organs. Freeze-dried kits containing the inactive reagents have been developed for the preparation of these complexes [3, 4]. Scanning reagents are therefore always available when required, as the lyophilized products can be readily reconstituted with sodium pertechnetate solution obtained from ^{99m}Tc generators.

The AAEC's Isotope Division uses ionizing radiation for terminal sterilization of freeze-dried reagents. In a laminar flow cabinet, bulk solutions of the reagents are purged with nitrogen, membrane filtered (0.22 μm) and dispensed in 1-ml quantities into vials. After lyophilization, the vials are fully stoppered, tested for leaks and sterilized by 2.5 Mrad at -70°C in a ^{60}Co rig in the AAEC storage pond.

Stannous chloride is the basis for two gamma-sterilized lyophilized reagents routinely produced by the AAEC for use with ^{99m}Tc. Combined with diethylenetriamine pentaacetic acid (DTPA), the product, known as PENTASTAN RM5, has a refrigerated shelf-life of at least 12 months and is used for brain or renal scans. Renal scans are also carried out using complexes of ^{99m}Tc with stannous chloride and calcium gluconate (RENAL REAGENT RM6). The possibility of freeze-dried calcium gluconate being degraded by a radiation dose of 2.5 Mrad delivered at -70°C was investigated by thin-layer chromatography using carbon-14 labelled calcium gluconate. Plates were developed in five solvent systems of various polarities. No significant differences were found between irradiated and non-irradiated materials.

Two further radiation-sterilized products for use with ^{99m}Tc are being developed. Clinical trials have been completed on a lyophilized mixture of stannous chloride and sodium pyrophosphate (SKELTEC II), used for bone scanning [5]. Human serum albumen microspheres (used for lung scanning) have also been radiation sterilized, and preliminary results indicate that absorption is enhanced after irradiation.

Sterilization of biological tissues

The Royal Victorian Eye and Ear Hospital has been using irradiated cadaver fascia lata since 1967. The fascia lata is mostly used in retinal detachment surgery for eye encircling procedures although small 'patches' have been used in tympanic membrane repair operations. Under near aseptic conditions, the fascia lata is removed as a large sheet from cadavers. It is then transferred to an operating theatre where, in surgical procedure conditions, it is cut into strips of suitable sizes (e.g. 15 cm x 3 cm). Each strip is then placed into a sterilized Pyrex glass tube containing a drop of antibiotic solution. The tubes, sealed with nylon film and a rubber band, are sterilized by 2.5 Mrad and stored at -20°C until required. It is estimated that 1,500 pieces of fascia lata have been successfully treated by this method. The present rate of usage is about 500 pieces per annum.

Only minimal usage has been made of radiation for sterilization of heart valves, nerves, shin bone, and cartilage removed during sub-mucous resection operations.

REGULATORY ASPECTS

Australian Code of Good Manufacturing Practice for Therapeutic Goods

When the original Code of Practice for Radiation Sterilization of Medical Products formulated by the IAEA in 1966 was circulated in Australia, representatives of industry and government agreed there was a need for an authoritative specification for the sterilizing dose and a legal definition of the responsibilities of the manufacturer and the owner of an irradiation facility covering sterility of a product. The stress which the initial IAEA Code placed on the minimization of the initial microbial load, and on determining sterilizing doses on the basis of initial microbial loads and desired microbial inactivation factors, while laudable, was criticised as being good in theory but not the best in practice.

This criticism reflected a lack of confidence in methods for determining initial microbial loads, and the chance that they might be seriously underestimated both qualitatively and quantitatively. It was therefore considered preferable to recommend an arbitrary minimum sterilizing dose based on an acceptable inactivation factor for specific radiation-resistant organisms under environmental conditions which maximized their resistance. It was also agreed that specification of an arbitrary minimum dose should not preclude <u>lower</u> doses, provided these were approved by the appropriate authority from scientific evidence demonstrating adequacy of a lower dose. This latter consideration accepted the fact that some products might be amenable to a combination of treatments and that advantageous synergistic interaction could provide the required degree of sterility.

Since radiation processing of medical supplies was a rapidly expanding industry in Australia, it was decided to incorporate a code for radiation sterilization in a Code of Good Manufacturing Practice for Therapeutic Goods. The manufacturing code was to be the basis for initiating a system of government licensing to control standards for the manufacture and handling of therapeutic goods.

Federal and State Health Departments, the Australian Atomic Energy Commission, Industry, Universities, the National Council of Chemical and Pharmaceutical Industries, the Associated Chambers of Manufactures of Australia were invited in 1968 to nominate representatives to the drafting committee, and the Code was published in August 1971 [6].

As previously mentioned, the committee had considered that, despite the soundness of the principle of determining a sterilizing dose for each product based on types and maximum numbers of contaminating organisms prior to sterilization, Australian manufacturers did not have the facilities or microbiologists for such work. It would therefore be impracticable to operate in this way.

It was also recognised that manufacturers would not operate empirically. It was therefore preferable to specify that medical products should receive at least a certain minimal dose of radiation and that this dose should be sufficient to provide a degree of sterility equivalent to that achieved by autoclaving. Several years' experience in Australia and other countries had demonstrated the efficacy of 2.5 Mrad and this dose was specified in the Code. Consideration of the specifications for the manufacture of medical supplies which required strict microbiological control and monitoring of production areas led to the mandatory requirement that a minimum dose greater than 2.5 Mrad shall not be used without approval of the designated authority. This provision ensures that higher doses are not used arbitrarily to compensate for unhygienic conditions. It also inhibits overdosing of manufacturing materials which could have consequences more disastrous than the minute theoretical risk of contamination by pathogenic micro-organisms surviving irradiation.

The Code sets out provisions relating to premises, equipment, personnel, processes, sytems of quality control and product release, checking and recording of procedures, and special precautions necessary to achieve high-quality production consistency and to avoid contamination errors, omissions and mix-ups. It specifies the requirements for calibrating a radiation sterilization process, microbiological control, buildings and grounds for the radiation facility, radiation safety, irradiation process control dosimetry and the maintenance of records.

Industry expressed some concern, when the Code was published, that the restrictions and requirements might cause radiation sterilization to be disadvantaged commercially in comparison with gaseous sterilization. This has not proved to be the case. In fact, the two full-scale gamma irradiators have subsequently been constructed in conformity with the Code, while the reliability and adequacy of gaseous methods tend to be suspect for lack of agreed specifications.

Microbiological control

The Australian Code states that routine sterility testing is not needed for radiation-sterilized items, provided that the process has been validated and that conditions have not changed since validation.

The Code details requirements for validation and these have met with some resistance from industry, despite the incentive of a much reduced effort in routine sterility testing. A contributory factor is that most pharmaceutical companies in Australia are subsidiaries of overseas companies and are obliged, or compelled, to comply with their parent company's testing regulations which generally include routine sterility testing. In view of the well-known limitations of sterility testing, diversion of microbiological effort to validation of the process would appear justified. Validation of the radiation sterilization of AAEC lyophilized products will be carried out when a freeze-drying unit is installed.

There is some use of biological indicators (*B. pumilus*, E-601) for radiation processing control. One manufacturer routinely inserts biological indicators in cartons even though the product has been microbiologically validated for a radiation dose of 2.5 Mrad. Biological indicators are attached to every 200th carton passing through one of the commercial radiation plants.

DOSIMETRIC ASPECTS

In Australia, the ceric-cerous dosimeter is used in calibration and routine dosimetry in the megarad range. It has been used internationally for many years and is particularly well suited to the 2.5-megarad dose range. It is also an excellent dosimeter down to the tens of kilorad dose range.

Experimental difficulties in its use have been largely overcome. Alteration of the response of the system to trace impurities can be eliminated by adding cerous ions in concentration greater than 3 millimolar to the solution. When this concentration of cerous ions is present, quite satisfactory results are obtained using solutions prepared from singly distilled water, or even Sydney domestic supply water, in place of the triply distilled water formerly required [7]. It is also unnecessary to specially clean glassware used for solution preparation or ampoules in which the solution is irradiated.

<u>Measurement of dosimeters</u>

The measurement of dosimeters following irradiation has been simplified. The usual method with the traditional ceric dosimeter in the megarad range is spectrophotometric measurement at 320 nm after dilution of the solution to a concentration at which Beer's Law is obeyed. It has been shown [8] that the dose absorbed by a ceric-cerous solution is related to the potential difference, ΔE (millivolts), between the irradiated and unirradiated solutions by the following expression:

$$\underset{(\text{Mrad})}{\text{Dose}} = \frac{965.2}{\rho G(Ce^{3+})}\left[[Ce^{4+}]_u - \left[\frac{[Ce^{4+}]_u + [Ce^{3+}]_u}{1 + \frac{[Ce^{3+}]_u}{[Ce^{4+}]_u}\ \text{antilog}\ \frac{\Delta E}{59.16}}\right]\right] \qquad (1)$$

where ρ = the density of the ceric-cerous solution,

$G(Ce^{3+})$ = the number of ceric ions reduced for 100 eV radiation absorbed in the solution,

965.2 = $10^3/1.036$ = conversion factor for change in molar concentration to dose in megarad,

$[Ce^{4+}]_u$, $[Ce^{3+}]_u$ = the ceric and cerous ion molar concentrations in the unirradiated solution,

59.16 = 2.303 RT/F and R, T, and F have the usual meanings.

Once the ceric and cerous ion concentrations in the unirradiated dosimeter solution have been determined, it is only necessary to measure ΔE.

ΔE is measured by connecting the simple electrochemical concentration cell shown in Fig.1 to a millivoltmeter. A graph or table may be prepared from expression (1) relating dose to ΔE for a given solution composition, thereby avoiding repeat calculations.

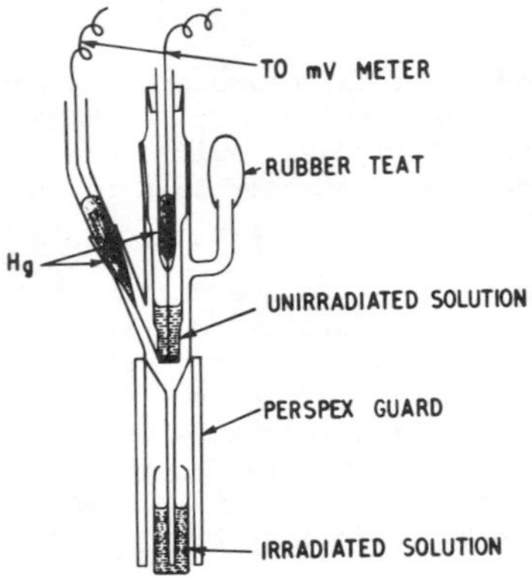

FIG.1. Dosimeter measuring cell.

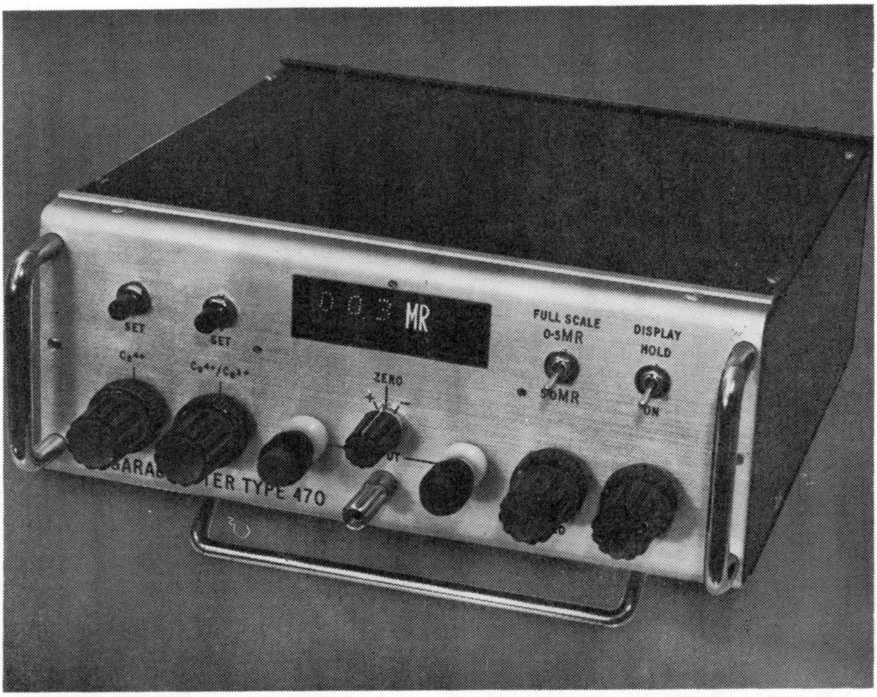

FIG.2. Direct-reading dose meter.

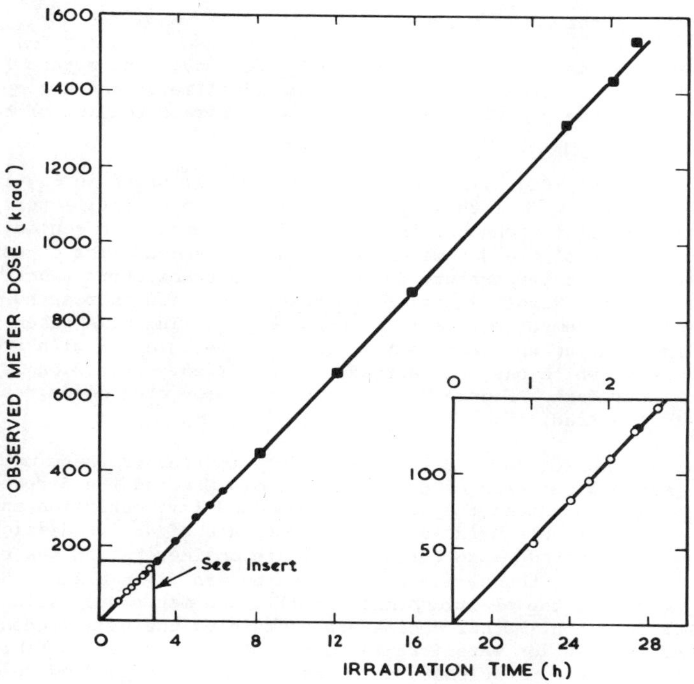

FIG.3. Observed meter dose versus irradiation time. Three different dosimeter solutions were used:
○ 4×10^{-4} M Ce^{4+}, 3×10^{-3} M Ce^{3+}; 50 to 145 krad
● 1×10^{-3} M Ce^{4+}, 3×10^{-3} M Ce^{3+}; 125 to 350 krad
■ 4×10^{-3} M Ce^{4+}, 9×10^{-3} M Ce^{3+}; 445 to 1535 krad.

TABLE II. COMPARISON BETWEEN DOSES GIVEN BY THE DIRECT-READING METER AND BY THE FERROUS SULPHATE METHOD

Dosimeter No.	$FeSO_4$ dose (krad)	Meter dose (krad)	Percentage difference
1	125.3	125	0.24
2	163.0	162	0.62
3	270.2	271	0.30
4	306.1	305	0.36
5	340.4	343	0.76

Direct reading meter

We have been using an alternative and more convenient method for several years in which an instrument containing a millivoltmeter and a comparatively inexpensive analogue computer is used to give a direct reading of total dose [9].

The instrument (Fig.2) is connected to the cell shown in Fig.1 and dose is read directly from the digital display. Two potentiometers are provided to correct for changes in the ceric and cerous ion concentration if the composition of the dosimeter solution is changed. The circuit is designed so that the temperature response of the transistors exactly compensates for the Nernst Equation increase in ΔE with increasing temperature, thus making temperature control unnecessary. The time taken for a dose reading is about the same as that for a pH measurement with a direct reading pH meter which has been already standardized. Precision is of the order of ± 1 per cent. Figure 3 shows the precision over doses ranging from 0.05 to 1.5 Mrad.

Table II compares doses determined by the ferrous sulphate method with doses determined with ceric-cerous sulphate solution and the direct-reading meter. In this experiment the solution contained 1 m$\underline{M}$ ceric ion and 3 m$\underline{M}$ cerous ion in 0.4 $\underline{M}$ sulphuric acid. The dose rate of an irradiation facility was determined using ferrous sulphate dosimeters and one of the dosimeters containing the ceric-cerous solution was given a known dose and introduced into the electrochemical cell; the meter was calibrated against this solution. Other dosimeters containing the same concentration of ceric-cerous solution were irradiated to the doses given in Table II according to the ferrous sulphate dose rate; the corresponding readings on the meter are also recorded in the table.

In later models of the direct-reading meter, an internal calibration reference voltage has been provided which eliminates the need to calibrate the instrument against the ferrous sulphate dosimeter although this can still be done as an additional check if required.

Calibration dosimetry service to industry

For several years the AAEC has provided a calibration dosimetry service to users of commercial gamma irradiators in Australia. Every two weeks, batches of six ceric-cerous dosimeters in glass ampoules sealed in plastic envelopes are mailed to the companies concerned for irradiation with the plastic dosimeters routinely used. After irradiation the ceric dosimeters are returned for measurement and the companies are informed of the result. In this way a check is maintained on the calibration of readings obtained with the plastic dosimeters.

One company uses the ceric-cerous system and the AAEC direct-reading meter exclusively and on a routine basis, using thousands of dosimeters each year. The method is found to be quicker and more reliable than the clear Perspex method it replaced.

Stability of ceric-cerous solutions

The solution used for calibration dosimetry at the 2.5-Mrad dose level is approximately 10 m$\underline{M}$ Ce^{4+}, 10 m$\underline{M}$ Ce^{3+} in 0.4 $\underline{M}$ sulphuric acid. After preparation it is allowed to age for 1 week before filling in 2-ml standard pharmaceutical ampoules. Checks on the ceric concentration of ampoules kept in the dark have shown less than 1 per cent change during 6 months.

ACKNOWLEDGEMENTS

The authors gratefully acknowledge Mr. V.E. Church and Mr. H.J. Fraser, Instrumentation and Control Division, Australian Atomic Energy Commission, for the design and construction of the direct-reading dose meter.

REFERENCES

[1] MURRAY, G.S., Industrial sterilization by gamma radiation, At. Energy Aust. 5 (1962) 4.

[2] O'LEARY, B., Gamma irradiated facilities using spent fuel elements, At. Energy Aust. 9 (1966) 7.

[3] DEUTSCH, M.E., REDMOND, M.L., MEAD, L.W., 'Freeze-dried kits for preparing radiopharmaceuticals from ^{99m}Tc', Radiopharmaceuticals and Labelled Compounds I (Proc. Symp. Copenhagen, 1973), IAEA, Vienna (1973) 189.

[4] BOYD, R.E., ROBSON, J., HUNT, F.C., SORBY, P.J., MURRAY, I.P.C., McKAY, J.W., ^{99m}Tc gluconate complexes for renal scintigraphy, Brit. J. Radiol. 46 (1973) 604.

[5] MORRIS, I.P.C., Clinical uses of bone scanning. Presented at 1st World Congress of Nuclear Medicine, Tokyo and Kyoto, 1974.

[6] AUSTRALIA, DEPARTMENT OF HEALTH, Code of Good Manufacturing Practice for Therapeutic Goods, National Biological Standards Laboratory, Canberra (1971).

[7] MATTHEWS, R.W., An evaluation of the ceric-cerous system as an impurity-insensitive megarad dosimeter, Int. J. Appl. Radiat. Isot. 22 (1971) 199.

[8] MATTHEWS, R.W., Potentiometric estimation of megarad dose with ceric-cerous system, Int. J. Appl. Rad. Isot. 23 (1972) 179.

[9] CHURCH, V.E., FRASER, H.J., MATTHEWS, R.W., A small analogue linearising instrument for ceric-cerous gamma ray dosimetry, 14th National Radio and Electronics Eng. Conv. (Proc. Conf. Melbourne, (1973)) 50.

DISCUSSION

K.S. AGGARWAL: I think that as the concentration of cerium in the dosimeter solution is increased, it becomes more and more energy-dependent. By adding cerium-3 ions you are in fact increasing the cerium concentration. Do you think this will adversely affect the dosimeter solution as far as its energy dependence is concerned? And if the energy dependence does increase, by how much would it be?

Pamela A. WILLS: The addition of cerium-3 ions does not adversely affect our dosimeter solution used for 2.5-Mrad dosimetry since the total cerium concentration does not exceed 0.02M. Mr. Chu[1] has calculated that at this concentration the build-up effect is 3.7% in an AECL JS-6500 plant.

[1] IAEA-SM-192/14, these Proceedings.

In practice, it is probably less. We tried to observe this effect in a JS-6500 plant, using 40 dosimeters with a cerium concentration of 0.045M. The results showed no evidence of a build-up factor in the irradiated solutions.

B.L. GUPTA: I note that you use commercial vials as containers for the dosimeter solution. Do you clean them or use them as such?

Pamela A. WILLS: Standard pharmaceutical ampoules are used, as supplied by the manufacturer. We do not clean them.

B.M. TERENT'EV: Can you give us some specific examples of the cost of processing medical products?

Pamela A. WILLS: The charges for commercial radiation sterilization (2.5 Mrad) are based on the annual throughput of the particular product. The basic charge is \$A1.50/ft^3 (0.028 m^3), which can be cut down to A\$1.20 for large volumes. The cost of radiation processing is independent of the density of the product.

LE FILM «TAC», DOSIMETRE PLASTIQUE POUR LA MESURE PRATIQUE DES DOSES D'IRRADIATION REÇUES EN STERILISATION

J.R. PUIG, J. LAIZIER
CEA, Centre d'études nucléaires de Saclay,
Gif-sur-Yvette, France

F. SUNDARDI
BATAN, Pasar Jumat Research Centre,
Jakarta, Indonésie

Abstract–Résumé

"TAC" FILM, A PLASTIC DOSIMETER FOR PRACTICAL MEASUREMENT OF IRRADIATION DOSES RECEIVED DURING STERILIZATION.

The characteristics of a dosimetric film ("TAC") recently placed on the market were investigated. Both for cobalt-60 gamma rays and for 0.3 - 6 MeV electrons, the absorption varies linearly with dose in the range 1 - 10 Mrad. The dose-rate effect is weak with gamma rays and nil with electrons. The response is stable in time and is not modified by heating for one hour at 80°C. This film therefore appears well suited for monitoring the radiosterilization process on an industrial scale, and also for many practical measurements, of which various examples are given.

LE FILM ≪TAC≫, DOSIMETRE PLASTIQUE POUR LA MESURE PRATIQUE DES DOSES D'IRRADIATION REÇUES EN STERILISATION.

Les caractéristiques d'un film dosimétrique (≪TAC≫) récemment mis sur le marché ont été étudiées. Aussi bien pour le rayonnement γ du cobalt-60 que pour les électrons de 0,3 à 6 MeV, la variation d'absorption est linéaire avec la dose dans le domaine compris entre 1 et 10 Mrad. L'influence du débit de dose est faible en γ et nulle sous électrons. La réponse est stable dans le temps et n'est pas modifiée par un chauffage d'une heure à 80°C. Ce film paraît donc bien adapté au contrôle industriel en radiostérilisation, mais également à de nombreuses mesures pratiques dont différents exemples sont donnés.

Qu'elles fassent appel à l'irradiation par des rayons γ ou par des électrons, les opérations industrielles de radiostérilisation impliquent la réalisation d'un nombre très important de mesures dosimétriques. Celles-ci sont nécessaires non seulement pour la fixation des conditions d'irradiation, mais surtout pour les contrôles de routine.

Les mesures dosimétriques constituent ainsi une part non négligeable du coût du traitement ; par ailleurs, la qualité de ce dernier en dépend directement, tout sous-dosage étant à proscrire, mais tout surdosage inutile étant à éviter.

Les films plastiques sont particulièrement bien adaptés à ce type de dosimétrie pratique, en particulier par leur faible coût et leur facilité d'emploi. Les produits disponibles présentent malheureusement plusieurs inconvénients comme, par exemple, la sensibilité au débit de dose et à la température, ou le manque de stabilité de la réponse dans le temps.

Le film dont les caractéristiques vont être décrites évite la plupart de ces inconvénients.

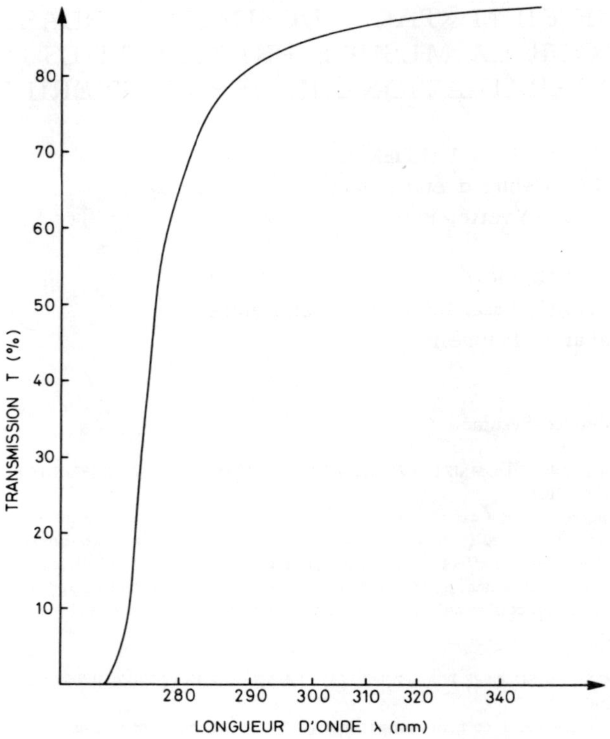

FIG.1. Courbe d'absorption du film "TAC" par rapport à l'air.

1. CARACTERISTIQUES DU DOSIMETRE "TAC"

Le dosimètre est un film[1] constitué de triacétate de cellulose et d'additifs convenables en teneur déterminée. Il est conditionné en galettes de largeur 8 mm, d'épaisseur 125 μm et de longueur variable.

La courbe d'absorption UV du film vierge (transmission en fonction de la longueur d'onde) est présentée à la figure 1. On y constate une variation continue de l'absorption en fonction de la longueur d'onde, le film étant pratiquement transparent vers 340 nm pour devenir opaque vers 270 nm.

L'irradiation accroît l'absorption du film, et cet accroissement est d'autant plus important que la longueur d'onde est faible. Les courbes transmission-longueur d'onde de films irradiés à différentes doses, mesurées par rapport à un témoin non irradié,

[1] Distribution en France et à l'étranger : Société NUMELEC, 2 Petite Place, 78000-Versailles, France.

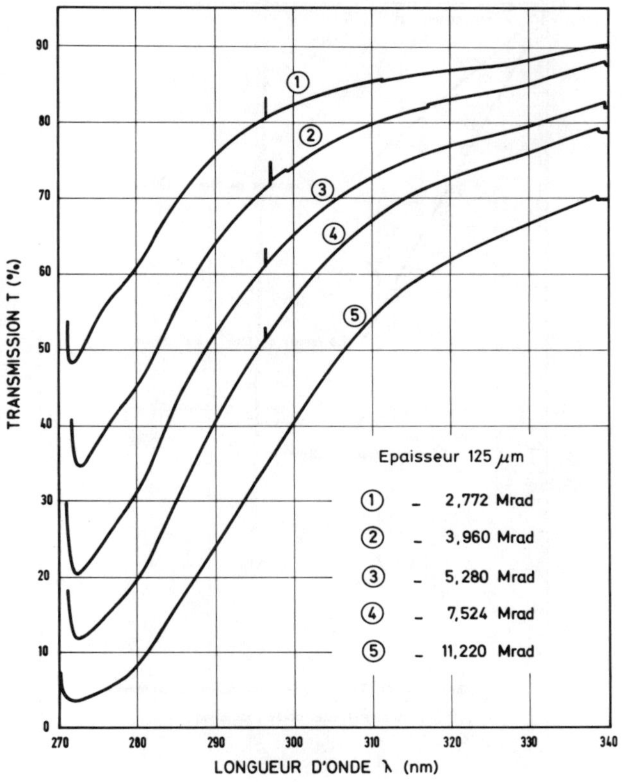

FIG.2. Absorption du film "TAC" irradié par rapport à un témoin.

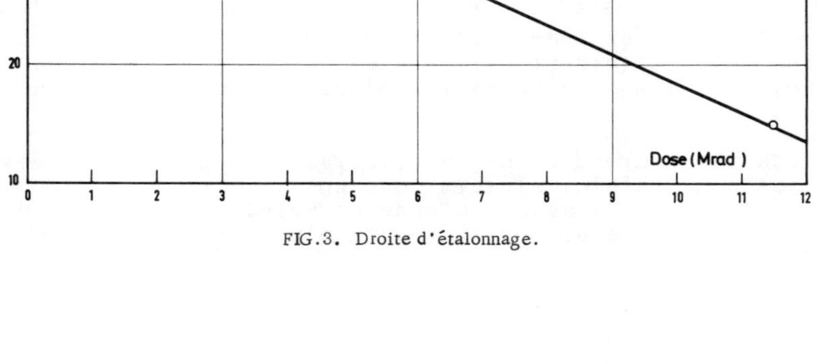

FIG.3. Droite d'étalonnage.

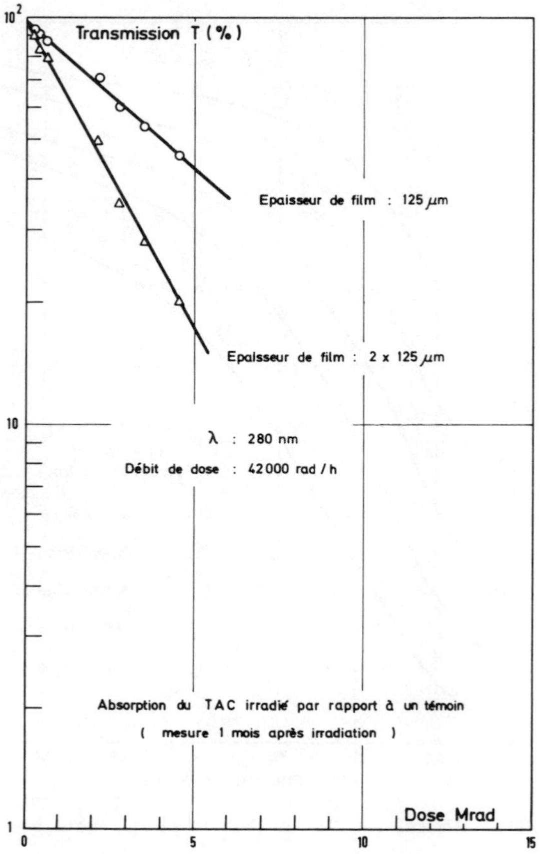

FIG.4. Influence de l'épaisseur du film.

sont présentées en figure 2. Leur allure générale est analogue à celle présentée en figure 1, mais il ne s'agit que d'une coïncidence. Au voisinage de 270 nm, la transmission propre du témoin étant pratiquement nulle, l'enregistrement spectrophotométrique fait apparaître un minimum : celui-ci n'est qu'un artefact, dû au système automatique de mesure. La longueur d'onde minimum pratiquement utilisable, qui conduit, pour une dose donnée, au maximum de variation d'absorption, est voisine de 280 nm.

Pour des longueurs d'ondes inférieures, la variation d'absorption est plus faible. Il sera donc possible, en choisissant convenablement la longueur d'onde de mesure, de mesurer des gammes différentes de doses, jusqu'à plusieurs dizaines de Mrad.

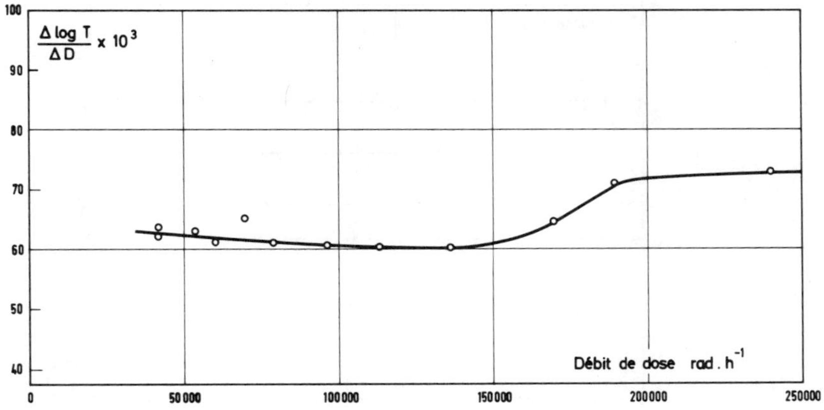

FIG.5. Variation de la pente de la droite d'étalonnage avec le débit de dose.

Cependant la majorité des doses délivrées dans le cadre des traitements industriels d'irradiation se situent dans un domaine compris entre 1 et 10 Mrad, et compris entre 2,5 et 5 Mrad pour les radiostérilisations. Une longueur d'onde de 280 nm sera donc préférentiellement choisie pour ce type de mesure.

2. ETUDE DU DOSIMETRE EN IRRADIATION γ DU Co^{60}

2.1. Réalisation des irradiations et des étalonnages

Les irradiations γ ont été réalisées dans "Pagure" et "Poséīdon", sources panoramiques au cobalt 60, d'activités respectives 20 et 200 kCi. Les dosimétries aux emplacements d'irradiation ont été effectuées par la méthode de Fricke, avec une précision d'environ 1 %.

2.2. Variation de la transmission avec la dose

Une courbe typique de variation de la transmission T avec la dose est présentée à la figure 3. Elle est linéaire en coordonnées semi-logarithmiques, ce qui revient à dire que la densité optique D.O. (ou absorption A) varie linéairement avec la dose D. On a donc :

$$\log T = 2 - \alpha D$$

ou $$(D.O.) = \alpha D$$

Selon la longueur d'onde de lecture, le coefficient α varie, et est d'autant plus élevé que la longueur d'onde est faible. Le film TAC permet donc la mesure de doses intégrées très élevées, moyennant le choix d'une longueur d'onde convenable. On doit cependant remarquer que pour des doses absorbées élevées (à partir d'environ 20-30 Mrad), le film devient cassant et ne permet plus une lecture continue dans un système de défilement.

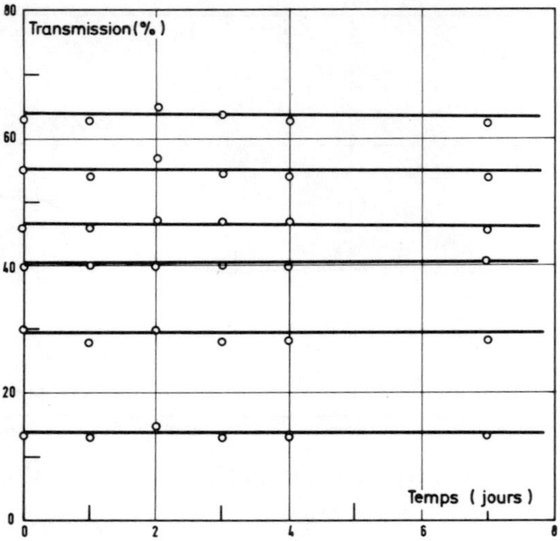

FIG.6. Stabilité de la réponse dans le temps.

Il est par ailleurs possible, dans une certaine mesure, d'utiliser deux épaisseurs de film pour accroître la sensibilité de la mesure. L'expérience (figure 4) montre alors que l'absorption suit la loi de Lambert-Beer et que $(DO)_{2x125\,\mu m} = 2\ (DO)_{250\,\mu m}$.

2.3. Influence du débit de dose

La variation du coefficient α avec le débit de dose est présentée à la figure 5.

On y constate qu'entre 4.10^4 et $1{,}5.10^5$ $rad \cdot h^{-1}$, la valeur de α est pratiquement constante, et égale à 0,0615.

Entre 1,5 et 2.10^5 $rad \cdot h^{-1}$, la valeur de α croît de façon relativement rapide pour passer à 0,0715, soit une variation de + 16%.

Le coefficient α demeure ensuite constant entre 2.10^5 et $3{,}34.10^6$ $rad \cdot h^{-1}$, cette dernière valeur représentant le débit de dose maximum étudié. On verra plus bas qu'aux plus forts débits de dose qui sont le fait de l'irradiation sous électrons, la valeur de α demeure inchangée.

2.4. Influence de la température

L'influence d'un chauffage après irradiation a été étudiée en irradiant à 1 et 1,7 Mrad ($I = 3{,}34$ $Mrad \cdot h^{-1}$) des lots de 15 dosimètres. La transmission de 8 d'entre eux était mesurée immédiatement tandis que les 7 autres étaient soumis à un chauffage d'une heure dans une étuve à 80 °C.

On n'a pas trouvé de différence statistiquement significative entre les deux séries de mesures.

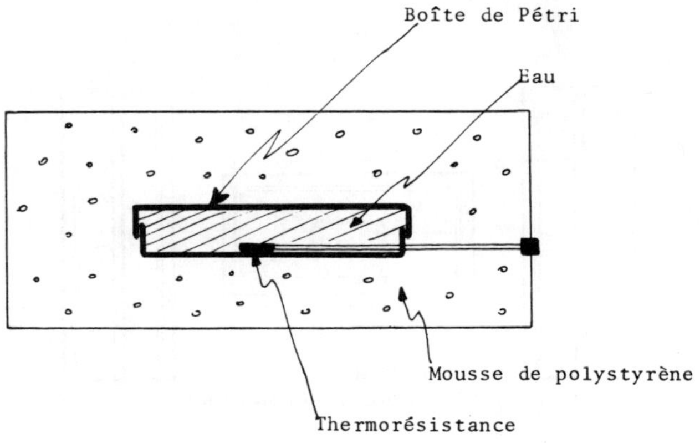

FIG.7. Calorimètre pour électrons de haute énergie.

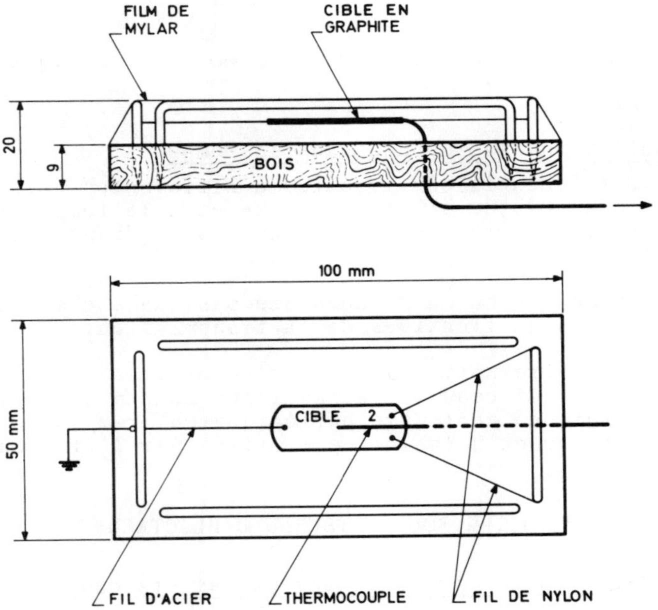

FIG.8. Calorimètre adiabatique n° 2 utilisé pour l'étalonnage du film "TAC" avec le faisceau d'électrons de 300 keV.

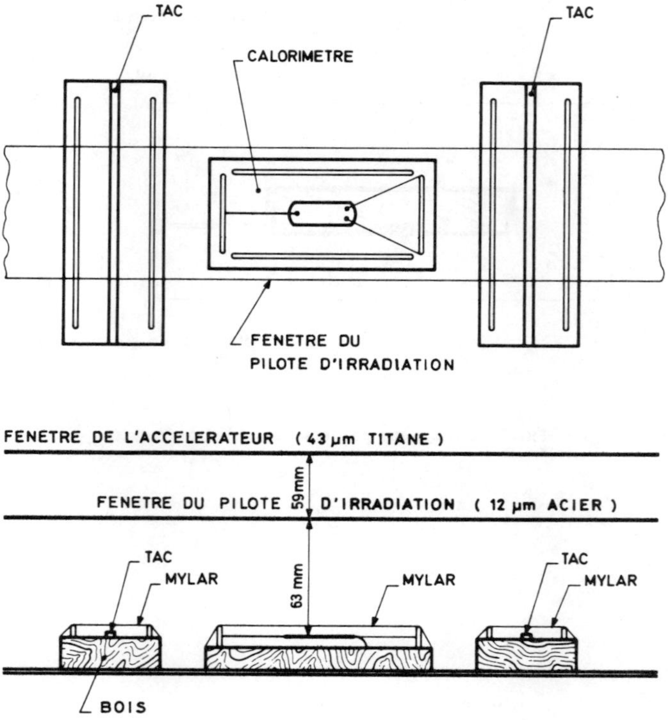

FIG. 9. Arrangement du calorimètre et du film "TAC" sous faisceau d'électrons de 300 keV.

2.5. Stabilité de la réponse dans le temps

Des échantillons irradiés à différentes doses ont été stockés pendant 7 jours à l'obscurité et à la température ambiante, des mesures périodiques de transmission étant effectuées dans l'intervalle.

La figure 6 montre qu'on ne constate pas, dans ces conditions, de variations significatives de la transmission.

On doit par contre noter que lorsqu'on utilise un spectrophotomètre à double faisceau , l'échantillon témoin non irradié utilisé comme référence doit être périodiquement remplacé, sa transmission se trouvant à la longue modifiée par l'irradiation UV subie.

3. ETUDE DU DOSIMETRE SOUS FAISCEAU D'ELECTRONS

3.1. Réalisation des irradiations et des étalonnages

Les irradiations sous faisceau d'électrons ont été réalisées en faisant appel à plusieurs types d'accélérateurs. Dans tous les cas la dosimétrie de référence est calorimétrique :

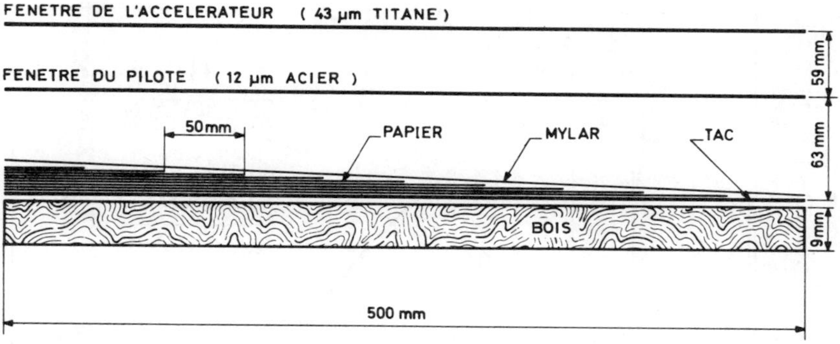

FIG.10. Dispositif expérimental utilisé pour obtenir les courbes pratiques profondeur-dose. Epaisseur du papier: 8,855 mg/cm^2. Epaisseur du film "TAC": 16,21 mg/cm^2.

a) A haute énergie (6,2 MeV), un accélérateur linéaire "CIRCE 10" (constructeur Thomson CSF - France) a été utilisé.

Cette machine délivre un faisceau pulsé et balayé dont l'énergie peut varier de 4 à 6,2 MeV, le courant moyen faisceau de 0,15 à 2,5 mA et la fréquence de répétition de 100 à 600 Hz. La puissance est comprise entre 1 et 9 kW.

Le dispositif calorimétrique utilisé pour ce type d'irradiation (figure 7) est constitué d'une boîte de Pétri en plastique, remplie d'eau et insérée dans une enveloppe isolante de mousse de polystyrène. La température est mesurée par l'intermédiaire d'une thermistance étalonnée. La valeur en eau du calorimètre est calculée d'après les éléments constitutifs, et vérifiée expérimentalement à l'aide d'une résistance chauffante.

On passe de la dose moyenne reçue par le calorimètre (épaisseur d'eau de 1 cm) à la dose de surface (reçue par le film) par l'intermédiaire d'un coefficient de proportionnalité. Il est calculé (valeur 0,86) à partir de la courbe expérimentale de pénétration des électrons dans l'eau.

Les irradiations sont réalisées par transport simultané du film et du calorimètre sous le faisceau grâce à un convoyeur.

Compte tenu de l'ensemble des conditions d'utilisation, la précision des mesures est évaluée à ± 10 %.

b) A basse énergie et pour une énergie de 500 keV, un accélérateur type Van de Graaff, d'énergie variable entre 500 keV et 3 MeV, de puissance maximum 3 kW (à 3 MeV) a été utilisé. Pour une énergie de 300 keV, on a fait appel à un accélérateur du type OFT de 30 kW.(Dans les deux cas, constructeur HVEC-USA).

Le dispositif calorimétrique utilisé dans les deux cas est présenté à la figure 8. Il est constitué d'une pastille de graphite suspendue dans l'air par des fils, protégée par une feuille mince de Mylar.

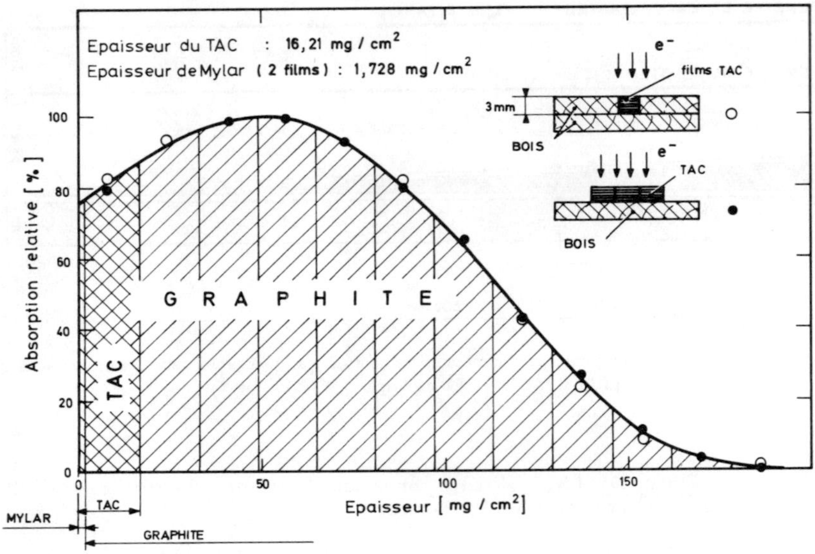

FIG.11. Courbe profondeur-dose pour un faisceau d'électrons de 500 keV.

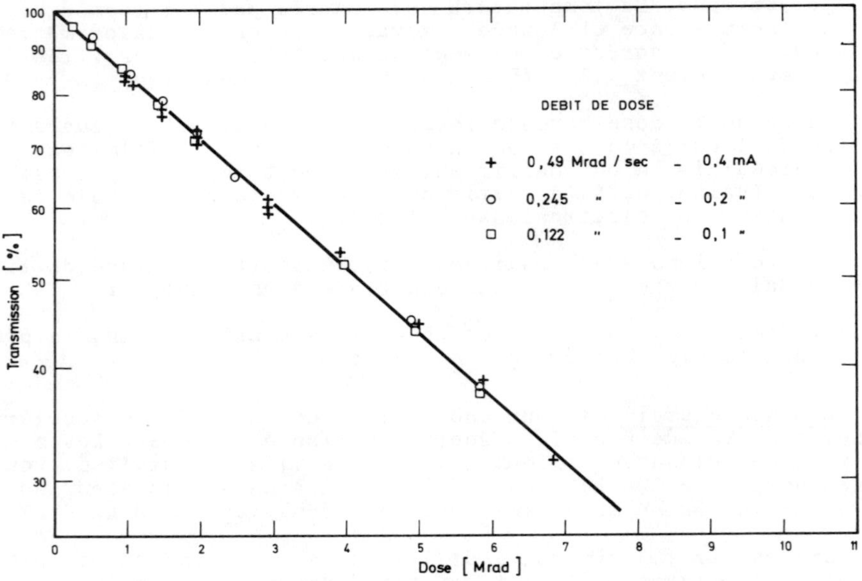

FIG.11bis. Essais en statique et en dynamique. Courbe d'étalonnage de l'absorption UV due à l'irradiation du film "TAC" ($\lambda = 280$ nm). Energie 500 keV, cible 2.

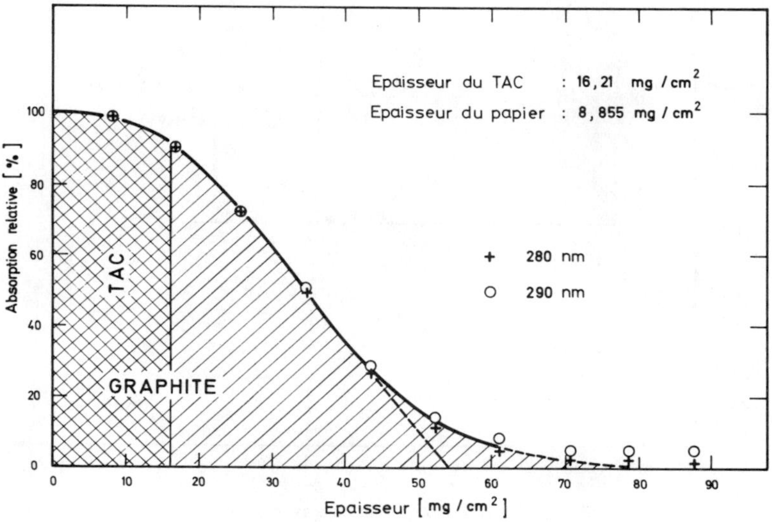

FIG.12. Courbe profondeur-dose pour un faisceau d'électrons de 300 keV.

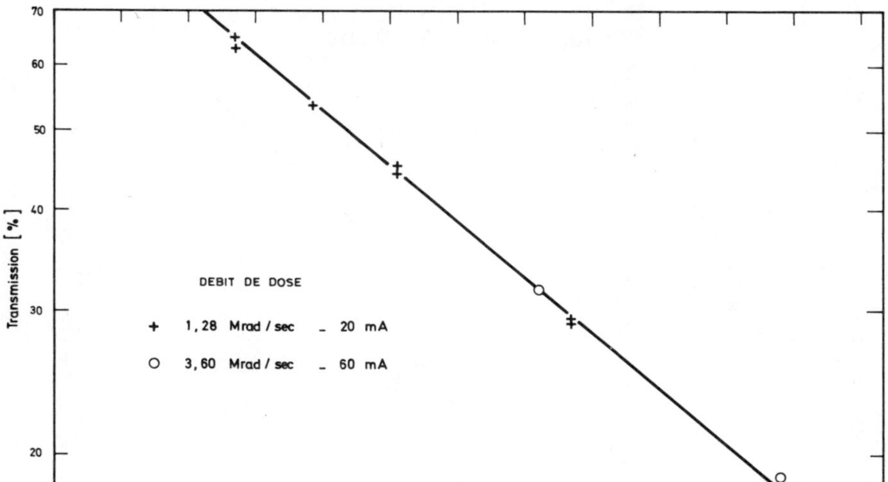

FIG.12bis. Essai en statique. Courbe d'étalonnage de l'absorption UV due à l'irradiation du film "TAC" (λ = 280 nm). Energie 300 keV, cible 2.

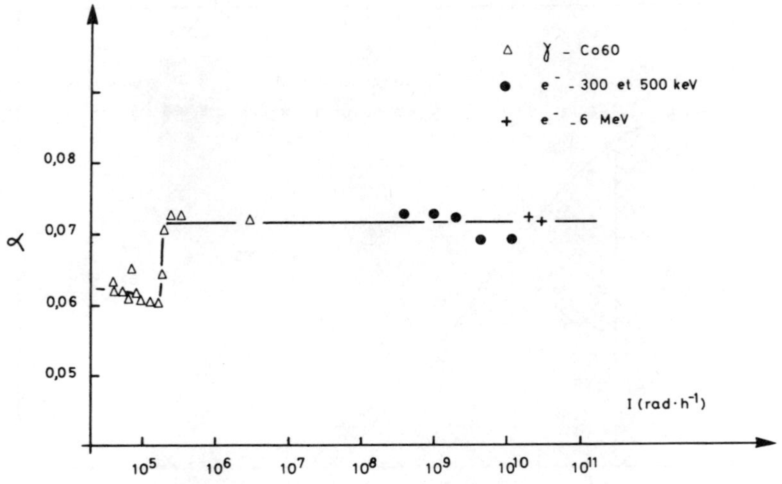

FIG.13. Variation de α avec I.

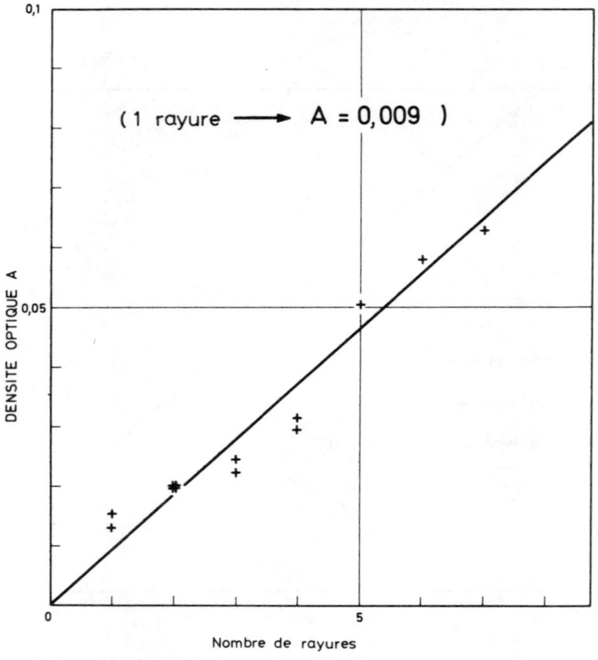

FIG.14. Influence des rayures.

Les irradiations ont été réalisées en statique sous le faisceau (figure 9 : disposition pour les irradiations à 300 keV). On a pu vérifier qu'il était également possible d'utiliser le même dispositif pour des irradiations sur un dispositif convoyeur avec de bons résultats.

On passe de la dose reçue par le calorimètre à la dose reçue par le film par l'intermédiaire d'un coefficient de proportionnalité comme dans le cas des hautes énergies. Il faut cependant remarquer qu'il est ici impératif d'utiliser les courbes expérimentales profondeur-dose, qui sont notablement différentes des courbes théoriques, du fait de la modification du faisceau due aux différents milieux traversés. La figure 10 montre le dispositif utilisé pour cette détermination à 300 keV. Les figures 11 et 12 présentent les courbes profondeur -dose correspondantes.

La précision des mesures dosimétriques est ici d'environ $\pm$ 3 %.

3.2. Variation de la transmission avec la dose

Comme dans le cas de l'irradiation γ, la courbe représentative de la variation de la transmission avec la dose est linéaire en coordonnées semi-logarithmiques ; les figures 11bis et 12bis présentent les courbes obtenues à 300 kV et 500 kV.

3.3. Influence du débit de dose

La figure 13 rassemble les résultats obtenus aux différents débits de dose sous électrons et les résultats obtenus par irradiation γ. On y constate que dans toute la gamme des débits de dose étudiés la valeur du coefficient α est constante aux erreurs expérimentales près. Cette valeur est identique à la valeur obtenue en γ pour la zone des forts débits de doses. Il semble donc qu'au-delà de 2.10^5 rad$\cdot$h^{-1}, la réponse du dosimètre est indépendante de la nature du rayonnement et de l'énergie.

4. UTILISATION PRATIQUE DU DOSIMETRE

4.1. Exactitude

Dans le cadre d'une utilisation pratique du dosimètre, l'exactitude des mesures dépend en premier lieu de l'exactitude de l'étalonnage, mais également de la longueur d'onde réelle de lecture. On constate, en effet, à la figure 2, que les mesures se font dans une zone où la transmission varie de façon continue avec la longueur d'onde. Une intercomparaison entre deux spectrophotomètres, et des essais consistant en une répétition des mesures en reprenant les réglages d'un appareil donné (Beckman DK 2) paraissent montrer que ce facteur ne diminue pas de façon appréciable l'exactitude.

4.2. Erreurs sur les mesures

L'incertitude sur les mesures, par ailleurs, a pour principales causes les hétérogénéités locales du film et les détériorations qu'il peut subir au cours des manipulations.

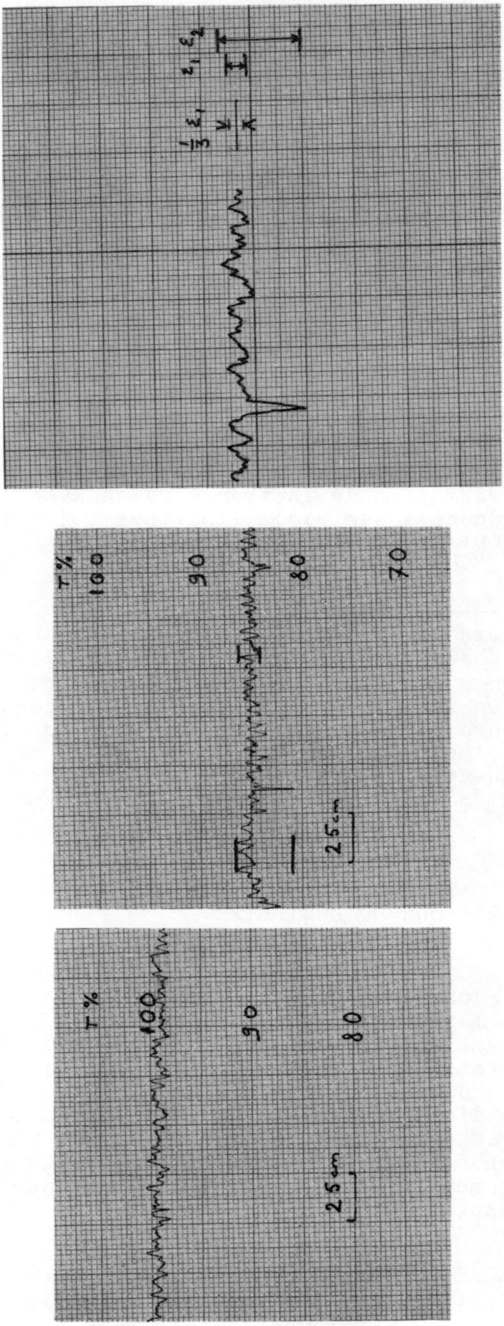

FIG. 15. Dosimétrie au film "TAC": enregistrements typiques.

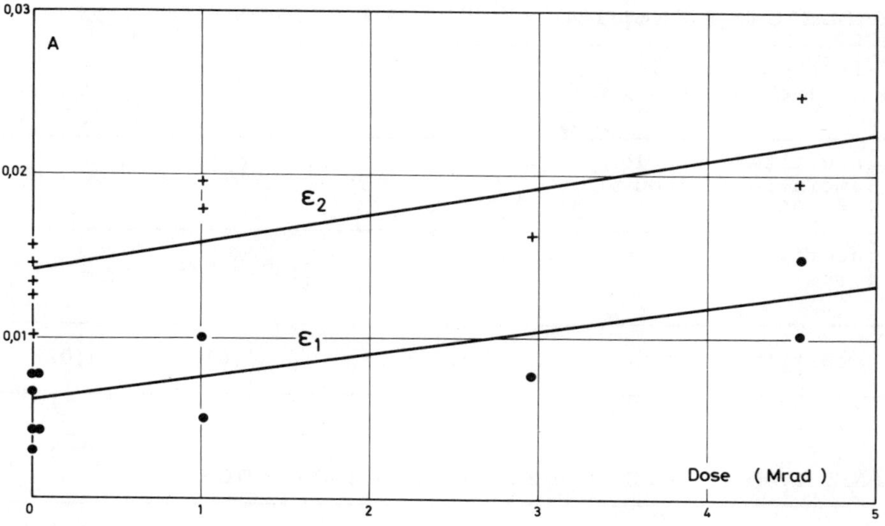

FIG.16. Variation des erreurs avec la dose.

L'utilisation dans le cadre d'un traitement industriel implique qu'il soit possible de manipuler commodément le film sans pour autant en détériorer les caractéristiques. On pourrait craindre qu'une lecture en ultra-violet entraîne une forte influence des rayures, poussières et salissures diverses.

Pour vérifier ce point, des rayures ont été effectuées, à l'aide d'une pointe de compas et sur toute la largeur du film, en nombre variable et croissant. La variation correspondante de densité optique (ou absorption A) a été mesurée. On constate à la figure 14 qu'elle demeure tout à fait limitée en dépit de la sévérité d'un tel traitement. Une rayure entraîne un accroissement de la densité optique équivalant, en moyenne, à 0,13 Mrad. Des essais d'utilisation pratique dans différentes conditions de manipulation (très soigneuse avec des gants, sans précautions particulières etc...) ont confirmé cette conclusion.

Les enregistrements de la transmission sur une certaine longueur de film présentent, par ailleurs, des fluctuations aléatoires. La figure 15 présente un exemple de lecture (15a : film non irradié : réglage du 100 % du spectrophotomètre ; 15b film irradié). Ces fluctuations sont la source la plus importante d'incertitude.

Leur influence dépend évidemment du mode de mesure adopté. Selon, en effet, la longueur de film dont on enregistre la transmission, leur incidence est plus ou moins grande. Cette dernière sera plus faible lorsqu'il est possible de faire une mesure moyenne sur une grande longueur d'enregistrement que lorsqu'on désire une mesure ponctuelle.

TABLEAU I. INCERTITUDES SUR LES MESURES D'ABSORPTION

Longueur de film lue (cm)	50	10	mesure pratiquement ponctuelle
Incertitude sur le Témoin (A = 0% ou T = 100 %)	$\frac{1}{3}\ \varepsilon_1 = 0,003$	$\frac{1}{3}\ \varepsilon_1 = 0,003$	$\frac{1}{3}\ \varepsilon_1 = 0,003$
Incertitude sur le film irradié	$\frac{1}{3}\ \varepsilon_1 = 0,005$	$\varepsilon_1 = 0,015$	$\varepsilon_2 = 0,025$
Incertitude totale	0,008	0,018	0,028

TABLEAU II. INCERTITUDES SUR LES MESURES DE DOSE

Longueur de film lue (cm)	50	10	mesure pratiquement ponctuelle
Incertitude sur le Témoin (Mrad) (T = 100 %)	0,04	0,04	0,04
Incertitude sur le film irradié (Mrad)	0,07	0,21	0,35
Incertitude totale (Mrad)	0,11	0,25	0,40

TABLEAU III. PRINCIPALES CARACTERISTIQUES DU LECTEUR DE FILMS NUMELEC

Longueur d'onde : Bande passante voisine de 280 nm (obtenue par lampe UV + convertisseur en quartz + filtre).

Vitesse de lecture : 62,5 cm/min.

Pouvoir séparateur : 5 mm (obtention de deux pics séparés en réponse à deux échelons opaques de 1 mm de large sur le film, si ceux-ci sont distants de plus de 5 mm).

Stabilité :

Zéro : pas de modification après 24 h

100 % : - pas de modification après 1 h
- passe à 95 % après 24 h

Pour évaluer l'importance de l'incertitude dans les différents cas, les hypothèses suivantes ont été prises (figure 15c).

a) Le réglage du 100 % est toujours effectué en enregistrant une grande longueur (1 m) de film irradié.

b) Dans le cas d'une mesure ponctuelle, on considère que l'ampleur de l'incertitude est égale à la plus grande variation constatée sur l'enregistrement d'une grande longueur de film (environ 4 m), soit ε_2 sur la figure 15c.

c) Dans le cas d'une mesure utilisant une longueur courte, environ 10 cm, de film, l'incertitude correspond à l'amplitude des fluctuations moyennes observées, soit ε_1 sur la figure 15c.

d) Dans le cas d'une mesure faisant appel à une grande longueur de film, de l'ordre de grandeur de 50 cm, il est possible de déterminer une moyenne, et l'incertitude sur cette moyenne a été prise égale à $(1/3)\,\varepsilon_1$.

On a donc enregistré les courbes obtenues pour des longueurs d'environ 4 m de films irradiés de façon homogène à différentes doses entre 0 et 5 Mrad, et mesuré les valeurs de ε_1 et ε_2.

Les résultats, présentés à la figure 16, montrent que ε_1 et ε_2 varient très peu avec la dose d'irradiation. On peut donc les considérer comme constants et prendre comme valeurs de base pour l'évaluation celle qui est déterminée pour 5 Mrad. Ceci revient à considérer que la <u>valeur absolue de l'incertitude est constante</u>, et que la <u>valeur relative de l'incertitude décroît avec la dose</u>.

Les Tableaux I et II présentent les différentes valeurs, en absorptions, et en doses correspondantes, calculées à partir de $\alpha = 0,71$.

On y constate en particulier que pour une longueur lue de 10 cm, la dose est déterminée à ± 0,13 Mrad près, soit, si la dose est de 2,5 Mrad, à ± 5 ou 6 % près. Dans le cas d'une mesure quasi ponctuelle, l'incertitude devient ± 0,2 Mrad, soit, pour la même dose, ± 8 %.

Dans le cas des irradiations γ, et en particulier lorsqu'elles sont réalisées à débit de dose variable, et lorsque ceux-ci se situent dans la zone où α lui-même varie, l'incertitude se trouve accrue de celle qui peut résulter de l'approximation faite sur α.

4.3. Dispositif de lecture des films

Un dispositif particulièrement simple et fiable de lecture de film existe sur le marché[2], qui en permet une exploitation commode.

---

[2] Société NUMELEC.

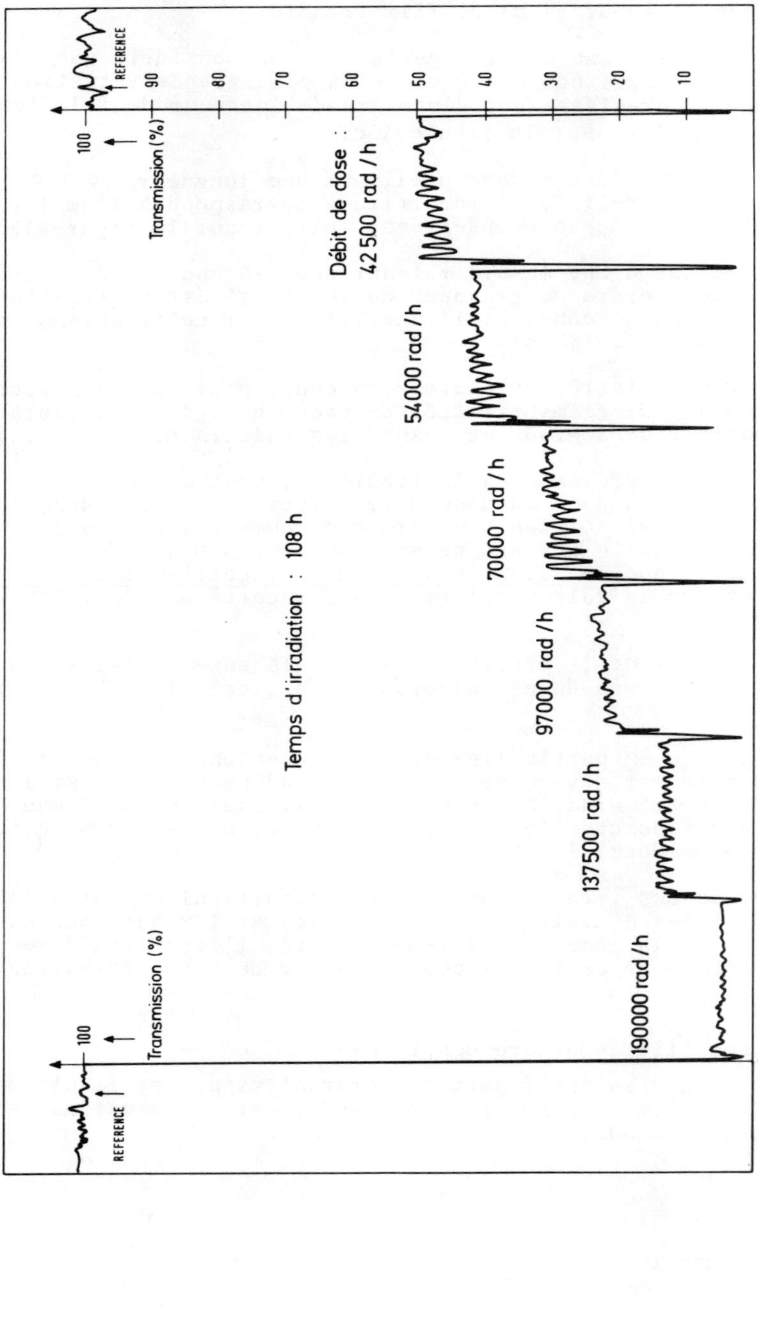

FIG. 17. Lecture de films "TAC" irradiés, enroulés et placés dans des tubes de verre.

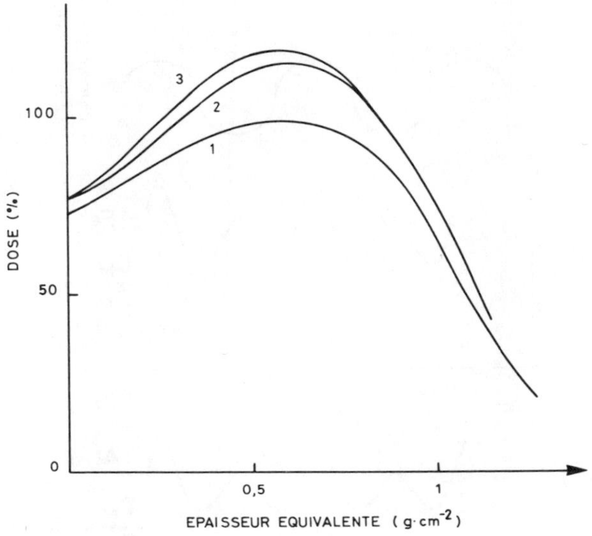

FIG.18. Courbes profondeur-dose (Plexiglas, 3 MeV).
1 - Poly-méthylméthacrylate (Plexiglas); 2 - idem, sur aluminium; 3 - idem, sur fer.

Celle-ci est en effet particulièrement efficace lorsqu'on peut lire le film de façon continue ; mais elle n'impose pas nécessairement l'utilisation d'un spectrophotomètre, puisqu'une seule longueur d'onde au voisinage de 280 nm est suffisante dans la majorité des cas.

L'appareil, dont les principales caractéristiques sont données au Tableau III, est constitué d'un densitomètre UV et d'un système de défilement du film. Le densitomètre UV utilise une bande étroite de longueurs d'onde au voisinage de 280 nm, qui est obtenue par un barreau convertisseur en quartz et un filtre. De cette façon, l'inconvénient résultant de la mesure sur la pente de la courbe transmission-longueur d'onde se trouve éliminé. Un étalonnage particulier de chaque appareil doit, par contre, être effectué. La stabilité des réglages de l'appareil dans le temps est particulièrement bonne.

5. EXEMPLES DE MESURE

Quelques exemples permettront de mieux évaluer les possibilités de dosimétrie offertes par le film, tant en γ qu'en électrons.

5.1. Hétérogénéités d'irradiations sous rayonnement γ

Des petites bobines de film ont été réalisées, introduites dans des tubes de verre (Ø 15 mm), et irradiées en position horizontale, à différentes distances d'une source de Co^{60}.

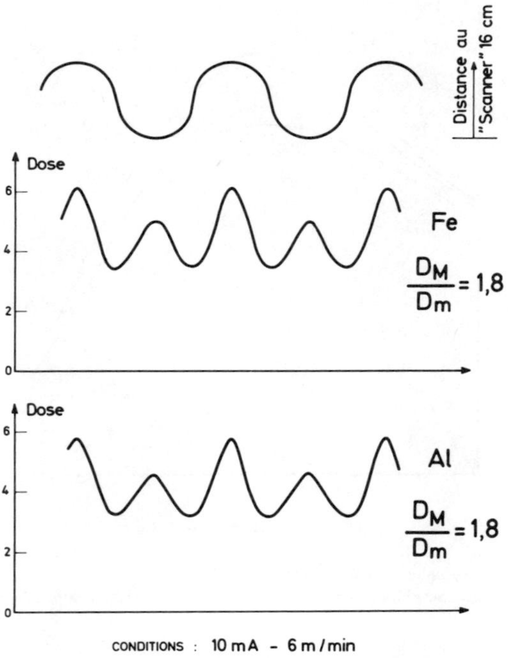

FIG.19. Irradiateur de profils courbes (profondeur creux-crête = 10 cm).

On peut prévoir, dans ces conditions, que la face avant de chaque spire recevra une dose légèrement supérieure à la face arrière, et que la différence ira décroissant de l'extérieur à l'intérieur de la bobine, les distances à la source se faisant de plus en plus voisines. L'enregistrement présenté à la figure 17 met clairement en évidence ces variations cycliques et leur amortissement.

5.2. Courbes profondeur-dose en rayonnement β

Les courbes expérimentales profondeur-dose, pour une machine donnée et dans un milieu donné, sont un des éléments principaux du choix des conditions de fonctionnement ou du choix du conditionnement dans les traitements de radiostérilisation par faisceaux électroniques. Leur détermination est rendue très rapide par la possibilité d'enregistrement continu de la réponse du film. Il suffit, en effet, de placer ce dernier entre un écran en forme de coin et un support épais du matériau considéré pour obtenir de façon quasi directe la courbe désirée (figure 18). Si le support est constitué d'un matériau différent, en particulier d'un métal formant écran de rétrodiffusion, l'augmentation de dose résultant de cette dernière peut être également mesurée (figure 18).

Dans le cas des faisceaux de basse énergie, la détermination des courbes profondeur-dose est également possible, soit en utilisant des écrans, formés d'empilements de films minces, soit en utilisant des empilements du film dosimétrique lui-même (figures 11 et 12).

5.3. Hétérogénéités de dose en rayonnement β de basse énergie

Dans le cas où des profils courbes sont irradiés par des électrons de basse énergie, d'importantes hétérogénéités de doses peuvent être attendues, du fait des différences de distance, outre les différents points du profil et la fenêtre de l'accélérateur, et des différences d'angle d'incidence.

Un exemple d'une telle irradiation est présenté à la figure 19, dans le cas d'un profil sensiblement sinusoïdal. De façon assez surprenante, on constate l'existence de maximums secondaires de la dose aux points les plus bas du profil, c'est-à-dire les plus éloignés de la fenêtre, et de minimums aux points d'inflexion.

On a pu, cependant, mesurer de façon distincte l'influence de la distance à la fenêtre, et celle de l'angle d'incidence sur la dose reçue. Le calcul, à partir de ces données, de la dose le long du profil donne des résultats qui sont en excellent accord avec les résultats expérimentaux de la figure 19. Cette vérification expérimentale directe montre qu'il est possible de prévoir la dose reçue aux différents points d'un profil complexe irradié par des électrons de basse énergie, et l'hétérogénéité correspondante.

6. CONCLUSION

Le film TAC présente un ensemble de caractéristiques qui en rendent l'utilisation particulièrement commode : linéarité de la réponse avec la dose, influence faible ou nulle de paramètres de l'irradiation comme le débit de dose ou la température, possibilité de lecture continue, épaisseur faible. Ces caractéristiques le rendent bien adapté aux mesures dosimétriques diverses. Jointes à son faible coût et à sa fiabilité, elles le rendent particulièrement utile pour le contrôle dosimétrique des irradiations de type industriel, dont les radiostérilisations.

DISCUSSION

R. EYMERY: Was the film described used to measure the variation in dose in direct proximity to any of the metal parts which radiosterilized medical items sometimes contain? Tests of this kind are very worth while, I think, especially in the case of electron beam irradiation.

J. LAIZIER: Yes, some tests of this kind have been carried out. Figure 18 of the paper, for example, illustrates the different dose increases for plexiglas, aluminium and iron.

The data we have accumulated are not really sufficient for us to draw any general conclusions or compare the extent of the respective effects when using electrons or gamma radiation, but they do show that there may be considerable increases in dose – as much as 20%.

Eh.I. SEMENENKO: Can you say anything about the chemical composition of the film?

J. LAIZIER: No, I'm afraid I can't give you the details at the present time.

Eh.I. SEMENENKO: How does the error increase when measuring doses of less than 2.7 Mrad? Your Fig.1 shows the slope of the absorption curve to be only very slight in such cases.

J. LAIZIER: Although the absolute error increases slightly with dose, the variation is small. We may therefore assume for practical purposes that the absolute error is constant, and we have taken the value corresponding to a dose of 5 Mrad. Under these conditions the relative error increases as the dose decreases. This fact does of course limit the use of the dosimeter for low doses.

EFFECTS OF STERILIZING RADIATION DOSE ON THE CONSTITUENTS OF MEDICAL PRODUCTS

(Session III)

Chairman: K.H. MORGANSTERN (United States of America)

ДОЗИМЕТРИЧЕСКОЕ ОБЕСПЕЧЕНИЕ ПРОЦЕССА РАДИАЦИОННОЙ СТЕРИЛИЗАЦИИ

В.Г.ХРУЩЕВ, В.А.БАЛИЦКИЙ, М.П.ГРИНЕВ,
Л.М.МИХАЙЛОВ
Институт биофизики
Министерства здравоохранения СССР,
Москва,
Союз Советских Социалистических Республик

Abstract–Аннотация

MONITORING OF THE RADIATION STERILIZATION PROCESS.

A system for monitoring the radiation sterilization of medical products is described, which consists of (1) measuring the basic radiation characteristics of a facility at the time of commissioning and after overhauls, (2) periodic checking of the basic parameters of the radiation fields, and (3) direct monitoring of the sterilizing dose. The dosimetric systems and methods used to perform these measurements at each stage are examined. The operation of the proposed system is illustrated by specific examples, namely the calibration of radiation fields during commissioning and routine operation of two facilities — a ^{60}Co facility with an activity of 126 kCi and a facility using a linear electron accelerator with a rated energy of 8 MeV and beam power of 5 kW.

ДОЗИМЕТРИЧЕСКОЕ ОБЕСПЕЧЕНИЕ ПРОЦЕССА РАДИАЦИОННОЙ СТЕРИЛИЗАЦИИ.

Рассмотрена система дозиметрического контроля процесса радиационной стерилизации изделий медицинского назначения, которая включает в себя: 1. Измерение основных радиационных характеристик установки при ее пуске в эксплуатацию и после модернизации. 2. Периодический контроль основных параметров полей излучения. 3. Непосредственный контроль и индикация величины стерилизующей дозы. Рассмотрен набор дозиметрических систем и методов, обеспечивающих эти измерения на каждом этапе. Применимость предложенной системы дозиметрического контроля подтверждена на примере калибровки полей излучения при пуске и при повседневной эксплуатации двух установок — изотопной с излучением активностью 126 кКи ^{60}Co и использующей линейный ускоритель электронов с номинальной энергией 8 МэВ и мощностью пучка электронов 5 кВт.

Промышленное использование метода радиационной стерилизации изделий медицинского назначения требует непрерывного дозиметрического контроля величины стерилизующей дозы. Система дозиметрического контроля при радиационной стерилизации хотя и обладает рядом специфических отличий является частью общей системы дозиметрического обеспечения радиационных процессов и базируется на принятой в СССР системе обеспечения единства измерений физических величин.

Объем дозиметрического контроля, методы и средства измерений величины поглощенной дозы и других характеристик поля излучения установки определяются, с одной стороны, характером ее работы – ввод в эксплуатацию (модернизация) или повседневная работа по облучению продукции – и, с другой стороны, спецификой производства изделий медицинского назначения – величиной стерилизующей дозы и ее допустимым перепадом по объему упаковки, масштабностью и номенклатурой производства и т.д.

В общем виде система дозиметрического обеспечения включает в себя: 1. Измерение основных радиационных характеристик установки при ее пуске и после модернизации (ремонтных работ) с помощью абсолютных методов измерения. К их числу относятся калориметры для определения величины поглощенной дозы для всех типов установок;

цилиндр Фарадея для измерения тока пучка и магнитный анализатор для измерения энергии электронов для ускорителей. Погрешность этих методов не должна превышать 1-2 % для всех величин и для всего диапазона измерений. Этот этап заканчивается аттестацией установки и передачей ее в эксплуатацию. 2. Периодический контроль основных радиационных параметров с использованием приборов и методов, обеспечивающих погрешность измерений в пределах 3-6 % – так называемые вторичные стандарты. 3. Непосредственный контроль и индикация величины стерилизующей дозы, который осуществляется с помощью проверки стабильности параметров установки в целом и с помощью дешевых детекторов малых размеров, прикрепленных непосредственно к каждой упаковке облучаемой продукции. К их числу относятся пленочные дозиметры, цветовые индикаторы и другие химические системы.

Кроме радиационных характеристик установки постоянному контролю подлежит скорость движения конвейера (на изотопных установках требуется контроль стабильности скорости конвейера, а на ускорителях электронов обычно осуществляется коррекция скорости конвейера в зависимости от изменения интенсивности излучения).

Рассмотрим более подробно дозиметрические системы, применяемые на различных стадиях дозиметрического обеспечения процесса.

Наиболее широко при контроле процесса стерилизации используются химические дозиметры, позволяющие непосредственно определять поглощенную дозу при высокой интенсивности излучения, это дает возможность обеспечить надежный и экономичный контроль практически на всех стадиях за исключением абсолютных измерений поглощенной дозы, выполняемых калориметрическим методом.

Химические дозиметры, используемые в рассматриваемом процессе, можно условно разделить на три группы в зависимости от точности и особенностей применения систем.

К первой группе относятся жидкостные химические системы, позволяющие определять поглощенную дозу с погрешностью ±5-6 % и используемые в качестве вторичных стандартов на стадии паспортизации установки, а также при периодическом контроле воспроизводимости условий облучения. При выборе вторичного стандарта должны учитываться такие характеристики системы как диапазон измеряемых доз, чувствительность к примесям, зависимость показаний от мощности дозы и температуры при облучении, метод анализа радиационных превращений в системе. При массовом применении немаловажное значение приобретает и возможность выпуска таких систем в централизованном порядке. Наиболее полно предъявляемым к ним требованиям удовлетворяют глюкозный дозиметр и дозиметр Фрике.

Однако несоответствие диапазонов измеряемых с их помощью доз с интервалом, представляющим интерес при стерилизации, затрудняет их использование при решении ряда задач. Этого недостатка лишена система хлорбензол-этанол при использовании осциллометрического метода определения HCl без вскрытия ампул с дозиметрическим раствором. Применение этого метода позволяет быстро проводить анализ большого числа детекторов, система проста в приготовлении, не требует использования реактивов особой чистоты и пригодна для массового выпуска в централизованном порядке в виде запаянных ампул с дозиметрическим раствором. Диапазон измерения доз 0,1-10,0 Мрад.

При выполнении отдельных измерений иногда применяются ферро-купрумсульфатная или цериевая системы, обладающие неплохими дозиметрическими характеристиками, однако весьма чувствительные к примесям, что делает невозможным их массовое применение.

Следующая большая группа химических дозиметрических систем — рабочие дозиметры, используемые в качестве дозиметров сопровождения.

Она объединяет целый ряд систем на основе прозрачных полимерных пленок, меняющих свои характеристики под действием облучения, что регистрируется спектрофотометрически на определенной длине волны. Ошибка в определении дозы с помощью таких систем ±10 %.

При выборе рабочего дозиметра особое внимание следует уделять воспроизводимости показаний для детекторов, изготовленных из различных партий полимерного материала, зависимости показаний системы от мощности дозы вследствие влияния процесса диффузии на окислительную деструкцию и другие процессы в полимерах и постэффект. Заслуживает также внимания вопрос корректной калибровки пленочных дозиметров.

Цветные индикаторы играют особую роль в процессе дозиметрического контроля радиационной стерилизации изделий медицинского назначения. Несмотря на невысокую точность измерений дозы, цветовые индикаторы позволяют визуально оценить степень облучения продукции, отличить облученные изделия от необлученных или двухкратнооблученных, гарантируют потребителю использование изделия, подвергнутого стерилизации, а в комплексе с другими дозиметрическими мероприятиями обеспечивают полную надежность процесса в целом.

К цветовым индикаторам мы предъявляем следующие требования.

1. Контрастность цветовых переходов, наличие не менее 4-х индивидуальных окрасок системы в зависимости от степени облучения, что позволяет реализовать четырехцветный метод контроля степени облучения. Например:

Начальная окраска — изделие не облучено	— синий;
Недооблучение	— зеленый;
Облучение в пределах нормы	— желтый;
Конечная окраска — переоблучение	— красный.

2. Погрешность при количественной оценке дозы с помощью цветовой шкалы около 30 % в диапазоне 0,4-4,0 Мрад.

3. Стабильность окраски облученного и необлученного индикаторов во времени под действием факторов окружающей среды.

4. Независимость цветового перехода от условий облучения.

5. Возможность надежного крепления индикаторов к облучаемым объектам.

В настоящее время в промышленном контроле используется три типа индикаторов, отличающихся принципами получения цветовых переходов. Это, во-первых, индикаторы, работающие по классическому принципу изменения окраски кислотно-основных индикаторов, введенных в галогено-содержащий полимер, индикаторы на основе полистрола с люминофором, и, наконец, системы, работающие по принципу смешения цветов красителей, имеющих различный радиационно-химический выход обесцвечивания. Технология получения индикаторов также различна.

Рассмотренный порядок проведения дозиметрических исследований был применен на двух экспериментально-промышленных установках —

изотопный с излучателем активностью 126 кКи ^{60}Co (ЭПГУ-200) и использующей ускоритель электронов с номинальной энергией 8 МэВ и мощностью 5 кВт (ЛУЭ-8-5).

ИЗОТОПНАЯ УСТАНОВКА ЭПГУ-200

Изотопная установка имеет источник цилиндрической формы, вокруг и вдоль которого движутся с помощью конвейерной системы ящики с облучаемой продукцией. Облучение происходит в 12 положениях – ящик движется вдоль образующей и вращается вокруг источника. Время облучения задается выдержкой ящика в фиксированных положениях.

После завершения всех монтажных работ и зарядки источников было проведено полное исследование распределения поглощенных доз в рабочих ящиках. Измерения проводились в воздухе и в средах, моделирующих по плотности объекты, подлежащие радиационной стерилизации (плотность изменялась от 0,1 до 1,5 г/см^3).

Эти исследования были выполнены с помощью различных химических (ферросульфатный, глюкозный и т.д.) дозиметров и пленочных (винипроз) дозиметрических систем. Дозиметры равномерно распределялись по объему ящика. При обработке показания дозиметра в геометрическом центре объема принимались за 100%. Погрешность относительных измерений в зависимости от типа дозиметра и величины поглощенной дозы составляла от 5 до 15%.

Относительным измерениям в фантомах предшествовали измерения абсолютной величины поглощенной дозы в воздухе в центре ящиков. Эти измерения были выполнены с помощью специальной наперстковой ионизационной камеры, предназначенной для работы в мощных полях ионизирующих излучений. Аналогичная камера использовалась для измерения распределения мощности дозы вдоль образующей цилиндрического источника. Погрешность измерений с помощью ионизационных камер составляла не более 5%.

Наконец, с помощью калориметра локальной поглощенной дозы была измерена абсолютная величина мощности дозы в центре ящика. На основании этих измерений было определено время облучения, необходимое для набора требуемой стерилизующей дозы. На наших установках эта величина составляет 2,5 Мрад с допустимым перепадом +20%.

Время облучения задается с помощью реле времени в цепи управления конвейерной системы. Знание величины мощности дозы в одной точке при наличии относительных дозных распределений достаточно для определения мощности дозы излучения в любой точке.

Контроль стабильности работы установки в процессе эксплуатации осуществляется с помощью механических датчиков, свидетельствующих о приходе источников в рабочее положение. Кроме того, контроль стерилизующей дозы, полученной объектом, осуществляется с помощью дозиметров-свидетелей, имеющихся на каждой упаковке изделий.

Коррекция времени облучения вследствие распада ^{60}Co производится один раз в месяц на основании вычислений. Погрешность величины отпускаемой дозы из-за распада источника не превышает таким образом 1%.

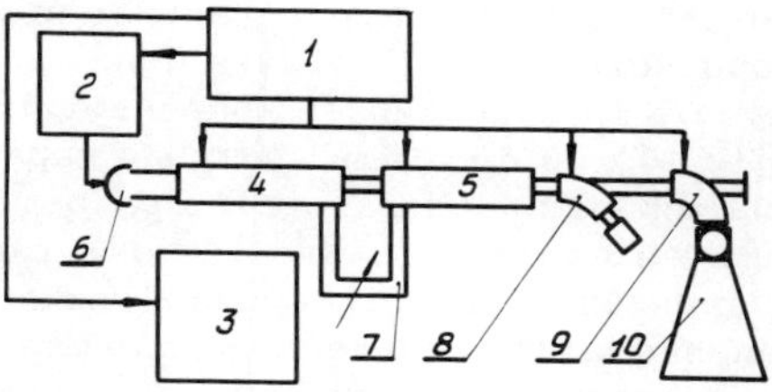

Рис.1. Блок-схема ускорителя ЛУЭ-8-5: 1. Система управления; 2. Модулятор источника; 3. Блок генератора высокой частоты; 4. Первая секция; 5. Вторая секция; 6. Электронная пушка; 7. Направленный ответвитель; 8. Анализирующий магнит и цилиндр Фарадея; 9. Поворотный магнит; 10. Камера развертки.

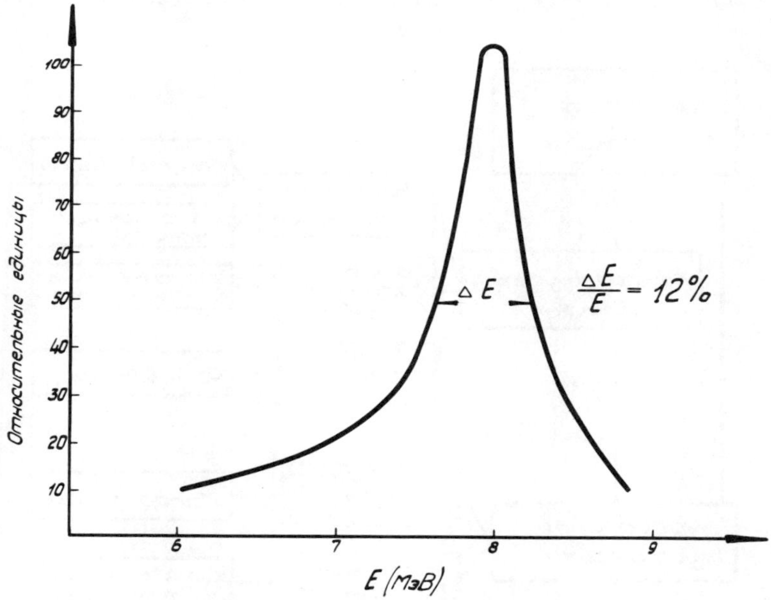

Рис.2. Спектр ускоренных электронов.

ЛИНЕЙНЫЙ УСКОРИТЕЛЬ ЭЛЕКТРОНОВ ЛУЭ-8-5

В установке используется линейный ускоритель электронов на бегущей волне, работающий в S-диапазоне. Конструктивно он выполнен в виде двух секций, запитываемых от одного источника высоко-частотной мощности, что позволяет регулировать энергию на выходе машины от 4 до 12 МэВ. Номинальным режимом работы ускорителя являются: энергия – 8 МэВ, мощность пучка – 5 кВт, частота посылок – 500 1/с и длительность импульса тока – 3,0 мкс. Схема ускорителя приведена на рис.1.

На выходе ускорителя установлено специальное выходное устройство, состоящее из поворотного (на 90°) магнита и скепирующего магнита, позволяющего получать из поверхности конвейера пучок электронов в виде полосы размером 10 × 50 см. Для контроля параметров пучка ускоритель снабжен магнитным анализатором с разрешением по импульсу 0,5 % и индукционным датчиком тока. Кроме того, на выходе камеры развертки установлен монитор вторичной эмиссии.

При наладке установки были проведены детальные измерения выходных параметров пучка ускоренных электронов таких, как спектральное распределение и распределение интенсивности по ширине развертки. Для калибровки датчиков тока использовался цилиндр Фарадея полного поглощения (эффективность собирания электронов 99,5 % при энергии 10 МэВ). Первоначально были измерены спектральные распределения ускоренных электронов и составлены карты режимов работы ускорителя. Один из образцов спектра ускоренных электронов приведен на рис.2.

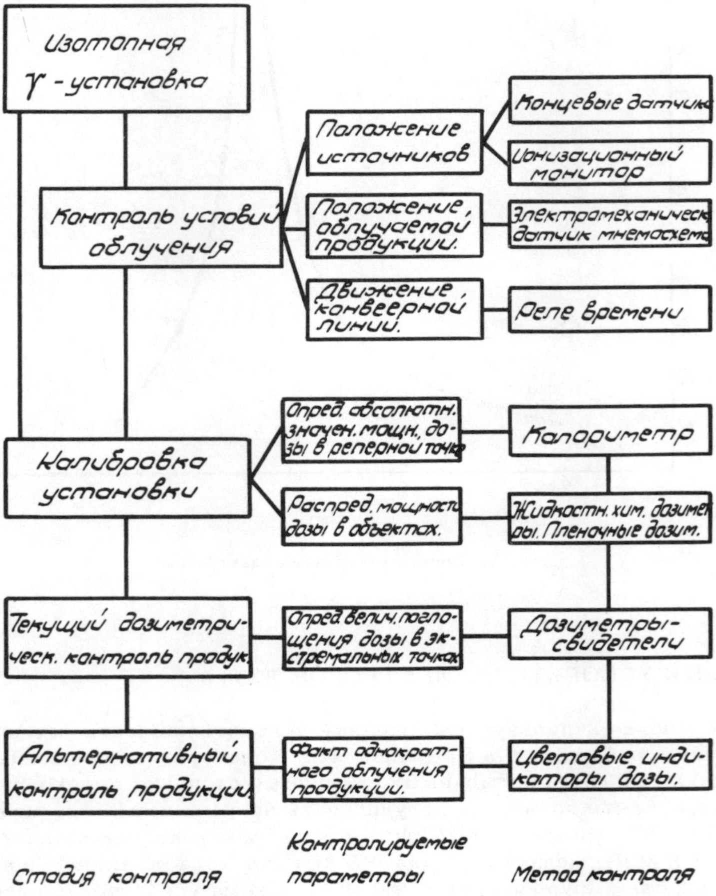

Рис.3. Схема контроля процесса радиационной стерилизации на изотопных γ-установках.

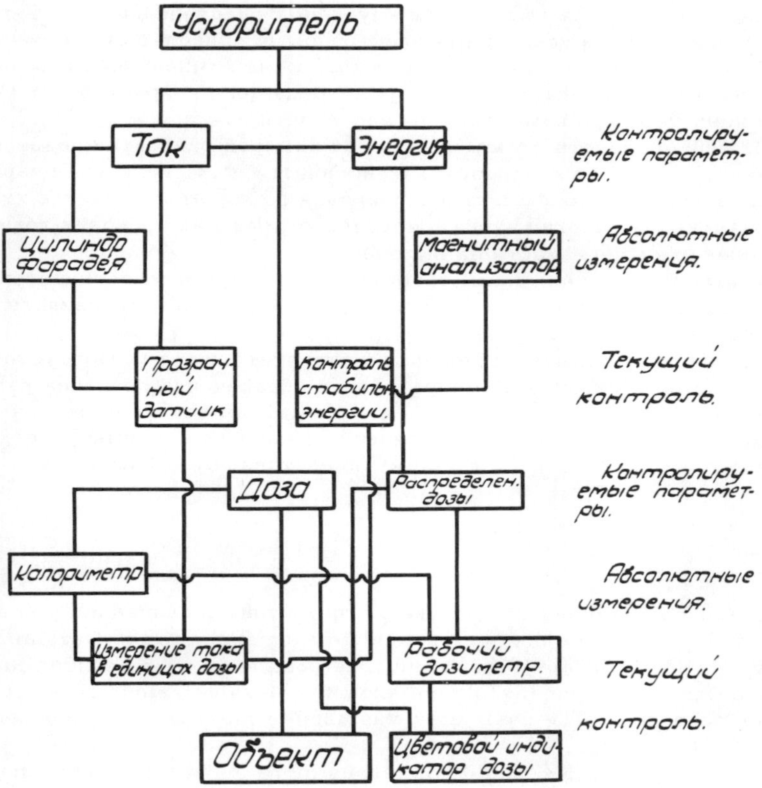

Рис.4. Схема контроля процесса радиационной стерилизации на ускорителе электронов.

Измерения равномерности распределения интенсивности по длине развертки было выполнено с помощью термопары, дистанционно перемещаемой вдоль развертки, и с помощью пленочных дозиметров (винипроз, окрашенный ПВХ). Оказалось, что максимальные колебания интенсивности при развертке пучка не превышают ± 3 %, что является вполне допустимым для целей радиационной стерилизации.

Поскольку процесс стерилизации на ускорителях электронов проходит при фиксированном значении энергии пучка, величина мощности дозы строго пропорциональна среднему току пучка, а точнее плотности тока (мкА/см^2) на поверхности образца. Поэтому показания монитора вторичной эмиссии на выходе развертывающего устройства были прокалиброваны с помощью калориметра локальной поглощенной дозы в единицах мощности дозы, принимая приближенно, что мощность дозы на единицу плотности тока не зависит от энергии.

В дальнейшем также как и для изотопной установки были определены дозные распределения в различных фантомах. Однако, измерялось главным образом одномерное распределение дозы по глубине.

Как показали наши измерения глубинных распределений, небольшие флуктуации энергии ускоренных электронов не приводят существенно к изменению величины поглощенной дозы, а для компенсации изменений тока пучка в схеме установки предусмотрена цепь корректировки скорости конвейера при изменении интенсивности пучка.

Наличие поворотного магнита перед камерой развертки позволяет ограничиться только контролем интенсивности, т.к. величина энергии задана магнитным полем и стабилизирована с хорошей степенью точности, а изменения энергии пучка в небольших пределах приводят только к изменению интенсивности на выходе.

Контроль величины дозы, сообщенной объекту, осуществляется по выходным параметрам ускорителя и с помощью дозиметров-свидетелей, укрепляемых на каждой упаковке.

На основании опыта калибровки и эксплуатации этих двух установок предложены схемы дозиметрического обеспечения процесса радиационной стерилизации, включающие в себя методы контроля основных параметров излучения и их соподчинения в процессе дозиметрического исследования установок. Эти схемы представлены на рис.3 и 4.

DISCUSSION

G. O. PHILLIPS: I should like to know what measurement you actually made with the glucose dosimeter, and how it relates to the G-value.

V. A. BALITSKIJ: With the glucose dosimeter we measured the rotation of the polarization plane. I do not know the G-value, since it is not required for our purposes. The dosimeter was used to measure the absorbed dose in the 10.0-Mrad range.

K. H. CHADWICK: Could you tell us more about the colour-change indicators you mentioned and perhaps comment on the sensitivity of this system to humidity, temperature and light exposure. Also, how dose-rate dependent was this system?

V. A. BALITSKIJ: The dosimeters we used showed the following colour changes according to dose: initial tone = blue; 1.0 - 1.5 Mrad = green; 2.0 - 3.0 Mrad = yellow; higher than 3.5 Mrad = red.

These systems are intended for routine work at ambient temperature of -10°C to 30°C, relative humidity 40-90%, and dose-rates up to 10^5 rad/s.

K. H. CHADWICK: You stated that the accuracy of your plastic films was ± 10% and that there were batch-to-batch reproducibility problems. You also mentioned the name Viniproz, which suggested to me that the films might be the PVC type. Could you enlarge on this?

V. A. BALITSKIJ: Yes, Viniproz, the name I mentioned, is a type of PVC. The dose measurements were made by the normal procedure of recording the variation in optical density after annealing at 60°C. But the film was chiefly used for relative measurements.

POLYETHYLENE PLASTICS AS CONTAINERS FOR WATER FOR INJECTION AND AS MATERIAL FOR DISPOSABLE MEDICAL DEVICES STERILIZED BY RADIATION

Nazly HILMY, S. SADJIRUN
Pasar Jumat Research Centre,
National Atomic Energy Agency,
Jakarta, Indonesia

Abstract

POLYETHYLENE PLASTICS AS CONTAINERS FOR WATER FOR INJECTION AND AS MATERIAL FOR DISPOSABLE MEDICAL DEVICES STERILIZED BY RADIATION.

Investigations on polyethylene plastic manufactured in Indonesia have been carried out in order to study the possibilities of using it for containers for pharmaceutical products and as material for medical devices sterilized by irradiation. The investigations involved biological experiments on systemic toxicity, skin irritation, pyrogen, haemolysis, and contamination prior to sterilization; and physico-chemical experiments such as the determination of hydrogen peroxide, oxidizable matter, the absorption of u.v. spectra, non-volatile residue, determination of pH and of heavy metals, and weight loss during storage. Biologically the results of these experiments could be accepted, but physico-chemically no conclusion could be reached because of the non-existence of maximum limits of hydrogen peroxide, oxygen, and also because of the possible problems of radicals in water for injection.

INTRODUCTION

Plastic is composed of an unlimited number of substances with a resin of a molecular weight as high as the basic material [1]. Depending on its composition plastic may contain other substances such as additives, fillers, plasticisers, etc. Additives in plastic are any minor constituents introduced in a plastic material in small quantities to achieve the most desirable final product. Some of the additives are colours, anti-oxidants, heat-resistant substances, u.v.-resistant substances, etc. [2].

Although no definite standard exists for plastic used as containers for water for injection and eye-drops, and for medical devices, a regulation for this has already been made [3].

The development of the radiosterilization technique or cold sterilization, which uses gamma rays or an electron accelerator, makes it possible to mass-produce plastic which is not heat resistant but inert enough for medical devices and pharmaceutical containers.

Plastic polyethylene is one of the well-established plastics which is already used for these purposes, but the quality of the plastic depends considerably on the additive used and the manufacturing method [4].

Biological and physico-chemical experiments have been carried out to predict the possibility of using polyethylene plastics made in Indonesia for medical devices sterilized by gamma radiation. Such research on polyethylene plastic has already been carried out in Australia [5]. This type of plastic is chosen because it is mass-produced in Indonesia.

This research aims at stimulating the manufacturers to produce these gamma-sterilized devices, since the future of the manufacturing of plastics in Indonesia is more ensured.

MATERIALS AND METHODS

1. Research material

Plastic bottles of low-density polyethylene B.A.S.F. Lupulen 1810 E were used as research material. These bottles were blow-moulded by P.T. Berlina in Indonesia, with the following measurements for each bottle excluding the cap: volume about 20 ml, height about 5 cm, wall thickness about 0.5 mm, and weight about 4.5 g. Type I glass vials made in Australia were used as glass containers. These vials had a volume of 100 ml. The water used was freshly prepared for injection. A total of 650 bottles were used. These bottles were collected at random by the manufacturer from several batches. A ^{60}Co Gamma Cell 220, made in Canada with a dose-rate of 0.37×10^6 rad/h, was used as a gamma-ray source.

2. Sample preparations

Two kinds of samples were prepared:

(1) Samples used to examine water for injection in plastic and glass containers sterilized by ^{60}Co gamma rays:
Before the bottles were used, they were washed in doubly distilled water and afterwards in water for injection.

The samples prepared were as follows:
(a) Unsterilized water for injection in plastic and glass containers;
(b) Water for injection in plastic and glass containers irradiated at 2.5×10^6 and 5×10^6 rad.
(c) Water for injection only in glass vials, autoclaved.

Physico-chemical experiments were done after storing the solution for 0, 1, 2, and 3 months, and biological experiments after 0 and 3 months. Special processing for biological experiments was as follows: before experiments were carried out, into each bottle containing 20 ml of water for injection was injected 1 ml solution of 18.9% sodium chloride so that the final concentration would be 0.9%. Storing was done at a temperature of 30 ± 3°C (room temperature) and a relative humidity of about 70 - 90% (room condition).

(2) Samples employed in the examination of plastics for medical devices, especially those used outside the human body such as syringes, infuse devices, petri dishes, etc.:
The preparation carried out was based on the combination method of the United States Pharmacopeia [3] and Gopal [6]: 10 g of plastic was chopped and washed in water for injection, dried and dropped into a 100-ml glass vial, and then irradiated under 2.5×10^6 and 5×10^6 rad gamma rays. For biological treatment, after irradiation 50 ml sterile

saline solution was injected into each vial, and for physico-chemical examinations water for injection was employed. Afterwards these vials were stored away for 4 h at a temperature of 37°C. This extract was tested biologically and physico-chemically after being stored for 0 and 3 months at a temperature of 30 ± 3°C and a relative humidity of about 70 - 90%.

3. Biological experiments

3.1. Toxicity test

This test was made by combining the methods of the United States Pharmacopeia [3] and Moore [5].

Injections were given intravenously and intraperitoneally to 15 mice each weighing 17 - 23 g. An injection of 0.5 ml solution was introduced intravenously to each mouse, while 2 ml of solution was introduced intraperitoneally. Sterile saline solution in a glass container was used as a negative control. The animals were observed 0, 4, 24, 48, 72 and 96 h after injection. During observations the mortality rate and change in behaviour or any other noticeable deviations were investigated.

3.2. Skin-irritation test

This test was carried out by combining the methods of the United States Pharmacopeia [3], Lawrence et al. [10] and Banziger [7].

In this test the possibility of tissue reaction such as erythema and oedema was observed after an intradermal injection at the clipped back part of a rabbit. 0.2 ml of solution was injected in 10 places, while saline solution in a glass vial was used as a negative control (0). This solution was sterilized in an autoclave, henceforth referred to as solution A. For positive control, 20% ethanol in saline solution was used, (3^+) for erythema and (1^+) for oedema. This solution will henceforth be called solution C, and the solution to be examined solution B. The reactions were observed 24, 48 and 72 h after injection. Each test was designed to be applied to three rabbits.

3.3. Pyrogen test

This test was carried out according to the British Pharmacopeia [9]. Each sample involved 10 rabbits. Observation was done only on samples with a dose of 2.5×10^6 rad and after being stored for 0 month.

3.4. Haemolysis test

This test was designed to test the possibility of haemolysis forming in the rabbit's blood caused by samples 1 and 2. This test employed the Lawrence method [8]. Saline solution was used as a negative control (solution A) and a solution of 1% sodium carbonate in saline served as the positive control (solution B). This solution resulted in 100% haemolysis. Solution C was the sample solution. Three test tubes of 16 × 150 mm were prepared, 10 ml of solution A were dropped into the first test tube, 10 ml of solution B into the second and 10 ml of solution C into the third. The tubes were allowed to stand for 30 min at 37°C. Then to each solution 0.2 ml oxalated rabbit blood was added. The solution was allowed to stand for 60 min at 37°C, then centrifuged. The supernatant was examined using an u.v. spectrometer at 540 nm.

The per cent of haemolysis was calculated as follows:

per cent haemolysis =

$$\frac{\text{absorbance of test sample minus absorbance of negative control}}{\text{absorbance of positive control}} \times 100\%$$

3.5. Contamination prior to sterilization

Twenty plastic bottles were taken at random from all samples received. The bottles were opened aseptically and 2 ml sterile saline solution were transferred into each bottle. The bottles were shaken for 5 min and allowed to stand for 30 min. From each bottle 1 ml of the solution was taken and injected immediately on to a nutrient agar plate. This agar plate was allowed to stand for 24 h at a temperature of 37°C. The number of colonies grown on this plate was counted.

4. Physico-chemical experiments

4.1. Determination of hydrogen peroxide

Into each of 5 measuring 25-ml glasses were dropped 5 ml of 10% potassium iodide solution and 5 ml of 0.1M dihydro potassium phosphate. Into the first measuring glass water for injection was added up to 25 ml. This solution was used as a blank. 10 ml of the sample solution was dropped into each of the other four measuring glasses. Then water for injection was added up to 25 ml. This mixture was allowed to stand for 1 h. Absorption of the iodine produced was determined at 350 nm using a Beckman D.U. 2 spectrophotometer. In the same way, a calibration curve was made using an iodine 2×10^{-4}N standard solution. Hydrogen peroxide was only measured on sample 1 [6].

4.2. Determination of oxidizable matter

The result was determined based on Moore [5].

To a 20-ml sample solution was added 20 ml 0.005N potassium permanganate and 0.8 ml 2N sulphuric acid. This mixture was heated in such a way that it would boil in 3 min. The heating was continued for 6 min and then cooled to 25°C; then 1 ml of 10% potassium iodide was added and titrated with 0.005N sodium thiosulphate. The amount of 0.005N potassium permanganate consumed per 20 ml sample was then determined.

4.3. Determination of non-volatile residue

The determination of the non-volatile residue of the 20-ml sample solution was done, based on the United States Pharmacopeia [3].

4.4. Determination of weight loss

The bottles were weighed monthly for 3 months.

4.5. Determination of heavy metals

Heavy metals such as calcium, magnesium, and lead were determined using a Beckman Atomic Absorption Spectrophotometer 448.

4.6. pH determination

The pH was determined by using a Leed and Northrup pH meter.

4.7. Ultra-violet absorption spectra

The absorption was determined by using a Beckman D.U. 2 spectrophotometer from 200 to 400 nm. Doubly distilled water was used as a blank. The determination was only done on samples which had been stored for 3 months.

RESULTS AND DISCUSSIONS

The results of the toxicity test on samples 1 and 2 can be seen in Table I. The results show that there were no deaths among the mice, nor were there visible behaviour changes.

The results of the skin-irritation test, given in Table II, showed no formation of erythema or oedema which should be caused by solution B on samples 1 or 2. The difference in the formation of both oedema and erythema on the skin of each rabbit caused by solution C, which is marked in the Table as 3^+ and 1^+, might be based on the difference of the sensitivity of each of the rabbits to the solution.

The pyrogen test, as seen in Tables IIIA and IIIB, shows a satisfactory result according to the British Pharmacopeia 1968.

The haemolysis test gave the result of 0% haemolysis, in other words the results show that samples 1 and 2 did not form haemolysis in the rabbits' blood.

The maximum contamination prior to sterilization resulting from the test on 20 bottles, was 4 colonies of microorganisms in each bottle. The volume of each bottle was 20 ml, so that the contamination prior to sterilization was then one colony per 5-ml solution.

Of the microorganisms found in dust 99% had a degree of sterility of about 10^{-12} at 2.5×10^6 rad, the initial contamination being = 1. Suppose that all microorganisms which contaminated these samples belong to this 99%, then the safety margin was about 10^{-11}, while the maximum safety margin was 10^{-6} [11].

So, all the samples had fulfilled the requirement of being sterilized at 2.5×10^6 rad.

The result of the physico-chemical experiments can be seen in Tables IV, V and VI, which show the results of the determination of hydrogen peroxide, oxidizable matter, and the pH of the solution. These Tables show that the difference in the containers, the time of storage and the

TABLE I. SYSTEMIC INJECTION TEST

Manner of injection	No. of samples	Treatment (Mrad)	Storage time (months)	Number of mice	Mortality rate of mice after injection (h) 0	4	24	48	72	96	Visual changes in mice
Intraveneous	1	0	0	15	-	-	-	-	-	-	-
			3	15	-	-	-	-	-	-	-
		2.5	0	15	-	-	-	-	-	-	-
			3	15	-	-	-	-	-	-	-
		5	0	15	-	-	-	-	-	-	-
			3	15	-	-	-	-	-	-	-
	2	0	0	15	-	-	-	-	-	-	-
			3	15	-	-	-	-	-	-	-
		2.5	0	15	-	-	-	-	-	-	-
			3	15	-	-	-	-	-	-	-
		5	0	15	-	-	-	-	-	-	-
			3	15	-	-	-	-	-	-	-
Intraperitoneal	1	0	0	15	-	-	-	-	-	-	-
			3	15	-	-	-	-	-	-	-
		2.5	0	15	-	-	-	-	-	-	-
			3	15	-	-	-	-	-	-	-
		5	0	15	-	-	-	-	-	-	-
			3	15	-	-	-	-	-	-	-
	2	0	0	15	-	-	-	-	-	-	-
			3	15	-	-	-	-	-	-	-
		2.5	0	15	-	-	-	-	-	-	-
			3	15	-	-	-	-	-	-	-
		5	0	15	-	-	-	-	-	-	-
			3	15	-	-	-	-	-	-	-

TABLE II. SKIN-IRRITATION TEST

Manner of injection	No. of samples	Treatment (Mrad)	Storage time (months)	Number of rabbits	Erythema: Solution A 24	48	72	Erythema: B 24	48	72	Erythema: C[a] 24	48	72	Oedema: A 24	48	72	Oedema: B 24	48	72	Oedema: C[a] 24	48	72
Intra-dermal	1	0	0	3	0	0	0	0	0	0	3	3	3	0	0	0	0	0	0	1	1	1
			3	3	0	0	0	0	0	0	3	3	3	0	0	0	0	0	0	1	1	1
		2.5	0	3	0	0	0	0	0	0	3	3	3	0	0	0	0	0	0	1	1	1
			3	3	0	0	0	0	0	0	3	3	3	0	0	0	0	0	0	1	1	1
		5	0	3	0	0	0	0	0	0	3	3	3	0	0	0	0	0	0	1	1	1
			3	3	0	0	0	0	0	0	3	3	3	0	0	0	0	0	0	1	1	1
	2	0	0	3	0	0	0	0	0	0	3	3	3	0	0	0	0	0	0	1	1	1
			3	3	0	0	0	0	0	0	3	3	3	0	0	0	0	0	0	1	1	1
		2.5	0	3	0	0	0	0	0	0	3	3	3	0	0	0	0	0	0	1	1	1
			3	3	0	0	0	0	0	0	3	3	3	0	0	0	0	0	0	1	1	1
		5	0	3	0	0	0	0	0	0	3	3	3	0	0	0	0	0	0	1	1	1
			3	3	0	0	0	0	0	0	3	3	3	0	0	0	0	0	0	1	1	1

[a] Observed after 24, 48, 72 h after having been injected.

TABLE IIIA. PYROGEN TEST OF SAMPLE No. 1[a]

Number of rabbits	Weight of rabbits (kg)	Mean initial temperature (°C)	Time after being injected (min)						Max. temp. (°C)	Temp. diff. (°C)	Room temp. (°C)
			30	60	90	120	150	180			
			Temperature (°C)								
1	1.515	39.6	39.5	39.6	39.6	39.6	39.5	39.5	39.6	0	24 ± 1
2	2.2	39.3	39.2	39.4	39.5	39.4	39.4	39.2	39.5	0.2	
3	1.8	39.3	39	39.2	39.4	39.4	39.3	39.1	39.4	0.1	
4	2.2	39	38.8	39	39.2	39.2	39.2	39	39.2	0.2	
5	2.3	38.9	39	39.2	39.4	39.4	39.4	39.2	39.4	0.5	
6	2.3	38.8	38.5	38.4	38.4	38.7	38.8	38.6	38.8	0	
7	1.86	39.4	39.3	39.2	39.2	39.3	39.3	39.2	39.3	0	
8	2.1	39.1	39	39	39.1	39.1	39.1	39.1	39.1	0	
9	1.64	39	39	39	39	39.2	39.2	39.3	39.3	0.3	
10	1.85	38.9	38.6	39.2	39.4	39.4	39.4	39.4	39.4	0.5	
Summed response of the difference in temperature										1.8°C	
Maximum limit of summed response for 9 rabbits = 4.45°C (British Pharmacopeia 1968)											

[a] Treatment: 2.5×10^6 rad.
Storage time: 0 month.

TABLE IIIB. PYROGEN TEST OF SAMPLE No. 2[a]

Number of rabbits	Weight of rabbits (kg)	Mean initial temperature (°C)	Time after being injected (min)						Max. temp. (°C)	Temp. diff. (°C)	Room temp. (°C)
			30	60	90	120	150	180			
			Temperature (°C)								
1	2.2	38.8	38.4	38.8	38.9	38.6	38.8	38.8	38.9	0.1	24 ± 1
2	1.8	38.9	39	38.6	38.4	38.4	38.4	38.4	39	0.1	
3	2.4	39	39.1	39.3	39.2	39.1	39.3	39.2	39.3	0.3	
4	2.0	39	39	38.8	39	38.6	38.8	39	39	0	
5	1.5	38.9	38.6	38.6	38.6	38.4	38.4	39	39	0.1	
6	1.9	38.7	38.6	38.4	38.6	38.8	38.8	38.8	38.8	0.1	
7	2.3	38.9	39	38.8	38.8	38.6	38.8	38.8	39	0.1	
8	2.1	38.7	38.2	38	37.8	39	38.8	38.8	39	0.3	
9	2.2	39	38.8	39	39.2	39	39	39	39.2	0.2	
10	1.8	38.8	38.8	38.6	38.7	38.7	38.8	38.8	38.8	0	
Summed response of the difference in temperature										1.3°C	
Maximum limit of summed response for 9 rabbits = 4.45°C (British Pharmacopeia 1968)											

[a] Treatment: 2.5×10^6 rad.
Storage time: 0 month.

TABLE IV. HYDROGEN PEROXIDE

Number of samples	Containers	Treatment (Mrad)	Storage time (months)	Hydrogen peroxyde (ppm)[a]	% D[b]
1	Plastic	2.5	0	0.4628	0-15
			1	0.4450	0-15
			2	0.3400	0-15
			3	0.2924	0-15
		5	0	0.5474	0-15
			1	0.4665	0-15
			2	0.4590	0-15
			3	0.3128	0-15
	Glass	2.5	0	3.5666	0-10
			1	3.1756	0-10
			2	3.1382	0-10
			3	1.649	0-10
		5	0	3.9338	0-10
			1	3.400	0-10
			2	3.196	0-10
			3	3.1382	0-10

[a] Each figure is the mean of six replicate determinations.

[b] $D = \sqrt{\frac{\epsilon (\bar{x} - x_{1-n})^2}{n-1}} \rightarrow \% D = \frac{D}{\bar{x}} \times 100\%.$

radiation dose could give a different concentration of hydrogen peroxide, oxidizable matter and the pH solution.

Figure 1 indicates the u.v. absorption spectra of sample 1, and Fig.2 those of sample 2, after 3 months storage. In Fig.1 it can be seen that water for injection in plastic containers, which have been sterilized at 2.5×10^6 and 5×10^6 rad, contains a little impurity from the leaching of the plastic as a result of radiation. The sample had a peak of around 230 nm, but has not yet been identified.

The non-volatile residue test yielded a maximum result of 3 ppm, while the test to determine heavy metals yielded less than 1 ppm.

Loss of weight caused by the penetration of water vapour through the plastic bottles was less than 0.1% in 3 months.

No change in the appearance of any of the plastic bottles was noticed after irradiation and storage.

TABLE V. OXIDIZABLE MATTER

Number of samples	Containers	Treatment (Mrad)	Oxidizable matter[a]				% D[b]
			0 month	1 month	2 months	3 months	
1	Plastic	Blank	1.06	1.05	1.00	1.08	0 - 10
		0	1.285	2.015	2.100	2.225	0 - 10
		2.5	3.14	3.0	3.91	3.05	0 - 10
		5	3.19	3.7	3.87	3.04	0 - 10
	Glass	Blank	1.01	1.03	1.05	1.04	0 - 10
		0	1.06	1.05	1.07	1.0	0 - 10
		2.5	2.15	1.71	1.22	1.01	0 - 10
		5	2.21	1.71	1.21	1.05	0 - 10
2	Glass	Blank	1.01	-	-	1.03	0 - 7
		0	1.01	-	-	1.53	0 - 6
		2.5	1.41	-	-	1.83	0 - 7
		5	1.92	-	-	2.33	0 - 5

[a] ml of 0.005N potassium permanganate consumed per 20 ml of sample, each figure is the mean of six replicate determinations.

[b] $D = \sqrt{\frac{\epsilon(\bar{x} - x_{1-n})^2}{n-1}} \rightarrow \% D = \frac{D}{\bar{x}} \times 100\%.$

TABLE VI. pH

Number of samples	Containers	Treatment (Mrad)	pH[a]				% D[b]
			0 month	1 month	2 months	3 months	
1	Plastics	0	6.12	5.85	5.7	5	0 - 7.3
		2.5	4.87	4.17	4.4	4.46	0 - 5.7
		5	4.8	4.31	4.27	4.45	0 - 5
	Glass	0	6.5	6.9	6	6.1	0 - 7.5
		2.5	6.4	6.5	6	6.55	0 - 5
		5	6.4	6.4	6.25	6.425	0 - 7
		Autoclave	6.6	6.7	6.6	6.5	0 - 7
2	Glass	0	5.58			5.1	0 - 6
		2.5	4.90			5.33	0 - 5
		5	4.63			4.72	0 - 5

[a] Each figure is the mean of six replicate determinations.

[b] $D = \sqrt{\frac{\epsilon(\bar{x} - x_{1-n})^2}{n-1}} \rightarrow \% D = \frac{D}{\bar{x}} \times 100\%.$

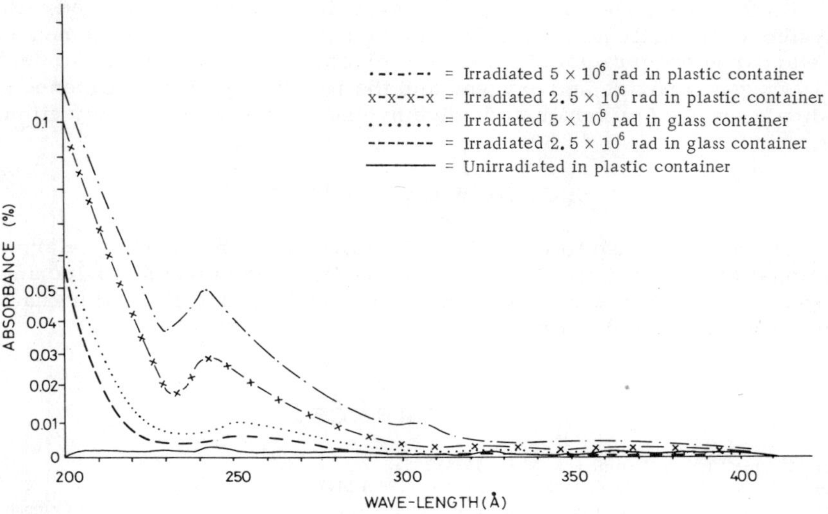

FIG. 1. Typical ultra-violet absorption spectra of sample 1.

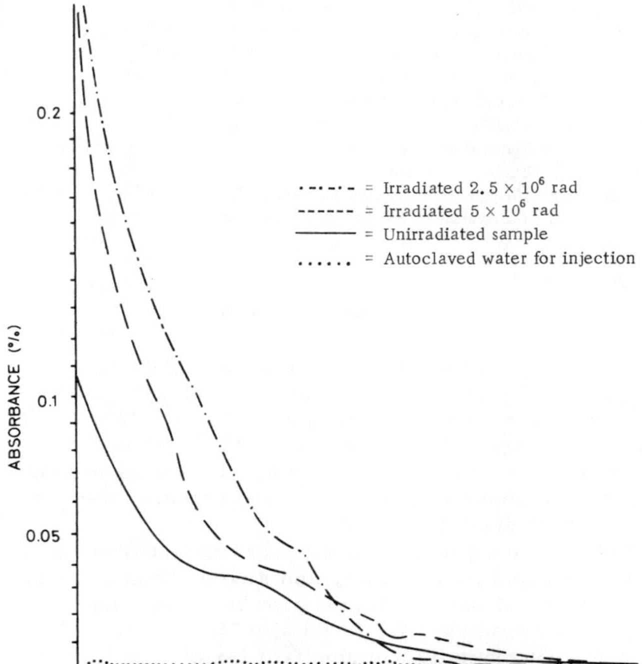

FIG. 2. Typical ultra-violet absorption spectra of sample 2.

Biologically, the result of these experiments can be accepted. But physico-chemically no conclusion can be reached because of the non-existence of standards for irradiated water for injection, such as the maximum limits of hydrogen peroxide and oxygen, and the possibility of the existence of radicals, caused by irradiated polyethylene, in the water for injection.

ACKNOWLEDGEMENTS

The authors wish to thank Mr. B. Zulkarnain and Lembaga Farmasi Nasional (National Institute of Pharmacy) for their valuable advice and co-operation; Mrs. C. Hendratno for her advice; and P.T. Berlina & Co., for co-operation and information.

REFERENCES

[1] AUTIAN, J., J. Pharm. Sci. 53 11 (1964) 1289.
[2] HAWARD, J., J. Chem. Industry, August 15 (1964) 1442.
[3] The United States Pharmacopeia (Board of Trustees, Rev.), Container, Mack Publishing Company, Easton, Pa (1970).
[4] CRAWFORD, C.G., "Material and packaging aspect", Sterile Single Use Medical Products, Nord - Emballage, Maret (1970).
[5] MOORE, P.W., Evaluation of polyethylene ampoules as containers for radiation sterilised solution, A.A.E.C/T M.444, Australian Atomic Energy Commission, Lucas Heights (1968).
[6] GOPAL, N.G.S., Unpublished Data, Isotope Division, Bhabha Atomic Research Commission, Bombay (1972).
[7] BANZIGER, R., POOL, W., "Safety testing of pharmaceuticals", Quality Control in the Pharmaceutical Industry 1 (MURRAY, S.C., Ed.), Academic Press, New York (1972).
[8] LAWRENCE, W.H., DILLINGHAM, E.O., TURNER, J.E., AUTIAN, J., J. Pharm. Sci. 61 1 (1972) 19.
[9] British Pharmacopeia (General Medical Council), Test for Pyrogens, Pharm. Press, London (1968).
[10] LAWRENCE, W.H., DILLINGHAM, E.O., TURNER, J.E., AUTIAN, J., J. Pharm. Sci. 61 11 (1972) 1712.
[11] CHRISTENSEN, E.A., "Hygienic requirements, sterility criteria, and quality and sterility control", Manual on Radiation Sterilization of Medical and Biological Materials, Techn. Rep. Ser. 149, IAEA, Vienna (1973).

DISCUSSION

J. FLEURETTE: I have some questions regarding toxicity. Did you perform acute toxicity tests, for example, on animals after injection of high doses of fluids in which plastics had been irradiated? Second, did you perform chronic toxicity tests, for instance, on animals that had been given 2 or 3 daily injections, for 2 or 3 months, with the same type of fluids? And, third, did you make conjunctival toxicity tests, which are better than skin tests, to detect irritation products?

Nazly HILMY: We carried out a preliminary test on acute toxicity with 10 doses, using evaporated extract, and also a chronic toxicity test with one daily injection for one month only, the investigations having been continued for three months. We found similar results. For certain technical reasons, we did not make the conjunctival toxicity test, but I agree that it is more sensitive than the skin irritation test.

R. KLEN: Do you think that the tissue culture technique could be used for the biological testing of certain irradiated materials? Have you had any experience in this area?

Nazly HILMY: Tissue culture techniques could be helpful in elucidating some of the pharmacological effects which I have described in my paper. Besides being fairly sensitive, such techniques need certain experience and equipment. However, I believe that in in-vitro experiments they are very valuable from the scientific point of view, and may provide useful information. On the other hand, since in-vitro results cannot be extrapolated to in-vivo conditions, I prefer to concentrate on in-vivo experimentation.

РАДИАЦИОННО-ХИМИЧЕСКИЕ ПРЕВРАЩЕНИЯ В ПОЛИМЕРАХ, ПРИМЕНЯЕМЫХ ДЛЯ УПАКОВКИ ЛЕКАРСТВЕННЫХ СРЕДСТВ

А.И.ИВАНОВ, С.И.ПОНОМАРЕНКО,
В.М.КОВАЛЬЧУК, Э.И.СЕМЕНЕНКО,
В.И.МЫШКОВСКИЙ
Всесоюзный научно-исследовательский институт
медицинских полимеров,
Москва,
Союз Советских Социалистических Республик

Abstract–Аннотация

RADIOCHEMICAL CHANGES IN POLYMERS USED FOR PACKAGING DRUGS.

Polymers are now widely used for packaging drugs sterilized with ionizing radiation. Sterilizing doses of radiation can cause changes in the structure and characteristics of polymers. Gas chromatography methods were used to study the composition of gaseous products formed during the radiation sterilization of low- and high-pressure polythene, polymethyl methacrylate (PMMA), PVC, and divinylstyrene and isoprenestyrene rubbers (DST-30 and IST-30), and it is shown that radiation sterilization causes complex radiochemical changes. Infra-red spectroscopy was used to investigate the effect of radiation sterilization on the structure of poly-4-methylpentene-1, low- and high-pressure polythene, PMMA, DST-30, IST-30 and PVC, and it is shown that gamma radiation causes radiochemical changes in polymers leading to the formation of ketones, aldehydes, acids, spirits, esters, unsaturated compounds, etc. Effective constants for the rate of formation of ketones and acids are determined and it is found that they are practically identical and independent of the thickness of the specimen within the range 30 to 150 μm. It is shown that radiation sterilization causes insignificant changes to the ageing mechanism of DST-30 in air and water. The investigation showed that the greatest resistance to the effects of radiation sterilization and aqueous solutions of insulin is offered by dimethyl siloxane rubber, amongst the rubbers, and polythene amongst the polyolefines.

РАДИАЦИОННО-ХИМИЧЕСКИЕ ПРЕВРАЩЕНИЯ В ПОЛИМЕРАХ, ПРИМЕНЯЕМЫХ ДЛЯ УПАКОВКИ ЛЕКАРСТВЕННЫХ СРЕДСТВ.

Для упаковки лекарственных средств, стерилизуемых ионизирующим излучением, в настоящее время широко применяются полимеры. Стерилизующие дозы радиации могут привести к изменениям структуры и свойств полимеров. Методами газовой хроматографии изучен состав газообразных продуктов, образующихся при радиационной стерилизации полиэтилена низкого и высокого давления (ПЭНД и ПЭВД), полиметилметакрилата (ПММА), поливинилхлорида (ПВХ), дивинилстирольного и изопропенстирольного каучуков (ДСТ-30 и ИСТ-30) и показано, что при радиационной стерилизации происходят сложные радиационно-химические превращения. Методом ИК-спектроскопии изучено влияние радиационной стерилизации на структуру поли-4-метилпентена-1 ПЭНД, ПЭВД, ПММА, ДСТ-30, ИСТ-30, ПВХ и показано, что в полимерах при γ-облучении происходят радиационно-химические превращения, приводящие к образованию кетонов, альдегидов, кислот, спиртов, эфиров, непредельных соединений и др. Определены эффективные константы скорости образования кетонов и кислот и найдено, что они практически одинаковы и не зависят от толщины образца в пределах от 30 до 150 мк. Показано, что радиационная стерилизация вносит несущественные изменения в механизм старения ДСТ-30 в воздухе и в воде. Проведенное исследование показало, что наиболее стойким к воздействию радиационной стерилизации и водных растворов инсулина является диметилсилоксановый каучук среди каучуков и полиэтилен среди полиолефинов.

Для упаковки лекарственных средств, стерилизуемых ионизирующим излучением, в настоящее время применяются полимерные материалы разнообразного строения. Стерилизующие дозы радиации могут привести к изменению структуры и свойств полимеров [1]. Однако эти изменения относительно невелики, а большинство применяемых для их определения методов (физико-механические испытания, поглощение кислорода, выделение газообразных продуктов и др.) не отличается высокой чувствительностью. В связи с этим механизм радиационно-химических превращений остается неизученным.

При действии радиации на полимерные материалы возможно протекание следующих процессов:

1. Образование газообразных продуктов радиолиза и их диффузия в контактную среду;

2. Образование в полимерной матрице низкомолекулярных продуктов радиолиза, способных мигрировать в окружающую среду;

3. Изменение степени кристалличности и разветвления, приводящие к изменению диффузии контактных сред внутрь полимерного материала, а также к изменению взаимодействия полимера с контактной средой и др. [2].

Радиационно-химические процессы, протекающие в период стерилизации, и постэффекты могут привести к необратимым изменениям свойств материала упаковки: изменение цвета и прозрачности, механической прочности, уменьшение формоустойчивости, ухудшение санитарно-химических характеристик.

Эти процессы могут ускорить старение полимерных упаковок, а параметры старения определяют сроки хранения лекарственных средств, упакованных полимерными материалами.

Настоящая работа посвящена исследованию радиационно-химических превращений полимеров, применяемых в упаковке лекарственных средств: полиэтилен высокого давления (ПЭВД), полиэтилен низкого давления (ПЭНД), поли-4-метилпентен-1 (П4МП1), поливинилхлорид (ПВХ), полиметилметакрилат (ПММА), дивинилстирольный и изопренстирольный термоэластопласты(ДСТ-30 и ИСТ-30), натуральный и силиконовый каучук (НКиСК), а также изучению влияния радиационной стерилизации на процессы старения стерилизованных материалов упаковки в атмосфере воздуха, в воде и при контакте с некоторыми лекарственными препаратами.

Для проведения исследования были разработаны селективные и высокочувствительные методы газовой хроматографии [1] и использованы приемы спектроскопии нарушенного полного внутреннего отражения [3]. Объектом исследования служили промышленные образцы и пленки толщиной от 30 до 200 мк.

Облучение проводили на установке с изотопом ^{60}Co дозами от 2,5 до 120 Мрад при мощности 1 и 3 Мрад/ч.

Ускоренное старение стерилизованных материалов в воздухе, в воде и водном растворе инсулина проводили термостатированием пленок при температурах от 37° до 80°C до и после стерилизации, наблюдая за изменениями в ИК-спектрах.

Методом газовой хроматографии [1] обнаружено выделение летучих продуктов радиолиза при гамма-облучении стерилизующей дозой 2,5 Мрад (табл.I).

Из табл.I видно, что при дозе облучения 2,5 Мрад происходит деструкция полимерных материалов. По углеводородному составу можно судить о процессах, протекающих в материалах при действии на них радиации.

ТАБЛИЦА I. КОЛИЧЕСТВА ЛЕТУЧИХ ПРОДУКТОВ РАДИОЛИЗА, ОБРАЗУЮЩИХСЯ ПРИ ОБЛУЧЕНИИ ДОЗОЙ 2,5 МРАД В АТМОСФЕРЕ ВОЗДУХА (мг/мл·10^4)

№№ п/п	Продукты радиолиза	ПЭНД-А	ПЭНД-В	ДСТ-30	ИСТ-30	ПММА	ПВХ
1.	CH_4	0,8	0,59	0,09	1,14	2,25	1,05
2.	C_2H_6	1,1	0,82	-	0,08	0,11	4,1
3.	C_2H_4	-	-	0,5	0,15	0,11	1,13
4.	C_3H_8	0,23	0,20	-	0,03	-	1,10
5.	C_3H_6	0,026	0,021	-	0,13	-	0,15
6.	i-C_4H_{10}	0,28	0,30	-	-	-	0,45
7.	n-C_4H_{10}	0,18	0,20	0,46	0,09	-	1,14
8.	транс-C_4H_8	0,27	0,22	0,90	0,78	0,10	0,40
9.	n-C_5H_{12}	-	-	0,10	-	-	1,40
10.	i-C_5H_{12}	-	0,23	-	-	-	0,60
11.	H_2	34,0	18,0	8,0	27,0	3,0	5,0
12.	CO	3,0	3,0	1,0	6,0	52,0	63,0
13.	CO_2	55,0	56,0	15,0	43,0	80,0	102,0

Примечание: ПЭНД-А – сополимер этилена с небольшой добавкой пропилена, ПЭНД-В – гомополимер этилена, ПВХ пластифицирован диоктилфтататом (40 %).

При радиолизе ИСТ-30 образуется в 12 раз больше метана, чем при радиолизе ДСТ-30. Это обусловлено тем, что в ИСТ-30 происходит отрыв от основной цепи боковых метильных групп, входящих в звенья изопрена. Многообразие углеводородов в продуктах деструкции указывает на сложность радиационно-химических превращений. Наряду с отрывом боковых групп происходит также отрыв конечных участков полимерной цепи и разрыв основной цепи. Об этом свидетельствует появление в газовой фазе ненасыщенных углеводородов (этилена и трансбутена), которые образуются при рекомбинации свободных радикалов:

$$2CH_3-CH_2^{\cdot} \rightarrow CH_2=CH_2+CH_3-CH_3$$

$$2CH_3CH_2CH_2^{\cdot} \rightarrow CH_3-CH=CH-CH_3+CH_3-CH_3$$

Ненасыщенные соединения образуются не только в летучих продуктах радиолиза, но и в самих полимерах. Например, при облучении ПЭВД образуются, а также трансформируются трансвиниленовые, винильные и

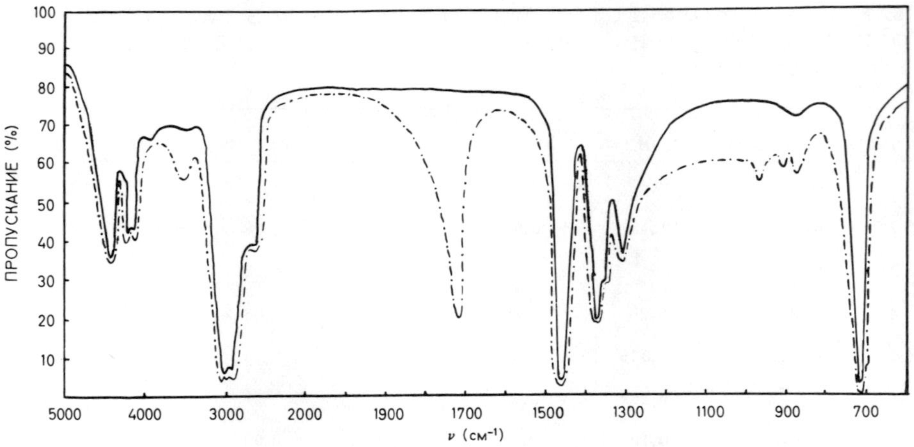

Рис.1. ИК-спектры ПЭВД: 1 - исходного, 2 - облученного дозой 30 Мрад, l = 150 μ.

винилиденовые двойные связи. Об этом свидетельствуют данные ИК-спектров, в которых появляются или меняют интенсивность полосы в области 970, 910 и 890 $см^{-1}$, характерные для внеплоскостных деформационных колебаний связей С-Н в трансвиниленовой, винильной и винилиденовой группах (рис.1). Это согласуется с известными литературными данными [4].

Методом ИК-спектроскопии [5] показано, что при γ-облучении происходит уменьшение степени разветвления полиэтилена.

Уменьшение степени разветвления полиэтилена происходит за счет радиационного отрыва боковых ответвлений, а отрыв боковых ответвлений приводит к образованию летучих продуктов. Другими словами, данные ИК-спектроскопии подтверждают данные, полученные методом газовой хроматографии.

В ИК-спектрах полимеров, облученных на воздухе, появляются полоса поглощения в области 1720 $см^{-1}$, соответствующая валентным колебаниям С = 0 групп, и полоса поглощения в области 3450 $см^{-1}$, которая является полосой валентных колебаний ОН-групп (рис.1). Появление и рост этих полос свидетельствует о реакциях радиационного окисления и деструкции [2, 6].

Для разных полимеров при толщине образца — 50 мк эти полосы проявляются при разных дозах облучения: ПЭВД — 10 Мрад, ПЭНД — 20 Мрад, П4МП1 — 5 Мрад, ПВХ — 10 Мрад, ПММА — 10 Мрад, ДСТ-30 — 5 Мрад, НК — 20 Мрад, СК — 100 Мрад.

Эти дозы являются относительной мерой радиационной стойкости указанных полимеров.

Наиболее стойким к радиационному окислению полимером оказался силиконовый каучук, наименее стойким — дивинилстирольный блок-сополимер. Из литературы [7] известно, что дивинилстирольные каучуки являются более радиационностойкими полимерами, чем резины

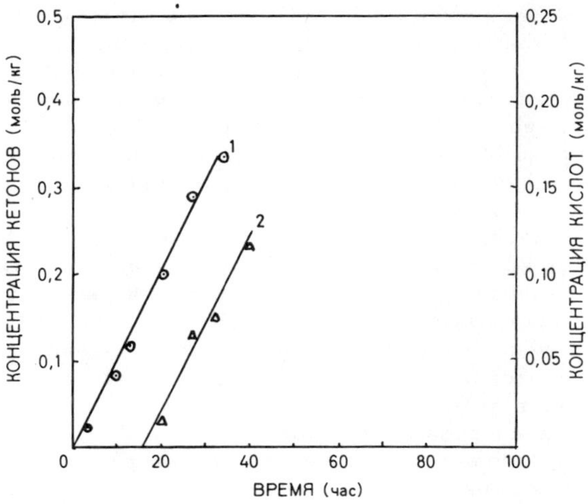

Рис.2. Кинетика накопления кетонов и кислот, образующихся при облучении пленок ПЭВД (l = 60 μ).

на основе бутадиенового каучука. Это объясняется тем, что внутримолекулярный перенос энергии электронного возбуждения от звеньев бутадиена к молекулам радиационностойкого стирола обеспечивает радиационную защиту в сополимере [8]. Низкая радиационная стойкость дивинилстирольных блоксополимеров, по-видимому, обусловлена особенностями блоксополимеров, в которых электронное возбуждение концентрируется на блоках дивинила или на границе блоков дивинила и стирола, но не передается к молекулам радиационностойкого стирола. В результате подвергаются усиленной деструкции блоки дивинила. Об этом же свидетельствует состав газовой фазы при радиолизе, в которой не удается обнаружить стирол или ароматические соединения.

В связи с тем, что при дозе 2,5 Мрад не удается обнаружить заметные изменения в ИК-спектрах, исследование проводили при более высоких дозах с целью экстраполяции изменений в интервале стерилизующих доз.

Интересно отметить, что интенсивность C=0 и O-H-групп, образующихся при облучении, пропорциональна толщине пленки и дозе облучения при толщинах до 150 мк. Этот факт свидетельствует о том, что при γ-облучении полимеров толщиной до 150 мк окисление происходит по всей глубине и не лимитируется диффузией кислорода воздуха, что находится в согласии с данными работы [9].

Проведена спектрофотометрическая идентификация карбонилсодержащих соединений, образующихся при радиационном окислении ПЭВД [9,10] по методике [11].

На рис.2 представлена кинетика накопления кетонов и кислот, образующихся при облучении пленок ПЭВД.

Из рис.2 видно, что зависимость концентрации от времени линейна. По наклону прямых в предположении нулевого порядка реакции рассчитаны эффективные константы скорости образования кетонов и кислот при радиационном окислении ПЭВД. Оказалось, что константы скорости образования кетонов и кислот примерно одинаковы $1{,}6 \cdot 10^{-4}$ мин$^{-1}$ и $1{,}3 \cdot 10^{-4}$мин$^{-1}$ и практически не зависят от толщины пленки в пределах от 30 до 150 мк. Это подтверждает вывод о том, что при облучении пленок за время, необходимое для набора нужной дозы (~1 час для стерилизующей дозы 2,5 Мрад), диффузия кислорода воздуха обеспечивает протекание радиационно-окислительных реакций по всей толщине образца при изученных толщинах пленок.

Интересно отметить, что при увеличении мощности дозы в три раза эффективные константы скорости тоже возрастают в три раза.

Указанные наблюдения имеют большое практическое значение. Для того, чтобы предотвратить быстрое развитие окислительных процессов по всей глубине пленки при лучевой стерилизации, можно рекомендовать для изготовления упаковок полиэтиленовые пленки толщиной более 150 мк, т.к. в этом случае окисление будет лимитироваться скоростью диффузии кислорода внутрь образца.

Из рис.2 в предположении линейного характера экстраполируемой части кинетической кривой определена концентрация кетонов, образующихся при стерилизующей дозе облучения 2,5 Мрад. Она оказалась значительной и равна $5 \cdot 10^{-3}$ моль на 1 кг полиэтилена.

Такая концентрация одного из продуктов окисления ПЭВД, образующихся при стерилизующей дозе облучения, может влиять на процессы старения облученного полиэтилена, а также на процессы термической и термоокислительной деструкции.

Действительно, в работе [12] показано, что изменения структуры, вызванные облучением, повышают скорость взаимодействия полиэтилена с кислородом воздуха. Аналогичный эффект обнаружен нами при

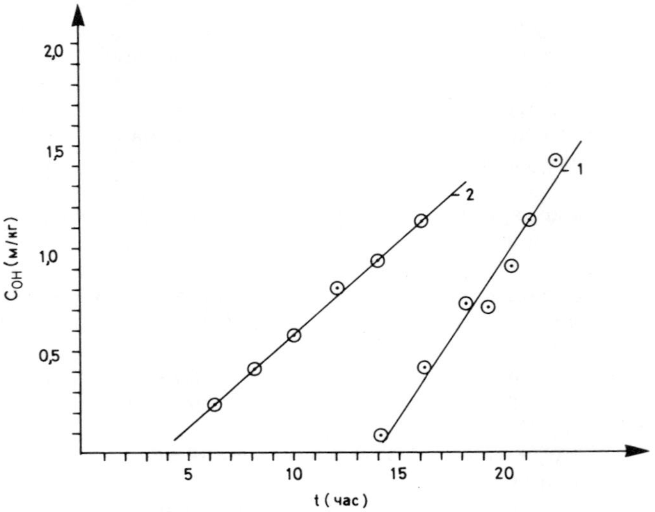

Рис.3. Кинетика возрастания концентрации ОН-групп при ускоренном старении (80°С) ДСТ-30: 1- исходного, 2- стерилизованного дозой 2,5 Мрад.

исследовании ускоренного старения ДСТ-30 в атмосфере воздуха и в воде.

Ускоренное старение ДСТ-30 в воздухе и воде проводили термостатированием пленок при температуре 60°, 70°, 80°C до и после стерилизации, наблюдая за изменениями, в ИК-спектрах.

При старении стерилизованных пленок ДСТ-30, как стабилизированных, так и нестабилизированных в ИК-спектрах появляются новые интенсивные полосы в области 1700 $см^{-1}$ и 3450 $см^{-1}$ (рис.1), соответствующие валентным колебаниям карбонильных и гидроксильных групп, соответственно.

Зависимость возрастания концентрации гидроксильных и карбонильных групп со временем при старении стерилизованных пленок на воздухе и в воде является прямолинейной (рис.3).

По наклону прямых рассчитаны эффективные константы скорости старения в предположении нулевого порядка реакции. Из графика зависимости логарифма эффективной константы скорости от обратной температуры определены энергии активации процесса старения.

Полученные результаты приведены в табл. II.

Из табл. II видно, что стерилизация дозой 2,5 Мрад уменьшает индукционный период старения нестабилизированного ДСТ-30 при 60°C в 8 раз, а облучение дозой 5 Мрад — при той же температуре в 56 раз.

Введение антиоксиданта резко увеличивает индукционный период старения нестерилизованных образцов при 60°C с 168 до 790 часов. Однако, стерилизация приводит к тому, что индукционный период старения стабилизированного ДСТ-30 при той же температуре уменьшается с 790 до 38 часов, т.е. примерно в 20 раз; а облучение дозой 5 Мрад уменьшает τ до 5 часов, т.е. в 158 раз. Следует отметить, что при t = 80°C индукционный период старения стерилизованного ДСТ-30 уменьшается до 4 часов, а старение облученного дозой 5 Мрад ДСТ-30 при той же температуре протекает без индукционного периода.

При увеличении дозы от 2,5 до 5 Мрад в 4 раза возрастает эффективная константа скорости старения при 60°C как для стабилизирован-

ТАБЛИЦА II. ЭФФЕКТИВНЫЕ КОНСТАНТЫ СКОРОСТИ ($k_{эф}$), ИНДУКЦИОННЫЕ ПЕРИОДЫ (τ) И ЭФФЕКТИВНЫЕ ЭНЕРГИИ АКТИВАЦИИ ($E_{эф}$) ОБРАЗОВАНИЯ ГИДРОКСИЛЬНЫХ ГРУПП В НЕОБЛУЧЕННЫХ И ОБЛУЧЕННЫХ ДСТ-30 В ПРОЦЕССЕ СТАРЕНИЯ В ВОЗДУХЕ

Доза ДСТ-30 (Мрад)		60°C		70°C		80°C		$E_{эф}$ (ккал/моль)
		$k_{эф}10^4$ $мин^{-1}$	τ (час)	$k_{эф}10^4$ $мин^{-1}$	τ (час)	$k_{эф}10^4$ $мин^{-1}$	τ (час)	
0	без антиоксид.	6,0	168	13	20	28	14	27
	с антиоксид.	2,8	790	9,0	160	18	63	35
2,5	без антиоксид.	3,1	20	11	16	18	4	26
	с антиоксид.	1,2	38	5,5	24	12	14	27
5,0	без антиоксид.	12	3	25	2	50	0	27
	с антиоксид.	5,4	5	12	3	21	1	27

Примечание: в качестве антиоксиданта применяли 4-метил-2,6-ди-трет-бутил-фенол.

ных, так и для нестабилизированных образцов. Это может быть связано с тем, что при облучении дозой 5 Мрад образуются такие продукты радиационно-химических превращений, которые ускоряют процесс старения ДСТ-30.

Облучение в интервале исследованных доз практически не влияет на энергию активации нестабилизированного ДСТ-30. Введение антиоксиданта увеличивает энергию активации от 27 до 35 ккал. Однако, облучение дозами 2,5 и 5 Мрад приводит к тому, что энергия активации уменьшается до 27 ккал для обеих доз. Исходя из этих данных можно предположить, что стерилизация вносит несущественные изменения в механизм реакций старения ДСТ-30.

Представляло интерес сравнение эффективных констант скоростей образования гидроксил- и карбонилсодержащих соединений как в атмосфере воздуха, так и в воде. Эти данные приведены в табл. III.

Из табл. III видно, что $k_{эф}^{C=O}$ в воздухе больше, чем в воде. Это можно объяснить тем, что карбонилсодержащие соединения образуются при взаимодействии полимера с кислородом [9], а концентрация кислорода в воздухе на несколько порядков больше, чем в воде. $k_{эфф}^{OH}$ в воде больше, чем в воздухе, это можно объяснить тем, что гидроксилсодержащие соединения образуются при участии воды в процессе старения, а концентрация воды в воздухе ограничена.

ТАБЛИЦА III. ЭФФЕКТИВНЫЕ КОНСТАНТЫ СКОРОСТИ ОБРАЗОВАНИЯ ГИДРОКСИЛСОДЕРЖАЩИХ ($k_{эф}^{OH}$) И КАРБОНИЛСОДЕРЖАЩИХ ($k_{эф}^{C=O}$) СОЕДИНЕНИЙ ПРИ СТАРЕНИИ СТЕРИЛИЗОВАННОГО ДОЗОЙ 2,5 МРАД ДСТ-30 С АНТИОКСИДАНТОМ ПРИ t = 70°C

среда	$k_{эф}^{OH} \cdot 10^4$	$k_{эф}^{C=O} \cdot 10^4$
воздух	5,5	3,3
вода	12,6	1,6

Из табл. III видно, что как при старении в атмосфере воздуха, так и при старении в воде образуются карбонилсодержащие и гидроксилсодержащие продукты с близкими эффективными константами скоростей образования.

Это свидетельствует о том, что старение в атмосфере воздуха происходит под действием кислорода и воды воздуха, а старение в воде — под действием растворенного кислорода и самой воды.

Образование различных карбонил- и гидроксилсодержащих соединений при γ-облучении можно объяснить [13] мономолекулярным или биомолекулярным распадом перекисных радикалов — промежуточных продуктов радиолиза:

$$RCH_2OO^{\bullet} \rightarrow RCHO + OH^{\bullet}$$

$$RR'CHOO^{\bullet} \rightarrow RR'C=O + OH^{\bullet}$$

$$\searrow RCHO + R_1O^{\bullet}$$

$$ROO^{\bullet} + R'OO^{\bullet} \rightarrow RCO^{1} + R^{II}OH + O_2$$

Возможны реакции перекисных радикалов с гидроперекисями

$$ROO^{\bullet} + R^{1}OOH \rightarrow RCOR^{1} + H_2O + O_2,$$

а также реакции рекомбинации

$$ROO^{\bullet} + ROO^{\bullet} \rightarrow ROOR^{1} + O_2$$

$$ROO^{\bullet} + R^{\bullet 1} \rightarrow ROOR^{1}$$

Кроме этого, карбонилсодержащие соединения могут образоваться при непосредственном окислении двойных связей, имеющихся в полимере до облучения [6]

$$CH_2=CHR \xrightarrow{O_2} CH_2O + RCHO$$

$$CH_2=CR_1R_2 \xrightarrow{O_2} CH_2O + R_1CO-R_2$$

$$R_1CH=CHR_2 \xrightarrow{O_2} R_1CHO + R_2CHO$$

При облучении на воздухе альдегиды легко превращаются в кислоты

$$RCHO \xrightarrow{O_2} RCOOH$$

Для исследования старения стерилизованных упаковок из натурального и силиконового каучуков при контакте с раствором инсулина применена спектроскопия нарушенного полного внутреннего отражения (НПВО), позволяющая изучать как прозрачные пленки, так и непрозрачные композиции и изделия из этих продуктов.

На рис.4 приведены ИК-спектры отражения стерилизованных натурального и силиконового каучуков, подвергнутых старению при 37°С в условиях контакта с водным раствором инсулина в течение 1 месяца.

Из рис. 4 видно, что структура натурального каучука существенно изменяется. Об этом свидетельствуют новые интенсивные полосы поглощения в области 1720 и 3450 $см^{-1}$, ответственные за изменение структуры. В ИК-спектре силиконового каучука эти полосы не появляются даже при контакте с раствором инсулина в течение 4-х месяцев.

Следует отметить, что в спектрах как натурального, так и силиконового каучуков появляются полосы поглощения инсулина в области 1510 и 1610 $см^{-1}$. Это говорит о том, что происходит сорбция инсулина на поверхности этих материалов. Показано, что кривые сорбции имеют характер насыщения.

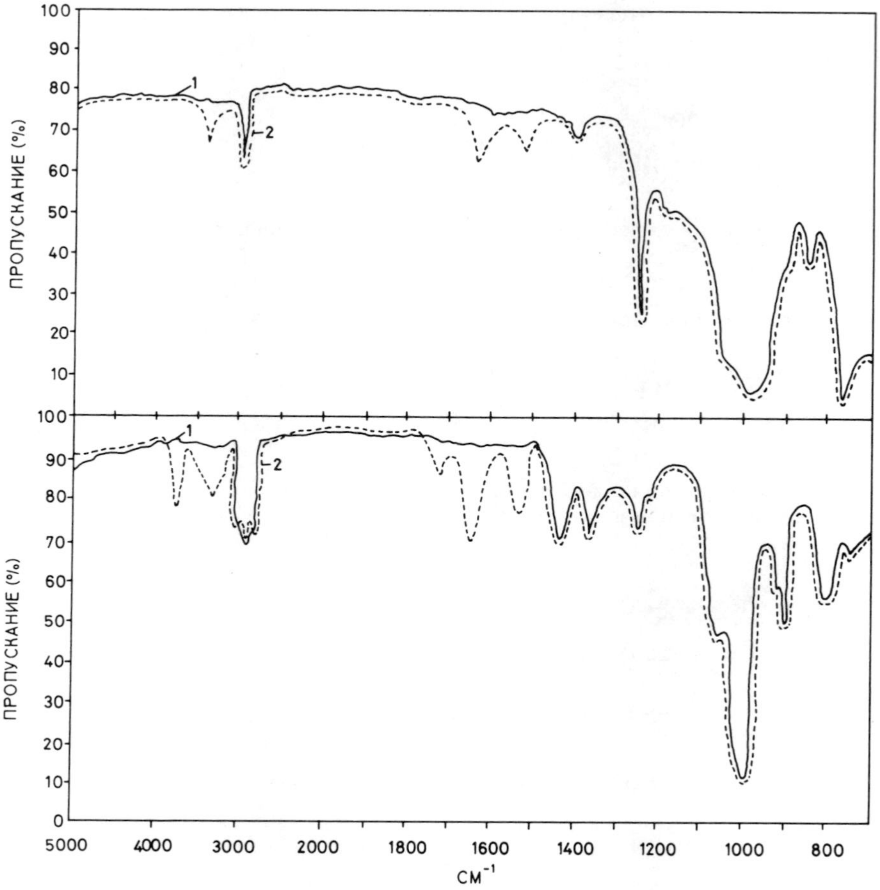

Рис.4. ИК-спектры отражения стерилизованных натурального и силиконового каучуков: 1 - исходных, 2 - подвергнутых старению при 37°С в растворе инсулина в течение 1 месяца.

Проведенное исследование показало, что наиболее стойким к воздействию радиационной стерилизации и водных растворов лекарственных средств является диметилсилоксановый каучук среди каучуков и полиэтилен среди других изученных полимеров.

ЛИТЕРАТУРА

[1] СЕМЕНЕНКО, Э.И., МЫШКОВСКИЙ, В.И., ВОБЛИКОВА, В.А., ВЕНДИЛЛО, В.П., Пластмассы 5 (1973) 30.

[2] ЧАРЛЬЗБИ, А., В кн. Радиационные Эффекты в Физике, Химии и Биологии, М., Атомиздат, 1965 г.

[3] ХАРРИК, Н., В кн. Спектроскопия Внутреннего Отражения, М., "Мир", 1970 г.

[4] DOLE, M., J.Am.Chem.Soc. 80 (1958) 1980.

[5] LUONGO, J.P., Appl. Polym. Symp., 10 (1969) 121-129.
[6] БРАГИНСКИЙ, Р.П., ФИНКЕЛЬ, Э.Э., ЛЕЩЕНКО, С.С., В кн. Стабилизация Радиационно-модифицированных Полиолефинов, М., "Химия", 1973 г.
[7] ANDERSON, H.R., J. Appl. Polym. Sci. 3 (1960) 316.
[8] КУЗЬМИНСКИЙ, А.С., ФЕДОСЕЕВА, Г.С., МАХЛИС, Ф.А., В кн. Радиационная Химия Полимеров, М., "Наука", 1973 г.
[9] GIBERSON, R.C., J. Polym. Sci., A-2 (1964) 4965.
[10] БЕЛЛАМИ, Л., Новые Данные по ИК-Спектрам Сложных Молекул, М., "Мир", 1972 г.
[11] COOPER, Y.D., PROBER, M., J. Polym. Sci. 44 (1960) 397.
[12] ПАВЛОВА, Ж.Д., ЛЕЩЕНКО, С.С., ЕГОРОВА, З.С., КАРПОВ, В.Л., ВМС A16 1 (1974) 111.
[13] ДОЛГОВА, Р.В., ИЛЬИЧЕВА, З.Ф., АХВЛЕДИАНИ, И.Г., СЛОВОХОТОВА, Н.А., РОЗЕНБЛЮМ, Н.Д., ВМС A16 (1974) 220.

ВЛИЯНИЕ РАДИАЦИОННОЙ СТЕРИЛИЗАЦИИ НА САНИТАРНО-ХИМИЧЕСКИЕ СВОЙСТВА ПОЛИМЕРНЫХ МАТЕРИАЛОВ, КОНТАКТИРУЮЩИХ С МЕДИЦИНСКИМИ ПРОДУКТАМИ

Э.И.СЕМЕНЕНКО, В.А.ПОЛУНИН, М.А.МАРКЕЛОВ
Всесоюзный научно-исследовательский институт медицинских полимеров,
Москва,
Союз Советских Социалистических Республик

Abstract–Аннотация

INFLUENCE OF RADIATION STERILIZATION ON THE MEDICOCHEMICAL PROPERTIES OF POLYMER MATERIALS COMING INTO CONTACT WITH MEDICAL PRODUCTS.

In view of the ever-increasing use of radiation for sterilizing medical products it is of interest to study the effect of ionizing radiation on the medicochemical properties of these substances, in particular polymer materials that come into contact with biological and medicinal media. In this paper the authors study how radiation affects the migration of monomers from such widely used polymers as polystyrene and copolymers of methyl methacrylate. A direct relationship is established between the concentration of monomers in polymers and the extent of their migration into standard media. It is shown that both gamma radiation and accelerated electrons tend to reduce the unpolymerized monomer content of a polymer. A mechanism for the process is proposed, based on the relationship established between the monomer content of polymers and the dose-rate.

ВЛИЯНИЕ РАДИАЦИОННОЙ СТЕРИЛИЗАЦИИ НА САНИТАРНО-ХИМИЧЕСКИЕ СВОЙСТВА ПОЛИМЕРНЫХ МАТЕРИАЛОВ, КОНТАКТИРУЮЩИХ С МЕДИЦИНСКИМИ ПРОДУКТАМИ.

В связи с все более широким распространением радиационного способа стерилизации изделий медицинского назначения, представляет интерес изучение влияния ионизирующего излучения на санитарно-химические свойства этих изделий, в частности, полимерных материалов, предназначенных для контакта с биологическими и лекарственными средами. Настоящая работа посвящена изучению влияния радиационного воздействия на миграцию мономеров из таких широко распространенных полимеров, как полистирол и сополимеры метилметакрилата. Выявлена прямолинейная зависимость между концентрацией мономеров в полимерах и величинами их миграции в модельные среды. Показано, что при действии как γ-излучения, так и ускоренных электронов, содержание незаполимеризовавшегося мономера в полимере уменьшается. Предложен механизм процесса, исходя из обнаруженной зависимости содержания мономеров в полимерах от мощности дозы.

Метод радиационной стерилизации медицинских продуктов в последнее время получает широкое распространение. Высокая производительность, максимальная степень автоматизации, относительная простота и другие ценные качества делают этот метод весьма перспективным технологическим приемом массового производства медицинских изделий. Однако, до сих пор остаются невыясненными многие вопросы действия ионизирующего излучения на полимерные материалы в случае стерилизации этим способом пластмассовых изделий.

Одной из важнейших характеристик полимерного материала, применяемого для медицинских целей (в частности, для упаковки лекарственных средств и других медицинских продуктов), являются его санитарно-химические свойства, т.е. способность выделять в контактную среду низкомолекулярные соединения различной природы и состава. Такими соединениями могут быть как составные части композиции полимерного материала, остаточные мономеры и разнообразные добавки, так и продукты деструкции, образовавшиеся под действием факторов внешней среды (температура, кислород воздуха, контакт с активными жидкостями) или стерилизация.

Исследование процессов миграции низкомолекулярных веществ из полимерных материалов в контактные среды предусматривает использование селективных и высокочувствительных методов их определения, т.к. появление в контактной среде даже малых концентраций указанных веществ может оказаться нежелательным.

В настоящем сообщении излагаются результаты экспериментальных исследований влияния радиационной стерилизации полимерных материалов на основе стирола, акрилатов и их сополимеров на санитарно-химические свойства этих материалов.

Для анализа полимерных материалов на содержание в них низкомолекулярных примесей, а также для определения их миграции в контактирующие с полимером жидкие среды (в частности, в воду) применяли разработанные авторами газо-хроматографические методики [1]. Содержание остаточного стирола и метилметакрилата (ММА) определяли с чувствительностью 0,01 и 0,005%, соответственно. Чувствительность определения стирола и ММА в воде составляла 0,05 и 0,1 мг/л, соответственно.

Исследовали сополимеры ММА с 4% бутилакрилата (Дакрил-4Б), полистирол блочный (ПС), тройные сополимеры стирола с ММА и дивинильным каучуком (25% стирола, 70% ММА, 5% СКД). Все указанные материалы созданы для медицинского назначения. Образцы материалов использовались в экспериментах в виде стандартных пластин толщиной 0,5 и 1,1 мм. Материалы облучали в кобальтовом облучателе при мощностях 270 и 830 рад/с и на линейном ускорителе электронов дозами от 0,1 до 10 Мрад.

В табл.I представлены экспериментальные данные по изменению содержания остаточного мономера (ММА) в образцах материала Дакрил-4Б, в зависимости от дозы γ-облучения и мощности дозы (интенсивности облучения). Как видно из этой таблицы, содержание остаточного мономера резко снижается в зависимости от интегральной поглощенной дозы и интенсивности облучения. Причем облучение в вакууме приводит к еще более заметному уменьшению концентрации мономера в полимере вплоть до полного исчезновения. Для большей наглядности эта картина представлена на рис.1.

Падение содержания мономера в полимере можно объяснить его полимеризацией в процессе облучения стерилизующими дозами ионизирующего излучения. Анализ кинетических кривых изменения концентрации мономера в зависимости от времени облучения с различной интенсивностью при наборе необходимой интегральной дозы показал, что скорость уменьшения концентрации мономера пропорциональна интенсивности облучения и подчиняется известному соотношению [2]: $V = kI^n$, где I – интенсивность облучения.

При разнице величин интенсивности γ-облучения образцов в 3 раза начальные скорости изменения концентраций ММА в полимере составили

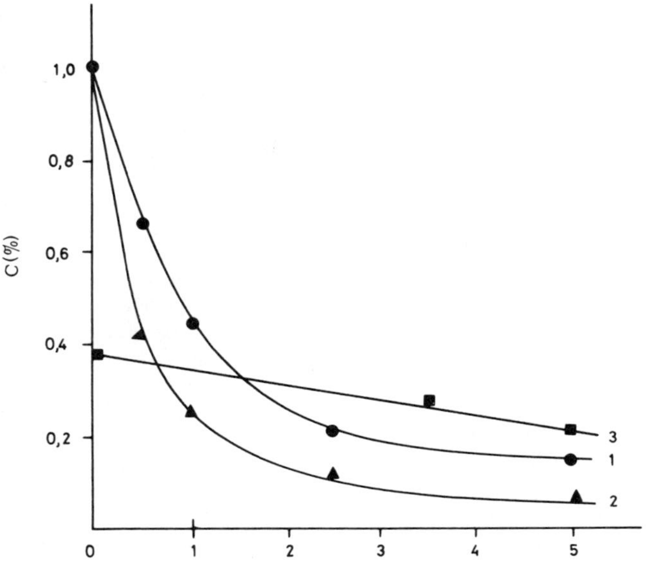

Рис.1. Зависимость процентного содержания остаточных мономеров в образцах материалов Дакрил-4Б и блочного полистирола марки Д от дозы и интенсивности γ-излучения. 1 – Дакрил-4Б, 270 рад/с ; 2 – Дакрил-4Б, 830 рад/с ; полистирол Д, 270 рад/с.

ТАБЛИЦА I. ИЗМЕНЕНИЕ СОДЕРЖАНИЯ ОСТАТОЧНОГО МОНОМЕРА В ОБРАЗЦАХ МАТЕРИАЛА ДАКРИЛ-4Б В ЗАВИСИМОСТИ ОТ ДОЗЫ γ-ОБЛУЧЕНИЯ И МОЩНОСТИ ДОЗЫ

Толщина образца (мм)	Доза (Мрад)	Мощность дозы (Мрад/ч)	Концентрация ММА в полимере (%)	
			Облучение на воздухе	Облучение в вакууме ($4 \cdot 10^{-4}$ мм Hg)
0,5	0,5	1	0,59	-
0,5	0,5	3	0,57	-
1,1	0,5	1	0,71	-
1,1	0,5	3	0,42	-
0,5	1,0	1	0,40	0,27
0,5	1,0	3	0,32	0,06
1,1	1,0	1	0,46	0,21
1,1	1,0	3	0,25	0,06
0,5	2,5	1	0,25	0,18
0,5	2,5	3	0,20	0,04
1,1	2,5	1	0,21	0,11
1,1	2,5	3	0,13	0,03
0,5	5,0	1	0,14	≤ 0,005
0,5	5,0	3	0,09	≤ 0,005
1,1	5,0	1	0,15	≤ 0,005
1,1	5,0	3	0,06	≤ 0,005

для образцов толщиной 1,1 мм – 1,4 и 4,2 $мин^{-1}$, а для образцов толщиной 0,5 мм – 2,0 и 6,6 $мин^{-1}$, соответственно. Экспериментально обнаруженное значение n в приведенном выше уравнении, близкое к единице, может служить подтверждением свободно-радикального протекания полимеризации остаточного мономера в полимере. Скорость процесса полимеризации пропорциональна интенсивности инициирования, т.е. числу начальных активных центров.

Полимеризацию остаточного мономера в полимере можно рассматривать как полимеризацию в твердой фазе со всеми присущими ей особенностями (ограниченная подвижность полимерных молекул, затрудненная диффузия молекул мономера и т.д.). Пропорциональность скорости полимеризации первой степени интенсивности облучения указывает на то, что обрыв цепей происходит при взаимодействии радикалов одного трека и увеличение I приводит к возрастанию количества изолированных элементарных объемов, в каждом из которых протекает полимеризация [3].

Обнаруженное в данной работе замедление конверсии мономера в присутствии кислорода также может служить подтверждением радикального механизма радиационной полимеризации [4]. Дать объяснение увеличению конверсии мономера в более толстых образцах Дакрила-4Б с возрастанием интенсивности облучения, а также более высоким начальным скоростям полимеризации в тонких образцах из имеющихся экспериментальных данных пока не представляется возможным.

В табл.II приведены результаты анализа Дакрила-4Б на содержание остаточного мономера после облучения образцов ускоренными электронами.

Общая тенденция изменения содержания мономера в материале в этом случае та же, что и при облучении γ-лучами. В образцах с бóльшей толщиной конверсия мономера выше. Это может быть связано с повышением скорости полимеризации за счет разогрева образцов в период облучения и существенно меньшим теплоотводом в более толстых образцах. Однако глубина конверсии мономера в случае облучения ускоренными электронами ниже по сравнению с γ-облучением.

ТАБЛИЦА II. ИЗМЕНЕНИЕ СОДЕРЖАНИЯ ОСТАТОЧНОГО МОНОМЕРА В ОБРАЗЦАХ МАТЕРИАЛА ДАКРИЛ-4Б В ЗАВИСИМОСТИ ОТ ДОЗЫ ОБЛУЧЕНИЯ УСКОРЕННЫМИ ЭЛЕКТРОНАМИ. МОЩНОСТЬ ДОЗЫ 7,35 · 10^4 рад/с

Толщина образца (мм)	Доза (Мрад)	Концентрация ММА в полимере (%)
0,5	0,1	0,88
1,1	0,1	0,75
0,5	1,0	0,74
1,1	1,0	0,49
0,5	5,0	0,25
1,1	5,0	0,13
0,5	10,0	0,06
1,1	10,0	0,08

На рис.1 показано изменение концентрации остаточного мономера в блочном полистироле. Вид этой кривой отличен от кривых изменения содержания ММА в Дакриле-4Б.

В этом случае мы также имеем дело с полимеризацией мономера. Прямолинейная зависимость для радиационной полимеризации стирола в твердой фазе является типичной [5,6]. Кроме того, присутствующий в качестве примеси в полистироле этилбензол в концентрации 0,29% остается после облучения в первоначальном количестве, что свидетельствует о полимеризации стирола, а не уменьшении его содержания в материале, например, в результате испарения за счет радиационного разогрева образца.

Уменьшение содержания остаточных мономеров при стерилизующих дозах радиации отмечено и в тройных сополимерах ММА со стиролом и СКД. Бруски из такого материала содержали в исходном состоянии 0,335% ММА и 0,016% стирола. После γ-облучения образцов дозой 2,5 Мрад концентрация ММА снизилась до 0,167%, а стирола — до нуля. В иглах для прокола пробок флаконов с медицинскими препаратами, изготовленных из этого сополимера методом литья под давлением, содержание остаточных мономеров составляло: 0,371% ММА и 0,012% стирола. После облучения 2,5 Мрад концентрация ММА снизилась до 0,185%, а стирол не обнаружен.

В обоих случаях содержание этилбензола оставалось неизменным. Несомненно, что в описанных выше случаях наряду с реакциями полимеризации протекают также и процессы радиационной деструкции, окис-

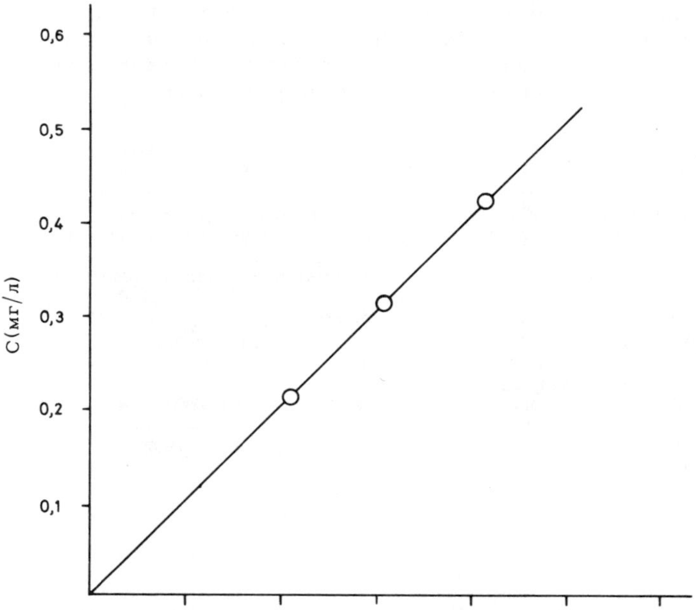

Рис.2. Зависимость концентрации стирола в водной вытяжке, приготовленной настаиванием пластин из блочного полистирола в воде, от содержания мономера в полимере.

ления и сшивки [7]. Картина совокупного механизма этих процессов весьма сложна, тем более, что полимеры на основе акрилатов в основном подвергаются деструкции, а полистирол относится к классу сшивающихся полимеров [8]. Однако, вероятно, в условиях облучения материалов стерилизующими дозами ионизирующего излучения заметной деполимеризации с образованием свободных монометров не происходит. Поэтому основным процессом с участием молекул мономера является радиационная полимеризация.

Наличие определенного количества низкомолекулярных соединений (в частности мономеров) в полимерном материале неизбежно приводит к опасности их миграции в контактирующую с материалом среду. Причем величина миграции пропорциональна содержанию мигрирующих веществ в полимере.

На рис.2 приведена зависимость величины миграции стирола в воду из блочного полистирола от концентрации остаточного мономера в полимере. Аналогичная зависимость получена нами и для случая миграции ММА из Дакрила-4Б. Обнаруженная зависимость дает возможность оценки санитарно-химических свойств материала по величине остаточного содержания мономера.

Подводя итог изложенному, следует отметить, что действие стерилизующих доз ионизирующего излучения на полимерные материалы на основе акрилатов, стирола и их сополимеров не приводит к ухудшению их санитарно-гигиенических характеристик за счет увеличения миграции наиболее токсичных компонентов – мономеров. Более того, радиационная стерилизация в этом смысле благоприятна и носит "облагораживающий" с санитарно-гигиенической точки зрения характер.

Результаты проведенных исследований позволяют рекомендовать указанные материалы для производства изделий, контактирующих с медицинскими продуктами, подлежащими радиационной стерилизации.

ЛИТЕРАТУРА

[1] МАРКЕЛОВ, М.А., СЕМЕНЕНКО, Э.И., Пластмассы 3 (1973) 65.
[2] ПШЕЖЕЦКИЙ, С.Я., Механизм Радиационно-химических Реакций, М., Изд-во "Химия", 1968 г.
[3] БАГДАСАРЬЯН, Х.С., Теория Радикальной Полимеризации, М., Изд-во "Наука", 1966 г.
[4] CHAPIRO, A., J. Chim.Phys. 47 (1950) 747, 764.
[5] CHAPIRO, A., STANNET, V., J. Chim. Phys. 57 (1960) 35.
[6] HAYASHI, K., OKAMURA, T., Proc. Int. Symp. on Radiation Industr. Polymeric, Rep. AEC, TJD-7643, 1962, p. 150.
[7] СЕМЕНЕНКО, Э.И., МЫШКОВСКИЙ, В.И., ВОБЛИКОВА, В.А., ВЕНДИЛЛО, В.П., Пластмассы 5 (1973) 30.
[8] ЧАРЛЗБИ, А., Ядерные Излучения и Полимеры, М., ИЛ, 1962 г.

DISCUSSION

on papers IAEA-SM-192/42, 43.

K. D. KEHR: I don't think that you mentioned the use of polymers in the packaging of metal products, such as blades, needles and so on. Do you consider that the toxic effect of the degradation of polymers has any relevance for that range of products? A further point is that besides the actual toxicity, there is also the problem of corrosion, particularly in the case of PVC, PVA and similar polymers. What in your opinion is the best method of counteracting it?

Eh. I. SEMENENKO: I would certainly agree that polymer degradation products may be toxic and that if they are sorbed on the surfaces of medical products, they could find their way into the organism. It should be remembered, however, that the toxic product concentrations are usually fairly small and that in most cases they do not therefore constitute a genuine hazard.

The experimental results I have reported suggest a reaction mechanism at the molecular level in polymers when sterilized by irradiation; this is important for the selection of the most suitable material for packaging or producing new materials to specification.

When using polymer materials that may give off, during radiosterilization, aggressive chemicals, such as HCl gas in the case of PVC, there is a risk of the packaged instrument being corroded. But it should be borne in mind that the total amounts of chemically active substances present are relatively small, and that the medical instruments are usually made of corrosion-resistant steels. So that in practice this risk, too, is negligible.

DISCUSSION

CHEMICAL AND BIOLOGICAL EFFECTS OF RADIATION STERILIZATION OF MEDICAL PRODUCTS

B. L. GUPTA
Division of Radiological Protection,
Bhabha Atomic Research Centre,
Trombay, India

Abstract

CHEMICAL AND BIOLOGICAL EFFECTS OF RADIATION STERILIZATION OF MEDICAL PRODUCTS.

Radiation is extensively used for the sterilization of plastic materials, pharmaceuticals and biological tissue grafts. The pharmaceuticals may be solid, liquid, or suspension in a liquid or a solution. Cobalt-60 gamma radiation, generally used for sterilization, primarily interacts with these materials through the Compton process. The resulting damage may be direct or indirect. In aqueous systems the primary species produced compete for interaction among themselves and the dissolved solutes. The nature, the G-values and the reactions of the primary species very much depend on the pH of the solution. The important chemical changes in plastic materials are gas liberation, change in concentration of double bonds, cross-linking, degradation and oxidation. These chemical changes lead to some physical changes like crystallinity, specific conductivity and permeability. The reactions in biological systems are very complex and are influenced by the presence or absence of water and oxygen. Water produces indirect damage and the radiation effect is generally more in the presence of oxygen. Most microorganisms are relatively radioresistant. Various tissues of an animal differ in their response to radiation. Catgut is not stable to irradiation. Lyophilized human serum is stable to irradiation whereas, when irradiated in aqueous solutions, several changes are observed. Generally, pharmaceuticals are considerably more stable in the dry solid state to ionizing radiations than in aqueous solutions or in any other form of molecular aggregation.

INTRODUCTION

The types of radiations which are encountered in the study of radiation effects are: X, gamma, alpha, beta, neutrons, deuterons and protons. In the sterilization of medical products gamma rays and high-energy electrons are the only ones in common use [1-6]. Table I gives a summary of the interaction of various radiations with matter.

The overall process of producing chemical transformations by the use of high-energy radiation starts with the bombardment of a material system by radiation and terminates with the re-establishment of the thermodynamic equilibrium [7]. This process can be divided into three stages as shown in Table II. When living organisms are involved, there may be delayed effects. Table III gives a time scale of events in radiation effects.

The primary reactive species formed in the physicochemical stage present a specific spatial distribution which depends upon the quality of the radiation used and the medium [8]. They then proceed to diffuse according to macroscopic diffusion laws and to react with themselves or with species already existing in the medium prior to irradiation. The products of these reactions may be chemically unstable and participate in further reactions, some of which may involve the primary reactive species. Table IV gives a summary of some physical and physicochemical reactions [9].

TABLE I. INTERACTION OF VARIOUS RADIATIONS WITH MATTER

- Radiation
  - Electromagnetic
    - x, γ
    - Photoelectric effect, Compton scattering, pair production
  - Particle
    - α, β, n, d, p
      - n
        - (n,γ), (n,p), (n,α), (n,2n), fission
      - α, β, d, p
        - Excitation
        - Ionization

TABLE II. KINETICS OF RADIATION CHEMICAL TRANSFORMATIONS

Physical stage (Energy deposition)	Physicochemical stage (Thermal equilibrium)	Chemical stage (Diffusion and chemical equilibrium)

TABLE III. TIME-SCALE OF EVENTS IN RADIATION EFFECTS

Energy absorption	10^{-16} s
Radical formation	10^{-13} s
Hydration of electrons	10^{-11} s
Molecular changes	10^{-5} s
Early physiological effects	seconds to hours
Mutations	minutes to hours
Delayed somatic effects	hours to years

TABLE IV. SUMMARY OF PHYSICAL AND PHYSICOCHEMICAL REACTIONS

1. Excitation
2. De-excitation
3. Energy transfer
4. Free radical formation
5. Dissociation into molecular products
6. Ionization
7. Neutralization

From the point of view of radiation chemistry, gaseous systems are simple because their low density greatly reduces the effect of LET; for example, alpha particles and gamma rays produce practically the same product yields, and because the active species are not confined in the particle tracks where they are formed [10].

In liquids or solids, the active species are formed close together and are hindered in moving apart by closed packed surrounding molecules [11,12]. Radicals formed in an organic liquid will predominantly react with the liquid if this reaction is at all facile [12]. Thus, track effects may be unimportant in the radiolysis or organic compounds because of such radial solvent reactions; for instance, product yields may be independent of LET. This is not invariably true and some organic materials give product yields which are dependent upon LET. In the radiolysis of water and aqueous solutions, the primary reactive species mostly either react among themselves or with the dissolved solutes [13-15].

The interaction of high-energy radiation with the network of a solid body may result in the formation of a number of defects: (a) vacancies, (b) interstitial atoms, (c) impurity atoms, (d) collisions resulting in substitutions of atoms, (e) thermal spurs and resulting thermal dislocations, and (f) ionizing effects. However, as far as the mechanism of radiation chemical processes in gamma and electron irradiated solid bodies is concerned, it is the ionization effects which are the most important [16].

The ionizing radiation may react with the biological systems in two ways — (a) direct effects, and (b) indirect effects [17-20]. The direct action model assumes energy to be absorbed by the molecule, which determines the biologically observable damage. There may be energy transfer within the molecule of one cellular structure and the excited target molecule may react with cellular constituents. The indirect action model assumes that inactivation of the biological unit is brought about by energy absorbed in the material surrounding the target molecule. This leads to chemical intermediates, which convey the damage to the target molecule.

The chemical effects of radiation are related in terms of chemical yield and dose. The units of dose normally used in sterilization are the electron volt (eV), erg and rad. One rad is 100 erg of energy absorbed per gram of any substance. The chemical yield is defined in G-values and it is the chemical change produced per 100 eV of energy absorbed.

In radiation sterilization, plastics are extensively used for making of single-use containers and syringes [5]. They are also used for intravenous feeding tubings, catheters and prostheses such as vascular grafts, supportive cartilage replacements, breast prostheses and pace makers. Several pharmaceuticals such as eye drops, vitamins, saline water and vaccines are also sterilized by this method [6]. Biological tissue grafts for transplant therapy are the other items sterilized by radiation. It is therefore of interest to examine the chemical, physical and immunological alterations taking place in such materials by sterilizing radiation doses [21].

AQUEOUS SYSTEMS

Table V gives a summary of reactions in irradiated water [22, 23]. The action of ionizing radiation on water results in the formation of ions and excited states close to its path or track [13-15]. The electrons so

TABLE V. SUMMARY OF REACTIONS AND RATE CONSTANTS IN IRRADIATED WATER

Reaction			Rate constant ($M^{-1} \cdot s^{-1}$)
H_2O	$\rightsquigarrow$	$H_2O^+ + e^-$	
H_2O	$\rightsquigarrow$	H_2O^*	
$e^-_{aq} + H_2O$	$\rightarrow$	$H + OH^-$	16.0
$H_2O^+ + H_2O$	$\rightarrow$	$H_3O^+ + OH$	
$H_3O^+ + e^-$	$\rightarrow$	H_3O	
H_3O	$\rightarrow$	$H_2 + OH$	
H_2O^*	$\rightarrow$	$H + OH$	
$H + H$	$\rightarrow$	H_2	2.0×10^{10}
$OH + OH$	$\rightarrow$	H_2O_2	0.6×10^{9}
$H + OH$	$\rightarrow$	H_2O	2.0×10^{10}
$e^-_{aq} + e^-_{aq}$	$\rightarrow$	$H_2 + 2OH^-$	0.6×10^{10}
$e^-_{aq} + H$	$\rightarrow$	$H_2 + OH^-$	2.5×10^{10}
$e^-_{aq} + OH$	$\rightarrow$	OH^-	3.0×10^{10}
$e^-_{aq} + H_3^+O$	$\rightarrow$	$H + H_2O$	2.0×10^{10}
$H + OH^-$	$\rightarrow$	e^-_{aq}	1.8×10^{7}
OH	$\rightarrow$	$O^- + H^+$	pka = 11.9 ± 0.2
$H + O_2$	$\rightarrow$	HO_2	
$e^-_{aq} + O_2$	$\rightarrow$	O_2^-	2.0×10^{10}
HO_2	$\rightarrow$	$H^+ + O_2^-$	pka = 4.5 ± 0.2

formed have sufficient energy to ionize a few other molecules of water. The clusters of ions thus produced are called 'spurs'. A few other molecules of water, located a little further away, become excited. The secondary electron, after losing its energy is solvated some distance away, forming an atom of hydrogen. The mother ion also reacts with water molecules with the formation of an OH radical. The direct dissociation of excited water molecules also produces H and OH radicals.

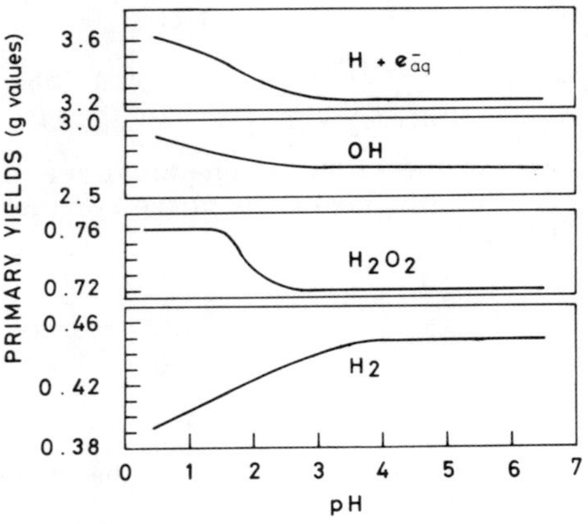

FIG. 1. Primary species yields at different pH.

The H and OH formed react with each other in those regions where their concentration is high. When the LET of the radiation increases, the re-combination of radicals among themselves increases, leading to a decrease in radicals and an increase in H_2 and H_2O_2 production. It is believed that a part of H_2 is produced directly through a pseudo first-order reaction [25]. Thus, the effect of ionizing radiation on water can be written as

$$H_2O \rightsquigarrow H,\ e^-_{aq},\ OH,\ H_2,\ H_2O_2$$

In acidic solutions the hydrated electrons are converted to hydrogen atoms. In alkaline solutions, hydrogen atoms are converted to hydrated electrons and there is ionization of hydroxyl radicals. When oxygen is present in the solution, H and e^-_{aq} react very rapidly with it. In neutral and alkaline solutions HO_2 also ionizes. The yields of primary species for any radiation depend on the pH of the solution and Fig.1 gives the primary yields at a different pH in acidic and neutral solutions [25].

In aqueous solutions, the primary species either interact among themselves or with the dissolved solute as mentioned earlier. The reactions of primary species with solutes are (1) Oxidation; (2) reduction; (3) addition; (4) abstraction; (5) electron transfer. The reactions with the inorganic solutes are generally oxidation-reduction reactions, whereas the reactions with the organic solutes are addition, abstraction and electron transfer. As seen from the Fig.1, the nature and yield of primary species depends upon the pH of the solution; therefore, the initial chemical reactions are also dependent on the pH [26]. The initial products from the reaction of primary radiolytic species with dissolved solutes may further undergo various types of reactions depending upon the type of the system being irradiated. These reactions may be different in the presence and absence of oxygen. For example, the abstraction of a hydrogen atom by OH radical from an organic molecule RH produces a radical R which may revert to RH in the absence of oxygen or it may form a peroxide in the presence of oxygen:

$$RH + OH \rightarrow R + H_2O$$

$$R + H \rightarrow RH$$

$$R + O_2 \rightarrow RO_2$$

Though the mechanisms of the radiation chemical reactions of simple organic molecules are well studied, the reactions of the complex molecules are not known in much detail [27].

POLYMERS

General

The irradiation of high polymers produces the following general reactions [28-30]:

(1) Formation of ionized molecules
(2) Formation of excited molecules
(3) Direct decomposition of ionized molecules
(4) Formation of free radicals as a result of decomposition of excited molecules

Chemical reactions

The free radicals can produce a variety of reactions:

(1) Gas liberation
(2) Formation and disappearance of double bonds
(3) Exchange reactions
(4) Migration of the free electron
(5) Cross-linking
(6) Degradation
(7) Oxidation

Physical changes

The profound chemical changes observed in polymers under the influence of radiation result in very substantial changes of the following physical properties of the material:

(1) Crystallinity
(2) Specific gravity
(3) Coefficient of thermal expansion
(4) Specific conductivity
(5) Modulus of elasticity
(6) Permeability to gases

Plastic materials

Polyvinyl chloride, the most commonly used material in the production of transfusion assemblies is a quite unattractive material from a toxicological point of view, because it contains a wide range of various compounds that may leak into fluids in contact with it [31]. The risk of sterilization-induced toxicity is increased with the chemical complexity of the material. It has been reported that the patients developed tracheal stenosis following tracheostomy in which a polyvinyl chloride cuffed tracheostomy tube was used [32]. The polyvinyl chloride tubes had been sterilized with gamma rays and subsequently sterilized with ethylene oxide. The chemical irritants that were involved were ethylene oxide, ethylene glycol produced by the interaction of moisture with ethylene oxide, and ethylene oxide with chloride ions released from polyvinyl chloride by gamma irradiation.

All nylons contain a –CO–NH– group that can be hydrolysed, and the breakdown products produce a greater tissue reaction [33]. However, radiation shows no significant change in nylon up to about 2.5 Mrad [34]. Some products made of polypropylene become fragile after the irradiation according to their material composition, and those made of teflon and fluoroethyl propylene are unsuitable for radiosterilization. Those made of polystyrene are generally suitable. Polyethylene shows little damage on irradiation. Polyester is resistant to radiation and no change in strength up to 2.5 Mrad has been reported. Segmented polyurethanes and fluorocarbons, though resistant to radiation, have poor strength. Gamma-irradiated Tygon tubing has been found to have certain advantages over other methods of sterilization for use in heart-lung by-pass machines [35]. The use of a radiation dose of 2.5 Mrad has been shown to sterilize a Starr-Edwards valve prosthesis. At this dose there is little change in Teflon, although some reduction in tensile strength occurs.

Plastic irradiation containers for liquids carry the risk of introducing chemical contamination which might lead to false results [36-46]. Different specimens of what is nominally the same plastic material may contain different impurities and hence may differ in their reliability. Ideally, plastic irradiation containers should only be used when experiments have demonstrated their suitability. The risk of contamination by polyethylene from any source, when suitably cleaned, is small.

BIOLOGICAL SYSTEMS

Macromolecules

Proteins are, in general, radiosensitive. All the proteins are denatured by ionization [18, 19, 47]. Many other physical effects have been observed, such as changes in optical rotation, refractive index, surface tension, electrophoretic mobility and electrical conductivity.

All the enzymes are inactivated when irradiated in solution [1]. Many parts of the molecule can react without, however, producing a change in biological activity. In such an aqueous medium, enzyme radiosensitivity is affected by solute concentration, the presence of other molecules, pH, temperature and enzyme conformation [47]. The inactivation of enzymes in dilute solution by indirect action is not affected by the presence of oxygen.

When DNA is irradiated in solution, the viscosity decreases [18, 19, 48, 49]. When DNA is irradiated in dry state, cross-linking and degradation both take place. In the absence of oxygen, cross-linking predominates and DNA is converted into gel. In the presence of oxygen only degradation is seen.

It is often possible to measure functional changes in molecules when they are not a part of a living system [19]. It is not always correct to extrapolate these findings to a living situation where the molecule may be in a different chemical form and may be surrounded by other molecules with differing radiosensitivities and protective capacities. Water participates in the development of cellular radiation damage [50] and for many lesions the radiosensitivity towards sparsely ionizing radiations is some two to three times greater in air than in the complete absence of air.

Microorganisms

Most microorganisms are relatively radioresistant, especially in the vegetative stages [28]. The lethal doses depend on the experimental conditions. For example, oxygen presence or absence influences the radiosensitivity of microorganisms. Radiosensitivity also depends on the stage of the growth cycle of the microorganisms at the time of irradiation. This apparent variation in the sensitivity at different stages may be related to repair ability.

The radiation doses required to produce measurable changes in the common metabolic processes are higher than those necessary to decrease survival significantly. Many of the examples of altered metabolic function are probably related to changes in cell membranes.

In general, it is accepted that viruses are more resistant than bacterial spores, resistance increasing with decreasing particle size, and in turn spores are more resistant than vegetative organisms, yeasts and moulds. Chemicals like hydrogen sulphide, aliphatic alcohols, glycerol, dimethyl sulphoxide and thiourea reduce the lethal effects of radiation. The presence of a pharmaceutical can alter the effect of radiation on microorganisms.

Tissues

Various tissues of an animal differ in their response to radiation. Part of this variation in radiosensitivity may be attributed to differences in the sensitivities of the cells which make up the tissues. In addition, apparent tissue sensitivity is influenced by such factors as normal turnover time of the cells, interaction between the various types of cells within a tissue, the ability of the tissue to replace or to repair damaged cells, and the reserve capacity of the tissue [19].

The heart and the large arteries and veins are relatively radioresistant. The endothelium of capillaries is radiosensitive; occlusion of capillaries and small arteries is produced by moderate doses of radiation. An increase in permeability of capillaries has been demonstrated after moderate doses of radiation. It has been postulated that this is due to a depolymerization of the mucopolysaccharide "cement substance".

Non-growing portions of bone are relatively radioresistant, but the epiphyseal-diaphyseal plate in growing long bones can be damaged by moderate radiation doses. If large doses of radiation are given to a growing bone, most cartilage cells degenerate, osteoblasts are destroyed, and an acellular, avascular bone-like substance may be formed from the cartilage. This will gradually be resorbed and replaced by bone to form a solid plate which is incapable of growth. The result will be shortened bone. Irradiation of formed bone may result in a derangement of the synchronization of resorption and new bone deposition. This may produce either excessive absorption or an overgrowth of bone. In irradiated bone, fractures are common owing either to structural damage to the bone matrix or to defective bone mineralization.

Sutures

Catgut, the main absorbable suture consists essentially of collagen and is derived from sheep or goat intestinal submucosa or bovine intestinal serosa [34]. Structurally, collagen consists of three polypeptide chains bound together in the form of a triple helix, primarily through intramolecular hydrogen bonds. Collagen irradiated dry is degraded by the direct action of radiation resulting in random scission of the polypeptide chain [51, 52]. As a result of this degradation collagen becomes more easily attached by non-specific proteases, its shrinkage temperature is lowered, there is loss of crystallinity and its mechanical properties are also degraded. The damage is generally dose-related. However, even at high doses, there is little destruction of the amino acids. Irradiation in the wet state is less damaging because concurrent cross-linking tends to counteract the scission process. The presence of radical scavengers increases radiation damage by preventing the formation of these cross-links. The introduction of covalent cross-links prior to irradiation increases the radiation resistance of collagen. This is particularly important in the case of radiosterilization of sutures because most catgut and collagen sutures are treated with formaldehyde and basic chromium salts prior to irradiation.

Serum

Serum albumin, when irradiated in aqueous solutions undergoes a simultaneous polymerization and degradation. There is also denaturation. On irradiation in solid state only polymerization takes place. Lyophilized human serum is stable to irradiation even above 2.5 Mrad. The biological activity of sera with immunoglobulins and purified immunoglobulin is lost on irradiation in solution while it is retained when irradiated in lyophilized state. Physicochemical changes are observed under all conditions. Anti-A isoagglutinin is more radiosensitive than the anti-B isoagglutinin. Fibrinogen is more radiosensitive than serum albumin or the globulin fractions when irradiated either in an aqueous solution or in the lyophilized state. When measured in the blood coagulation system the biological activity of prothrombin and thrombin appreciably decreases after irradiation with doses of 1.5 to 2.0 Mrad, while the biological activity of thrombokinase does not decrease even after irradiation with 2.5 Mrad.

PHARMACEUTICALS

Generally, pharmaceuticals are considerably more stable in the dry solid state to ionizing radiations than in aqueous solutions or in any other form of molecular aggregation and only small losses in activity are found under these conditions [5]. While the drugs when irradiated alone or in a normal vehicle may themselves be stable to the sterilizing dose, there are frequently undesirable effects found as a result of chemical changes induced in the vehicle.

A number of ointment formulations containing hydrocortisone and neomycin and one containing hydrocortisone acetate and chloramphenicol have been found to be unchanged chemically toxicologically. Radiation severely degraded insulin preparations. Polypeptide and proteinaceous hormones are not as stable to radiation as those containing a steroid nucleus.

A variety of degradation products are formed in aqueous solution of monosaccharides by the effect of ionizing radiation. Irradiation, generally, induces the formation of acids and ring splitting, the latter yielding aldehydes with 2 and 3 carbon atoms. The degree of the splitting was found to be lower after irradiation in the absence of oxygen as compared with irradiation in the presence of oxygen. The initial concentration of hexose influences the degree of the ensuing changes. Solid-state D-glucose is extremely sensitive to gamma radiation owing to energy transfer processes. As a rule, the oxygen bridge is split upon irradiation of oligosaccharides such as maltose, lactose and raffinose.

When irradiating the cellulose molecule, dehydrogenation and decomposition lead to the release of gaseous hydrogen, carbon monoxide and carbon dioxide. Moisture and oxygen, when present during or after irradiation, affect the radiation products. The degradation products are more in the presence of oxygen. Cellulose that has reacted with aromatic groups is generally less radiosensitive.

Normal saline solutions, when sterilized in glass, polystyrene or polyethylene containers, show no appreciable pH drop or any other unfavourable effect [21]. However, in polyvinyl chloride containers, the pH drop is quite significant.

The effect of radiation-sterilization doses on pharmaceuticals is summarized as follows:

Dry pharmaceuticals	–	mostly stable
Ointments	–	mostly stable
Aqueous solutions of pharmaceuticals	–	mostly unstable.

ACKNOWLEDGEMENTS

The author is grateful to Dr. K.G. Vohra for his keen interest in this work.

REFERENCES

[1] IAEA, Radiosterilization of Medical Products, Pharmaceuticals and Bioproducts (Proc. Panel, Vienna, 1966), IAEA, Vienna (1967).

[2] Ionizing Radiation and the Sterilization of Medical Products, Taylor and Francies Ltd., Denmark (1964).

[3] IAEA, Radiosterilization of Medical Products (Proc. Symp. Budapest, 1967), IAEA, Vienna (1967).

[4] Bull. Inform. Ass. Tech. Energ. Nucl. 93 (1972).

[5] IAEA, Radiation Sterilization of Medical and Biological Materials, IAEA-159. Unpubl. Doc. (1973).

[6] BARC, Radiation Sterilization of Medical Products, Bhabha Atomic Research Centre, Bombay (1973).

[7] SCHWARZ, H.A., J. Phys. Chem. 73 (1969) 1928.

[8] APPLEBY, A., SCHWARZ, H.A., ibid., p.1937.

[9] SPINKS, J.W.T., WOODS, R.J., An Introduction to Radiation Chemistry, John Wiley and Sons, New York (1964).

[10] LIND, S.C., Radiation Chemistry of Gases, Reinhold, New York (1961).
[11] VERESHCHINSKII, I.V., PIKAEV, A.K., Introduction to Radiation Chemistry, Israel Program for Scientific Translations, Jerusalem (1964).
[12] SWALLOW, A.J., Radiation Chemistry of Organic Compounds, Pergamon Press (1960).
[13] ALLEN, A.O., The Radiation Chemistry of Water and Aqueous Solutions, Van Nostrand, Princeton, N.J. (1961).
[14] STEIN, G., Radiation Chemistry of Aqueous Systems, Weizman Science Press of Israel, Jerusalem (1968).
[15] DRAGANIĆ, I.G., DRAGANIĆ, Z.D., The Radiation Chemistry of Water, Academic Press, New York (1971).
[16] BURTON, M., SMITH, J.S.K., MAGEE, J.L., Comparative Effects of Radiation, John Wiley and Sons, Inc., New York (1960).
[17] LEA, D.E., Action of Radiation on Living Cells, Cambridge University Press, London, 2nd Edn (1955).
[18] ALEXANDER, P., BACQ, Z.M., Fundamentals of Radiobiology, Pergamon Press (1963).
[19] CASARETT, A.P., Radiation Biology, Prentice-Hall, Inc., Englewood, Cliffs, New Jersey (1968).
[20] ALTMAN, K.I., GERBER, G.B., OKADA, S., Radiation Biochemistry, Academic Press, New York (1970).
[21] GOPAL, N.G.S., RAJAGOPALAN, S., SHARMA, G., Ref.[6], p.105.
[22] HART, E.J., ANBAR, M., The Hydrated Electron, John Wiley and Sons, New York (1970).
[23] HART, E.J., Radiat. Res. Rev. 3 (1972) 285.
[24] FARAGGI, M., Int. J. Radiat. Phys. Chem. 5 (1973) 197.
[25] SEHESTED, K., BJERG BAKKE, E., FRICKE, H., Radiat. Res. 56 (1973) 385.
[26] BUXTON, G.V., Radiat. Res. Rev. 1 (1968) 209.
[27] GROSSWEINER, L.I., Radiat. Res. Rev. 2 (1970) 345.
[28] BOVEY, F.A., The Effects of Ionizing Radiation on Natural and Synthetic High Polymers, Interscience Publishers Inc., New York (1958).
[29] CHARLESBY, A., Atomic Radiation and Polymers, Pergamon Press, Oxford (1960).
[30] CHAPIRO, A., Radiation Chemistry of Polymeric Systems, Interscience Publishers, Inc., New York (1962).
[31] EMBORG, C., Ref.[5], p.191.
[32] LIPTON, B., GUTIERREZ, R., BLAUGRUND, S., LITWAK, R.S., RENDELL, B.L., Anesth. Analg. 50 (1971) 578.
[33] LITTLE, K., Ref.[5], p.199.
[34] ARTANDI, C., Ref.[5], p.173.
[35] ALLANDINE, M.F., GIBBONS, J.R.P., Ref. [3], p.285.
[36] GIBBONS, J.R.P., ALLANDINE, M.F., Ref.[3], p.289.
[37] HAYBITTLE, J.L., SAUNDERS, R.D., SWALLOW, A.J., J. Chem. Phys. 25 (1956) 1213.
[38] HALL, E.J., OLIVER, R., Brit. J. Radiol. 34 (1961) 397.
[39] KEENE, J.P., LAW, J., Phys. Med. Biol. 8 (1963) 83.
[40] DAVIES, J.V., LAW, J., Ibid., p.91.
[41] DUFFY, T.L., KASPER, R.B., Hlth Phys. 14 (1968) 45.
[42] PETTERSSON, C., HETTINGER, G., Acta Radiol. (Therapy) 6 (1967) 160.
[43] SEVENSSON, C., PETTERSON, C., HETTINGER, G., in Solid State and Chemical Radiation Dosimetry in Medicine and Biology (Proc. Symp. Vienna, 1966), IAEA, Vienna (1967) 251.
[44] LAW, J., REDPATH, A.T., Phys. Med. Biol. 13 (1968) 371.
[45] LAW, J., Phys. Med. Biol. 15 (1970) 117.
[46] MANFRED, P., in Proc. Annual Meeting of Radiation Protection Experts, Berlin, Fed. Rep. of Germany, 28-30 May 1969.
[47] LUSE, R.A., Radiat. Res. Suppl. 4 (1964) 192.
[48] KAPLAN, H.S., Actions Chim. Biol. Radiat. 12 (1968) 71.
[49] HUTCHINSON, F., Ibid., p.107.
[50] POWERS, E.L., TALLENTIRE, A., Ibid., p.3.
[51] CASSEL, J., J. Am. Leath. Chem. Ass. 54 (1959) 432.
[52] KUNTZ, E., WHITE, E., Fed. Proc. 20 (1961) 376.

DISCUSSION

H.M. ROUSHDY: You state in the section on macromolecules that all enzymes are deactivated by irradiation. This has not been correlated with radiation dose level, as is the case with all the other biological effects of radiation described. It might therefore be better if oxidative and digestive enzymes were considered independently in terms of their radiation dose dependence.

A second point is that you mentioned that radiation exposure causes complete dissolution of cartilage, which is then replaced by bone. Is this type of replacement a process of ossification, or bone formation, which necessitates an active proliferation of osteoblasts, and, if so, what could be the radiation effect on bone-forming cells and the latent time then required to recover from damage after exposure to the high radiation level causing complete dissociation of cartilage tissue?

B.L. GUPTA: What I said was that all the enzymes were deactivated in solution. At the high doses encountered in radiation sterilization, any differences between oxidative and digestive enzymes might not be very important in dilute solutions.

As for your second point, I have mentioned in my paper that cartilage cells and osteoblasts are both destroyed and there is no further growth, but I am not able to give you data on the exact processes involved since that is not my field.

M.A.R. MOLLA: Your statement to the effect that the heart, the large arteries and veins are radioresistant, and that the endothelium of capillaries is radiosensitive, strikes me as somewhat misleading. Experience has shown that all biological tissues are radiosensitive, though some are apparently more sensitive than others. The dividing cells and tissues show up the biological effect sooner after exposure than the non-dividing cells and tissues.

B.L. GUPTA: Yes, I agree that most biological tissues are sensitive to radiation. What I really meant to say in my statement was that the sensitivity was relative.

H.M. ROUSHDY: I think it should be stressed that no living tissue can be regarded as absolutely radioresistant. Radioresistance is only a relative phenomenon in radiation biology. Any ambiguity in the use of the term could perhaps be resolved if the tissues termed "radioresistant" were referred to as "more radioresistant" than other reference tissues.

B.L. GUPTA: Thank you, Mr. Roushdy, for the comment. I agree with your suggestion.

E.P. PAVLOV: Perhaps I could add a brief comment on the question of what is termed the radioresistance of biomaterials. In the Soviet Union we have had a certain amount of experience with radiosterilized bone transplants. These materials are usually sterilized by γ-irradiation. Naturally, I am referring to non-living tissues. The homotransplants have been carefully investigated and now are being successfully applied in traumatology and orthopaedics.

РАЗРАБОТКА ПРИЕМОВ ПРОГНОЗИРОВАНИЯ СВОЙСТВ ПОЛИМЕРНЫХ МАТЕРИАЛОВ УПАКОВКИ, НЕ ВЫЗЫВАЮЩИХ ИЗМЕНЕНИЙ В ЛЕКАРСТВЕННЫХ ПРЕПАРАТАХ ПРИ ИХ СТЕРИЛИЗАЦИИ

А.И.КРИВОНОСОВ, М.В.ПОСТРИГАНЬ
Всесоюзный научно-исследовательский институт медицинских полимеров,
Москва,
Союз Советских Социалистических Республик

Доклад представлен Э.И.Семененко

Abstract–Аннотация

DEVELOPMENT OF METHODS OF FORECASTING PROPERTIES OF POLYMER PACKAGING MATERIALS WHICH DO NOT CAUSE ANY CHANGES IN MEDICINAL PREPARATIONS DURING STERILIZATION.

The authors present the results of an investigation of the effect of gamma radiation on the properties of low-molecular moulded polymethyl methacrylates (molecular weight distribution, physico-mechanical properties, resistance to the effects of aqueous media and physiological solution) in the form of copolymers of methyl methacrylate with methyl acrylate (MA) or butyl acrylate (BA) with the brand names Dacryl-4B, Dacryl-2M and Dacryl-4M (4% BA; 2 and 4% MA respectively) having a molecular weight of around 10^5. For comparison, block polymethyl methacrylate with a molecular weight of 3×10^6 was used. The results of experiments on the ageing of unirradiated and irradiated specimens in a heated store-room (t = 10 - 20°C) and at high temperatures (90°C) are also given. It is shown that irradiation with doses of 2.5 - 20 Mrad leads to radiolysis of the above copolymers, accompanied by destructive processes. Evidence of this is an increase in the polydispersity of the materials and a reduction in their mean molecular weight. These changes impair their physicomechanical characteristics to some extent. Irradiation with a sterilizing dose of 2.5 Mrad reduces the strength indices by around 15%, which is quite acceptable for polymer products having high initial physicomechanical characteristics. Irradiation has practically no effect on the molecular and strength characteristics of Dacryl products stored in a heated room, and the change in physicomechanical properties resulting from storage at high temperatures is practically the same for unirradiated and irradiated specimens. On the basis of these results the materials investigated can be recommended for the packaging of medicinal preparations.

РАЗРАБОТКА ПРИЕМОВ ПРОГНОЗИРОВАНИЯ СВОЙСТВ ПОЛИМЕРНЫХ МАТЕРИАЛОВ УПАКОВКИ, НЕ ВЫЗЫВАЮЩИХ ИЗМЕНЕНИЙ В ЛЕКАРСТВЕННЫХ ПРЕПАРАТАХ ПРИ ИХ СТЕРИЛИЗАЦИИ.

В докладе приводятся результаты исследования влияния γ-облучения на свойства низкомолекулярных литьевых полиметилметакрилатов (молекулярно-весовое распределение (МВР), физико-механические свойства, стойкость к воздействию водной среды и физиологического раствора), представляющих собой сополимеры метилметакрилата с метилакрилатом (МА) или бутилакрилатом (БА) марок Дакрил-4Б, Дакрил-2М и Дакрил-4М (4%БА; , 2 и 4 %МА, соответственно) с молекулярным весом около 10^5. Для сравнения взят блочный полиметилметакрилат (ПММА) с молекулярным весом $3 \cdot 10^6$. Приводятся также результаты старения необлученных и облученных образцов в условиях отапливаемого склада (t = 10 ÷ 20°C) и при повышенных температурах (90°C). В докладе показно, что при облучении дозами 2,5 ÷ 20 Мрад происходит радиолиз указанных сополимеров, сопровождаемый процессами деструкции. Об этом свидетельствует увеличение полидисперсности исследованных материалов и снижение их среднего молекулярного веса. Эти изменения приводят к некоторому ухудшению физико-механических характеристик. При облучении стерилизующей дозой 2,5 Мрад снижение прочностных показателей составляет около 15 %, что вполне допустимо при эксплуатации изделий из полимеров, имеющих высокие исходные физико-механические показатели. Облучение Дакрилов практически не сказывается на молекулярных и прочностных характеристиках, при хранении в условиях отапливаемого склада, а при хранении при повышенных температурах одинаково как для необлученных, так и облученных стерилизующей дозой образцов. Результаты работы позволяют рекомендовать исследованные материалы для упаковки лекарственных препаратов.

Необходимость стерилизации лекарственных препаратов ставит задачу изучения воздействия γ-излучения на свойства материала упаковки. Наиболее дешевым, удобным и перспективным видом упаковки является упаковка лекарственных препаратов в изделия из полимерных материалов.

При γ-облучении в полимерах происходят определенные изменения, которые могут влиять как на свойства лекарственных препаратов, вследствие появления новых функциональных групп и низкомолекулярных фракций, способных вымываться в лекарственные среды, так и на физико-механические характеристики полимерной упаковки. Радиационно-химические превращения в полимерах при интегральной дозе облучения 2,5 Мрад, как правило, невелики, однако, при длительном хранении облученных изделий постэффекты могут приводить к ряду изменений физико-химических и физико-механических свойств в полимерах, нежелательных с точки зрения эксплуатации.

Вопрос прогнозирования свойств полимеров и изделий из них является одним из наиболее сложных и мало изученных вопросов общей проблемы исследования старения полимеров. Предлагаемый ряд аналитических выражений носит в большинстве случаев эмпирический характер и является справедливым для какого-либо одного или нескольких полимеров.

В настоящее время отсутствует единый научно-обоснованный подход к решению этого вопроса, что, в частности, затрудняет сопоставление результатов, полученных различными авторами при исследовании одного и того же полимера.

Обычно в тех случаях, когда при хранении полимерные материалы или изделия из них подвергаются только тепловому воздействию, а действием других внешних факторов (нагрузок, влажности воздуха, солнечной радиации, химических сред и т.д.) можно пренебречь, с целью прогнозирования сроков службы полимеров проводят их ускоренное старение при повышенных температурах, используя принцип температурно-временной эквивалентности.

Вопрос разработки приемов прогнозирования изменения свойств полимерных материалов медицинского назначения осложняется необходимостью изучения радиационно-химических превращений при лучевой стерилизации и их влияния на процесс старения при длительном хранении.

Объектами настоящего исследования являются низкомолекулярные литьевые полиметакрилаты, представляющие собой сополимеры метилметакрилата с метилакрилатом (МА) или бутилакрилатом (БА) марок Дакрил-4Б, Дакрил-2М и Дакрил-4М (4 % БА ; 2 и 4 % МА, соответственно) с молекулярным весом около 10^5 . Для сравнения взят блочный полиметилметакрилат (ПММА) с молекулярным весом $3 \cdot 10^6$. Эти полимеры представляют интерес для упаковки лекарственных препаратов вследствие хороших санитарно-гигиенических свойств, высокой прозрачности, ценных физико-механических характеристик и т.д. Полимеры в виде гранул диаметром около 2 мм или стандартных образцов (лопатки и бруски) подвергали облучению в присутствии кислорода воз-

духа на изотопном источнике ^{60}Co дозами от 2,5 до 50 Мрад при мощности дозы 1 Мрад/ч. Исследовано влияние γ-облучения на молекулярно-весовое распределение (МВР), молекулярный вес (M_v), физико-механические свойства и устойчивость к тепловому старению.
МВР оценивали по данным турбидиметрического титрования в системе ацетон-вода при 25 ± 0,1 °С. Дифференциальные кривые распределения (ДКР) по растворимости получали численным дифференцированием по Тайхгреберу [1] и вычисляли фактор формы ДКР [2], характеризующий полидисперсность полимера

$$F = \frac{\gamma_{max}}{\lg \frac{\Delta\gamma}{h/2}}$$

где γ_{max} – объемная доля осадителя, соответствующая максимуму ДКР: $\Delta\gamma$ – полуширина кривой на высоте $h/2$.

Вискозиметрический молекулярный вес определяли из измерений характеристической вязкости раствора $[\eta]$ в хлороформе при 25 °С по одной точке (концентрация раствора 0,3 %). Молекулярный вес вычисляли по уравнению $[\eta] = 4{,}8\cdot 10^{-5}\cdot M_v^{0,8}$; $[\eta]$ раствора концентрации С определялась по уравнению: $[\eta] = \frac{\sqrt{2}}{C}\sqrt{\eta_{уд} - \ln \eta_{отн}}$, где $\eta_{уд}$ и $\eta_{отн}$ соответственно удельная и относительная вязкость раствора. Физико-механические свойства определяли в соответствии с существующими ГОСТами. Доверительная вероятность результатов испытаний $\alpha = 0{,}95$. Старение полимеров проводили в условиях отапливаемого склада при температуре 10 ÷ 30 °С и относительной влажности около 65 %, а также в термокамере при температуре 90 ± 2 °С. Перед испытанием стандартные образцы кондиционировали.

На рис.1 приведены ДКР по растворимости исследуемых полимеров, цифры у кривых указывают дозу облучения. Из рисунка видно, что необлученные Дакрилы имеют близкое распределение, причем более широкое, чем распределение ПММА. Облучение приводит к снижению среднего молекулярного веса (сдвиг максимума ДКР в сторону больших значений γ) и увеличению полидисперсности (рис.2). Последнее объясняется тем, что, как показано для деструктирующихся при радиолизе полимеров, их МВР стремится к наиболее вероятному [3]. Изменение содержания метил- или бутилакрилатных звеньев от 2 до 4 % не влияет на характер изменения полидисперсности Дакрилов от дозы облучения. Облучение Дакрилов и ПММА стерилизующей дозой не вызывает существенного увеличения суммарной весовой доли низкомолекулярных фракций, заметное возрастание которых обнаруживается при дозе 20 Мрад и резко возрастает при дозе 50 Мрад.

Характер изменения M_v Дакрилов от дозы облучения (рис.3) свидетельствует о том, что как и в случае высокомолекулярного ПММА, при радиолизе Дакрилов разрыв макромолекул происходит по закону случая [3]. Отклонение представленной зависимости от линейной в области малых доз облучения свидетельствует о том, что исходные полимеры не имеют наиболее вероятного МВР.

Уменьшение среднего молекулярного веса и возрастание полидисперсности приводит к ухудшению их прочностных характеристик (табл.1). При этом предел прочности при растяжении является линейной функцией фактора формы F (рис.4), что позволяет предсказывать прочность изделий из этих полимеров на основании исследования молекулярных

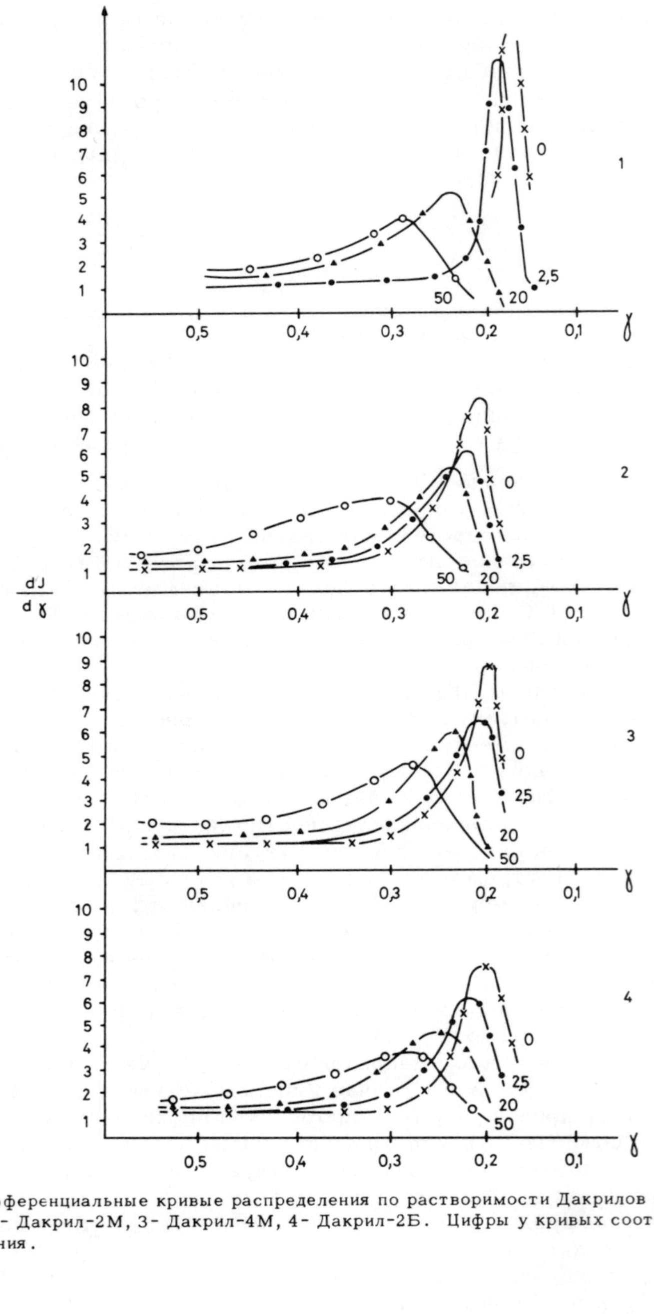

Рис.1. Дифференциальные кривые распределения по растворимости Дакрилов и ПММА: 1- ПММА, 2- Дакрил-2М, 3- Дакрил-4М, 4- Дакрил-2Б. Цифры у кривых соответствуют дозе облучения.

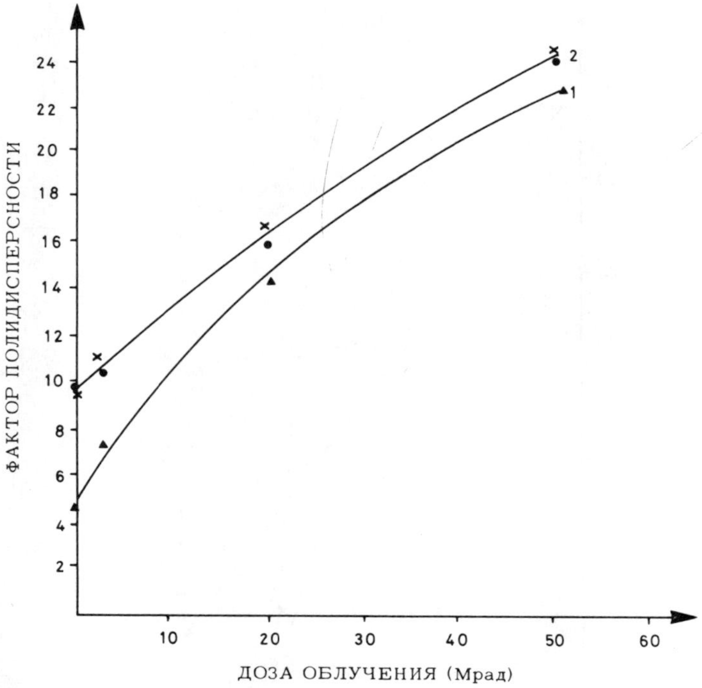

Рис.2. Влияние дозы облучения на полидисперсность ПММА (1) и Дакрилов (2).

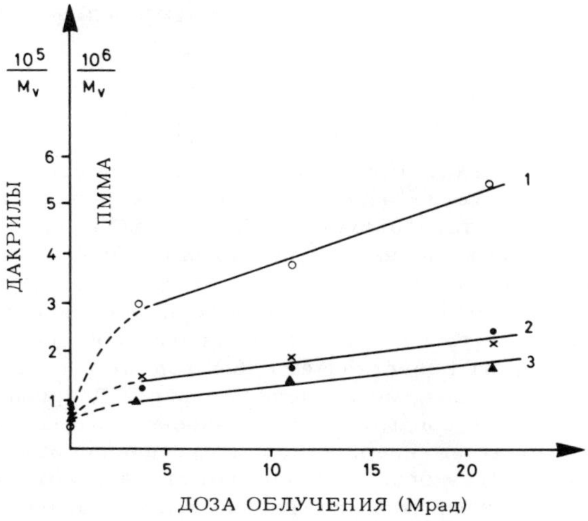

Рис.3. Изменение молекулярного веса под действием облучения: 1- ПММА, 2- Д-2М,Д-4Б, 3- Д-4М.

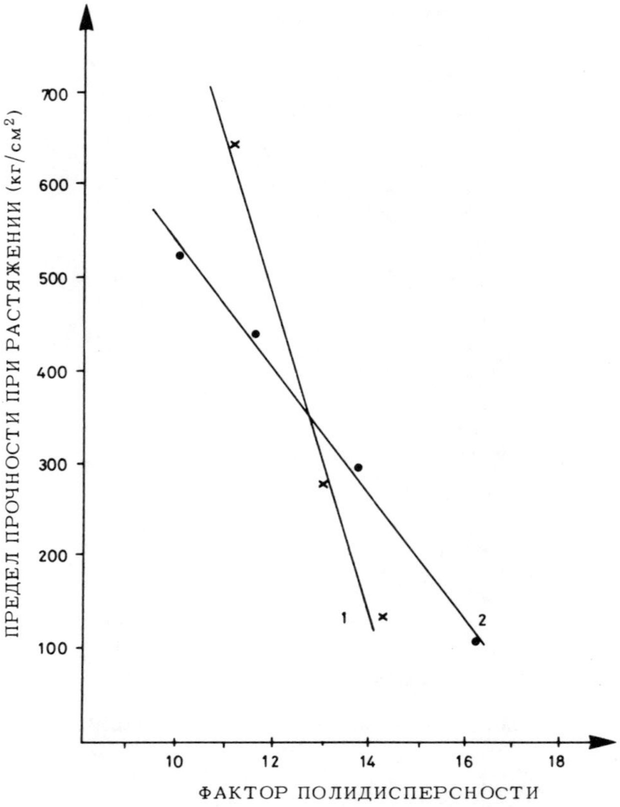

Рис.4. Зависимость прочности при растяжении Дакрил-2М(1) и Дакрил-4Б(2) от полидисперсности.

характеристик материала. При облучении Дакрилов стерилизующей дозой 2,5 Мрад уменьшение значений приведенных в табл.I прочностных характеристик составляет около 15 %, что вполне допустимо при эксплуатации изделий из полимеров, имеющих высокие исходные физико-механические показатели.

В связи с тем, что эксплуатация полимерных материалов упаковки происходит в условиях воздействия лекарственной среды и значительная часть лекарств представляет собой водные растворы, была произведена оценка физико-механических свойств Дакрила после 40 суток воздействия водной среды и физиологического раствора (табл.II). Показано, что водопоглощение полимера по истечении этого срока составляет около 1 %. Прочность при разрыве, статическом изгибе и удельная ударная вязкость как необлученных, так и облученных стерилизующей дозой образцов, после 40 суток выдерживания в воде и физиологическом растворе практически не меняется.

ТАБЛИЦА I. ИЗМЕНЕНИЕ ФИЗИКО-МЕХАНИЧЕСКИХ СВОЙСТВ ДАКРИЛОВ ПОД ДЕЙСТВИЕМ ОБЛУЧЕНИЯ

Полимер	Свойства	Доза облучения (Мрад)						
		0	1,5	2,5	3,5	5,0	10	20
Дакрил-2М	Предел прочности при растяжении (кгс/см^2)	645	550	542	480	450	280	140
	Предел прочности при статическом изгибе (кгс/см^2)	1295	1100	1100	1050	1030	770	-
	Удельная ударная вязкость (кгс·см/см^2)	17,6	-	16,4	16,0	15,5	11,6	-
Дакрил-4Б	Предел прочности при растяжении (кгс/см^2)	530	495	440	402	392	295	142
	Предел прочности при статическом изгибе (кгс/см^2)	1190	-	1100	1090	1050	790	-
	Удельная ударная вязкость (кгс·см/см^2)	16,0	15,9	15,8	14,6	13,8	12,6	-

ТАБЛИЦА II. ИЗМЕНЕНИЕ ФИЗИКО-МЕХАНИЧЕСКИХ СВОЙСТВ ДАКРИЛ-4Б ЧЕРЕЗ 40 СУТОК ВОЗДЕЙСТВИЯ ВОДНОЙ СРЕДЫ

Свойство	Доза облучения (Мрад)	Среда		
		воздух	дистиллированная вода	физиологический раствор
Предел прочности при растяжении (кгс/см^2)	0	615	600	620
	2,5	520	525	513
Предел прочности при статическом изгибе (кгс/см^2)	0	1048	1115	1080
	2,5	920	1090	1100
Удельная ударная вязкость (кгс·см/см^2)	0	13,4	12,0	11,9
	2,5	12,0	12,2	12,4
Привес (%)	0	-	0,99	1,02
	2,5	-	1,00	1,10

ТАБЛИЦА III. ИЗМЕНЕНИЕ НЕКОТОРЫХ СВОЙСТВ ДАКРИЛОВ ПРИ СТАРЕНИИ*

Свойства	Исходные	Через 2,5 года хранения в условиях отапливаемого склада	Через 120 суток при 90°C
Дакрил-2М			
Молекулярный вес $M_v \cdot 10^{-3}$	76,6/67,5	75,2/66,3**	73,9/66,5
Полидисперсность (F – фактор)	11,1/12,0	-	11,1/12,3
Предел прочности при растяжении (кгс/см^2)	640/510	638/510	-
Предел прочности при статическом изгибе (кгс/см^2)	1195/1105	1295/1103	950/498
Удельная ударная вязкость (кгс·см/см^2)	17,5/16,3	17,7/16,5	11,4/11,4
Дакрил-4Б			
Молекулярный вес $M_v \cdot 10^{-3}$	76,4/67,5	75,0/66,5**	73,5/64,3
Полидисперсность (F -фактор)	10,3/10,8	-	10,8/11,2
Предел прочности при растяжении (кгс/см^2)	530/440	500/460	-
Предел прочности при статическом изгибе (кгс/см^2)	1190/1070	1069/1000	750/870
Удельная ударная вязкость (кгс·см/см^2)	16,0/15,5	16,0/15,1	10/10

Примечание: * числитель – необлученный полимер, знаменатель – полимер, облученный дозой 2,5 Мрад; ** через 1 год хранения в условиях отапливаемого склада.

Исследование старения необлученных и облученных стерилизующей дозой Дакрилов как в условиях отапливаемого склада, так и при длительном воздействии повышенных температур показало, что облучение Дакрилов стерилизующей дозой практически не сказывается на среднем молекулярном весе и полидисперсности (табл.III). Практически неизменными остаются величины, характеризующие физико-механические свойства через 2,5 года хранения в условиях отапливаемого склада. Изменение прочностных характеристик при воздействии повышенных температур объясняется, возможно, изменениями в структуре исследованных полимеров. Величина относительного изменения предела прочности при статическом изгибе и удельной ударной вязкости как для исходных, так и для облученных стерилизующей дозой Дакрилов примерно одинакова.

ВЫВОДЫ

1. Показано, что радиолиз низкомолекулярных сополимеров метилметакрилата с 2 и 4% метил- или бутилакрилата сопровождается процессами деструкции, также как и радиолиз ПММА.

2. Изменение среднего молекулярного веса и МВР под действием облучения приводит к ухудшению физико-механических свойств Дакрилов.

3. Уменьшение значений прочностных характеристик Дакрилов, облученных стерилизующей дозой, составляет около 15%, что допустимо при эксплуатации изделий из полимеров, имеющих высокие исходные физико-механические показатели.

4. Хранение в условиях отапливаемого склада в течение 2,5 лет как необлученных, так и облученных дозой 2,5 Мрад Дакрилов практически не сказывается на их молекулярно-весовых характеристиках и прочностных свойствах.

5. Ухудшение физико-механических свойств Дакрилов при тепловом старении (90°C, 120 суток) практически одинаково как для необлученных, так и облученных стерилизующей дозой образцов.

ЛИТЕРАТУРА

[1] TEICHGRÄBER, Faserforschung und Textiltechn, 19 (6), (1968) 249.
[2] Ph., SCHERER, B.P.Rouse, Rayon and Synthetic Textiles, 30, 11, (1949) 42-44; 30, 12, (1949) 47-49.
[3] ЧАРЛЗБИ, А., Ядерные Излучения и Полимеры, Изд-во ИЛ, М., 1962 г.

APPLICATION OF RADIATION STERILIZATION TO MEDICAL PRODUCTS OF BIOLOGICAL ORIGIN

(Session IV)

Chairman: V.G. KHRUSHCHEV (USSR)

COMPARISON OF SOME PROPERTIES OF TISSUE GRAFTS STERILIZED BY COLD SHOCK AND IONIZING RADIATION*

R. KLEN, J. PÁCAL
Faculty Hospital, Tissue Bank and Medical School,
Department of Histology,
Hradec Králové,
Czechoslovakia

Abstract

COMPARISON OF SOME PROPERTIES OF TISSUE GRAFTS STERILIZED BY COLD SHOCK AND IONIZING RADIATION.

The paper compares some essential properties of grafts prepared for clinical grafting by a common procedure (sterilizing by cold shock, preservation by freeze-drying) with those experimental grafts sterilized by 2.5 Mrad gamma rays from a ^{60}Co source before or after preservation by freeze-drying. In rib-bone grafts, the course of re-hydration and the mechanical properties were tested. The results indicate that radiation sterilization is not a suitable method for bone grafts. In rib-cartilage grafts, the rates of weight and volume increasing during re-hydration were examined and the differences compared. The content of potassium and mucoproteins released into the eluate was measured during re-hydration in order to obtain some insight into the type of radiation damage. Sterilization by irradiation was again found unsuitable. In skin, chorion and fascia grafts, the permeability was evaluated. For skin and fascia grafts, the prospect of using ionizing radiation for sterilization purposes seems promising. More data are still needed to reach a definite conclusion.

There are three conceivable areas of application for ionizing radiation in the processing of tissue grafts. First is the so far rarely used radiation-induced modification of the graft, meaning a controlled change in its properties [1]. Secondly, preservation in the narrow sense, meaning an inhibition of enzymes, has not yet been used with tissue grafts prepared for clinical transplantation, though this method has been successful in increasing the storage time of some kinds of food. Finally, since even the most rigorous observance of asepsis during the dissection of tissues cannot guarantee perfect sterility of the grafts and since infection is 'Enemy No.1' in transplantation, sterilization of the graft is imperative. Ionizing radiation is a matter of choice [2]. However, the very fact that rather moderate doses of ionizing radiation are capable of inducing changes in some properties of the exposed tissue indicates that doses which are orders of magnitude higher, such as are necessary for sterilization purposes, would affect not only the contaminating microorganisms, but also the graft tissue itself. It is therefore essential to investigate the effect of sterilization doses of ionizing radiation upon those properties of the grafts that are decisive for the success of transplantation.

This paper briefly summarizes the results of our laboratory studies and compares certain basic properties of different types of grafts processed during preparation for clinical use either by conventional methods or by

* This work was partly supported by an IAEA grant.

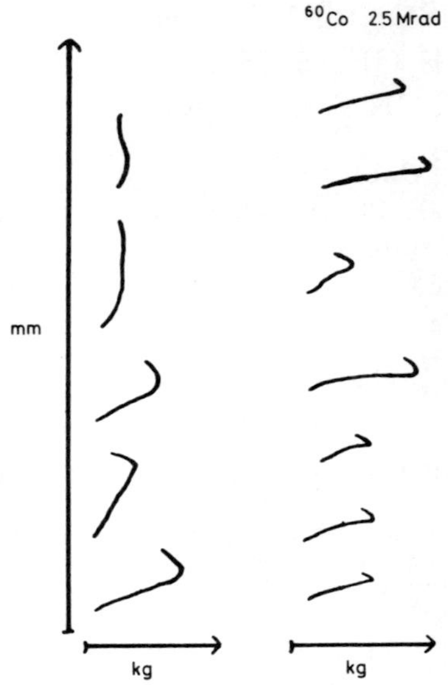

FIG. 1. Resistance to bending of the cancello-cortical bone grafts.

ionizing radiation. The grafts were sterilized by the so-called low-temperature shock [3], and the controls were gamma-irradiated with a 2.5-Mrad dose from a ^{60}Co source [4]. Indications and the technique of tissue removal, its further processing and conservation of the grafts by means of freeze-drying, were identical in both groups. Only substantial facts are given here since details have already been published elsewhere (see References).

1. CANCELLO-CORTICAL BONE GRAFTS

1.1. Weight changes

Since information on the degree of re-hydration of freeze-dried grafts is important when choosing the appropriate surgical and operational approach, we have studied the course of re-hydration by determining the rate of weight gain. The data of the measurements were evaluated statistically. It was found that re-hydration of grafts irradiated after freeze-drying proceeds at a slower rate than that of non-irradiated freeze-dried grafts. The irradiated grafts require approximately twice as long to reach the same degree of re-hydration as the non-exposed grafts.

1.2. Mechanical properties (Fig.1)

Mechanical properties of the grafts are important for the expected functioning of the transplant; besides, they can greatly influence the surgical technique used. We have therefore tested the resistance to bending by means of a tearing apparatus. The characteristics of the resulting curves indicate a decrease in the elasticity of irradiated grafts.

1.2.1. X-ray diffraction analysis (Fig.2)

To shed some light on the structural cause of the mechanical changes observed, investigations were carried out on the crystalline structure of collagen, which is the major constituent of fascia and a component critically determining most of the mechanical properties of bone. No difference was found in the diffraction patterns of irradiated and non-irradiated fasciae.

1.2.2. Electron-optic examination

Since X-ray diffraction analysis furnished no explanatory information on the observed mechanical changes, both irradiated and non-irradiated fasciae were also examined by electron microscopy. However, also no differences were detectable by this method.

The discrepancy between the positive finding of a change in mechanical properties and the negative results of X-ray diffraction and electron microscopy can best be explained by assuming that the irradiation induces a deformation of the lattice of collagen molecules, resulting in a change of the physico-chemical and mechanical properties of collagen, which is irreversibly stabilized [5].

2. CARTILAGE GRAFTS

Since the role of a cartilage graft is different from that of a bone graft, we have followed the course of re-hydration in more detail: directly, by measuring the changes in weight and volume, and indirectly by biochemical methods determining the content of potassium and mucoproteins in the solution used for re-hydration of the graft. All measurements were carried out in native grafts, freeze-dried grafts, and grafts irradiated either before or after freeze-drying.

2.1. Weight changes

Owing to freeze-drying, the cartilages lose 60% of their weight; in 30 min re-hydration they regain it. If the grafts are irradiated prior to freeze-drying, their initial weight level is restored in 2 h. Thereafter, the weight continues to increase significantly. The grafts irradiated after freeze-drying reach their initial weight in only 6 h. The weight increase again continues and the difference between the mean values after 6 and 20 h is significant. The described weight changes are more evident if one compares the rate with which the initial weight is restored. Grafts irradiated before freeze-drying re-hydrate four times more slowly and grafts irradiated after freeze-drying twelve times more slowly than non-irradiated freeze-dried grafts. Irradiation after freeze-drying prolongs re-hydration threefold in comparison with the effect of radiation applied before freeze-drying.

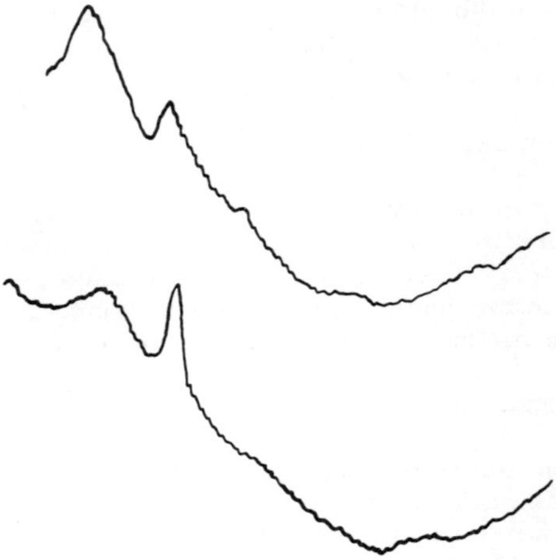

FIG.2. X-ray diffraction analysis of the fascia.

Thus, summarizing the data on weight changes of irradiated cartilages, one can conclude that (a) irradiation, particularly if it follows freeze-drying, increases the re-hydration time considerably; (b) irradiation, particularly if it precedes freeze-drying, significantly increases the weight of a cartilage graft re-hydrated for 20 h.

2.2. Volume changes

Grafts irradiated prior to freeze-drying reach their initial volumes in approximately 1 h; the volume increases further, but the difference is insignificant because of a large scattering of values. The original volumes of grafts irradiated after freeze-drying are restored in about 4 h; further increase in volume is again insignificant. The volume changes are more evident if one compares the rate with which the grafts re-hydrate to their initial volumes. The rates found for grafts irradiated after and before freeze-drying were eight and two times lower, respectively, than those of the untreated freeze-dried grafts. Comparing the grafts irradiated either before or after freeze-drying, a fourfold delay of re-hydration was found it irradiation took place afterwards.

Grafts irradiated prior to freeze-drying have a significantly greater volume compared with only freeze-dried grafts. If the increase in volume is plotted against re-hydration time, the resulting curve pertaining to grafts irradiated before freeze-drying has two peaks. The first phase ends after about 2 h, the second after about 6 h. In grafts irradiated after freeze-drying, the volume continues to rise up to 20 h, i.e. for the whole period of observation in this particular experiment. The volume of these grafts increases in the three successive phases, by 33, 21 and 13%, respectively, of the initial volume. On the other hand, if preceded by irradiation, the grafts increase after 2 h by 77%, after 6 h by 46%, and after 20 h by 1% of the initial volume.

Thus, irradiation after freeze-drying induces changes in the first and second phases that are reduced to approximately one-half those observed if irradiation is applied before freeze-drying. It is assumed that the first phase represents re-hydration of the superficial younger layer having an immediate contact with the re-hydration solution, and absorbing water much more readily than the deeper older layer which begins its re-hydration only after the superficial layer has been completely soaked up. This assumption is confirmed by macroscopic observation made 20 h after the onset of re-hydration of grafts irradiated before freeze-drying. The image is characterized by laceration of the outer layer and transformation of the inner core to a substance resembling spawn both in colour and consistency.

Thus, summarizing the data on volume changes in freeze-dried cartilages, one can conclude that (a) irradiation, particularly if it follows freeze-drying, results in a considerable prolongation of re-hydration time; (b) irradiation (the difference between the exposures before and after freeze-drying was insignificant owing to a large scattering of values) significantly increases the volume changes of cartilage grafts re-hydrated for 20 h. The increase in volume progresses in two phases.

2.3. Comparison of weight and volume changes

Since 1 ml of re-hydration solution corresponds to 1 g, the changes in weight and volume should run parallel, provided all volume changes were exclusively due to a transfer of liquid. The weight of freeze-dried grafts and of grafts irradiated after freeze-drying decreased by 60%, but their volume decreased only by 45%. In grafts irradiated before freeze-drying, the decrease in weight and volume amounted to 53 and 45%, respectively. The difference in the degree of changes in weight and in volume can most probably be attributed to the fact that the frozen and successively dried tissue is inflexible and does not collapse. Neither does it collapse when the drier drum is filled with an inert gas at a pressure of 1 atm. Some cavities inside the tissue formed by the sublimation of frozen water are occupied by the gas which, as re-hydration progresses, is gradually replaced by the liquid. The time course of volume changes in grafts irradiated after freeze-drying suggests, to a certain extent, that the replacement of gas trapped within the tissue by the liquid is slow. The time course of changes observed in grafts irradiated before freeze-drying can only be explained as a consequence of alterations induced by radiation in normal hydrated cartilage.

2.4. Biochemical examination of the eluate (Table I)

An estimate of the content of potassium and mucoproteins in the eluate during re-hydration provides indirect information on the changes occurring in the tissue. Potassium was chosen because it leaks from the cells into the re-hydration solution when the cell membranes are damaged. It was shown in a previous study that the increase in potassium content is a more sensitive indicator than the total protein content [5]. Mucoproteins, on the other hand, are substances that are represented in substantial amounts in the matrix, which has a decisive role in the ultimate fate of the transplanted cartilage graft [6]. Measurements were made at three time intervals: 1, 6, and 24 h after the onset of re-hydration. If the grafts are irradiated

TABLE I. AVERAGES TESTED BY MEANS OF t-TEST (ASSOCIATED POPULATION) OF THE POTASSIUM (K) AND MUCOPROTEIN (MP) CONTENT IN THE ELUATE OF RE-HYDRATED CARTILAGES

N = natural
L = freeze-dried
LCo = irradiated after freeze-drying
CoL = irradiated before freeze-drying

	(h)	N Ø	N 1 - 6	N 6 - 24	N 1 - 24	L Ø	L 1 - 6	L 6 - 24	L 1 - 24	LCo Ø	LCo 1 - 6	LCo 6 - 24	LCo 1 - 24	CoL Ø	CoL 1 - 6	CoL 6 - 24	CoL 1 - 24
K	1	0.095	0	0	0	0.04	++	0	0	0.1	++	++		0.05	++	++	
	6	0.12				0.07				0.27				0.13			
	24	0.14				0.07				0.71				0.33			
MP	1	0.14	0	0	0	0.7	0	0	++	0.03	0	+		1.1	/	/	0
	6	0.24				0.9				0.05				/			
	24	0.31				1.5				0.12				1.3			

0 = $p > 0.05$ + = $p < 0.05$ ++ = $p < 0.01$

TABLE II. AVERAGES TESTED BY MEANS OF t-TEST OF THE POTASSIUM (K) AND MUCOPROTEIN (MP) CONTENT IN THE ELUATE OF RE-HYDRATED CARTILAGES AFTER DIFFERENT TIME INTERVALS

		1 h	6 h	24 h
K	N : L	N	N	N
	L : LCo	LCo	LCo	LCo
	L : CoL	CoL	CoL	CoL
	LCo : CoL	0	LCo	LCo
MP	N : L	0	L	L
	L : LCo	0	L	L
	L : CoL	0	/	L
	LCo : CoL	CoL	/	CoL

N = natural; L = freeze-dried; LCo = irradiated after freeze-drying; CoL = irradiated before freeze-drying

TABLE III. SEQUENCE OF POTASSIUM AND MUCOPROTEINS AFTER DIFFERENT PROCEDURES IN THE ELUATE

Sequence according to level of K	Graft processing	Sequence according to amount of mucoproteins
1	Irradiated after freeze-drying	5
2	Frozen	4
3	Irradiated before freeze-drying	2
4	Original	3
5	Freeze-dried	1

before freeze-drying, the release of potassium is highly significant at the first and second intervals. The amount of released mucoproteins is already very high after 6 h. However, when compared with the value obtained at the 24-h interval, the difference is not significant. It may be due to an indirect effect of water radicals. If grafts are irradiated after freeze-drying, the release of potassium is significantly increased at the first two intervals. Mucoproteins, on the other hand, are significantly increased only after 6 h.

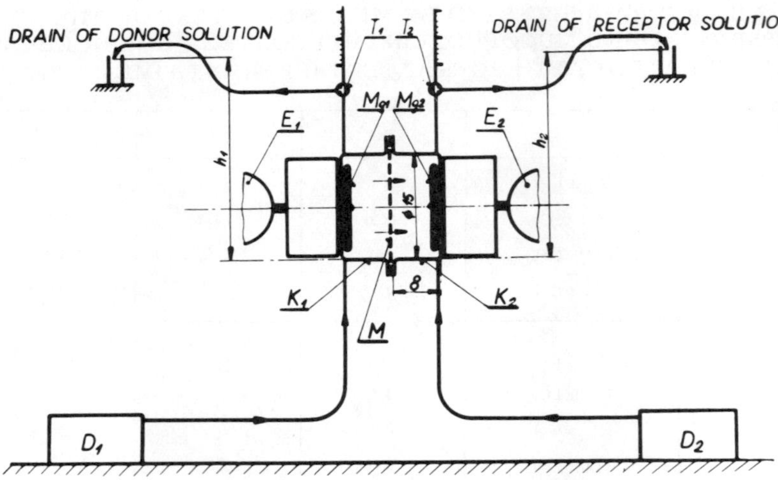

FIG.3. Scheuplein's and Blank's apparatus modified by Pácal and Sokol.

To elucidate the causal relationship between the observed changes and a particular treatment, only those results are compared that are obtained with grafts differing in one single procedure of the processing scheme (Table II). If grafts are irradiated before freeze-drying, the release of potassium is significantly higher at all three time intervals, when compared with untreated freeze-dried grafts. As to mucoprotein release, no significant difference is found at the first interval, while at the third interval the release is significantly higher from non-irradiated freeze-dried grafts; no data are at hand for the second interval. If grafts are irradiated after freeze-drying, the rate of potassium release, compared with merely freeze-dried grafts, significantly increases at all three time intervals, while that of mucoproteins is significantly lower at the 6- and 24-h intervals. Comparing the effect of irradiation either before or after freeze-drying, significant differences are observed at two time intervals, namely after 6 and 24 h. An exposure following the freeze-drying results in a high release of potassium while mucoproteins are released at a significantly higher rate if the exposure precedes freeze-drying.

Thus, summarizing the results of biochemical examinations, one can conclude that (a) irradiation, particularly if it follows the freeze-drying, raises the release of potassium, suggesting a damage inflicted primarily to chondrocytes; and (b) irradiation before freeze-drying causes a higher release of mucoproteins than an exposure after freeze-drying. Since the amount of mucoproteins released into the eluate is highest with grafts irradiated before, and lowest with grafts irradiated after freeze-drying, the effect of radiation upon the matrix evidently differs.

If the cartilage grafts processed in different ways are arranged into a sequence according to the amount of released potassium, the order, with the exception of original grafts, is the reverse of the sequence order obtained if they are arranged according to the amount of mucoproteins released (Table III).

TABLE IV. DERMO-EPIDERMAL GRAFTS

	Non-irradiated	Irradiated
No. of measurements	9	5
Mean value of the coefficient of mass permeability 10^6 k $\left[\frac{cm}{s}\right]$	3.25	1.73
Maximum and minimum value of the coefficient	7.30 - 0.61	2.0 - 0.75
Standard error	± 0.6	± 0.25
Time in which steady diffusion is reached τ [min]	20 - 90	15 - 60

TABLE V. CHORIO-AMNIOTIC GRAFTS

	Non-irradiated	Irradiated
No. of measurements	4	4
Mean value of the coefficient of mass permeability 10^4 k $\left[\frac{cm}{s}\right]$	1.43	0.88
Maximum and minimum value of the coefficient	1.96 - 0.58	1.24 - 0.48
Time in which steady diffusion is reached τ[min]	30 - 60	30 - 60

From the viewpoint of radiation biology, one finding seems particularly important, namely that irradiation is more damaging to chondrocytes if it follows the freeze-drying than if it precedes it, while the effect upon the matrix is the opposite [7].

3. PERMEABILITY STUDIES

For certain types of grafts, knowledge of permeability is very important; this applies not only to dermo-epidermal grafts and chorio-amniotic grafts [8] used for local treatment of burns, but also to fascial grafts used for plastic surgery of dura mater. In the following series of experiments, the permeability of grafts prepared for clinical use and processed in a routine way

TABLE VI. FASCIAL GRAFTS

	Non-irradiated	Irradiated
No. of measurements	6	5
Mean value of the coefficient of mass permeability $10^4 k \left[\frac{cm}{s}\right]$	1.21	0.83
Maximum and minimum values of the coefficient	1.65 - 0.60	0.99 - 0.73
Standard error	± 0.12	± 0.04
Time in which steady diffusion is reached τ[min]	5	10 - 30

already described was compared with that of control grafts, differing only in the sterilization technique. Instead of the cold shock, the control grafts were sterilized by irradiation applied after freeze-drying. Before starting the measurements, the grafts were re-hydrated for 10 min in isotonic NaCl solution.

3.1. Method

Since the methods used in all experiments discussed so far are standard, their descriptions are omitted. For our permeability studies, however, a new method has been worked out which is not commonly known, and it seems appropriate to describe it briefly.

An apparatus designed by Scheuplein and Blank [9] was modified by adding a flowing system with stirring on both sides of the membrane [10]. At the speed of 500 rev/min, the conditions on the membrane are very close to ideal; therefore, the transmission resistance on the membrane surface can be neglected. The method utilizes the first Fick's law, expressing mathematically the conditions obtaining in passive diffusion.

First Fick's Law [1]

$$m_D = k(c_1 - c_2)$$

m_D	penetration flux	$\left[\frac{mol}{cm^2 \cdot s}\right]$
c_1, c_2	concentration of donor and receptor solution	$\left[\frac{mol}{cm^3}\right]$
k	coefficient of the mass permeability	$\left[\frac{cm}{s}\right]$

If the amounts of the substance in donor and recipient solutions are known, the coefficient of mass permeability k can be calculated. The basic element of the apparatus consists of a diffusion chamber having two identical cells K_1 and K_2. The membrane to be tested is stretched so that it separates the two cells. One cell is flushed with water labelled with tritium (specific activity 2 - 10 μCi/ml), the second one with distilled water. Dermo-epidermal grafts were spread with their epidermal surface, chorio-amniotic grafts with their chorionic surface facing the cell filled with the donor solution. No particular attention was paid to the orientation of fascial grafts. After a certain time period, a steady concentration gradient, independent of time, is established. This interval is called stabilization time of a stationary process. A constant flow is secured by two dosers (D_1, D_2). The pressure difference on the membrane surface is measured by two manometers (T_1, T_2) and can be adjusted by pre-setting the height of the overflow (h_1, h_2). Magnetic stirrers (Mg_1, Mg_2) are driven by means of two electric motors (E_1, E_2) adapted to a gradual changing of rotation speed. At appropriate time intervals, samples of the donor and recipient solutions are collected at the outlet and their ^3H-activity measured by means of Nuclear Chicago Mark I scintillation counter (Fig.3).

The results presented here are tentative and experiments are at present in progress to verify them.

3.2. Dermo-epidermal grafts

Five irradiated and nine non-irradiated grafts were tested. The coefficient of mass permeability fluctuates greatly, particularly in the group of non-irradiated grafts. The mean value is equal to 3.25×10^{-6} cm/s for non-irradiated grafts, against 1.73×10^{-6} cm/s for irradiated specimens. The time required to reach the steady state ranged in non-irradiated grafts from 20 - 90 min, in irradiated grafts from 15 - 60 min (Table IV).

3.3. Chorio-amniotic grafts

Four irradiated and four non-irradiated grafts were tested. The coefficient of mass permeability varies about equally in both groups. The mean values found in non-irradiated and irradiated grafts amounted, respectively, to 1.43×10^{-4} and 0.88×10^{-4} cm/s. The time necessary to establish the steady state was 30 - 60 min for both groups (Table V).

3.4. Fascial grafts

Six non-irradiated and five irradiated samples were tested. The values of the coefficient of mass permeability vary comparatively little, in particular among irradiated grafts. The mean value for non-irradiated samples was 1.21×10^{-4} cm/s, for irradiated grafts 0.83×10^{-4} cm/s. The time necessary to reach the steady state was 5 min for all non-irradiated grafts and rose to 10 - 30 min owing to irradiation (Table VI).

As already stated, the number of measurements performed so far is not sufficient in all groups to allow complete statistical evaluation. Nevertheless, some preliminary conclusions seem justified:

(a) The radiation dose used reduced the scatter of measurement values in dermo-epidermal and fascial grafts;
(b) Permeability of dermo-epidermal grafts decreased to about one half;
(c) Permeability of dermo-epidermal grafts was by two orders of magnitude lower than that of chorio-amniotic and fascial grafts. There seems to be no difference between the latter two types of grafts;
(d) There was practically no difference in permeability between irradiated and non-irradiated fascial grafts, although the time necessary to reach steady-state diffusion was substantially greater in irradiated grafts;
(e) If further, more extensive experiments confirm the findings that radiation induces a decrease in permeability of dermo-epidermal grafts, it will become necessary to submit these favourably modified grafts to further tests, particularly in respect of their compatibility.

4. GENERAL CONCLUSIONS

Our experiments did not aim at an explanation of the mechanisms of the effects observed. This will more appropriately be studied by radiobiologists and biochemists. We have merely attempted to decide whether sterilization by means of ionizing radiation is a suitable method for processing tissue grafts prepared for clinical use. A sterilization dose of 2.5 Mrad given under the specified conditions must be considered unsuitable for cancello-cortical bone grafts and for cartilage grafts. On the basis of our preliminary results, however, this method seems promising for sterilization of dermo-epidermal and fascial grafts. More data are needed to confirm this finding.

REFERENCES

[1] Melsunger Medsche Mitt. 46 (1972) 116.
[2] IAEA, Sterilization and Preservation of Biological Tissues by Ionizing Radiation (Proc. Panel Budapest, 1969), IAEA, Vienna (1970).
[3] KLEN, R., Czechoslovak patent No. 92515.
[4] KLEN, R., HEGER, J., JEŽEK, K., Contribution to the rehydration of freeze-dried grafts. I. Re-hydration of rib-bone grafts, Acta Chir. Plast. 9 (1967) 67.
[5] KLEN, R., KLENOVÁ, V., PAZDERKA, J., Use of the anterior chamber of the eye for selection and preservation of cornea, Am. J. Ophthalmol. 60 (1965) 879.
[6] KLEN, R., JIRÁSEK, J., Fate of the cartilage homograft deprived of a part of the matrix (in Russian), Ortop., Travmatol. Prot. 4 (1964) 14.
[7] KLEN, R., HEGER, J., Re-hydration of the cartilage, Acta Chir. Plast. 9 (1967) 146.
[8] KLEN, R., "Preparation of chorion and/or amnion grafts used in burns", Research in Burns (MATTER, P., BARCLAY, T.L., KONIČKOVÁ, Z., Eds), Hans Huber, Bern (1971) 235.
[9] SCHEUPLEIN, R.J., BLANK, J.H., Permeability of the skin, Phys. Rev. 51 (1971) 702.
[10] PÁCAL, J., SOKOL, D., Permeability of the normal and freeze-dried skin, Paper read at the Int. Meeting Czech. Soc. for Low Temp. Biol., Opočno, 1974.

EFFECTS OF STERILIZING DOSES OF IONIZING RADIATION ON THE MECHANICAL PROPERTIES OF VARIOUS CONNECTIVE TISSUE ALLOGRAFTS

N. TRIANTAFYLLOU, P. KARATZAS
Human Tissue Bank,
Nuclear Research Center Democritos,
Athens, Greece

Presented by G.O. Phillips

Abstract

EFFECTS OF STERILIZING DOSES OF IONIZING RADIATION ON THE MECHANICAL PROPERTIES OF VARIOUS CONNECTIVE TISSUE ALLOGRAFTS.

Ionizing radiations are being utilized to sterilize in our laboratory hard and soft connective tissues. With the sterilizing dose of 2.5 Mrad, deterioration is found in the mechanical properties of the tissues. For bone the deterioration is mainly in the organic collagen matrix. It has been observed that the freeze-drying procedures utilized as a preliminary procedure before radiation treatment contributes significantly to the overall mechanical deterioration, and is further accentuated by the radiation treatment. Experiments with dura mater and pericardium show that not only collagen, but the fibroelastic structure is sensitive towards the sequence of treatments usually accorded to radiation-sterilized transplant materials.

It is now accepted [1-5] that one of the most widely used sterilization methods for the preparation of tissues for allo-transplantation is gamma irradiation. This is usually provided by ^{60}Co sources [6]. In our laboratory we use both a ^{60}Co source and a 5-MW open pool reactor [4, 7]. In the latter the samples are introduced to the core following shutdown.

The standard dose given to our products is between 2.5 and 3 Mrad [4, 6, 7], which is found adequate, as seen from microbiological tests. Although from the sterility aspect, gamma irradiation is very good, problems arise with various tissues. The main problem is the deterioration of the mechanical qualities of the grafts.

The allografts supplied by our tissue bank are as follows: bone, dura mater, fascia latta, tendons, cartilages, skin, peripheral nerves, and chorion and/or amnion. The alterations of the mechanical properties of the bone and dura mater allografts are discussed here.

The bone matrix consists of two phases, the inorganic and the organic. The inorganic phase consists of calcium phosphate, which is about 65% of the dry, fat-free weight of the bones. The remainder is the organic phase. The basic structure of the inorganic phase is considered to be a mixture of the mineral hydroxyapatite, of which the prototype is

$$CA_{10}(PO_4)_6(OH)_2 \qquad (1)$$

and of amorphous calcium phosphate. The organic phase of the bone matrix consists mainly of collagen. In the adult beef bone, collagen comprises about

88-89% of the whole dry weight of the organic matrix, while the remaining is the non-collagenous fraction.

It is well established [4, 8] that the bone allografts exhibit lower mechanical strength (tensile strength and elasticity). The cause of this change is attributed to the different stages of the procedure used for preparing the allografts.

As a rule the allografts are subjected to deep freezing (DF), freeze-drying (or lyophylization) (LP), and irradiation (Ir). Of these, deep freezing is a standard treatment following sample collection. Thus, the remaining two stages are examined, while the values of (DF) are taken as basis.

To make a quantitative evaluation of the changes possible, we proceeded in measuring the tensile strength and the modulus of elasticity of the bovine tibia. The pieces were cleaned and all soft tissues including periosteum removed. They were then cut into pieces of 8 × 0.5 × 0.5 cm. Care was taken to use animals of the same age as donors, and also to remove bone pieces from the same part of the tibial diaphysis. All the pieces were then checked by X-ray radiography and those found to have fractures or gaps were excluded. The remaining pieces were then kept at -35°C for three days, following which they were subjected to the tests. Figure I shows the set-up of the experiment and Table I the results obtained from the bovine tibial pieces.

All tests were carried out at room temperature. The re-hydration of the freeze-dried samples was done with normal saline solution at 20°C for two hours. Each bone beam was subjected to an increasing concentrated load, applied in the middle of the beam, until failure occurred. The maximum tensile strength σ at failure was determined by means of the equation:

$$\sigma = \frac{M}{b\ h^2/6}\ (\mathrm{kg/cm^2}) \qquad (2)$$

where:

M = Maximum bending moment in the middle of the beam = $\frac{Pl}{4}$ (simply supported beam)
P = Load at failure (kg)
l = Distance between supports = 3.70 cm
b = Width of the beam = 0.5 cm
h = Height of the beam = 0.5 cm

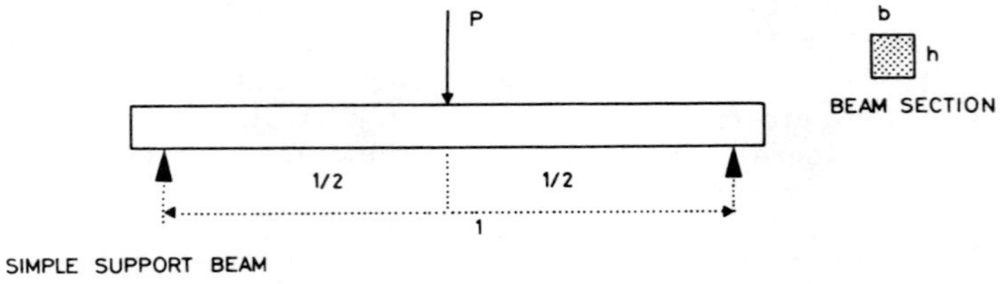

FIG.1. Set-up for strength measurement.

TABLE I. RESULTS OBTAINED FROM BOVINE TIBIAL PIECES

Treatment	Donors																	
	A			B			C			D			E			F		
	P	σ	f	P	σ	f	P	σ	f	P	σ	f	P	σ	f	P	σ	f
DF	39.8	1.8	0.75	62	2.7	1.1	51	2.8	0.8	69	3	1.2	63	2.8	1	68	3	0.9
DF + LP	34.8	1.5	0.5	37	1.6	0.4	34	1.5	0.5	55	2.4	1	29	1.3	0.6	30	1.3	0.5
DF + Ir	14.6	0.65	0.2	35	1.6	0.5	30	1.4	0.4	47	2.1	0.6	47	2.0	0.8	49	2.2	0.6
DF + LP + Ir				15	0.64	0.4	7.6	0.4	0.5	11	0.49	0.7	9.5	0.42	0.15	19	0.86	0.7

P: Average values of failure load (kg). σ: Tensile strength at failure (kg/cm^2). f: Deflection at failure (mm). DF: deep-freeze; LP: lypholization; Ir: irradiation.

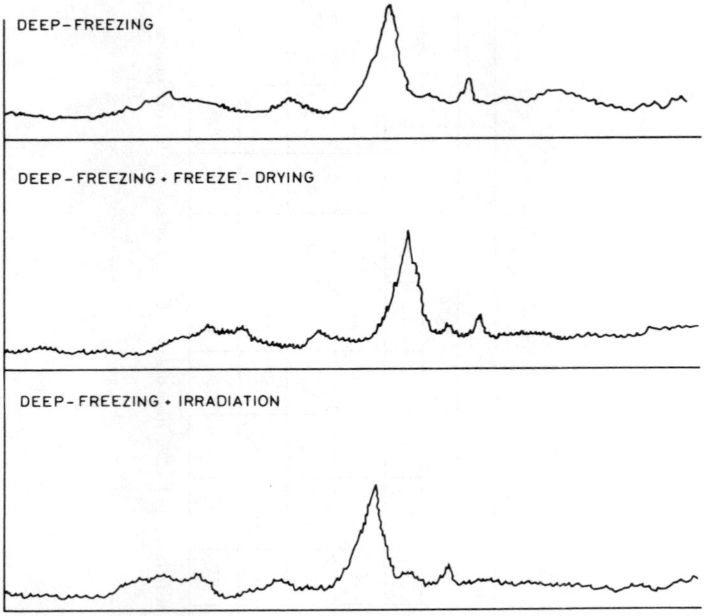

FIG.2. X-ray patterns for hydroxyapatite.

Following the above findings the question arises: which parts of the bone suffer because of the treatments. The main bone constituents were therefore examined, i.e. the crystalline hydroxyapatite and collagen. Since there is no evidence for gross chemical changes in the inorganic material, we proceeded with an examination of the crystalline structure itself.

Bone powder from human samples was prepared and examined by X-ray diffraction, both before and after treatment. As can be seen from Fig.2, no difference was observed in the patterns, and this holds true even at a dose of 10 Mrad.

Therefore, the organic part of the bone, consisting mainly of collagen, had to be examined.

It is very difficult to separate intact collagen from the bone and therefore we decided to use human dura mater and pericardium as experimental models.

The main constituent of dura mater is collagen with some elastic fibres present, whereas the pericardium [9] consists mainly of elastic fibres. Therefore, by testing these two connective tissue membranes, we have a model for the effect of freeze-drying and irradiation on the mechanical properties of the collagen fibrils and the fibroelastic network.

Dura mater was taken from the frontal area of the cranium of human donors within 8 h of death. Pericardium was also taken from human donors

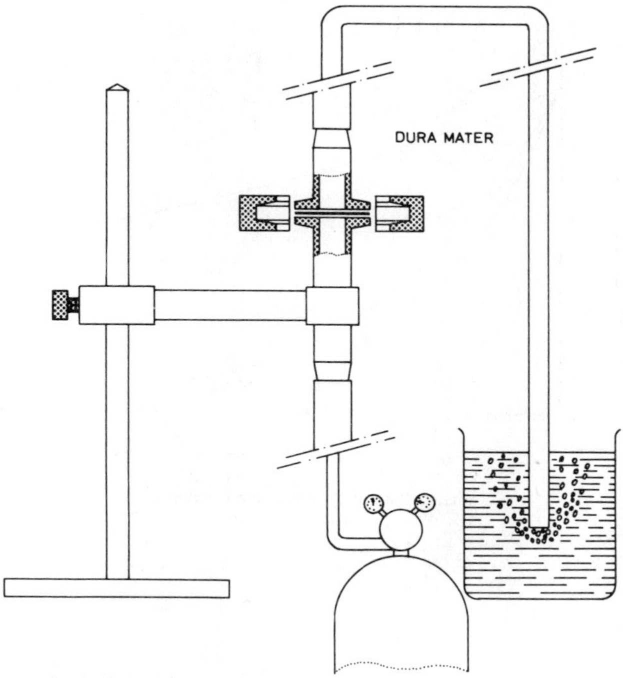

FIG.3. Set-up for the measurements of the dura mater.

TABLE II. PRESSURE THRESHOLD (PT) AT WHICH AIR DIFFUSION BEGINS. PRESSURE AT WHICH SPECIMENS FAILED (PF)

	Deep-freezing (kg/cm^2)	Deep-freezing Freeze-drying (kg/cm^2)	Deep-freezing irradiation (kg/cm^2)	Deep-freezing Freeze-drying irradiation (kg/cm^2)
PT	4.5	0.37	0.22	0.24
PF	> 15	9-15	9-15	9-15

within 8 h of death. Special care was taken to collect the specimens only from donors without any pathological condition of the dura mater or the pericardium. Following collection, the specimens were kept at -40°C for three days.

The pieces of dura mater were cut in portions of approximately 4 × 4 cm. These were made air-tight and attached to the apparatus shown in Fig.3. The pressure from the air-tank was allowed to rise to 15 kg/cm^2, which was the maximum capacity of the manometer. The free end of the tube was immersed in a beaker containing water so that the beginning of diffusion of air through the membrane could be monitored from the bubbles.

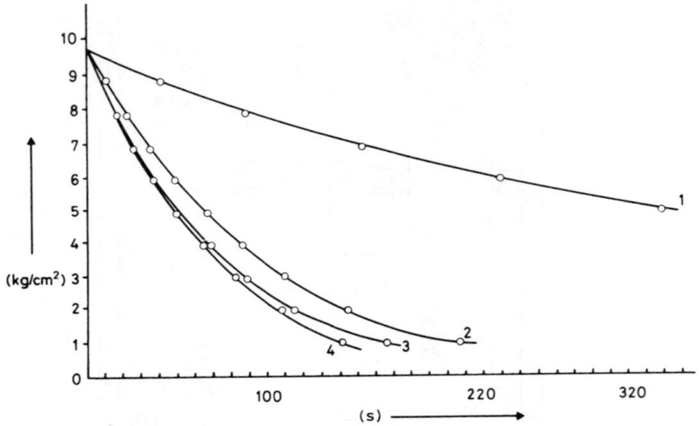

FIG.4. Rate of diffusion through dura mater.
1 = DF
2 = DF + Ir
3 = DF + LP
4 = DF + LP + Ir

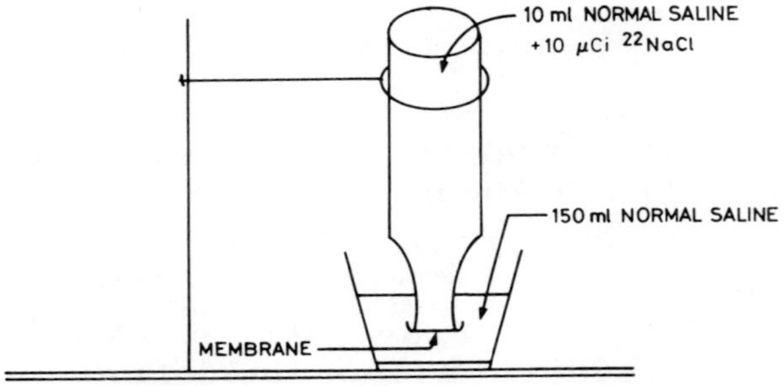

FIG.5. Set-up for measurement of $^{22}NaCl$ equilibrium through dura mater and pericardium.

Once the pressure was set as 10 kg/cm^2 the air supply was cut off and the pressure drop versus time was recorded. The results are shown in Table II and Fig.4.

The permeability of dura mater and pericardium was measured by a simple isotope dilution method. The set-up is shown in Fig.5. The membranes were tied at the bottom of a glass tube and inserted in 150 ml of normal saline. Then 10 ml of normal saline + 10 μCi of $Na^{22}Cl$ were poured into the inside of the tube, and samples taken from the beaker for radioactivity measurement. The results are shown in Fig.6 for dura mater and Fig.7 for pericardium. The experiment was based on the phenomenon of osmosis. It must be pointed out that the 2-ml samples for the measurement

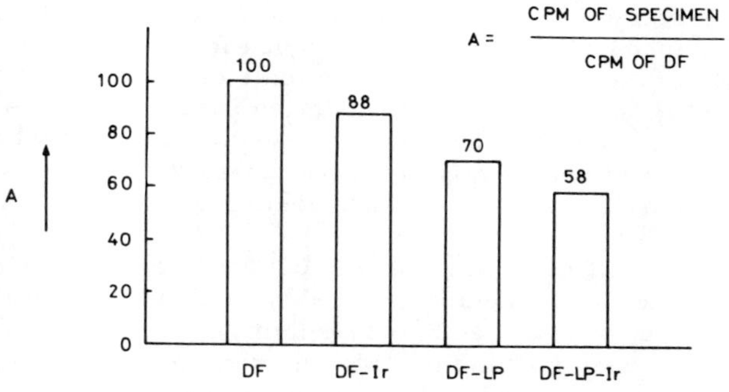

FIG.6. Permeability rates of dura mater.

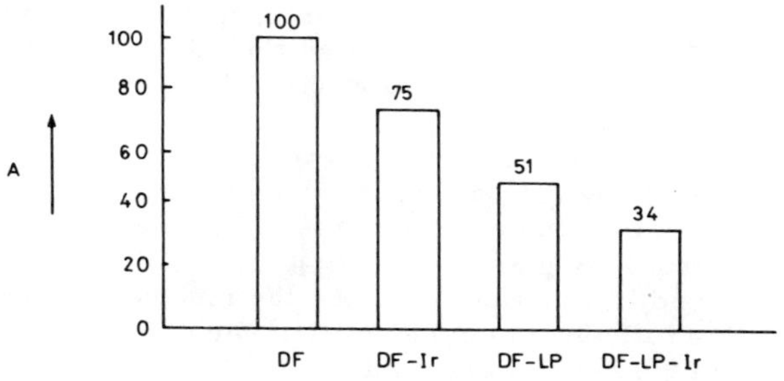

FIG.7. Permeability rates of pericardium.

were taken from the beaker after 30 min. Time was chosen so as to permit equilibrium between the ions, which were found in both sides of the membranes (entering and leaving). The samples were counted in a Packard gamma-counter for 2 min.

DISCUSSION AND CONCLUSIONS

These studies were undertaken and carried out in order to increase our knowledge of the causes of damage to the transplantable allografts produced by our tissue bank. This damage exhibits itself as a lowering of the mechanical properties of the processed allografts, therefore diminishing their value as transplantable materials [10, 11].

Since radiation damage is related to the solid phase of the bone matrix, i.e. the crystal lattice, we concentrated our efforts on detecting possible hydroxyapatite decomposition of different components like $CaO.H_2O$, as an effect of the gamma-ray sterilization of the bone. If the hypothesis of decomposition were true, then following the strongest reflection in the X-ray patterns, we should be able to distinguish an eventual decrease of intensity of that reflection and the appearance of new characteristic peaks of the new substances. This would be evidence in the effort to explain the loss in strength of the bones after irradiation sterilization. Our results were negative [10], even at gamma irradiation doses as high as 10 Mrad. We are not yet convinced that our negative results are critical and they have proved the stability of hydroxyapatite in the doses given. We believe that a further study must be undertaken that would lead to more conclusive results.

Because of the difficulty in removing collagen from the bone without destroying its structure we decided to use experimental models. For this purpose we chose dura mater and pericardium.

Of these, dura mater contains a high percentage of collagen in the form of fibrils. It has been shown by electron microscopy that these fibrils fuse at sterilizing doses of gamma irradiation and turn into an amorphous mass. Of course, it has still to be proved that this fusion of collagen fibrils is responsible for reducing the mechanical strength of bone and/or dura mater. It can certainly be held responsible for reducing the elasticity of the material. It must be stressed that in our permeation experiments we found that freeze-drying is the main cause of the observed alteration. Obviously an alteration of the collagen structure because of de-hydration is to be expected.

Although we feel it is rather early to draw final conclusions from our experiments, we can summarize our findings as follows:

(1) In all tissues examined the deterioration of the mechanical properties because of the processing and sterilizing techniques is significant. Freeze-drying and irradiation influence the mechanical properties of the grafts. This influence is evident after each treatment, and when both are applied the results are additive.

(2) The bone changes are most probably due to destruction of the collagen, and not of the hydroxyapatite [13].

(3) The experiments with dura mater and pericardium show that not only collagen, but also the fibroelastic structure, are sensitive to treatment usually applied to the transplantable materials.

To generalize, it could possibly be said that connective tissue as a whole suffers from the treatment.

REFERENCES

[1] BASSETT, C.A.L., HUDGINS, T.F., TRUMP, J.K., WRIGHT, The clinical use of cathode-ray sterilized grafts of cadaver bone, Surg. Forum 5 (1956) 549.

[2] COHEN, J., Cathode-ray sterilization of bone grafts, Arch. Surg. 71 (1955) 784.

[3] SWANSON, A.B., GLESSNER, J.R., Jr., BURDICK, H.W., MAHANEY, R.C., Seven years experience with irradiated bone graft material, J. Bone J. Surg. 45-A (1963) 1554.

[4] TRIANTAFYLLOU, N., "Observations on the clinical use of the freeze-dried, irradiated allografts during bone transplantation", Manual on Radiation Sterilization of Medical and Biological Materials, Tech. Rep. Ser. No.149, IAEA, Vienna (1973) Ch.24.

[5] WILBER, M.C., HYATT, G.W., Bone cysts: Results of surgical treatment in two hundred cases, Proc. Amer. Academy Orth. Surgeons, J.Bone Joint Surg. 42-A (1960) 879.

[6] OSTROWSKI, K., KOSSOWSKA, B., MOSKALEWSKI, S., KOMENDER, A., KURNATOWSKI, W., "Radiosterilization of tissues preserved for clinical purposes", Radiosterilization of Medical Products (Proc. Symp. Budapest, 1967), IAEA, Vienna (1967) 139-42.

[7] TRIANTAFYLLOU, N., CHRYSOCHOIDES, N., MITSONIAS, C., KIORTSIS, M., MARKETOS, D., LITSIOS, B., Méthode d'élaboration, stérilisation et conservation des Homogriffes Osseuses, Proc. Congr. Franco-Hellénique d'Orthop. et Traum. Athènes (1970) 507.

[8] TURNER, T., BASSETT, C.A.L., PATE, J.K., SAWYER, P.N., TRUMP, J.G., WRIGHT, K., Sterilization of preserved bone grafts by high-voltage cathode irradiation, J. Bone J. Surg. 38-A (1956) 862.
[9] HAM, A.W., Histology, Lippincott Co. (6th Edn) Philadelphis and Toronto (1965) pp. 519 and 614.
[10] TRIANTAFYLLOU, N., PHILLIPPAKIS, S., The effects of the radiation sterilization on human transplantable bone, IAEA Progr. Rep. (1971).
[11] TRIANTAFYLLOU, N., SOTIROPOULOS, E., TRIANTAFYLLOU, J.N., The mechanical properties of the lyophilized and irradiated bone grafts, 3rd Symp. Biomécanique Osseuse, C.I.B.O., Brussels (1974).
[12] TRIANTAFYLLOU, N., ANDONADOS, D., Studies on the influence of high doses of irradiation to the bone's components, Proc. J. Meeting Assoc. Radiat. Research and UK Panel, Salford University, December, 1973.

DISCUSSION

R. KLEN: We have not found that freeze-dried grafts are as badly damaged as those you describe. We therefore consider the irradiation to be the more dominant factor. Compared with other methods which sterilize only the surface, irradiation affects the whole of the graft. The changes attributed to freeze-drying may actually be due to the use of inappropriate freeze-drying techniques. I don't think that freeze-drying, if properly carried out, badly damages the graft and I consider it to be a good fixation method for histologic examination.

J. FLEURETTE: At what temperature were the tissues frozen? I would think the best temperature for preserving living cells would be that of liquid nitrogen. This is the temperature used for bacteria, viruses, cell cultures, and certain organs.

G.O. PHILLIPS: The tissues were taken out of the container at -40°, and placed directly in the freeze-drier.

ORGANIZATION AND IMMUNOLOGICAL QUALITIES OF LYOPHILIZED AND RADIOSTERILIZED EXTRASYNOVIAL DOG-TENDON PREPARATIONS*

I. FEHÉR, S. PELLET Jr., E. UNGER, Gy. PETRÁNYI
Frédéric Joliot-Curie National Research Institute
for Radiobiology and Radiohygiene,
Budapest,
Hungary

Abstract

ORGANIZATION AND IMMUNOLOGICAL QUALITIES OF LYOPHILIZED AND RADIOSTERILIZED EXTRASYNOVIAL DOG-TENDON PREPARATIONS.

In experiments the organization and immunological properties of lyophilized radiosterilized dog-tendon preparations after implantation were studied. According to the experimental results the lyophilized dog-tendon preparations seem to be of the same value for surgical use as the fresh tendon. The preparations can be easily stored and transported and if implanted they are quickly rebuilt to living tendon tissue. According to the pilot experiments the method of preparation does not influence the immunological qualities of the human tendon; this preparation does not behave as antigen.

Repair of injuries to tendon tissue is nearly routine in reconstructive surgery. For a long time, reconstruction of seriously or totally damaged tendon tissue has been carried out by implanting autologous or allogenic tendon tissue. Allogenic tendon tissue prepared for implantation is generally not used when fresh, but is stored in a preservative until use. The widespread use of these methods made it possible to establish banks of tendon tissue.

In our experiments we tried to elaborate a tendon-preserving method to enable us to prepare preparations which can be treated more easily than previously and can be stored for an almost unlimited period.

Preserving the tendon tissue by lyophilization and sterilizing the dry preparations by irradiation with ^{60}Co gamma-rays proved to be suitable methods. The preparations were prepared from Achilles tendon of 1-2 year-old mongrel dogs. Twenty-four hours after removal the tendons were lyophilized on a plate at 10°C for 2 d. The lyophilized tendon preparation was packed in a double-walled polyethylene bag, then irradiated with 2.5 Mrad at +4°C in a Noraton device (240 rad/s). The preparations treated in this way can be stored at room temperature, are transportable and can be stored for almost an unlimited period. The mechanical qualities of the lyophilized tendon are satisfactory; the tendon is not breakable, but it has to be bathed in physiological saline for a few hours before transplantation.

Figure 1 shows a comparison of the histological pattern of both normal and lyophilized radiosterilized tendon. The histological pattern of the preparation prepared by this method does not differ from that of the control.

* Work supported by IAEA Research Agreement No.889.

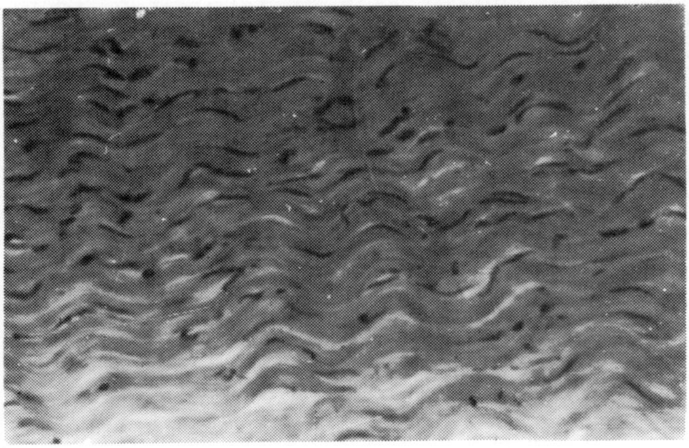

(a)

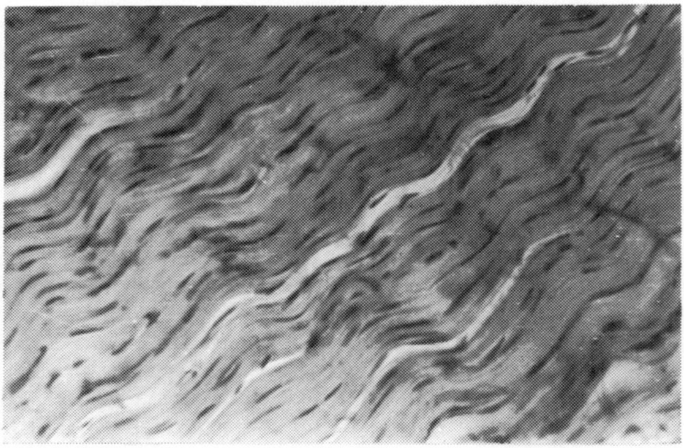

(b)

FIG.1. Comparison of the histological pattern of both normal and lyophilized, radiosterilized tendon. (a) Normal; (b) lyophilized.

The significance of our dog-tendon preparations is that they may serve as models for human extrasynovial tendon. Preparations originating from human cadavers were also prepared by the same method. The physical qualities of these human tendons seemed to correspond to that of the dog-tendon preparations. However, these have been used only for immunological studies; no implantation has been carried out up to now.

This paper reports the experiments in which the organization of lyophilized, radiosterilized dog-tendon preparations was studied after implantation. Then a brief account is given of the results of the immunological experiments performed so far.

The tendon preparations were implanted into 1-2 year-old mongrel dogs. The operation was performed under intranarcon narcosis and sterile conditions. Figure 2 shows the area of operation. From the middle part

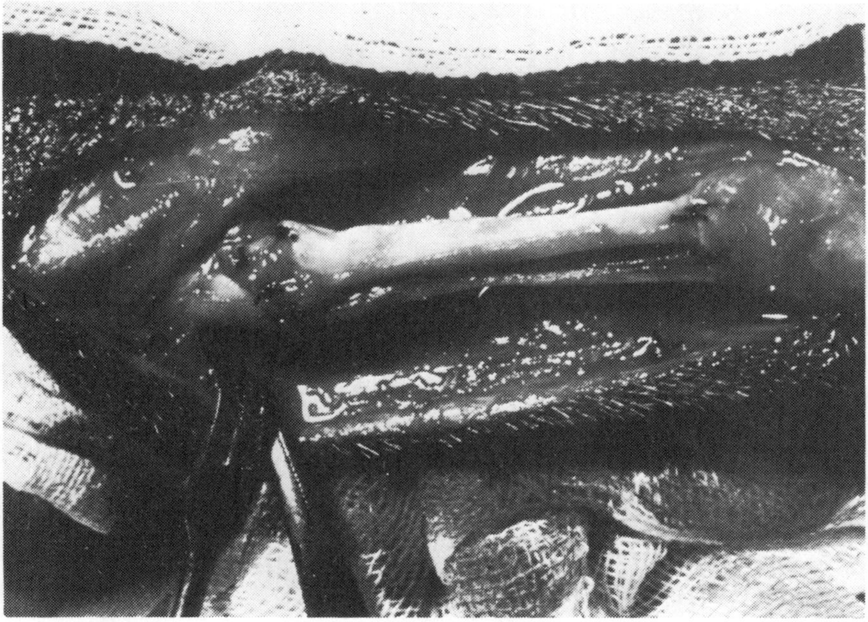

FIG.2. Area of operation.

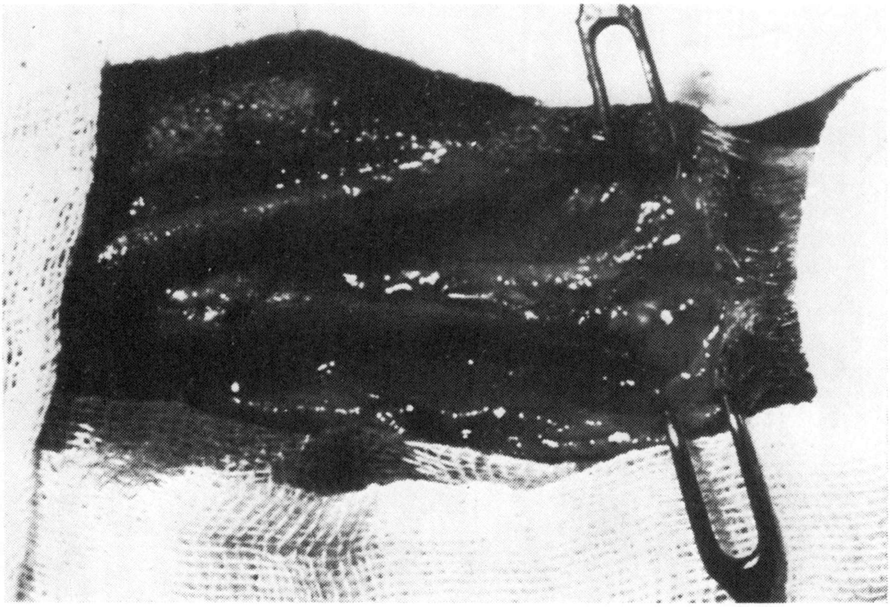

FIG.3. Implant distinguishable from recipient's own tendon.

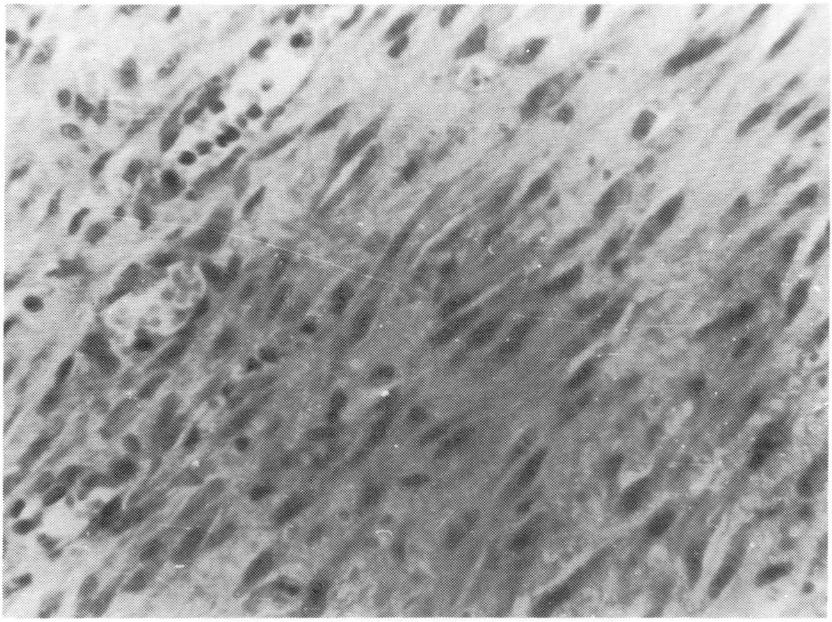

FIG.4. Visible inflammatory infiltrated granulation tissue.

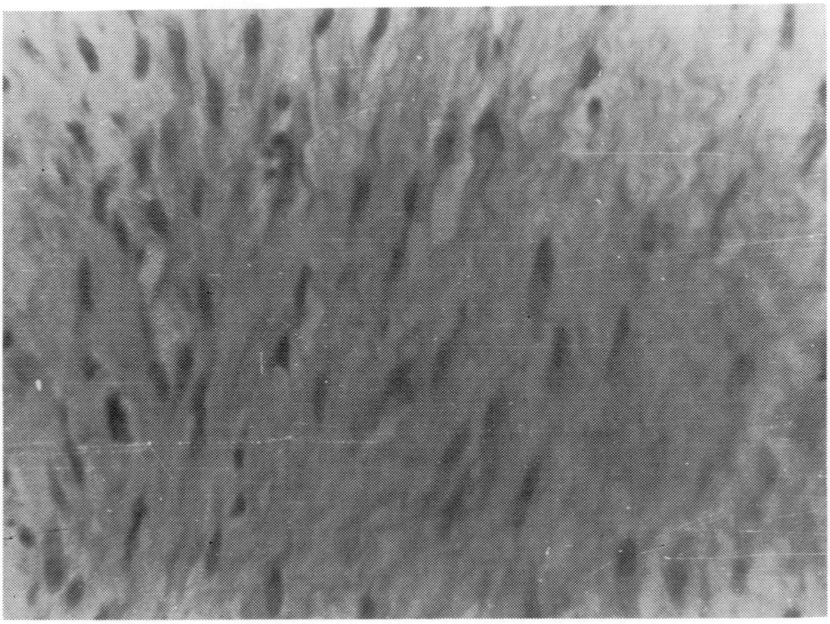

FIG.5. Irregular spots of young, cell-rich connective tissue showing initial fibre formation.

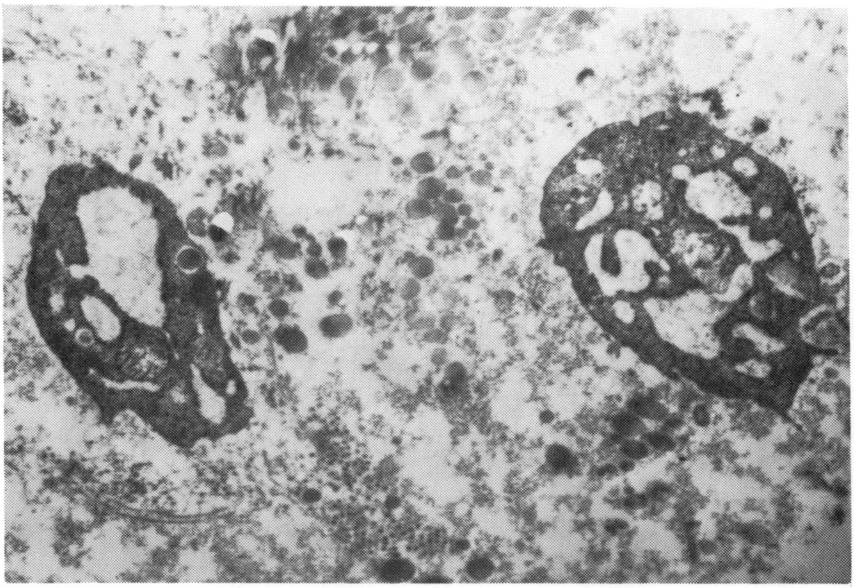

FIG.6. Visible thin, young fibres of the implant.

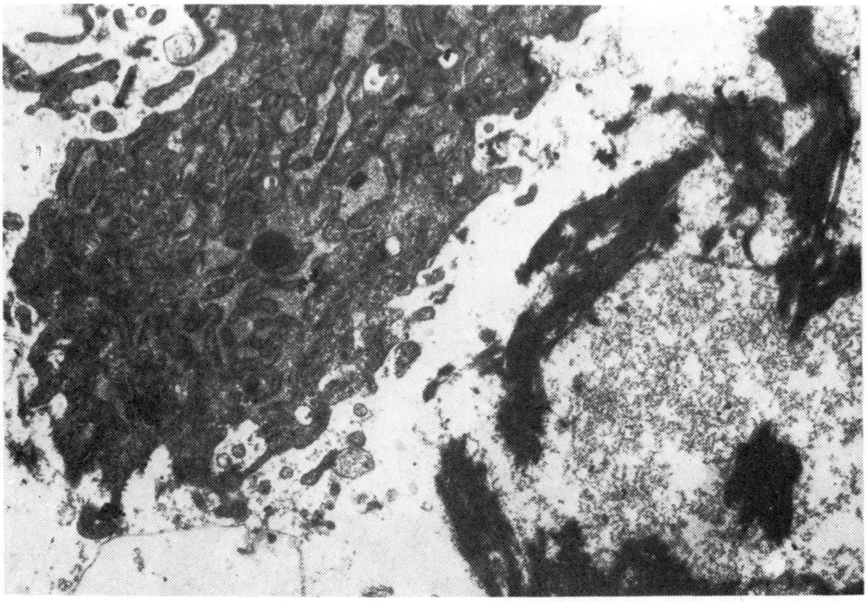

FIG.7. Renewal of collagenous fibrils.

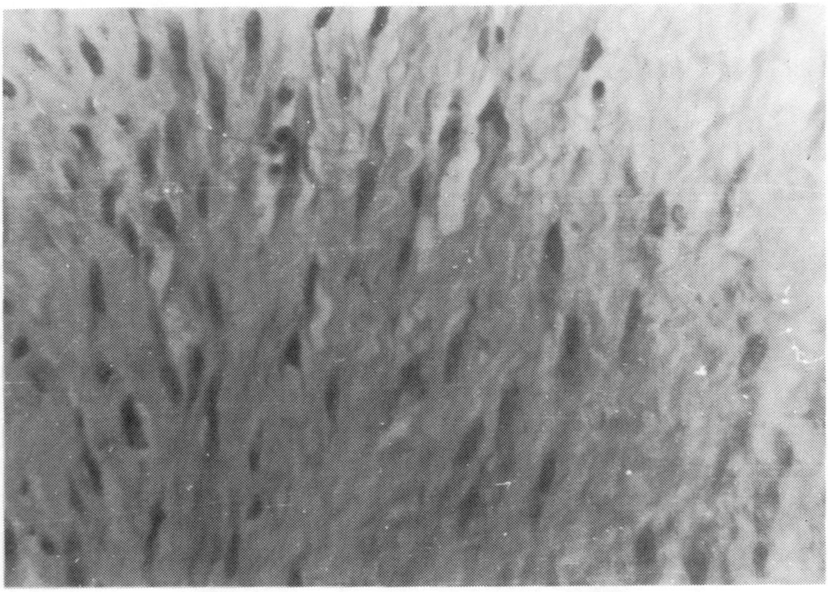

FIG.8. Part of implant surrounded by granulation tissue.

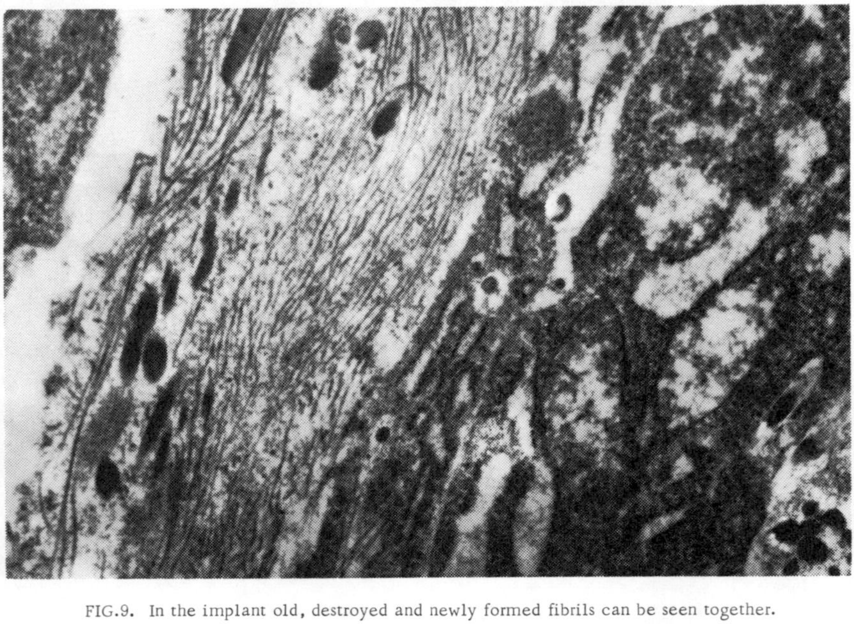

FIG.9. In the implant old, destroyed and newly formed fibrils can be seen together.

of the Achilles tendon a part 2-3 cm long was excised and a lyophilized tendon bathed in physiological saline was patched into the place of the removed tendon part by a modified Bunnel operation.

On the operated limb of the experimental animals a bandage was put which, however, did not inhibit mobility. During the 3 d after the operation they were given 10^6 units of penicillin daily.

Evaluation of the experiments was performed by examining the function of the limb by macroscopic inspection, as well as by histological and immunological examinations.

For 6-7 d after the operation a light tissular swelling could be detected on the operated limb. At this time the implant could be still distinguished macroscopically from the recipient's own tendon (Fig.3) but some relation between the implant and the original tendon was already detectable.

On the histological section of this part of the implanted tendon the structure of the original tendon shows a peculiar change: the proportion of the cells to the collagenous fibre bundles shifts in favour of the cells. The increased number of cells do not correspond to the characteristic, flattened tendon cells of the control tendon but rather they seem to be fibroblasts and young fibrocytes. This phenomenon is presumably connected with the increased regenerating activity of the original tendon.

In the implanted tendon a mosaic-like histological picture developed. While some parts still showed the structure of the tendon tissue, others showed nuclear staining. In some places inflammatory infiltrated granulation tissue can be seen (Fig.4), while elsewhere irregular spots of young, cell-rich connective tissue showing initial fibre formation could be detected (Fig.5).

The electron-microscopic preparation is prepared from the border zone of the implanted tendon near the original one. In the intracellular amorphous material can be seen both the original fibres originating from the implant and the newly formed fibres originating in the host. The fibres of the implant started to disintegrate. They are stained unequally; some can hardly be seen, but some still thin, young fibres appear in their place (Fig.6).

In Fig.7 one can see the renewal of the collagenous fibrils. When using a higher magnification, the laminosis material around the cells proves to consist of newly formed collagenous fibrils.

On the histological patterns obtained from the middle part of the implant, poorly stained, dead cell-nuclei and disintegrating collagenous fibres can be seen. This part is surrounded in each direction by granulation tissue (Fig.8).

The sections show that the renewal of the implant starts axially, from the border zone of the original tendon and the implant. From the original tendon, interstitial cells flow into the implant and the new fibrin fibres are formed from these cells. But — as is shown in the last section — in the middle parts of the implant, although only to a lesser extent, the formation of the new tendon tissue can be seen to encircle near the edges of the implant. These regeneration processes seem to take place rather rapidly, newly formed capillaries being detectable in the implant already 2-3 weeks after operation. Three weeks after operation the re-organization of the implant is still in progress. In the implant, both old, destroyed and newly formed fibrils can be seen together (Figs 9 and 10).

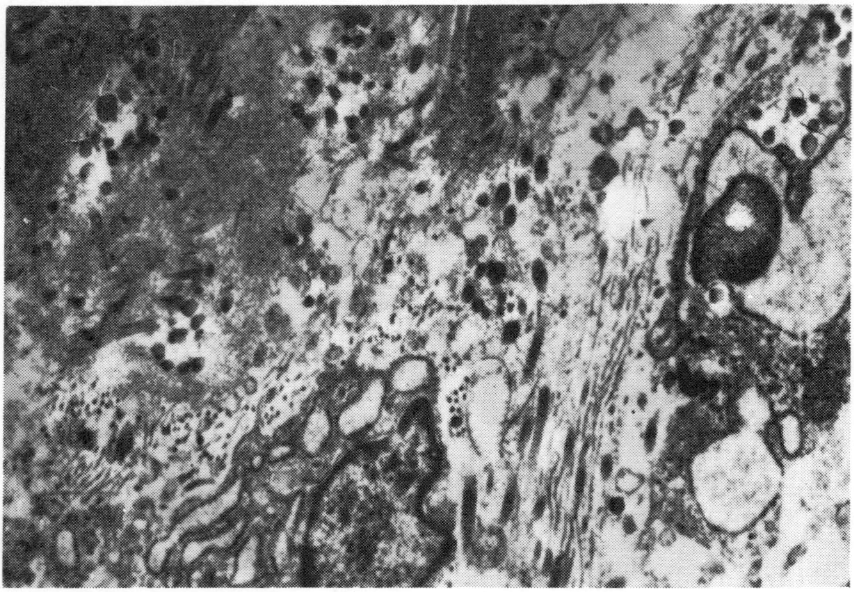

FIG.10. In the implant old, destroyed and newly formed fibrils can be seen together.

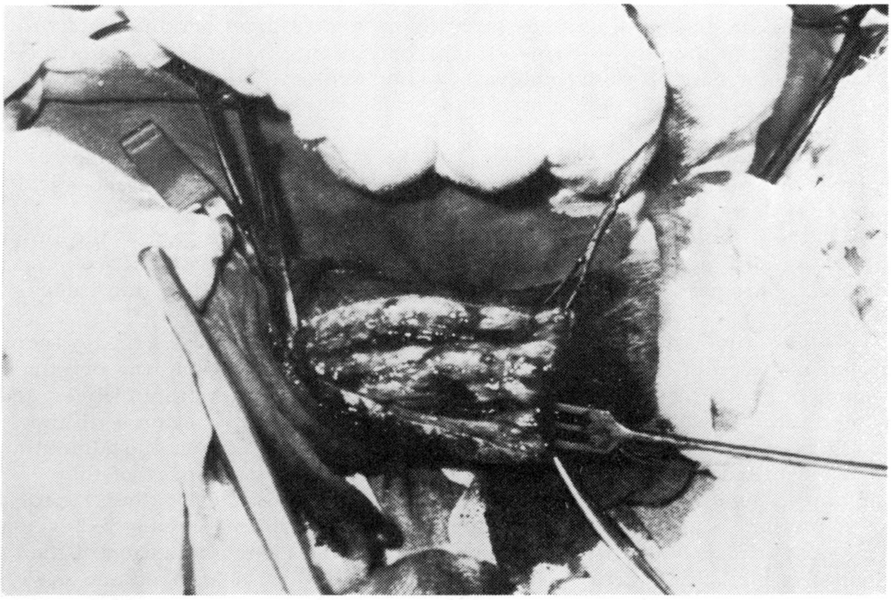

FIG.11. Strong connection between implant and recipient's tendon after 3 weeks.

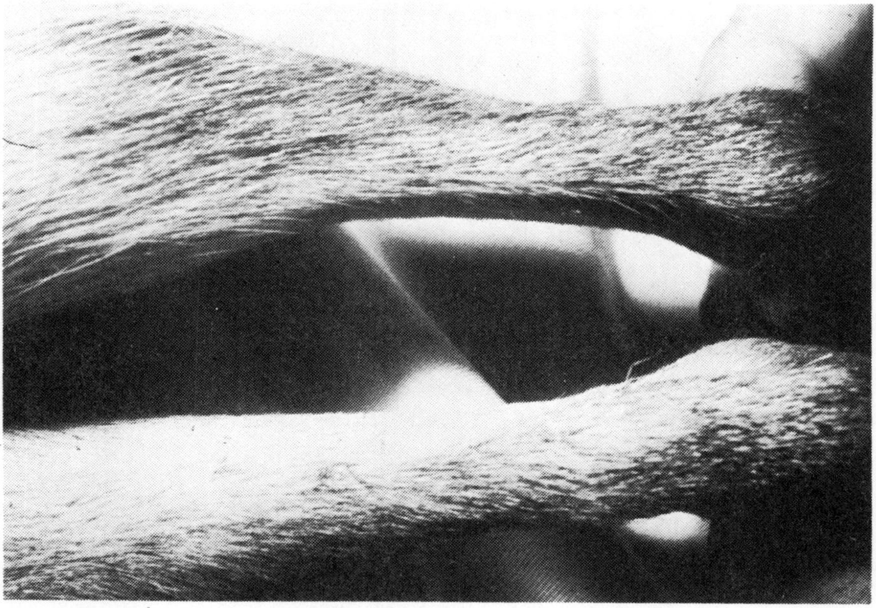

FIG.12. No macroscopic difference between implant and recipient's tendon.

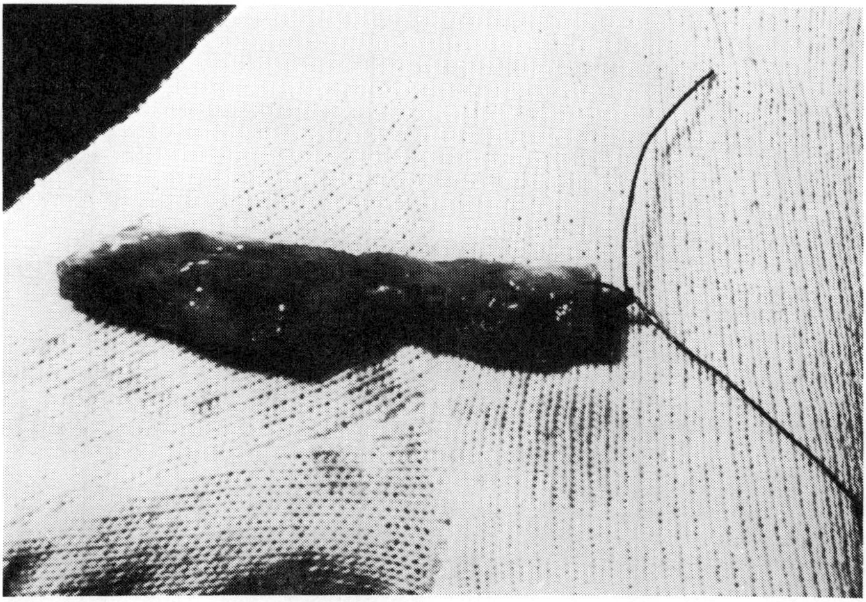

FIG.13. Significant vascularization on the surface of the implant.

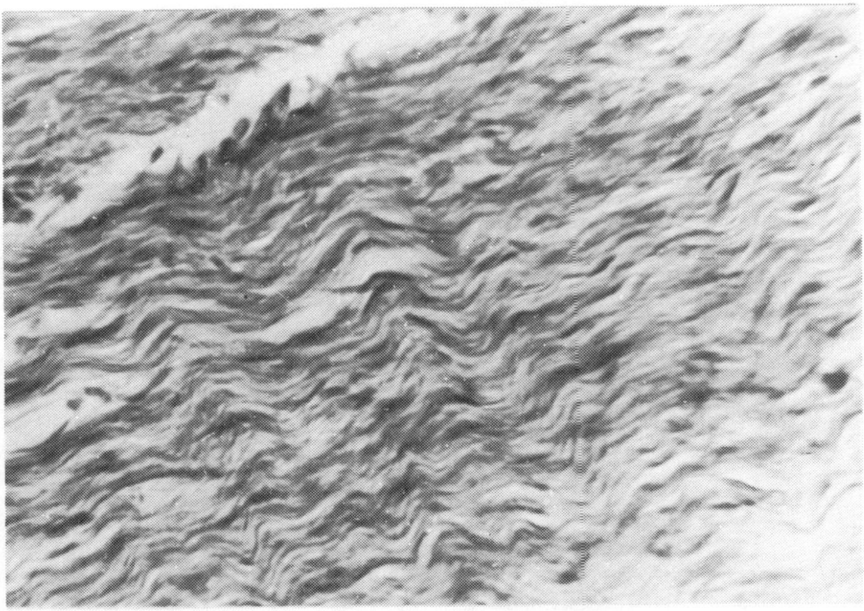

FIG.14. Capillaries run parallel to the longitudinal axis of the tendon but, as shown here, they also enter from the edges.

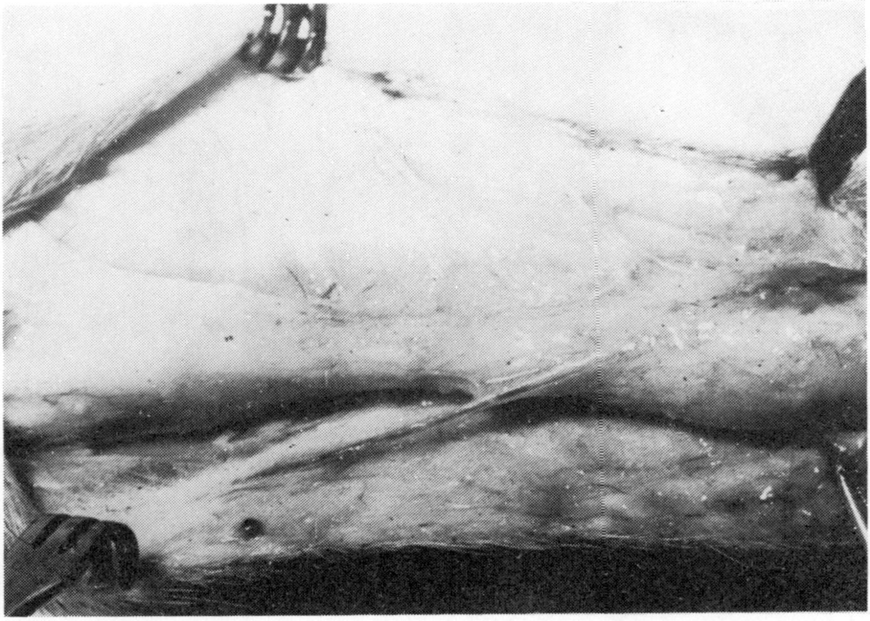

FIG.15. Macroscopically a slight thickening distinguishes implant from contralateral Achilles tendon.

Three weeks after implantation a strong connection between the implant and the recipient's own tendon can be observed macroscopically (Fig.11). A slight swelling can be seen in the implant. The implanted tendon can be easily distinguished from its environment. Strong inflammation reaction or signs of rejection cannot be detected.

Three to four weeks after operation the function of the operated limb recovered. The dogs used their operated limbs, although carefully. Eight to ten weeks after operation they could stand on their hind paws, and jump, which means that total functional recovery could be observed. At this time the operated limb could not be distinguished from the intact one by naked eye.

No difference can be seen macroscopically between the tendon implant and the recipient's own tendon (Fig.12). However, compared with the bradytrophic character of the tissue there is a significant vascularization observed on the surface of the implant (Fig.13). After two months the original and the implanted tendons show only slight differences histologically. The implanted tendon tissue is microscopically richer in cells than the normal one and is young in character; the majority of the cells are not typical, flattened tendon cells. The vascularization of the regenerated tendon is satisfactory. Some capillaries run parallel to the longitudinal axis of the tendon, but capillaries enter also from the edges (Fig.14). Later the swelling of the implant decreases and the increased vascularization ceases six months after the operation. Macroscopically, only a slight thickening distinguishes the implant from the contralateral Achilles tendon (Fig.15).

After six months the histological pattern of the implanted tendon does not differ from that of the control tendon.

From the point of view of the applicability of the lyophilized, radio-sterilized tendon preparations, the study of the immunological qualities of the preparation is of great importance.

The data in the literature indicate that the tendon tissue is not immunogenic. These data, however, refer to native tendon tissue, and it may happen that the tendon becomes antigenic owing to some eventual changes of the collagenous structure induced by lyophilization and irradiation with a high dose. This induced us to study our preparations also by immunological methods. The experiments reported in this paper were performed on mongrel dogs, consequently histo-compatibility between the donors and recipients was hardly to be expected. Despite this, no signs of rejection or any other immuno-reactions could be observed. Almost two years have passed since we implanted our first preparation and the dog is still healthy. In six cases we implanted tendons from two different donors into the same recipient within a year and, in one case, we repeated the operation four times with tendon preparations originating from different dogs. From immunological considerations these repeatedly operated dogs had already been sensitized to tendon tissue. Even if the dog tendon had proved to be a weak antigen, the second graft should have been rejected. But no rejection occurred in any of the cases.

During our studies, model experiments with mouse tendon were performed and it was found to be a strong antigen. Accordingly, the antigeneity of the tendon tissue is species-dependent. Thus, there is the possibility that the results obtained with dog tendon cannot be applied to human tendon. Therefore, we are now studying the antigeneity of lyophilized, radiosterilized human tendons in human lymphocyte cultures.

According to the results we have obtained so far, our human tendon preparations have no antigenic properties in this test system.

CONCLUSIONS

The lyophilized, radiosterilized dog-tendon preparations seem to be of the same value as fresh tendon for surgical use. The preparations can be easily stored and transported and, if implanted, are quickly rebuilt to living tendon tissue. According to our pilot experiments, the method of preparation does not influence the immunological qualities of the human tendon; human tendon preparations do not behave like antigens.

DISCUSSION

H.M. ROUSHDY: You did not mention any histological changes induced in radiosterilized dog-tendon preparations or comment on the functional status of the tendon preparation after implantation. Have you carried out any studies to evaluate the physiological activity of the tendon preparations after implantation and also their stability with time?

I. FEHÉR: Though the lyophilized radiosterilized preparations do not differ morphologically from the intact tendon, they cannot be considered as living tissue. Thus, we cannot speak of the physiological properties of the preparation. After implantation the preparation is rejected, parallel with the recovery of the new tendon tissue. The implant acts only as a matrix, necessary for the recovery of the new tendon tissue. The new tendon tissue develops from the host's own fibroblasts.

It was our aim to establish whether the absorbing preparation was antigenic or not, after which it was investigated. It was also necessary for the implant to be solid and not to tear until the new tendon tissue recovered. The physical properties of the preparation would appear to be satisfactory, since — as I have mentioned in the paper — 8 to 10 weeks after transplantation the animals used their operated limbs as easily as the contralateral ones.

J. FLEURETTE: I should just like to make a general comment in connection with the three papers presented at this session on the application of radiation sterilization to biological tissues. An uninformed listener might gain an unfavourable impression of the process inasmuch as it constantly gives rise to cell and tissue lesions. In actual fact, radiation sterilization has to be compared with other techniques to be seen objectively, and I should like, in this connection, to mention the experience we have had over the past four years with the sterilization of cardiac valves. Because of certain risks of infection, we rejected the technique consisting in non-sterilization while maintaining aseptic sampling conditions and turned to other recommended procedures. Sterilization by means of chemicals, for example, Formol or beta-propiolactone, was immediately discontinued on account of the serious tissue lesions produced. Then, having rejected treatment of the valves with antibiotics (which doesn't guarantee sterility), we tried radiation sterilization and, like other investigators, we observed lesions in the treated cells and tissues. Recent work has shown identical lesions in valves treated with antibiotics.

I do not think, therefore, that there is any sterilization method at the present time that does not harm the cells. We can only say that radio-sterilization is one of the least injurious. It should be our aim now to reduce the sterilizing dose. The microbiological tests we have carried out have enabled us to lower the dose to only 1.5 Mrad, as opposed to the previous 2.5 Mrad.

ACTIVITE BACTERIOSTATIQUE DE DIFFERENTS ANTIBIOTIQUES APRES IRRADIATION PAR RAYONS GAMMA*

J. FLEURETTE, S. MADIER, M.J. TRANSY
Laboratoire de bactériologie,
Hôpital cardiologique,
Lyon, France

Abstract–Résumé

BACTERIOSTATIC ACTIVITY OF VARIOUS ANTIBIOTICS AFTER GAMMA-RAY IRRADIATION.

The purpose of the work described was to discover whether the antibiotics used in medicine can be sterilized by gamma rays; in this preliminary study, only the antimicrobic activity – the principal criterion for this type of medicament – was evaluated. Thirty-three products belonging to the various families of antibacterial and antifungic antibiotics were studied. The substances were irradiated in the dry state and in an aqueous solution, using a caesium-137 irradiator. The antibacterial and antifungic activity before and after irradiation was investigated by the method of diffusion in gelose.

When irradiated in the dry state, 14 antibiotics preserve normal activity up to a dose of 10 Mrad; at doses between 5 and 10 Mrad, 15 other antibiotics are subject to a variable, but moderate, loss activity; and four register a slight loss of activity at a dose of 2.5 Mrad. In an aqueous solution all but two of the antibiotics suffer total loss of activity at a dose of 2.5 Mrad. As most commercial antibiotics are supplied in the dry state, gamma irradiation may be a useful sterilization process. However, preparations such as eye lotions, suspensions, ointments, etc. should be excepted.

ACTIVITE BACTERIOSTATIQUE DE DIFFERENTS ANTIBIOTIQUES APRES IRRADIATION PAR RAYONS GAMMA.

L'objectif du travail a été de rechercher si les antibiotiques utilisés en médecine pouvaient être stérilisés par les rayons gamma; dans ce travail préliminaire, on a évalué seulement l'activité antimicrobienne, critère essentiel pour ce genre de médicaments. Trente-trois produits appartenant aux différentes familles des antibiotiques antibactériens et antifongiques ont été étudiés. Les substances ont été irradiées à l'état sec et en solution aqueuse dans un irradiateur au césium-137. L'activité antibactérienne et antifongique avant et après irradiation a été recherchée par la méthode de diffusion en gélose.

Irradiés à l'état sec, 14 antibiotiques conservent une activité normale jusqu'à la dose de 10 Mrad; à des doses comprises entre 5 et 10 Mrad, 15 autres antibiotiques subissent une perte d'activité variable, mais modérée; à la dose de 2,5 Mrad, 4 subissent une légère perte d'activité. En solution aqueuse, tous les antibiotiques sauf deux subissent une perte d'activité totale à la dose de 2,5 Mrad. Etant donné que la plupart des antibiotiques sont délivrés commercialement à l'état sec, l'irradiation par les rayons gamma pourrait être un procédé de stérilisation valable. Cependant le cas des préparations de type collyres, suspensions, pommades, etc., doit être réservé.

INTRODUCTION

La stérilisation des médicaments par les rayonnements ionisants a été envisagée il y a déjà longtemps en raison de certains avantages: possibilité de stériliser des substances thermosensibles; stérilisation du produit dans son emballage définitif; pas de risque de contamination microbienne ultérieure; garantie de stérilité satisfaisante. Cependant, ce procédé de stérilisation n'est pas encore très utilisé pour de nombreuses raisons dont

* Travail effectué avec l'aide du Commissariat français à l'énergie atomique (contrat de recherche n° GR 751-746).

certaines ressortent du domaine médical; l'irradiation entraîne la formation de radicaux libres; on peut donc redouter que le médicament ne subisse une altération de ses caractères physiques et chimiques, une perte de son activité biologique; on peut aussi craindre la formation de corps nouveaux toxiques.

Un certain nombre de travaux ont été effectués sur ce problème et lors du Colloque organisé à Vienne en 1967 par l'AIEA [1-4], deux séances lui ont été consacrées.

Dans le domaine des antibiotiques, des publications anciennes existent; Controulis et coll. [5] en 1954 ont irradié au cobalt-60 des antibiotiques à l'état sec; cinq antibiotiques ont reçu une dose de 1,8 Mrad: pénicilline G (sel de potassium), chlorhydrate de streptomycine; auréomycine, terramycine et chloromycétine. Après remise en solution, l'activité biologique a été testée pendant une période allant de 3 jours à 90 jours. Les échantillons irradiés possèdent la même activité antibiotique que les échantillons non irradiés. A l'exception de la pénicilline G dont la solution, irradiée ou non, s'avère stable pendant 90 jours, les autres substances, qu'elles soient irradiées ou non, perdent progressivement leur activité antibiotique d'une manière identique. Les auteurs ont noté un léger changement de couleur (du blanc au gris) des échantillons irradiés de streptomycine et de chloromycétine; par contre, ils n'ont observé aucun changement de solubilité.

En 1957, Grainger et Hutchinson [6] ont irradié, au moyen de cobalt-60, de la pénicilline G en solution aqueuse aux doses de 1 et 2 Mrad, et à l'état sec à la dose de 1 Mrad sans observer de perte d'activité.

La même année, Colovos et Churchill [7] ont utilisé un accélérateur d'électrons de Van de Graaff pour irradier divers médicaments, parmi lesquels diverses formes de pénicilline G (à l'état sec, sauf une en suspension aqueuse). A la dose de $2 \cdot 10^6$ rad, il n'a pas été observé de perte d'activité antibactérienne. A des doses supérieures allant jusqu'à $20 \cdot 10^6$ rad, il y a destruction de la molécule de pénicilline.

De même, alors qu'il n'y a pas de phénomènes de toxicité avec l'antibiotique irradié à faible dose, il en est observé lorsque des doses de 8, 15 et $20 \cdot 10^6$ rad sont données.

Horne en 1956 [8] a irradié à l'état sec, à la dose de 2 Mrad, de la pénicilline G, de la streptomycine, du chloramphénicol, de la chlortétracycline et de l'oxytétracycline. Dans tous les cas les échantillons étaient stériles. Il n'a pas été observé de perte d'activité significative. De légers changements de couleur ont été notés. Le même auteur en 1958 [9] a irradié différentes formes, la plupart à l'état sec, de pénicilline, streptomycine, néomycine, polymyxine et bacitracine. Après une dose de 2,5 Mrad, certains échantillons acquièrent une faible coloration jaune, mais aucune perte d'activité n'est notée. Avec une dose de 25 Mrad, la coloration est plus marquée (brune pour certains échantillons), mais la perte d'activité est très faible.

Ogg en 1967 [2] a stérilisé à la dose de 2,5 Mrad une solution de framycétine pour usage ophtalmique; la solution irradiée devient acide et prend une coloration jaune paille, mais ne perd que faiblement son activité biologique; l'utilisation de cette solution dans 200 cas cliniques n'a révélé aucun signe de toxicité.

Hangay et coll. [3] ont stérilisé une pommade ophtalmique à l'acétate d'hydro-cortisone et au chloramphénicol, ainsi que les divers ingrédients séparément; deux doses ont été délivrées: 2,5 et 5 Mrad; la stabilité des produits a été testée après 3, 6, 12 et 24 mois de conservation. Le produit fini ne subit pas de changements importants. Le chloramphénicol prend une coloration jaune plus marquée qu'avant l'irradiation; il faut noter cependant que la conservation a le même effet sur le produit non irradié; par contre, il n'y a pas de perte d'activité antibiotique. Les auteurs ont trouvé après une dose de 2,5 Mrad des radicaux libres dans le chloramphénicol.

Holland et coll. [4] ont irradié trois tétracyclines: oxytétracycline, chlortétracycline et pyrrolidine-méthyltétracycline au moyen d'une source de cobalt-60 ayant un débit de dose de 0,7 Mrad/h, à l'air et à une température de + 4°C.

Ils ont étudié l'influence de l'irradiation sur les molécules d'antibiotiques par spectrophotométrie en u.v., chromatographie en couche mince et mise en évidence de radicaux libres, et ont effectué une étude bactériologique.

L'irradiation à l'état liquide entraîne une altération importante des molécules: les solutions aqueuses à 0,1 mg/ml perdent toute activité antibiotique avec une dose de 0,25 à 0,30 Mrad; des modifications importantes du spectre u.v. sont observées et il y a production de corps nouveaux non identifiés, décelables par chromatographie.

Il faut noter le pouvoir protecteur du mercaptoéthanol introduit dans la solution et la meilleure résistance des tétracyclines lorsqu'elles sont irradiées en solution dans le méthanol.

Par contre, irradiées à l'état sec à des doses allant jusqu'à 5 ou 10 Mrad, les tétracyclines ne subissent pas d'altération de leur activité antibiotique qui reste stable pendant la conservation. Il faut noter cependant la production de radicaux libres persistant 42 jours après l'irradiation.

En 1970, Pandula et coll. [10] ont étudié l'action des rayons gamma à la dose de 2,5 Mrad sur plusieurs médicaments, dont un antibiotique, le chloramphénicol; l'étude a été effectuée au moyen de procédés physico-chimiques (mesure du pH, spectrophotométrie u.v., polarographie). A l'état sec, il n'a pas été observé d'altération des substances; en solution aqueuse (à une concentration inférieure à 2%) au contraire, on constate dans les échantillons irradiés un abaissement du pH et des modifications du spectre d'absorption u.v.; cependant, avec des concentrations plus fortes (supérieures à 2%), l'altération est très faible. Les auteurs n'ont pas effectué d'étude microbiologique.

Crippa et coll. [11] en 1973 ont soumis de la rifampicine, des antibiotiques appartenant au même groupe des rifamycines et des noyaux chimiques apparentés, à une irradiation par le cobalt-60, à des doses progressives jusqu'à 2 Mrad; l'irradiation a été faite dans l'air et dans le vide. L'activité bactériostatique n'a pas été diminuée après irradiation à une dose de 0,5 Mrad. La rifampicine non irradiée émet un signal indiquant la présence de radicaux libres stables; après irradiation à 2 Mrad, le même signal a une intensité plus grande.

Plusieurs travaux ont donc été effectués sur des antibiotiques divers, irradiés à des doses différentes et dans des conditions différentes; par ailleurs, la plupart de ces publications sont anciennes et actuellement le nombre de substances antibactériennes et antifongiques utilisées couramment en médecine a beaucoup augmenté. Il nous a donc paru intéressant d'étudier l'activité bactériostatique d'un grand nombre d'antibiotiques modernes après irradiation à une dose usuelle de stérilisation; il nous semble, en effet, que le premier critère pour apprécier l'intérêt de la radiostérilisation sur les antibiotiques est celui de la conservation de l'activité antimicrobienne. C'est l'objectif essentiel de ce travail.

1. MATERIEL ET METHODES

1.1. Antibiotiques testés

Des représentants de chaque grande famille d'antibiotiques ont été étudiés:

Bêta-lactamines:
Groupe G: Pénicilline G
Groupe M: Méthicilline – oxacilline
Groupe A: Ampicilline – métampicilline – carbénicilline – ticarcilline
Groupe C: Céfalotine – céfaloridine

Aminosides:
Streptomycine
Néomycine – paromomycine – framycétine – kanamycine
Gentamicine
Tobramycine

Tétracyclines:
Tétracycline – minocycline

Chloramphénicol:
Chloramphénicol

Macrolides et antibiotiques apparentés:
Erythromycine – oléandomycine – lincomycine – virginiamycine – pristinamycine

Rifamycines:
Rifampicine (rifampine)

Antibiotiques polypeptidiques:
Colistine

Sulfamides et associations:
Triméthoprime + sulfaméthoxazole

Antibiotiques divers:
Vancomycine
Acide fusidique
Nitrofurantoïne
Acide nalidixique

Antibiotiques antifongiques:
Amphotéricine B
5-fluorocytosine.

Ces antibiotiques nous ont été livrés par les fabricants sous forme de poudre étalon titrée; ils ont été conservés à -20°C jusqu'à utilisation. Des solutions-mères titrées des différents antibiotiques sont préparées: solutions à 10 000 μg/ml réparties sous un volume de 1 ml et congelées à -20°C. Les solvants utilisés sont l'eau distillée stérile pour la plupart des substances, l'éthanol (chloramphénicol, tétracycline), le méthanol (rifampicine), le diméthyl-sulfoxide (amphotéricine B).

1.2. Préparation des échantillons à tester

Les antibiotiques ont été irradiés soit à l'état sec, soit à l'état liquide.

1.2.1. A l'état sec

On utilise soit des disques de papier du commerce (BBL) de 6 mm de diamètre, soit des disques de papier canson fabriqués au laboratoire, de même diamètre. Tous les disques sont stérilisés aux rayons gamma. Ils sont ensuite disposés sur un tamis métallique stérilisé situé sous une hotte à flux laminaire d'air stérile et imprégnés de 0,01 ml de la solution d'antibiotique diluée de manière à obtenir une concentration définie d'antibiotique sur le disque.

Les concentrations, variables avec chaque antibiotique, sont choisies de manière que la zone d'inhibition de la culture produite après diffusion dans le milieu gélosé ensemencé soit suffisamment large pour être mesurée facilement. Les concentrations utilisées sont celles généralement choisies dans la technique de l'antibiogramme. On laisse sécher les disques de 2 à 12 heures (selon le solvant) dans la hotte stérile à la température ambiante. Les disques sont conservés à -20°C en présence d'un dessiccateur jusqu'au moment du titrage.

1.2.2. A l'état liquide

Les mêmes solutions d'antibiotiques sont mises dans des tubes qui seront irradiés.

1.3. Irradiation

Nous avons utilisé l'irradiateur «Ammonite 407» au césium-137, fabriqué et mis à notre disposition par le Commissariat français à l'énergie atomique; le débit de dose est de 0,7 Mrad/h; les irradiations ont été effectuées à la température de +10°C, la chambre d'irradiation étant maintenue à cette température par une circulation de liquide réfrigérant.

Les doses suivantes ont été délivrées:

– antibiotiques à l'état sec: 1, 2,5, 5, 6, 7, 8 et 10 Mrad
– antibiotiques à l'état liquide: 2,5 et 5 Mrad.

1.4. Mesure de l'activité bactériostatique

Nous avons utilisé la méthode classique de diffusion en gélose. On utilise des plaques et des baguettes de verre délimitant une surface de 20 × 25 cm et stérilisées par la chaleur ou les rayons ultraviolets.

Le milieu de culture est du milieu de Mueller-Hinton dont le pH est ajusté en fonction du pH d'activité maximale des différents groupes d'antibiotiques: pH = 6 pour les bêta-lactamines, les tétracyclines, etc., pH = 8 pour les aminosides, pH = 7 pour les autres antibiotiques.

1.5. Souches microbiennes

Sarcina lutea ATCC 9341 est utilisée pour la plupart des antibiotiques. Une suspension stock contenant $4{,}5 \cdot 10^6$ germes/ml est préparée et conservée à +4°C pendant 1 à 2 mois; la suspension d'emploi est préparée pour chaque titrage par dilution au 1/5 de la solution stock.

Pseudomonas aeruginosa L 74/09/39 est utilisée pour la carbénicilline et la ticarcilline; dilution à 10^{-3} d'une culture en bouillon cœur-cervelle de 24 heures (environ 10^5 à 10^6 germes/ml).

Escherichia coli Monod, préparée dans les mêmes conditions, et utilisée pour la colistine, la nitrofurantoïne et l'acide nalidixique.

Candida albicans IP 646 est utilisée pour la 5-fluoro-cytosine et l'amphotéricine B (mise en suspension dans 20 ml de soluté salé isotonique d'une colonie ayant été cultivée pendant 3 jours sur milieu de Sabouraud, ce même milieu est utilisé pour le test).

1.6. Protocole

Le milieu gélosé en surfusion à 45°C est ensemencé avec 1 ml de la suspension microbienne convenable et est coulé dans les plaques de verre, sous un volume de 35 ml donnant une épaisseur de 3 à 4 mm.

Après solidification, on dispose à la surface du milieu:

– soit des disques de papier, imprégnés de l'antibiotique à tester, irradiés et non irradiés; il y a 9 disques par plaque dont 2 disques non irradiés (témoins),
– soit des cylindres d'acier stérile de 8 mm de diamètre intérieur, que l'on remplit au rhéomètre avec 0,2 ml de la solution d'antibiotique irradiée ou non irradiée (2 témoins).

Pour une dose de rayons gamma, trois mesures sont effectuées.

Les plaques sont ensuite portées à l'étuve à 37°C pendant 24 ou 48 heures selon les souches. La mesure des diamètres des zones d'inhibition est effectuée, soit au pied à coulisse, soit le plus souvent au lecteur optique.

1.7. Interprétation statistique des résultats

On effectue la moyenne des diamètres des zones d'inhibition produites par les témoins non irradiés et par les échantillons irradiés.

L'écart type σ par rapport à la moyenne des diamètres des zones d'inhibition des témoins est calculé; la variation du diamètre des échantillons n'est considérée comme significative (à 95%) que si l'écart par rapport à la moyenne est supérieur à 2 σ. Si la variation est significative, le pourcentage de diminution du diamètre d'inhibition de l'échantillon est calculé.

2. RESULTATS

2.1. Irradiation à l'état sec

La perte d'activité est indiquée sur les tableaux par un pourcentage de diminution du diamètre d'inhibition. Les tableaux I et II montrent les résultats globaux obtenus. Aucun des 33 antibiotiques testés ne subit de perte d'activité à la dose de 1 Mrad. A 2,5 Mrad, 4 d'entre eux sont légèrement inactivés (au maximum 5%). Entre 5 et 10 Mrad, il apparaît une perte d'activité pour 15 autres antibiotiques. 14 antibiotiques gardent une activité antibiotique intacte après avoir reçu 10 Mrad.

Les tableaux III à VII donnent les résultats regroupés par familles d'antibiotiques.

Parmi les bêta-lactamines (tableau III), on remarque que l'ampicilline, la métampicilline, la carbénicilline, et la ticarcilline, qui ont une structure chimique très voisine (groupe A), sont très stables. Par contre, dans le groupe M, il y a une différence de stabilité entre l'oxacilline et la méthicilline, la première étant plus sensible.

Quant aux céphalosporines, la différence est encore plus nette, la céfalotine montrant une perte d'activité dès 2,5 Mrad, alors que la céfaloridine est stable jusqu'à 10 Mrad.

Parmi les aminosides (tableau IV), la streptomycine et la kanamycine sont légèrement inactivées à partir de 5 Mrad. La néomycine, la paromomycine, et la soframycine sont stables jusqu'à 6 Mrad. La tobramycine est peu stable, alors que la gentamicine est totalement stable.

Les tétracyclines et la colistine (tableau V) sont remarquablement stables; parmi les macrolides (tableau VI), l'érythromycine et la lincomycine sont stables jusqu'à 10 Mrad, par contre la virginiamycine et la pristinamycine subissent une perte d'activité dès 2,5 Mrad.

Les antibiotiques divers (tableau VII) sont stables, jusqu'à 7 Mrad au moins, sauf l'acide fusidique qui subit une perte d'activité à partir de 5 Mrad.

2.2. Irradiation des antibiotiques en solution

Les résultats figurent sur les tableaux VIII et IX. Lorsqu'ils sont irradiés en solution, les antibiotiques sont rapidement inactivés puisque à 2,5 Mrad, 31 d'entre eux subissent une perte totale de leur activité. Deux antibiotiques ne sont pas dans ce cas, mais la tobramycine perd cependant 67% de son activité à cette dose. A 5 Mrad, tous sont inactivés, sauf l'érythromycine.

TABLEAU I. IRRADIATION A L'ETAT SEC
Nombre d'antibiotiques ne présentant aucune perte d'activité aux différentes doses

Dose (Mrad)	0	1	2,5	5	6	7	8	10
Nombre d'antibiotiques	33	33	29	25	25	22	19	14

TABLEAU II. IRRADIATION A L'ETAT SEC
Doses à partir desquelles les antibiotiques subissent une perte d'activité

2,5 Mrad	*	5 Mrad	*	7 Mrad	*	8 Mrad	*	10 Mrad	*
Céfalotine	2,3%								
Tobramycine	4,7%								
Virginiamycine	4 %								
Pristinamycine	2 %	Oxacilline	2 %						
		Kanamycine	5 %						
		Acide fusidique	3 %						
		Streptomycine	2,4%	Pénicilline	1,4%				
				Néomycine	8,4%				
				Paromomycine	12,3%	Soframycine	4,5%		
						Oléandomycine	1,2%		
						Rifampicine	1,6%	Méthicilline	1,7%
								Métampicilline	4,8%
								Triméthoprime + Sulfaméthoxazole	3,4%
								Vancomycine	1,7%
								5-fluorocytosine	2,2%

* Pourcentage de diminution du diamètre d'inhibition.

TABLEAU III. IRRADIATION A L'ETAT SEC
Perte d'activité des bêta-lactamines aux différentes doses

Antibiotiques	2,5 Mrad	5 Mrad	7,5 Mrad	10 Mrad
Pénicilline G	0	0	1,4%	6%
Méthicilline	0	0	0	1,7%
Oxacilline	0	2%	4,5%	10,5%
Ampicilline	0	0	0	0
Métampicilline	0	0	0	4,8%
Carbénicilline	0	0	0	0
Ticarcilline	0	0	0	0
Céfalotine	2,3%	5,7%	8,3%	13,5%
Céfaloridine	0	0	0	0

TABLEAU IV. IRRADIATION A L'ETAT SEC
Perte d'activité des aminosides aux différentes doses

	2,5 Mrad	5 Mrad	6 Mrad	7 Mrad	8 Mrad	10 Mrad
Streptomycine	0	2,4%	4%	5%	-	7,5%
Néomycine	0	0	0	8,4%	9,7%	15%
Paromomycine	0	0	0	12,3%	-	11%
Soframycine	0	0	0	0	4,5%	6%
Kanamycine	0	5%	6%	5,2%	6,3%	7,4%
Gentamicine	0	0	0	0	0	0
Tobramycine	4,7%	18,5%	-	26,3%	33,7%	37,4%

2.3. Contrôles de stérilité

Des contrôles de stérilité effectués sur les disques témoins et les disques irradiés à la dose de 2,5 Mrad ont tous été négatifs.

3. DISCUSSION

La méthode que nous avons utilisée exprime la perte d'activité après irradiation en pourcentage de diminution de l'aire d'inhibition de la culture

TABLEAU V. IRRADIATION A L'ETAT SEC
Perte d'activité des tétracyclines et des antibiotiques polypeptidiques aux différentes doses

	2,5 Mrad	5 Mrad	6 Mrad	7 Mrad	8 Mrad	10 Mrad
Tétracyclines:						
Tétracycline	0	0	0	0	0	0
Minocycline	0	0	0	0	0	0
Antibiotiques polypeptidiques:						
Colistine	0	0	0	0	0	0

TABLEAU VI. IRRADIATION A L'ETAT SEC
Perte d'activité des macrolides et apparentés aux différentes doses

	2,5 Mrad	5 Mrad	6 Mrad	7 Mrad	8 Mrad	10 Mrad
Erythromycine	0	0	0	0	0	0
Lincomycine	0	0	0	0	0	0
Oléandomycine	0	0	0	0	1,2%	2,1%
Virginiamycine	4%	11%	11,3%	11,3%	15,7%	18,8%
Pristinamycine	2%	5,3%	-	8,9%	7,4%	14,8%

TABLEAU VII. IRRADIATION A L'ETAT SEC
Perte d'activité des antibiotiques divers aux différentes doses

	2,5 Mrad	5 Mrad	6 Mrad	7 Mrad	8 Mrad	10 Mrad
Chloramphénicol	0	0	0	0	0	0
Vancomycine	0	0	0	0	0	1,7%
Rifampicine	0	0	0	0	1,6%	2,4%
Triméthoprime + sulfaméthoxazole	0	0	0	0	0	3,4%
Nitrofurantoïne	0	0	0	0	0	0
Acide nalidixique	0	0	0	0	0	0
Acide fusidique	0	3%	7%	9,7%	13,5%	15,8%
Amphotéricine B	0	0	0	0	0	0
5-fluorocytosine	0	0	0	0	0	2,2%

TABLEAU VIII. IRRADIATION EN SOLUTION
Nombre d'antibiotiques subissant une perte totale d'activité aux différentes doses

Dose (Mrad)	0	2,5	5
Nombre d'antibiotiques	33	31	32

TABLEAU IX. IRRADIATION EN SOLUTION
Perte d'activité de la tobramycine et de l'érythromycine

	2,5 Mrad	5 Mrad
Tobramycine	67%	100%
Erythromycine	9%	12%

bactérienne; il serait certes souhaitable, mais beaucoup plus long et onéreux, de juger l'efficacité thérapeutique des antibiotiques irradiés au cours des infections expérimentales. Cependant nos résultats paraissent devoir être considérés comme valables, car la technique de diffusion en gélose est couramment utilisée pour les titrages d'antibiotiques et l'efficacité de l'antibiothérapie en clinique est en accord avec les résultats de l'antibiogramme.

Les 33 antibiotiques étudiés irradiés à l'état sec se sont montrés relativement stables. A 2,5 Mrad, dose usuelle pour la radiostérilisation, 4 seulement subissent une légère perte d'activité ne dépassant pas 5%. Les différentes publications antérieures avaient montré que pénicilline G, streptomycine, néomycine, soframycine, différentes tétracyclines, chloramphénicol, polymyxine, rifampicine, irradiés à l'état sec à des doses comprises entre 0,5 et 2,5 Mrad, conservaient totalement leur activité; nous obtenons pour ces antibiotiques des résultats semblables.

Pour des doses plus fortes, Colovos et Churchill [7] avaient observé que la pénicilline G subissait une destruction (dose de l'ordre de 8 Mrad), par contre Horne [9] n'a pas trouvé de diminution sensible d'activité après une dose de 25 Mrad; nous obtenons une perte d'activité à partir de 7 Mrad, mais qui ne dépasse pas 6% à 10 Mrad. Horne [9] avait également observé que la streptomycine irradiée à 25 Mrad conservait son activité alors que nous notons une diminution d'activité à partir de 5 Mrad. Comme Holland et coll. [4] nous avons trouvé que les tétracyclines sont très stables à l'état sec.

Ces légères différences observées entre les publications sont peut-être liées à des différences techniques ou à des différences portant sur l'état d'hydratation plus ou moins important des antibiotiques; il semble bien cependant que, irradiés à l'état sec, les antibiotiques conservent leur activité antimicrobienne. Par contre, en solution, ils résistent mal à l'irradiation, puisque 31 sont totalement inactivés à 2,5 Mrad. Grainger et Hutchinson [6],

Colovos et Churchill [7], Horne [8] n'avaient pas observé de perte d'activité de la pénicilline G lorsqu'elle était irradiée en solution à la dose de 2 ou 2,5 Mrad et nos résultats sont en discordance avec ces travaux. Holland et coll. [4] avaient montré que les tétracyclines en solution sont totalement inactivées avec une dose de 0,25 à 0,30 Mrad. Ces résultats concordent donc avec les nôtres.

CONCLUSION

La stérilisation des antibiotiques à l'état sec par les rayons gamma à la dose de 2,5 Mrad nous paraît donc possible. Cependant, des recherches portant sur les éventuelles modifications physiques des molécules et la détection de produits de dégradation devront être menées à bien avant de pouvoir passer à l'utilisation clinique.

REFERENCES

[1] INTERNATIONAL ATOMIC ENERGY AGENCY, Radiosterilization of Medical Products (Proc. Symp. Budapest, 1967), IAEA, Vienna (1967) 458 p.

[2] OGG, A.J., «Gamma ray sterilization in ophtalmology», Radiosterilization of Medical Products (Proc. Symp. Budapest, 1967), IAEA, Vienna (1967) 49-54.

[3] HANGAY, G., HORTOBÁGYI, G., MURÁNYI, G., «Sterilization of hydrocortisone eye-ointment by γ-irradiation, I. Physical and chemical aspects», Ibid. p.55-62.

[4] HOLLAND, J., ANTONI, F., GALATZEANU, I., SCHULMAN, M., KOZINETS, G., «Effect of gamma irradiation (^{60}Co) on the tetracyclines», Ibid. p.69-81.

[5] CONTROULIS, J., LAWRENCE, C.A., BROWNELL, L.E., The effect of gamma radiation on some pharmaceutical products, J. Am. Pharm. Assoc. 43 (1954) 65-69.

[6] GRAINGER, H.S., HUTCHINSON, W.P., Irradiation of Penicillin, J. Pharm. Pharmacol. 9 (1957) 343.

[7] COLOVOS, G.C., CHURCHILL, B.W., The electron sterilization of certain pharmaceutical preparations, J. Am. Pharm. Assoc. 46 (1957) 580-83.

[8] HORNE, T., Sterilization by radiation, Pharm. J. (1956) 27-29.

[9] HORNE, T., «Biochemical applications of large radiation sources with special reference to pharmaceutical products», 2nd Int. Conf. Peaceful Uses At. Energy (Proc. Conf. Geneva, 1958) 26, UN, New York (1958) 338.

[10] PANDULA, E., FARKAS, E., NAGYKALDI, A., Untersuchung von strahlensterilisierten Arzneimitteln und ihren wässerigen Lösungen, Pharmazie 25 (1970) 254-58.

[11] CRIPPA, P.R., TEDESCHI, R., VECLI, A., The secondary effects of sterilization of the Rifampicin by gamma irradiation, Farmacologia 28 (1973) 226-32.

DISCUSSION

E.P. PAVLOV: We consider investigation of the antimicrobial activity of the irradiated antibiotics to be the first step towards the radiation sterilization of these substances on an industrial level. In addition to the physico-chemical studies mentioned, it is very important to conduct toxicological investigations based on official health requirements. What kinds of tests on the more promising antibiotics are necessary before the radiation sterilization of these substances in industry is permitted?

J. FLEURETTE: Any irradiated antibiotic is regarded in effect as a new drug and, in accordance with the French Pharmacopoeia, has to be subjected to analytical, pharmacological, toxicological and chemical investigations by experts.

N.G.S. GOPAL: Do the decomposition products of tetracycline, i.e. 4-epi-, anhydro-, and 4-epi-anhydro-tetracyclines, interfere with the microbiological assay of tetracycline? Do these products have any antibiotic activity?

J. FLEURETTE: We have had no problems on this score. The tetracycline products that we used, as supplied by the manufacturer, are guaranteed pure.

IRRADIATED COBRA (Naja naja) VENOM FOR BIOMEDICAL APPLICATIONS

S.R. KANKONKAR, R.C. KANKONKAR,
B.B. GAITONDE
Haffkine Institute,
Parel, Bombay

S.V. JOSHI
Biophysics Department,
Cancer Research Institute,
Bombay,
India

Abstract

IRRADIATED COBRA (Naja naja) VENOM FOR BIOMEDICAL APPLICATIONS.

Ionizing radiation is known to cause damage to proteins in aqueous solutions in a selective manner, thereby producing remarkable changes in their properties. Since venoms are very rich in proteins, it was felt that they would also show such changes upon irradiation. It was of interest to know if one could get rid of the toxicity and retain the immunogenicity of the venom by suitable choice of radiation dose and strength of venom solution. If so, the method could be profitably exploited for the rapid preparation of venom toxoid and this could be expected to have many applications in the biological sciences. Accordingly, laboratory investigations were undertaken on the effect of gamma radiation on cobra (Naja naja) venom. To avoid drastic changes, solutions of cobra venom having low protein content were irradiated with gamma radiation from a cobalt-60 source. The results obtained with 0.01 to 1.0% venom solutions are found to be encouraging. The solutions did not manifest any toxicity in mice. For the immunogenicity test, guinea pigs were immunized with varying doses of the irradiated cobra venom and the immunized guinea pigs were found to survive when challenged with as big a dose as 10 MLD (i.e. minimum lethal dose, approximately 1 mg). The paper describes the experimental details and the results of the observations.

1. INTRODUCTION

Even today snake-bites are known to take a heavy toll of human life in India. The number of deaths are reported to be around 20 000 to 30 000 per year (Ahuja and Singh [1]). The majority of these cases occur in the villages. Mostly the villagers go barefoot either through poverty or habit. The fields, farms and forests where they work are often infested with deadly venomous snakes. The victim has often to be carried quite long distances before even first-aid can be applied. Occasionally the envenomation is so rapid that antivenin therapy alone may fail to save the life of the victim. What is true of villagers is equally true of army personnel, especially when camping or fighting in jungle areas.

Venom toxoid aimed at active immunization would thus be a significant step towards saving a snake-bite victim. However, the quality of the venom toxoids prepared so far with the use of chemical reagents is far from satisfactory. Wiener [2] used formalin to prepare a toxoid of tiger snake (Notechis scutatus) venom. Sawai et al. [3-7] used formalin, dihydrothioctic acid (DHTA), and ethylene-di-amine tetra acetic acid (EDTA) to detoxify snake venoms. Fukuyama and Sawai [8] also employed formalin to detoxify cobra venom. Okonogi and Hattori [9] used alcohol to prepare venom toxoid. According to Sadahiro et al. [10], though the venom toxoids obtained by these methods were completely devoid of toxicity, they were not good immunogens.

The action of ionizing radiations on proteins in aqueous solutions is known to be highly selective (Alexander and Hamilton [11] and Pavlovskaya and Pasynskii [12]). Whereas carboxyl and sulphhydryl groups are most readily susceptible, the peptide bond in proteins is found to be relatively radiostable (Joshi [13]). The extent of damage by ionizing radiations can also be controlled much more easily than that by chemical methods. Moreover, while irradiation of proteins at lower concentrations would be mainly by indirect mechanism, direct hit effects come into play at higher concentrations Therefore, it is of interest to know if by suitable choice of radiation dose and venom concentration it is possible to eliminate the toxicity of the snake venom and yet retain its immunogenicity.

So far, not much work in this direction has been done on snake venoms although irradiation procedures have already yielded several efficient vaccines (Jovanovic [14] and Shibayeva et al. [15]).

Therefore, it was decided to irradiate snake venoms with gamma radiations and undertake investigations with a view to prepare a venom toxoid. Cobra being one of the main offenders in our country, and the symptomatology of its envenomation being well understood, its venom was the first chosen.

2. MATERIALS AND METHODS

Cobra venom

The venom was from milkings collected throughout the year from cobras procured from different parts of the country and was obtained in lyophilized form from the Zoonosis Department of Haffkine Institute.

The irradiation in these investigations was provided by a cage-type gamma chamber with a ^{60}Co source at its centre. Irradiation was carried out with gamma rays by exposing 20-ml aliquots of solutions taken in glass-stoppered cylindrical cuvettes made of Corning glass and placed at the central space of the chamber. The dose-rate as determined by Fricke's ferrous sulphate method [16] ranged from 1.3×10^5 to 0.85×10^5 rad/h during the course of investigation.

Venom solutions

Each time, fresh venom solutions of desired strength were prepared in sterile normal saline solution prepared in triple distilled water.

Venom irradiation

Starting from an arbitrary low dose, venom solutions of different strength were exposed to increasing doses of gamma radiations in the presence of air until the toxicity was almost lost. Aliquots of solutions kept aside before irradiation served as the control.

TABLE I. IMMUNIZATION OF GUINEA PIGS WITH CRUDE COBRA VENOM

Route	Subcutaneous
Schedule	Interval between first two injections: one month Thereafter weekly
Doses	(μg) First 2 injections = 50.0 Next 3 injections = 100.0
Total quantity of venom injected	400 μg

Note: The first injection was administered with Freund's complete adjuvant.

TABLE II. IMMUNOGENIC RESPONSE IN GUINEA PIGS AFTER IMMUNIZATION WITH CRUDE COBRA VENOM

No. of immunizing doses after which challenged	No. of MLDs of crude venom with which challenged	No. of survivals / No. of animals challenged	
		Control guinea pigs	Immunized guinea pigs
5	2	2/2	2/2
	3	0/2	2/2
	5	0/2	2/2

All the animals were challenged 8 d after the last, i.e. the 5th immunizing dose.

Toxicity determination in mice

The toxicity of venom solutions both before and after irradiation was determined by injecting 0.5 ml of solution intravenously (i/v) in albino mice (Kasauli strain), weighing between 18 to 20 g. Two mice were used for each dilution and the minimum amount of venom required to kill both the animals in 24 h was accepted as the minimum lethal dose (MLD) for the mouse.

Toxicity determination in guinea pigs

The minimum lethal dose (MLD) in guinea pigs was determined by injecting subcutaneously 1 ml solution containing different amounts of non-irradiated cobra venom into guinea pigs weighing 250-300 g. For each of the dilutions two guinea pigs were used and observed for 24 h. The smallest

TABLE III. IMMUNIZATION OF GUINEA PIGS WITH 0.1% IRRADIATED COBRA VENOM SOLUTION

Route	Subcutaneous
Schedule	Interval between first two injections: one month Thereafter weekly
Doses	(μg) First 4 injections = 50.0 Next 3 injections = 100.0 Next 2 injections = 200.0
Total quantity of venom injected	After 7 injections = 500.0 After 8 injections = 700.0 After 9 injections = 900.0

Note: The first injection was administered with Freund's complete adjuvant.

quantity of venom that killed both the guinea pigs was taken as the MLD for them. The determination of toxicity in guinea pigs was necessary for the study of immunogenicity as it was decided to immunize guinea pigs with a sub-lethal dose of venom to be subsequently challenged with a supra-lethal dose.

Immunization of guinea pigs with crude cobra venom

To compare the immunogenicity of irradiated cobra venom with that of normal venom, it was necessary to study the immunogenic response in the same species. For the present investigation, we decided to use guinea pigs of the same strain and weight as mentioned before. Throughout the investigation, immunizing doses and challenging doses were administered subcutaneously.

For the study of immunogenicity of normal crude venom, the animals received two injections of 50 μg each at an interval of one month. Thereafter, they received three more injections of 100 μg at weekly intervals. Except for the first injection, which was administered with Freund's complete adjuvant, all other injections were plain (Table I). Just prior to administering the next immunizing dose, the animals were bled and serum was collected. The serum so collected was tested for the presence of antibodies by using the micro-gel diffusion technique. Eight days after the administration of the last immunizing dose the animals were challenged with 2, 3 and 5 times the MLD of crude venom (Table II). Similarly guinea pigs that were kept as control, without the administration of any injection, were also challenged with corresponding doses of the crude venom. This was necessary to ascertain that the challenging doses were sufficient to kill these guinea pigs in spite of ageing and gaining weight during the period of immunization.

TABLE IV. IMMUNOGENIC RESPONSE IN GUINEA PIGS AFTER IMMUNIZATION WITH 0.1% IRRADIATED COBRA VENOM

No. of immunizing doses after which challenged	No. of MLDs of crude venom with which challenged	No. of survivals / No. of animals challenged			
		Control guinea pigs	Immunized guinea pigs		
			0.24 Mrad	0.33 Mrad	0.66 Mrad
7	3	0/2	0/2	1/2	2/2
	5	0/2	0/2	0/2	2/2
8	5	0/2	1/2	1/2	2/2
	7	0/2	0/2	1/2	1/2
9	7	0/2	1/2	2/2	2/2
	10	0/2	1/2	2/2	2/2

All the animals were challenged 8 d after the last immunizing dose.

TABLE V. IMMUNIZATION OF GUINEA PIGS WITH 0.5 AND 1.0% IRRADIATED COBRA VENOM SOLUTION

Route	Subcutaneous
Schedule	At weekly intervals
Doses	(μg) 1st injection = 75.0 Next 3 injections = 100.0 Next 1 injection = 150.0 Next 2 injections = 200.0 Next 3 injections = 250.0 Next 1 injection = 500.0
Total quantity of venom injected	After 6 injections = 725.0 After 7 injections = 925.0 After 9 injections = 1425.0 After 11 injections = 2175.0

TABLE VI. IMMUNOGENIC RESPONSE IN GUINEA PIGS AFTER IMMUNIZATION WITH 0.5 AND 1.0% IRRADIATED COBRA VENOM

No. of immunizing doses after which challenged	No. of MLDs of crude venom with which challenged	No. of survivals / No. of animals injected				
		Control	0.5% venom solution		1.0% venom solution	
			1.0 Mrad	1.5 Mrad	2.0 Mrad	2.0 Mrad
6	3	0/2	2/2	2/2	1/2	2/2
	5	0/2	2/2	2/2	1/2	1/2
7	5	0/2	2/2	1/2	1/2	1/2
9	5	0/2	1/2	2/2	2/2	1/2
	7	0/2	1/2	1/2	2/2	1/2
11	7	0/2	2/2	1/2	2/2	1/2

All the animals were challenged 8 d after the last immunizing dose.

Immunization of guinea pigs with irradiated cobra venom

In the case of irradiated cobra venom, three different solutions of 0.1% concentration detoxified by 0.24, 0.33 and 0.66 Mrad were used first for the immunization of guinea pigs. The first immunizing dose constituted 50 μg of irradiated venom with Freund's complete adjuvant. One month after the administration of the first injection all other injections were given at weekly intervals. The first three weekly doses were 50 μg, followed by three of 100 μg and two of 200 μg of the venom (Table III).

Animals at different stages of immunization were challenged with different doses of the crude venom. Thus animals which had received seven immunizing doses were challenged with 3 and 5 times the MLD of the crude venom, while those which received eight and nine doses were challenged with 5 and 7, and 7 and 10 times the MLD respectively, including the controls (Table IV).

Three solutions of 0.5% concentration, detoxified by 1, 1.5 and 2 Mrad of gamma radiations, as well as 1.0% solution detoxified by 2 Mrad, were taken for further immunogenic study. All the injections were administered at one-week intervals. The first dose of 75 μg was followed by three more doses of 100 μg each, one of 150 μg, two of 200 μg, three of 250 μg and one of 500 μg (Table V).

Animals thus immunized were challenged at different stages of immunization. Eight days after the administration of the last dose, animals that received six immunizing doses were challenged with 3 MLDs and animals immunized with 7 doses were challenged with 5 MLDs; those immunized with 9 doses were challenged with 5 and 7 times the MLD and the ones which received 11 doses were challenged with 7 times the MLD of the crude venom (Table VI).

TABLE VII. TOXICITY OF 0.01 AND 0.1% IRRADIATED COBRA VENOM SOLUTIONS

Irradiation doses (Mrad)	Quantity of venom administered i/v in mice (μg)	No. of survivals / No. of animals injected — Venom solution 0.01%	0.1%
0.03	50.0	2/2	0/2
0.06	50.0	-	0/2
0.12	50.0	-	2/2
	100.0	-	0/2
0.24	100.0	-	2/2
	250.0	-	2/2
	500.0	-	0/2
0.33	500.0	-	1/2
0.66	500.0	-	2/2

3. RESULTS

Venom toxicity

The MLD of crude venom in mice was 11 μg intravenously (i/v) while MLD in guinea pigs was 100 μg subcutaneously (s/c).

The results of the toxicity of cobra venom in concentrations ranging from 0.01 to 1.0%, irradiated with different doses of gamma radiations, are given in Tables VII and VIII. It can be seen from Table VII that a very small radiation dose was enough to detoxify venom in a very dilute solution such as 0.01%. Even with a dose of 0.03 Mrad, the 0.01% venom solution lost a considerable amount of toxicity. Thus, despite receiving 50 μg venom intravenously (i/v), that is a quantity equivalent to 5 times the MLD of crude venom, the mice survived. A slightly higher dose was required to detoxify 0.1% venom solution. Doses of 0.03 and 0.06 Mrad had a very slight effect on the toxicity of 0.1% venom solution. A dose of 0.12 Mrad was necessary to reduce the toxicity of 0.1% venom solution to the extent to which the toxicity of 0.01% venom solution was reduced by a 0.03-Mrad dose. A 0.24-Mrad dose was highly effective in detoxifying this solution and mice could survive a challenge with a dose of irradiated venom as high as 250 μg, i.e. 25 times the MLD of the crude venom. With still higher doses, namely 0.33 and 0.66 Mrad, the toxicity was reduced further, and an injection of 500 μg of venom solution irradiated with these doses produced no mortality in the animals tested (Table I).

For detoxification 0.5 and 1.0% cobra venom solutions needed higher radiation doses.

TABLE VIII. TOXICITY OF 0.5 AND 1.0% IRRADIATED COBRA VENOM SOLUTIONS

Irradiation doses (Mrad)	Quantity of venom administered i/v in mice (μg)	No. of survivals / No. of animals injected	
		Venom solution	
		0.5%	1.0%
1.0	50.0	2/2	0/2
	100.0	2/2	-
	250.0	2/2	-
	500.0	1/2	-
1.5	50.0	2/2	2/2
	100.0	2/2	0/2
	500.0	2/2	-
2.0	100.0	2/2	2/2
	150.0	2/2	2/2
	200.0	2/2	0/2
	1000.0	2/2	-
	2000.0	2/2	-
	2500.0	2/2	-

To detoxify 0.5% solution to a considerable extent 1 Mrad was required, although this dose was ineffective in extending the toxicity of 1.0% venom solution. At this high dose, 0.5% solution was completely detoxified and mice could withstand an intravenous dose as high as 2.5 mg (Table II).

These results clearly showed that irradiation very efficiently decreased the toxicity of cobra venom. Higher concentrations of venom solution required higher doses of radiation for detoxification.

Immunogenicity of irradiated venom vis-à-vis normal venom

Guinea pigs immunized with crude venom survived a challenge of 5 times MLD (see Table IV).

The immunogenic response in guinea pigs immunized with 0.1% irradiated cobra venom solution as described under Section 2 was determined by challenging the immunized animals at different stages of immunization with varying amounts of normal cobra venom. The results are summarized in Table VI. It can be seen from the Table that 7 immunizing doses of 0.1% venom solution, detoxified by 0.66 Mrad, were sufficient to protect animals thus immunized, when challenged with 5 times the MLD of the crude venom, while those immunized with the same concentration of venom but detoxified either by 0.24 or 0.33 Mrad, died. Challenge after the administration of eight immunizing doses showed that the state of immunity remained more or less the same as that of animals immunized with 7 doses. But immunization with 9 doses enhanced the immunity of animals, and the guinea pigs immunized with the venom solutions detoxified by 0.33 and 0.66 Mrad showed 100% survival when challenged with 7 times the MLD. The immunity conferred by 0.1% solution detoxified by 0.33 and 0.66 Mrad appeared to be more or less the same since animals challenged with 10 times the MLD of the crude venom also survived.

Immunity conferred by immunization with 0.5 and 1.0% irradiated cobra venom solutions is given in Table VIII. By comparing protection offered to animals immunized with 0.5 and 1.0% cobra venom solutions, it can be concluded that 0.5% cobra venom solution offered better protection and that there was not much difference in protection offered to animals immunized with 0.5% venom solution detoxified either by 1 or 2 Mrad. Comparison between animals immunized with 0.1% and those immunized with 0.5% irradiated venom solutions appeared to indicate that 0.1% venom solution detoxified by 0.33 and 0.66 Mrad offered rapid immunity. These results also show that irradiated venom can be safely used as a toxoid since there is a substantial detoxification while immunogenicity is hardly affected.

4. DISCUSSION

The results obtained in the present experiments appear encouraging. The irradiation dose appears to depend on the irradiation conditions, the higher concentrations needing a higher dose for detoxification. Despite high radiation doses, immunogenicity seems to remain unaffected.

The mechanisms of detoxification by radiation needs detailed investigation, which is in progress. Since low radiation doses are effective when dilute solutions of venom are used, it appears that gamma radiation achieves detoxifying action mainly by indirect mechanism. Such an indirect action in an aqueous system is mainly through short-lived radicals such as OH, H, HO_2 and free electrons and molecular systems such as H_2O_2 and O_2 (Alvin and Edvin [17]). Cobra venom toxins are reported to be polypeptides containing 61 to 63 or 72 amino acid residues (Yang [18]). There are hydrophilic regions in snake venom toxins. Of course, crude venom is a mixture of several toxic factors. In biological systems, the end effects of irradiation are often oxidative. Decarboxylation, de-amination and saturation of double bonds are most commonly observed effects at molecular level. It therefore seems likely that the part responsible for the toxicity of the venom is associated with above radiolabile group while the immunogenic part of the venom complex is located in a small restricted portion which is either immune to radiation doses or structurally shielded from it.

ACKNOWLEDGEMENTS

We are grateful to Dr. K.S. Korgaonkar, Head of the Biophysics Department of Cancer Research Institute, for his keen interest and some useful suggestions.

REFERENCES

[1] AHUJA, M.L., SINGH, G., "Snakebite in India", Venoms (BUCKLEY, E.E., PORGES, N., Eds), Publication No.44, Amer. Assoc. Advance. Sci., Washington (1956) 341-57.

[2] WIENER, S., Active Immunization of man against the venom of the Austrialian Tiger snake (Notechis scutatus), Amer. J. Trop. Med. Hyg. 9 (1960) 284-92.

[3] SAWAI, Y., KAWAMURA, Y., MAKINO, M., FUKUYAMA, T., SHIMIZU, T., HOKAMA, Z., LIN, Y.H., MIYAZAKI, S., MUTO, S., Studies on the immunization of borses by Habu venom treatment by Freunds Adjuvant or dihydrothioctic acid, Jap. J. Bact. 20 (1965) 271-73.

[4] SAWAI, Y., KAWAMURA, Y., MAKINO, M., FUKUYAMA, T., SHIMIZU, T., LIN, Y.H., KURIBAYASHI, H., ISHII, T., Studies on the Habu-snake (Trimeresurus flavoviridis) venom toxoid 1. On the antigenicity of toxoid, Jap. J. Bact. 21 (1966) 32-41.

[5] SAWAI, Y., KAWAMURA, Y., FUKUYAMA, T., KEEGAN, H.L., Studies on the inactivation of snake venom by dihydrothioctic acid, Jap. J. Exp. Med. 37 (1967) 121-28.

[6] SAWAI, Y., KAWAMURA, Y., FUKUYAMA, T., OKONOGI, T., Studies on the improvement of treatment of Habu (Trimeresurus flavoviridis) bites. 7. Experimental studies of the Habu venom toxoid by dihydrothioctic acid, Jap. J. Exp. Med. 39 (1969) 109-17.

[7] SAWAI, Y., CHINZEI, H., KAWAMURA, Y., OKONOGI, T., Studies on the improvement of treatment of Habu (Trimeresurus flavoviridis) bites. 9. Studies on the immunogenicity of the purified Habu venom toxoid by alcohol precipitation, J. Formosa, Med. Assoc. 71 (1972) 421-30.

[8] FUKUYAMA, R., SAWAI, Y., "Experimental study on cobra venom toxoid", in Proc. Symp. Toxins, Gunma, July, 1972. Summary reprinting from Jap. J. Med. Sci. Biol. 26 (1973) 32-33.

[9] OKONOGI, T., HATTORI, Z., "Attenuation of Habu-snake (Trimeresurus flavoviridis) venom treatment with alcohol and its effect as immunizing antigen", Jap. J. Bacteriol. 23 (1968) 137-44.

[10] SADAHIRO, S., KONDO, S., YAMAUCHI, H., KONDO, H., MURATA, R., "Studies on immunogenicity of toxoids from Habu (Trimeresurus flavoviridis) venom", Jap. J. Med. Sci. Biol. 23 (1970) 285-89.

[11] ALEXANDER, P., HAMILTON, L.D.G., Radiat. Res. 13 (1961) 193.

[12] PAVLOVSKAYA, T.E., PASYNSKII, A.G., Kolloid Zhur. 18 (1956) 583; quoted from Radiation Chemistry of Organic Compounds (SWALLOW, A.J., Ed.), Pergamon Press, London, 2 (1960) 215.

[13] JOSHI, S.V., Gamma irradiation studies with synthetic poly-L-lysine hydrobromide, Ind. J. Biochem. Biophys. 9 (1972) 341-44.

[14] JOVANOVIC, M., "The production of an irradiated vaccine against Dictyocaulus filaria", Production and Utilization of Radiation Vaccines Against Helminthic Diseases, Technical Reports Series No.30, IAEA, Vienna (1964) 25.

[15] SHIBAYEVA, I.V., IZVEKOVA, A.V., IVANOV, K.K., TUMANYA, M.A., Biological properties of antigens and chemical characteristics of specific polysaccharides obtained from S. typhi bacteria exposed to gamma irradiation, J. Hyg. Epid. Mic. Immunol. 17 (1973) 357-63.

[16] FRICKE, H., HART, E.J., Oxidation of ferrous to ferric by irradiation with X-rays of solution of ferrous solution in sulphuric acid, J. Chem. Phys. 3 (1935) 60.

[17] ALVIN, G., EDWIN, J.H., in Basic Mechanisms in the Radiation Chemistry of Aqueous Media, Proc. Conf. Nat. Acad. Sci., Nat. Res. Council of United States, Gatlinburg, Tenn., Radiat. Res. Suppl.4, Academic Press (1964).

[18] YANG, C C., Chemistry and evolution of toxins in snake venoms, Toxicon 12 (1974) 1-43.

TECHNOLOGICAL ASPECTS OF RADIATION-STERILIZATION FACILITIES

(Session V)

Chairman: V.K. IYA (India)

DIFFERENT METHODS OF STERILIZATION OF MEDICAL PRODUCTS

J.O. DAWSON
Ethicon Limited,
Edinburgh,
United Kingdom

Abstract

DIFFERENT METHODS OF STERILIZATION OF MEDICAL PRODUCTS.

The classical methods of sterilization - wet chemical, steam under pressure, dry heat, and gaseous methods, depend on time of exposure, temperature and moisture content. They are mainly batch processes, and scaling up leads to difficulties. In a properly designed irradiation facility the only variables are time of exposure and density of product. Subject to initial design the capacity can be increased. The process is applied to the final sealed container, no aseptic handling being required.

I suppose it can be said that the first method of sterilisation employed by man for medical purposes was fire, by incineration or cremation. Although used before micro-organisms were discovered, how often even nowadays do we find that when disaster strikes, and for example hundreds of people drown in floods, bodies are disposed of by fire to lessen the risk of epidemics. Incineration or flaming is the standard quick effective sterilisation method employed by the microbiologist in his day to day bench work.

However, as microbes came to be established as the causative factors of disease, methods introduced to destroy them were essentially wet chemical methods about which I will have more to say in my paper tomorrow. Suffice to say that in the field of medical items, Lister's carbolic acid treatment brought about a revolution in medical care. Carbolic acid was followed by various other chemical agents but in all cases, to be effective a wet chemical process depends on the concentration of the active agent, the temperature at which the process is carried out, the time of exposure to the process, and last but by no means least, the presence of water. This is something of a bête noire with me.

We probably all know now that the disinfectant action of alcohols is optimum at about 70% concentration. Lister, even with his carbolic oil treatment for catgut found that some of his contemporaries did not meet with success and had to publish communications pointing out that the essential ingredient which he called carbolic acid was in fact phenol liquefactum which in fact contains 20% of water and not pure phenol crystals. We all know the importance of humidity in gaseous ethylene oxide sterilisation.

However, over the years I have come across failures in chemical methods of sterilisation due to the presence of insufficient water. Most chemical methods require that the active reagent shall have access to the material to be sterilised so this requires what is in effect Good Manufacturing Practice to ensure that such material is clean, free from blood, mucous, grease, etc. Except for some sutures and other biological materials wet chemical methods have little part to play in commercial methods of sterilisation of medical products. They have their uses in hospitals, however, in the sterilisation and re-use of expensive instruments and even thermometers and so on.

With the acceptance of the germ theory of disease various methods of so-called sterilisation evolved, which whilst perhaps not perfect, represented a step forward. I was trained on the 1932 British Pharmacopoeia, and, amongst others, two particular methods of sterilisation were referred to therein, methods which may still be employed for specific purposes today. One was heating with a bactericide which was merely a wet chemical method employing say, chlorcresol or phenylmercuric nitrate in low concentration and in aqueous solution, but at an elevated temperature of say 100°C (boiling point).
The other was the process of Tyndallisation in which material essentially of a nutritious nature would be heated up to boiling point or less on three successive days, the idea being that the first heating would destroy vegetative organisms and overnight the spores remaining would germinate so becoming in themselves vegetative forms to be caught by the boiling process in the second day, and just to make sure, the process would be repeated on the third day. As with sterilisation by filtration, these were methods evolved to minimise the effect of the method of sterilisation on the active principles or on the materials involved.

The methods of steam sterilisation and of dry heat, however, were the methods of choice. Steam sterilisation involved saturated steam to convey heat or energy, the latent heat of vaporisation of water, to the product. In those days it was carried out at pressures of 5, 10, or 15 pounds above atmospheric pressure, nowadays at a pressure of 30 pounds or even higher for a shorter time. However, pressure is not the criterion. Temperature, heat energy is the active factor. It is far too easy to provide a pressure gauge which measures the pressure inside the vessel but if the pressure is that of a mixture of air and steam the resultant temperature will be too low, and sufficient energy will not be present to effect sterility. In the case of dry heat sterilisation higher temperatures are involved, and increased time. It was always emphasised that timing should start from the time at which the object or the material reached the required temperature. It can be seen that short-cuts are possible unless you have an automated piece of equipment. Basically, these two processes are batch processes. I have no doubt that with considerable capital expenditure it would be possible to evolve a continuous sterilisation process employing these methods and in fact I believe there is such a design for high pressure steam. Even so, how many present day products would withstand the rigorous treatment involved. Indeed, scaling up such a process with, for example larger batches, to ensure that every part of the sterilising load receives uniform treatment presents serious technological problems.

We must not overlook a method of sterilisation which evolved more or less contemporaneously with irradiation sterilisation. I refer of course to the method employing gaseous ethylene oxide, which is, of course, not the only gas used. Here we have a chemical process in which the method depends on the concentration of the active agent, the time and temperature of exposure, and the presence of adequate moisture in particular carrying out a pre-humidification cycle to saturate the products to be sterilised. Ethylene oxide is itself essentially a dangerous chemical. It can be used neat or can be diluted with carbon dioxide or Freon, in which case pressure increases are necessary to ensure an adequate concentration. I have used it neat in old oil drums and my shins are black and blue as a result of the lids blowing off at the end of the cycle. The problem is that even with sophisticated equipment, scaling up presents major technological problems to ensure that every part of the load receives uniform and adequate treatment.

So far we have discussed physical and chemical methods of effecting sterility. The physical wet or dry heat methods may be too rigorous for the material to be sterilised. The chemical methods may result in the introduction of undesirable agents to the body or require a decontamination treatment which itself could nullify the sterilisation process. In other words you would have to carry this decontamination out under aseptic conditions. But in all methods we should reduce the challenge to the process by starting with clean materials.

Now we turn to the present time to the subject of this week's symposium - radiation sterilisation. We employ energy, the energy of ionising radiations, be they electrons or gamma rays. The energy required is known, the field of a beam or a radio-iosotope source is known or can be calculated. Depending on its strength, the energy absorbed by a material of a known density can be calculated on the basis of time of exposure and geometry of the radiation field. Whilst machines operate on a continuous process radio-iosotope sources can operate as either batch or continuous processes. Indeed, subject to the physical design of the cell, they may be upgraded from a batch process to a continuous process. Again subject to shielding, such installations can be increased in capacity. Our own plant started with a 40 kilocurie source and is now operated to 200 kilocuries with an upper limit of perhaps 250 - 300 kilocuries. It started life with a cycle of approximately 72 hours, the current cycle is between 10 and 11 hours. But to be fair I am sure machines can also be upgraded, again subject to the shielding round about the irradiation cell.

The papers we are to hear this morning appear to me to cover a wide range of installations and will no doubt provoke considerable discussion from the floor. I know many people are interested in costs, but whilst radiation may appear at a disadvantage cost-wise when taken on its own, we must also take into consideration the cost reductions in materials and packaging, the avoidance of costly aseptic work since radiation sterilisation can be carried out in the final sealed container, and indeed the impossibility of treating some materials or active components by the more traditional methods. Finally, we must not forget the value of the materials which are to be sterilised.

DISCUSSION

K.H. CHADWICK: Could you comment on, first, the usefulness or applicability of the different sterilization techniques, second, the control methods for these various techniques, and, third, the possibility of gas remnants in medical products which are gas sterilized.

In the Netherlands it has recently been found that paper cartons used for milk which had been gas-sterilized while packed flat in "bricks" still contained some gas remnants. As a result, the gas sterilization of these products has been forbidden and the cartons are now being sterilized by radiation.

J.O. DAWSON: Briefly summarizing, I should say that heating with a bactericide is confined nowadays to extemporaneous preparation, while Tyndallization is used in microbiology. Dry heat is employed for bulk powders in thin layers and for oily products, while steam is suitable for sterilizing dressing drapes and infusions (a continuous process is possible).

The control methods are based on time and temperature, indicator tubes not being reliable. Fully automated equipment records the time, temperature, vacuum, etc. If use is made of non-automated equipment, then biological indicators are used. They are essential in the case of ethylene oxide.

Gas remnants are a real problem, and requirements are becoming more and more strict. A good evaluation system is needed. It is possible to convert the gas residues into ethylene glycol but the health authorities are now clamping down on this procedure.

APPRAISAL OF THE ADVANTAGES AND DISADVANTAGES OF GAMMA, ELECTRON AND X-RAY RADIATION STERILIZATION

K.H. MORGANSTERN
Radiation Dynamics, Inc.,
Westbury, N.Y.,
United States of America

Abstract

APPRAISAL OF THE ADVANTAGES AND DISADVANTAGES OF GAMMA, ELECTRON AND X-RAY RADIATION STERILIZATION.

Since radiation sterilization was first introduced as a viable technique for the sterilization of medical products, there has been considerable discussion of the relative merits of gamma rays versus accelerator produced electrons. Although the early sterilization work at Ethicon started with an electron device, the problems associated with these early generation electron devices convinced the researchers to switch to ^{60}Co gamma sources as their radiation choice. Over the intervening period of approximately two decades, considerable changes have taken place with respect to both ^{60}Co facilities and sources, and electron accelerators. The paper reviews and up-dates the status of both. The changes in source cost, construction techniques and production flow for ^{60}Co sterilization plants are examined. The advantages of ^{60}Co as it relates to simplicity and ease of maintenance are discussed as well as the disadvantages of long dwell time and facility complication. Typical plant efficiencies for both continuous flow and batch gamma plants are described, and the fact that radiation utilization (efficiency) factors with gamma facilities are still relatively low and have not changed radically over the period is reviewed. Finally, the economics of gamma-ray sterilization are examined for typical medical products.

With regard to electron accelerators, both linacs and d.c. generators have undergone considerable changes in both power levels and degree of process reliability. These improvements are described and the apparent acceptance afforded electron accelerators as production tools by industries other than medical disposables. In addition, the relationship of medical product packaging, product flow and point of sterilization in the production flow and the type of electron device most suitable are discussed. Further, the advantages of higher-powered electron accelerators (up to 150 kW) and their ability to generate X-ray fluxes comparable with those for ^{60}Co plaques makes it necessary to consider this possibility as well, in attempting to arrive at an optimum radiation sterilization facility design. The economics of medical product sterilization with electrons used directly or with secondary-produced X-rays are examined and compared with gamma-ray economics.

INTRODUCTION

The continuing growth of the world's population, in combination with improvements in medical care and techniques, has made the requirement for sterilization of medical disposables more important than ever. In the USA and Western Europe alone, it is projected that the medical disposable business will grow from the 1-billion-dollar level in 1970 to 3 billion by 1980. The most common items requiring sterilization are indicated in Table I. With this extensive business potential, it is not surprising that the economics of sterilization becomes an important consideration.

TABLE I. TYPICAL PRODUCTS BEING IRRADIATED

SUTURES (ABSORBABLE AND NON-ABSORBABLE)	DIALYSIS UNITS
PLASTIC HYPODERMIC SYRINGES	INFUSION SETS
HYPODERMIC NEEDLES	ARTERIAL PROSTHESES
PLASTIC AND RUBBER CATHETERS	ENDOTRACHEAL TUBES
PLASTIC EXAMINATION GLOVES	ACRYLIC POWDER
CANNULAE	PETRI DISHES
PLASTIC SCALPELS	SWABS
SURGICAL BLADES	MATERNITY TOWELS
PREPARATION RAZORS	VARIOUS PLASTIC BAGS AND CONTAINERS FOR PHARMACEUTICAL PREPARATIONS AND PATHOLOGICAL SAMPLES
BLOOD LANCETS	
DRESSINGS	

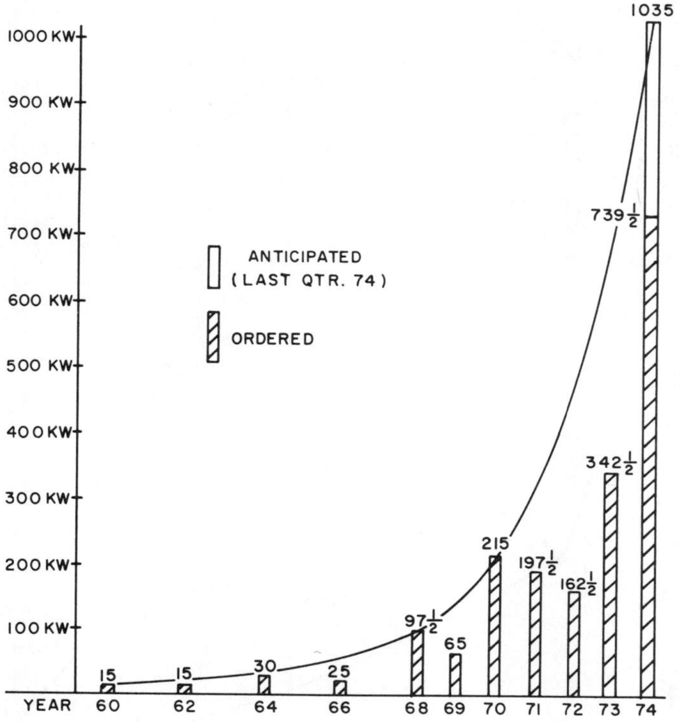

FIG. 1. Kilowatts of electron processing equipment ordered.

EARLY BACKGROUND

The first commercial application of radiation for medical product sterilization was started in the USA by Ethicon Inc. in 1956, with the intial experimentation dating back to 1949. Their early commercial activity made use of a linear accelerator; however, in 1964, the radiation sterilization programme was switched over to ^{60}Co. The initial products so treated were surgical sutures and then disposable hypodermic syringes and needles.

At about the same time (1960), the Package Irradiation Plant at Wantage Research Labs in the United Kingdom went into commerical operation. Using 300 000 Ci ^{60}Co, it has been used for sterilizing a wide variety of medical products.

On the continent, the first commercial radiation sterilization activity was launched by the Danish AEC in 1958 – using ^{60}Co. However, in 1961 the sterilization processing was transferred to a linear accelerator.

In the USSR, investigation and initial commercialization of radiation sterilization started in the early 1960s, again with gamma rays. In Czechoslovakia, the first commerical irradiation plant in 1964 made use of a d.c.-type electron accelerator.

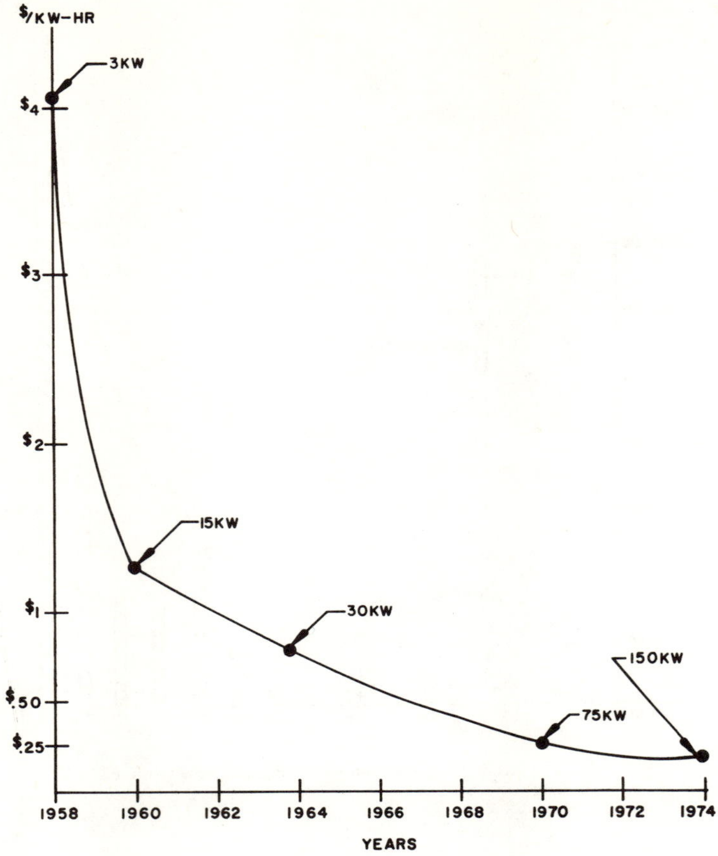

FIG. 2. Radiation costs versus time.

Although full commerical utilization of radiation in the Federal Republic of Germany did not take place until 1967, there was a service irradiation centre in operation using a 3-MeV Van de Graaff generator as early as 1957 and it was employed in 1958 to irradiate catheters.

What is evident from this brief historical resume is that radiation sterilization is certainly not new and, further, that both gamma rays and electrons have been employed in the process. However, as the process became more widely used, it is fair to state that ^{60}Co has dominated the scene. Today there are over 60 gamma-ray facilities in operation contrasted to perhaps less than a dozen accelerator systems. This bias in favour of cobalt appears to be unique to the radiation sterilization field when one considers the broader area of industrial radiation processing. To put this into proper perspective, we should consider the general radiation processing field.

TABLE II. ENERGY REQUIREMENT COMPARISON

	Heat	Radiation
1. Cross-linking polyethylene on 600 V - 4/0 wire	0.55 ¢/lb	0.28 ¢/lb (15 Mrad)
2. Vulcanization of sheet rubber	1.2 ¢/lb	0.1 ¢/lb (10 Mrad)
3. Curing reinforced polyester panels (3000 lb/h)	\$5.00/h	\$1.13/h
4. Curing paint (energy input for same product throughput)	\$4500/month (gas)	\$900/month (electricity)
5. Food preservation	0.4 ¢/lb/d (refrigeration)	0.01 ¢/lb (one time) (0.5 Mrad)
6. Medical disposable sterilization	2 ¢/lb[a] (steam)	0.05 ¢/lb (2.5 Mrad)

[a] Based on 26 ¢/ft^3 – density 0.2.

RADIATION PROCESSING

Since the end of World War II, numerous technical organizations, governmental, academic, and industrial, have investigated possible applications of radiation as an industrial processing tool. As has been the case with radiation sterilization, these activities go as far back as two decades. However, it has not been until rather recently that there has been a broadening of acceptance and an acceleration in the growth of industrial radiation process activities.

Figure 1 indicates the growth of electron beam kilowatts used for industrial processing as evidenced by our company's order bookings. As is shown from this figure, the growth-rate since 1970 has been very dramatic and shows every expectation of continuing its rapid climb.

The types of applications involved include the upgrading of insulation on wire and cable, the cross-linking of plastic film for heat shrinkable properties, the cross-linking of polyethylene foam, the vulcanization of elastomeric materials, the cross-linking of extruded and moulded plastics for improvement of temperature and stress-crack resistance, the curing of coating on a variety of substrates, the graft-polymerization of coatings on textile substrates, the upgrading of semiconductor materials.

In addition, a number of very exciting new horizons are opening up in areas such as pollution control (irradiation of sludge and stack gases) and the use of radiation as a process tool for the recovery of valuable by-products from the recycling of waste materials.

In the vast majority of the applications indicated above, electron accelerators have been the radiation mode employed. This has been due in the main to the recognition that electron beam power is less expensive than that from gamma sources and, further, that the electron power can be used more

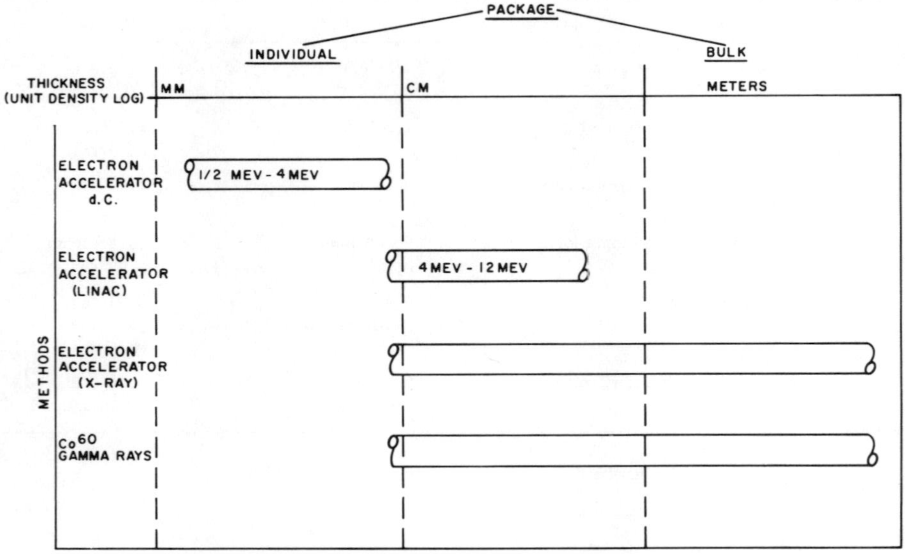

FIG.3. Summary of radiation sterilization.

efficiently. Figure 2 indicates the change in electron radiation costs that has taken place over the last 20 years. As is evident from this curve, radiation costs have come down approximately twenty-fold. As a result, many industrial process applications which one or two decades ago were technically feasible but not economical, have now crossed over into the realm of economic feasibility. Further, and of particular importance in today's energy crunch, is the fact that in most cases these radiation-processing techniques require less energy than do the more conventional and accepted techniques such as the application of heat or mechanical energy.

Table II shows the relative advantage of radiation in contrast to heat energy as it applies to half-a-dozen common industrial process areas. As a result of the substantial reduction in radiation costs, the recognition that radiation provides a technique for conserving energy and a general acceptance on the part of industry with respect to radiation as a viable process technique, we foresee very rapid growth in the application of radiation as a process tool in many areas, including sterilization.

RADIATION STERILIZATION

It would now be well to examine the various possibilities available to the industrial organization interested in using radiation as a sterilization technique.

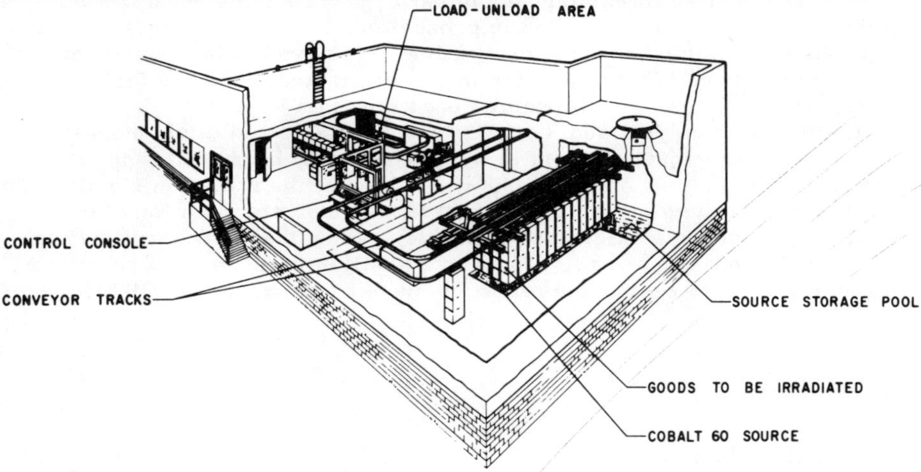

FIG. 4. Packaged goods continuous flow irradiator.

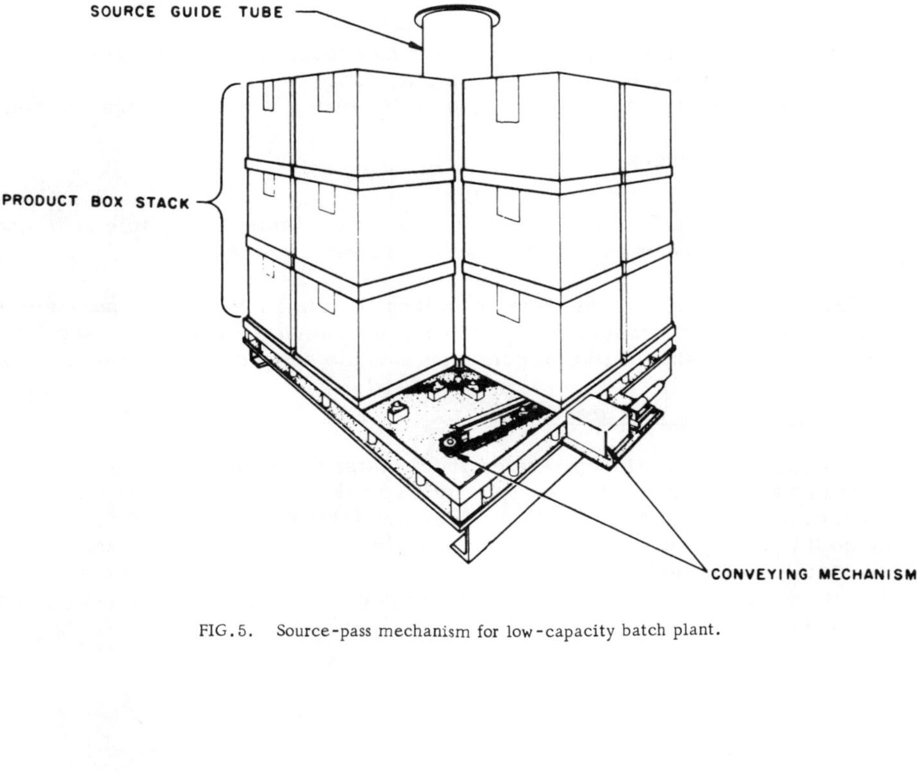

FIG. 5. Source-pass mechanism for low-capacity batch plant.

As was mentioned earlier, a number of radiation methods are available to the industrial user interested in product sterilization. Figure 3 attempts to put these possibilities into perspective. It is evident that the generalization that X-rays are suitable for thick objects and electrons for thin still holds. However, one must examine each situation in greater depth to make an intelligent value judgment as to the most suitable radiation process technique. Often that judgment is weighted heavily by the existing production flow chart rather than solely by the economics of the radiation sterilization system. For example, one might want to consider sterilization of unit packages, if the economics are attractive, rather than bulk sterilization of final shipment containers. To be in a position to make such judgments, it would be well to review the radiation systems at present available.

COBALT-60

General information

A ^{60}Co plant consists of the source, the biological shield, and the product conveyor or transfer mechanism. (Connected with the source is the obvious requirement for source replenishment and storage.) These plants are generally either automatic (continuous product flow) or batch. The automatic plants are generally classified as high (1 MCi or greater), intermediate (~ 500 000 Ci), or low capacity (100 000 Ci). Figures 4 and 5 illustrate typical automatic and batch-type facilities.

The basic advantages of ^{60}Co for sterilization are:

(a) Simplicity and reliability of radiation source
(b) Ability to match source strength to production throughput
(c) Broad experience gained in many plants
(d) Good flexibility in product selection owing to high penetration ability

The disadvantages of ^{60}Co are:

(a) Expensive source of radiation
(b) Complex product handling necessary to obtain reasonable efficiencies
(c) Source decay and source replenishment

Since previous authors have described in detail many of the parameters of plant design and product transfer (Brown, Baines, Ballantine — see Bibliography), I should like to consider now the economics of ^{60}Co sterilization

Economics of ^{60}Co sterilization

Cobalt-60 is reactor-produced from metal ^{59}Co, and as such has a finite production cost. The early sources in the 1950s generally had a cost of approximately \$2.00/Ci. With wider availability and greater demand, the cost per curie dropped to approximately 50 ¢ in the mid-60s and has essentially remained at that level. Because of the large investment in source capacity, it behoves the plant designer to strive for maximum plant efficiencies.

TABLE III. RANGE OF PERFORMANCE CHARACTERISTICS FOR FOUR TYPES OF PLANTS

	HIGH CAPACITY	INTERMEDIATE CAPACITY	LOW CAPACITY BATCH	LOW CAPACITY AUTOMATIC
PRODUCT BOX SIZE (in.)	12 x 15 x 21	18 x 18 x 18	16 x 16 x 16	16 x 16 x 16
PACKING DENSITY (g/cm^3)	0.10 to 0.25	0.10 to 0.25	0.10 to 0.25	0.10 to 0.25
DOSE MAX / MIN	1.19 to 1.25	1.30 to 1.35	1.27 to 1.29	1.27 to 1.29
EFFICIENCY % (1)	18.5 to 38.3	17.9 to 36.0	6.2 to 15.0	6.6 to 16.2
PROCESSING RATE (2) (ft^3/h-kCi Co60)	0.145 to 0.120	0.134 to 0.108	0.046 to 0.045	0.049
ANNUAL PRODUCTION RATE (3)	1.04 ft^3/curie	0.94 ft^3/curie	0.32 ft^3/curie	0.40 ft^3/curie

NOTES: (1) EFFICIENCY IS BASED ON SOURCE CONTENT AND MINIMUM DOSE.
(2) PROCESSING RATE IS BASED ON A MINIMUM DOSE OF 2.5 MEGARADS.
(3) ANNUAL PRODUCTION RATE IS BASED ON A MINIMUM DOSE OF 2.5 MEGARADS DELIVERED TO 0.2 g/cm^3 MATERIAL AND 8.000 h OPERATION.

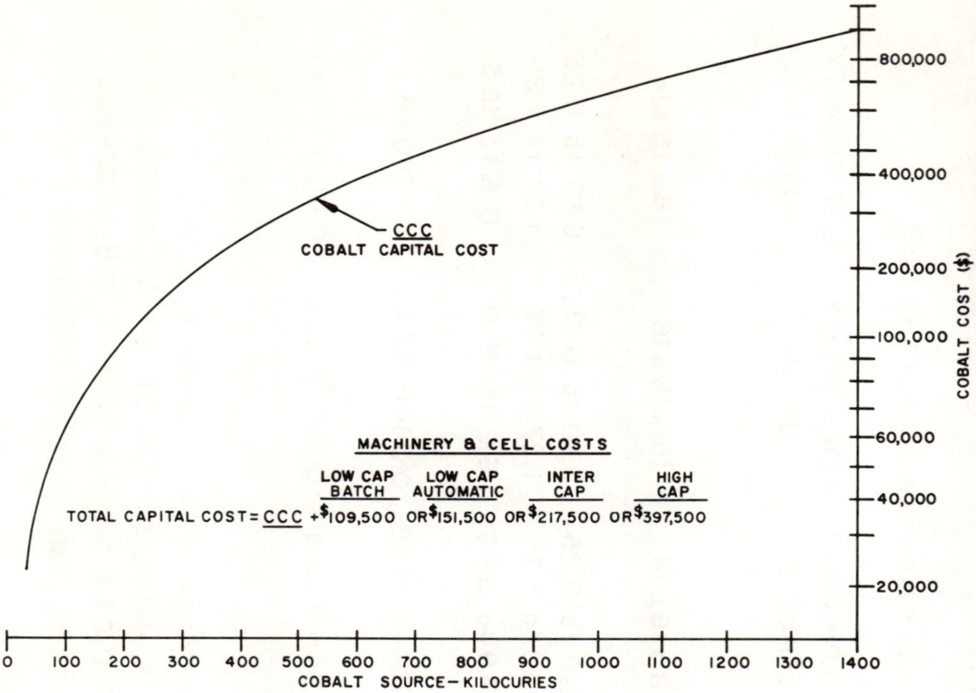

FIG. 6. Source cost for various size facilities.

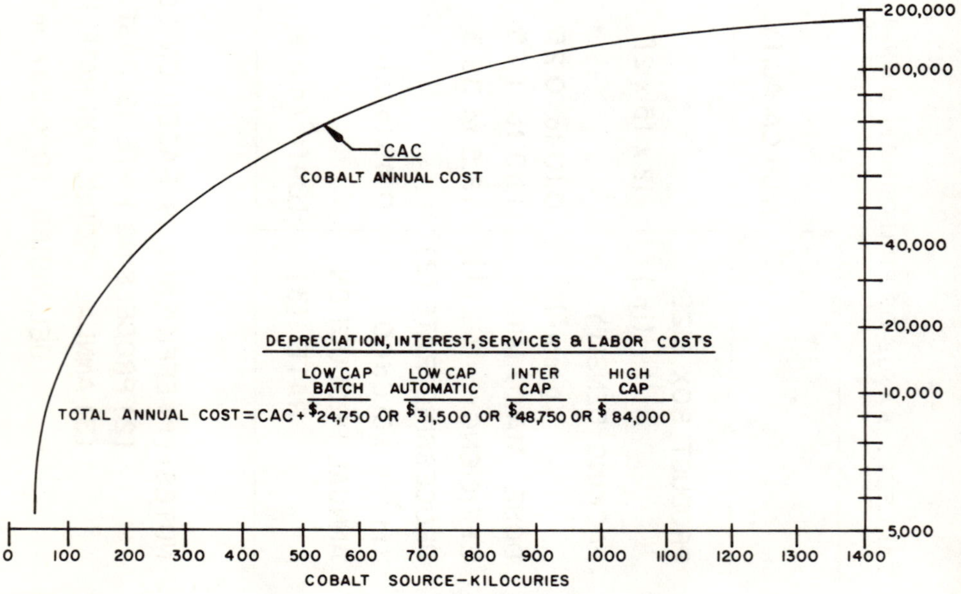

FIG. 7. Annual cobalt-60 facility operating costs.

FIG. 8. View of voltage-generating stack of typical dynamitron.

However, this can be somewhat difficult since the X-rays of 1.17 and 1.33 MeV from ^{60}Co are quite penetrating and are absorbed exponentially. (For example, it takes 27 cm of water to reduce the X-ray intensity by 50%.) The best in reported plant efficiency appears to be in the region of 30 to 40%, where efficiency is defined as the ratio of radiation absorbed by the product to the total emitted by the radiation source.

In 1967, Brown compared the costs, throughput, and efficiencies of various size plants, both batch and automatic. This data, summarized in Table III, for four types of plants still appears valid. His calculation of curie requirements and costs for various size plants are shown in Fig. 6. However, I have taken the liberty of updating the machinery and other costs based upon an estimated inflationary factor of 50% over the intervening seven years.

Using this cost input and choosing an intermediate plant of 500 000 Ci capacity, one has a capital cost of \$300 000 plus \$217 000 = \$517 000. Recognizing that 70 000 Ci is roughly equivalent to a kW, then this plant = 7.1 kW.

The annual operating costs, based on a ten-year amortization of the initial cobalt source would be for

6000-h operation

Amortization — source and facility	\$ 51 700
^{60}Co make-up (12 litres/2%/yr)	37 500
Labor (3 shifts) @ \$10.00/h	60 000
Power and water	2 000
Maintenance	5 000
Miscellaneous	2 000
	\$158 200
Hourly cost	\$ 26 400

Efficiency factor = 33%

Throughput = 0.33 × 7.1 kW × 800 M lb/h

= 1874.4 MR-lb/h at 2.5 MR ≈ 750 lb/h

$$\text{Cost/lb} = \frac{\$26.40}{750} = \underline{3.9\ ¢}$$

$$\text{Cost/ft}^3\ (\text{Density } 0.2) = \underline{47\ ¢}$$

Alternatively, using Brown's annual cost from Figure 7 (with a 50% inflationary factor), we have ≈ \$122 000/yr. From Table III, the intermediate plant has an annual capacity of 1.04 ft^3/Ci or 270 000 ft^3/yr (8000 h/yr). Normalizing to 6000 h, we have ≈ 200 000 ft^3/yr for a cost/ft^3 of $\frac{123\,000}{200\,000}$ = 61 ¢, or a relatively fair agreement between the two costs.

ELECTRON ACCELERATORS

An alternate choice to ^{60}Co for radiation sterilization is high-energy electron accelerators. The choice can be further divided depending upon the product thickness, to either d.c. generators or linear accelerators. In both cases, one has an electron gun, acceleration tube and a scanning device to spread the electrons over the desired target area. Figures 8 and 9 show typical d.c. and linear accelerators.

There are two major differences that must be considered in comparing gamma rays from an isotope source with electrons from a machine. The first relates to the penetration characteristics of each radiation and the latter to the intensity or energy flux from the two sources.

Figure 10 shows the penetration as a function of voltage for electrons in unit density ratios. By using two-sided irradiation, one can obtain two and a half times the penetration from one side (Fig. 11). Further, the penetration is inversely related to the product density; consequently, 3-MeV electrons, for example, could accommodate with two-sided irradiation approximately 2.5 cm of unit density material and 12 cm for density 0.2.

FIG. 9. View from electron gun end of 15-kW industrial processing Linac.

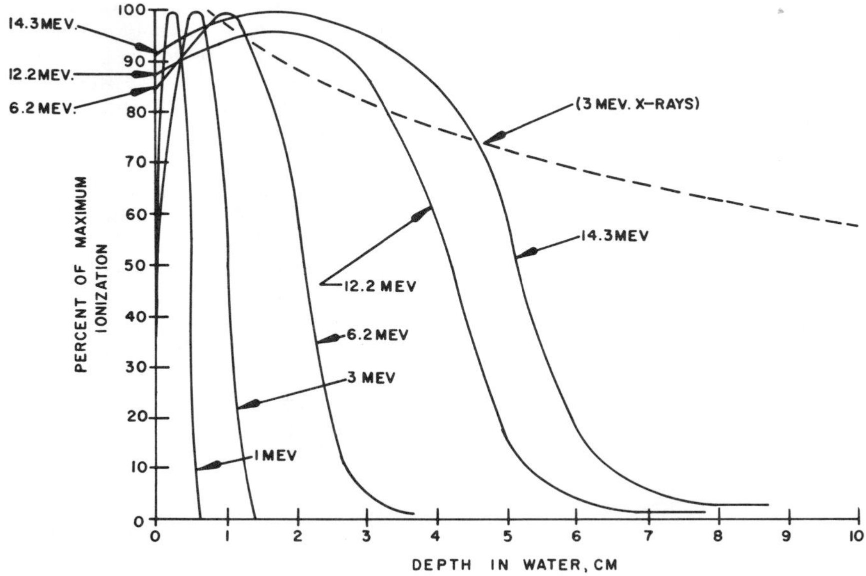

FIG. 10. Penetration of high-energy electrons in water.

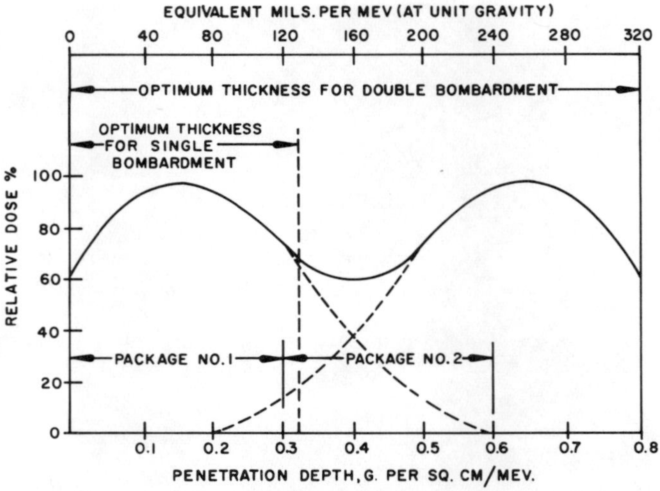

FIG. 11. Depth profile for fixed package thickness showing effect of irradiation from two directions.

The second difference between gamma and electron sterilization is the vast difference in rate of energy absorption or rate of sterilization. Most gamma-ray sterilization plants require 8-24 h to deliver the 2.5-Mrad sterilization dose or an average dose-rate of 0.1 - 0.3 Mrad/l. An electron accelerator can deliver the 2.5-Mrad sterilization dose in a second or less. There are then some rather obvious differences in product-handling rates and design of product-handling systems. Because of the brief irradiation time involved with the electron accelerator, there is a consequent necessity for very accurate monitoring for beam intensity and product speed through the radiation field. (A typical monitoring system employed by RDL with their industrial linear accelerators is shown in Fig. 12.) On the other hand, degradative effects on package materials are less severe at the higher dose-rate irradiations.

Further, because of the faster energy deposition, one must be aware of possible product-heating problems, although at 2.5 Mrad this will not be a real problem in most cases.

The conclusion of all these factors is that the thickness of target which can be irradiated by electrons from an accelerator is a function of the various factors of product density, electron energy, permissible max/min dose ratio and whether the product is irradiated from one or two sides. In general, the flexibility of product package size is more limited with electron irradiation than ^{60}Co gamma irradiation and the limitations become more severe at the lower electron energies and higher product densities.

ELECTRON STERILIZATION ECONOMICS

As was pointed out earlier, there have been relatively few changes in the economics of ^{60}Co sterilization over the last seven years. By contrast,

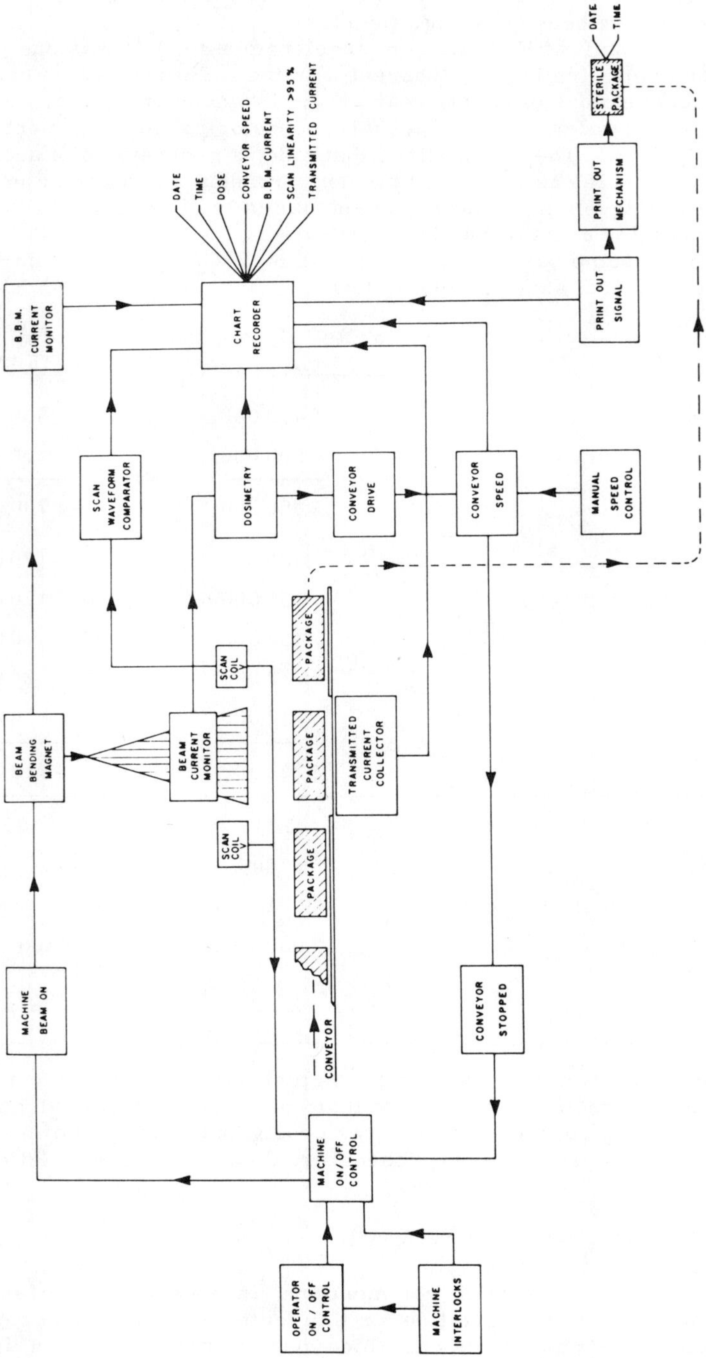

FIG. 12. Sterilization monitoring system for Linac.

the changes that have occurred with respect to electron accelerators in this same period have been quite significant.

The early shift at Ethicon from accelerators to ^{60}Co was due primarily to the much better reliability inherent with the radioisotopes as compared with the accelerators of twenty years ago. However, both d.c. generators and Linacs have undergone considerable technological improvement since those early days. The net result is that today's generations of accelerators are highly reliable, as evidenced by their broadening industrial applications, and the power levels have increased substantially, resulting in a lowering of their radiation costs. (See Fig. 2.)

Using the same basic cost analysis as for ^{60}Co, the radiation-sterilization cost for two typical accelerators follows:

	3-MeV Dynamitron 25 mA (75 kW)	10-MeV Linac (15 kW)
Capital cost	$450 000	$500 000
Building and conveyor	200 000	200 000
	$650 000	$700 000
Amortization	65 000	70 000
Labour (3 shifts) 2 people	120 000	120 000
Power and water	24 000	6 000
Maintenance	12 000	24 000
Miscellaneous	2 000	2 000
	$223 000	$212 000
Hourly cost (6000 h)	$37	$35
Efficiency factor	50%	50%
Thickness (2 sides) (0.2 density)	12 cm	40 cm
Pounds at 2.5 Mrad	12 000 lb/h	2 400 lb/h
Cost/lb	0.3 ¢	1.45 ¢

What is obvious from this comparison is that the 3-MeV accelerator is the least expensive radiation source but that, in order to utilize fully this inexpensive radiation source, one requires prodigious quantities of product. However, both accelerators represent a real cost saving over ^{60}Co.

ACCELERATOR PRODUCED X-RAYS

The generalization which was mentioned in the early part of this paper, namely that accelerators are fine for thin materials but gamma rays are required for thick objects, is one which must be re-evaluated in light of

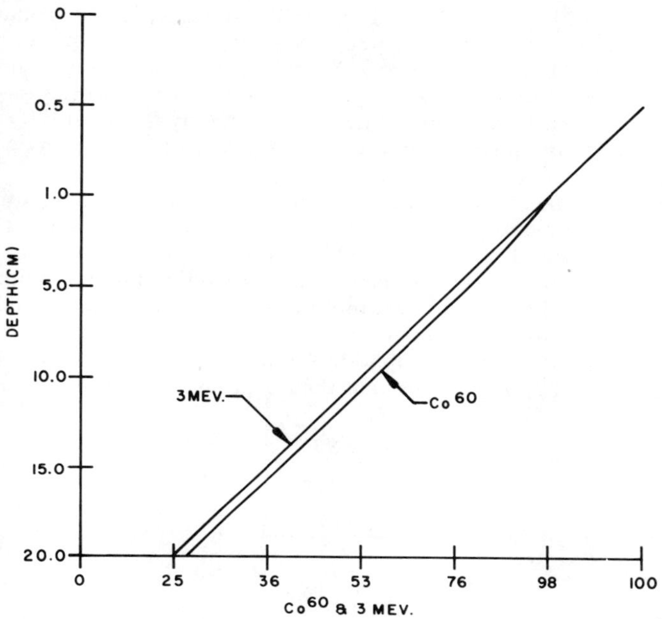

FIG. 13. The depth dose in tissue of cobalt-60 gammas and bremsstrahlung from a 3-MeV electron beam.

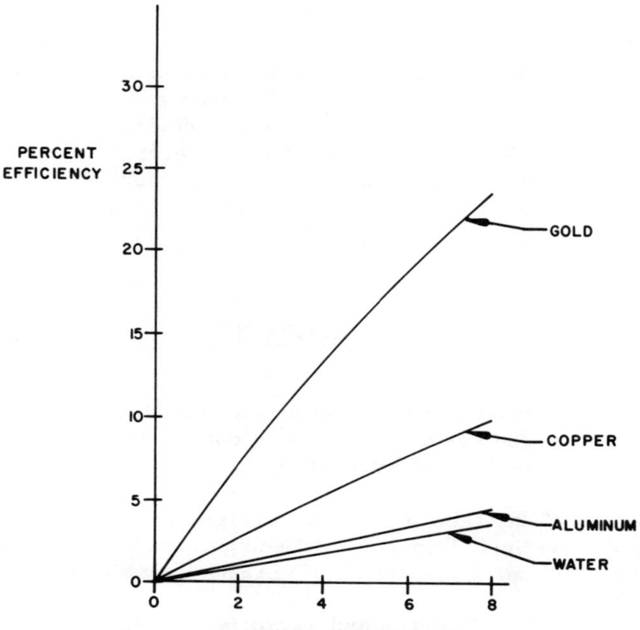

FIG. 14. Thick target conversion efficiency.

the availability of higher-powered, reliable accelerators. Today 3-MeV, 75-kW and 150-kW accelerators are available, and as such can produce copious supplies of X-rays. These 3-MeV X-rays are very similar in penetrating characteristics to the gamma rays from ^{60}Co (as shown in Fig. 13). Consequently, even though the generation of X-rays is basically inefficient, one should examine their relative suitability and economics as a radiation modality for sterilization.

From our in-house dosimetry on heavy "Z" targets, we can demonstrate an X-ray yield of approximately 7 kW, i.e. equivalent to approximately 500 000 Ci of ^{60}Co with a 75-kW, 3-MeV Dynamitron. (See Fig. 14.) Because of the spatial output of these X-rays, in contrast to the 4π geometry of ^{60}Co sources, one can obtain improvements in over-all efficiency and with a simpler product-transfer system.

Using the previous cost per hour for a 3-MeV, 25-mA accelerator generating 7 kW of X-rays, one can then calculate costs as follows:

Cost/h	=	\$37		
Efficiency	=	42%		
Pounds at 2.5 Mrad	=	$\frac{7 \times 800 \times 0.42}{2.5}$	=	940 lb/h
Cost/lb	=	\$37/940	=	3.8¢
Cost/ft³ (0.2 density)	=		=	45.0¢

SUMMARY

There are now four basic radiation systems available to the industrial user. Each has characteristic advantages and disadvantages, and we cannot a priori decide upon the best without a careful review of many factors. Certainly, economics is an important one; however, careful consideration must be given to the total production picture in order to make a meaningful value judgment.

BIBLIOGRAPHY

ARTANDI, C., "Present Status of Radiation Sterilization Technology and Experience", Proceedings of the U.S.P. Conf. on Radiation Sterilization, Washington, D.C. (1972) Tel Aviv (1973) 9.

BAINES, B.D., "The Relation of Plant Design to Product Cost", Radio-sterilization of Medical Products and Recommended Code of Practice, (Proc. of Symposium, Budapest, 1967) IAEA, Vienna (1967) 395.

BALLANTINE, D.S., "Sources and Facilities for Radiation Sterilization of Medical Supplies", Proceedings of the U.S.P. Conf. on Radiation Sterilization Washington, D.C. (1972) Tel Aviv (1973) 21.

BROWN, M.G., "The Relation of Design Parameters, Plant Capacity and Processing Costs in Cobalt-60 Sterilization Plants", Radiosterilization of Medical Products and Recommended Code of Practice, (Proc. of Symposium, Budapest, 1967) IAEA, Vienna (1967) 381.

CERNY, P., Report from the Czechoslovak Socialist Republic, Radiosterilization of Medical Products, Pharmaceuticals and Bioproducts, (Report of Panel Vienna, 1966) IAEA TRS 72 (1967) 81.

CHRISTENSEN, E., HOLM, N.W., JUUL, F., Report from Denmark, Radiosterilization of Medical Products and Recommended Code of Practice, (Proc. of Symposium, Budapest, 1967) IAEA, Vienna (1967) 60.

FARRELL, P., "The Bremsstrahlung Radiation Field of a Scanned Monoenergetic Beam", Tech. Meeting 9b 9/25, Int. Nuclear Industries Fair, Basle (1966).

HOLM, N.W., "Process Parameter Control, Dosimetry, and Operation in Radiation Sterilization Processing. Relative Merits of Cobalt-60 and Linear Accelerator Plants", Proceedings of the U.S.P. Conf. on Radiation Sterilization, Washington, D.C. (1972) Tel Aviv (1973) 71.

HOLM, N.W., Dosimetry, Ch. 9,Manual on Radiation Sterilization of Medical and Biological Materials, IAEA TRS 149 Austria (1973) 99.

JEFFERSON, S., CRAWFORD, C.G., "Development of Industrial Sterilization of Medical Products", Radiosterilization of Medical Products and Recommended Code of Practice, (Proc. of Symposium, Budapest 1967) IAEA, Vienna (1967) 361.

JEFFERSON, S., Facilities Required for Radiosterilization, Ch. 7, Manual on Radiation Sterilization of Medical and Biological Materials, IAEA TRS 149 Austria (1973) 89.

KISELEV, A.E., KOZINETS, G.I., Report from the Union of Soviet Socialist Republics, Radiosterilization of Medical Products, Pharmaceuticals and Bioproducts, (Report of Panel Vienna, 1966) IAEA TRS 72 (1967) 77.

LEY, F.J., CRAWFORD, C.G., KELSEY, J.C., Report from the United Kingdom, Radiosterilization of Medical Products, Pharmaceuticals and Bioproducts, (Report of Panel Vienna, 1966) IAEA TRS 72 (1967) 40.

MORGANSTERN, K. H., "X-Ray Radiation Sources", The Radiation Processing Industry, its Prospects and Problems, (ANS Washington Section Seminar (1964).

MORGANSTERN, K. H., "Radiation's Time Has Arrived for Plastics and Rubber", 1st Int. Conf. on Radiation Curing, Society of Manufacturing Engineers, (Atlanta, 1973).

SCHNELL, J., PETER, K. H., Report from the Federal Republic of Germany, Radiosterilization of Medical Products, Pharmaceuticals and Bioproducts, (Report of Panel Vienna, 1966) IAEA TRS 72 (1967) 82.

VAN WINKLE, W., Report from the United States of America, Radiosterilization of Medical Products, Pharmaceuticals and Bioproducts, (Report of Panel Vienna, 1966) IAEA TRS 72 (1967) 23.

DISCUSSION

K. KRISHNAMURTHY: You stated that with the increasing cost of energy, radiation treatment is more economical. However, in view of the poor conversion of electrical into radiation energy, I don't really see how it can be considered economical. Furthermore, on what grounds do you recommend electron and X-ray sources in preference to ^{60}Co sources, which do not require a large amount of external power, unlike X-ray machines and electron accelerators.

K.H. MORGANSTERN: The conversion efficiency from line power to beam power at 3 MW is approximately 30%. The point I was making was that radiation energy (electrons) is more efficient for most industrial processes when compared with more conventional energy sources, such as heat and mechanical mixing. The primary reason for this energy conservation is the fact that the radiation energy can be more efficiently transferred from space into the material being processed.

I was not trying to suggest that radiation energy from an accelerator is more efficient than that from ^{60}Co. Obviously, a cobalt facility, after installation, requires no electrical power to produce the radiation output. However, it would be interesting to calculate the energy required to produce the initial radioactivity level; I think we would find that the source decay was, in effect, comparable to energy consumption.

H.M. ROUSHDY: With regard to the new electron beam accelerator under development in the United States of America, enabling a switch-over from electrons to X-rays, I am wondering how long it takes to make the switch-over and what sort of geometry calculations are required. How far would the change-over interfere with the continuity of the irradiation processing for various items?

K.H. MORGANSTERN: The facility is only now being built. Consequently, I cannot give a precise answer. However, I would expect the switch-over from electrons to X-rays to take only a few minutes. The conveyor system is designed primarily for the X-ray mode, since it is under X-rays that total plant efficiency becomes of prime importance. The X-ray target is being designed to swing, by means of a mechanical drive, into the position for intercepting the electron beam. The target will be water-cooled and the dosimetry based on the target will be determined at the time of commissioning.

Consequently, during routine operation we will be able to change over from electrons to X-rays, with little effort — other than that of moving the heavy "Z" X-ray target into position.

TECHNOLOGIE DES INSTALLATIONS DE RADIOSTERILISATION
Quelques aspects intéressant les pays en voie de développement

R. EYMERY
CEA, Centre d'études nucléaires
de Grenoble, Grenoble, France

Abstract–Résumé

TECHNOLOGY OF RADIOSTERILIZATION INSTALLATIONS: SOME ASPECTS CONCERNING THE DEVELOPING COUNTRIES.

After briefly summarizing the conditions of interaction of accelerated electrons and gamma rays with matter, the author examines the most favourable geometries for irradiating medical material. Technical devices for irradiation are then examined, as is the mode of operation of the installation, which can be continuous or batch. It appears that only a detailed study of the present and future needs of the user can permit definition of a satisfactory type of conveyor and mode of operation. The possibility of using an accelerator is envisaged. The principal criteria that should be applied by the developing countries are, in order of importance: safety of the installation, its reliability, uniformity of the dose obtained, and, finally, versatility and yield.

TECHNOLOGIE DES INSTALLATIONS DE RADIOSTERILISATION: QUELQUES ASPECTS INTERESSANT LES PAYS EN VOIE DE DEVELOPPEMENT.

Après un bref rappel des conditions d'interaction des électrons accélérés et des rayonnements gamma avec la matière, l'auteur examine les dispositions les plus favorables, du point de vue géométrique, pour l'irradiation de matériel médical. Les solutions techniques permettant l'irradiation sont ensuite examinées, avec le mode de fonctionnement de l'installation: fonctionnement continu ou discontinu. Il apparaît qu'une étude approfondie des besoins actuels et futurs de l'utilisateur peut seule permettre de définir un type de convoyeur et un mode de fonctionnement satisfaisants. L'éventualité de l'emploi d'un accélérateur est envisagée. Les principaux critères qui doivent guider les pays en voie de développement sont, dans l'ordre: la sécurité de l'installation, sa fiabilité, l'uniformité de dose obtenue et, enfin, la souplesse et le rendement.

1. INTRODUCTION

C'est vers 1954 que la radiostérilisation est entrée dans la pratique industrielle. Depuis cette date, plusieurs dizaines d'installations ont été mises en service un peu partout dans le monde.

Nous nous proposons ici de rappeler les principes qui sont à la base des études d'installations d'irradiation, de montrer l'importance de certains paramètres et de décrire les principales solutions techniques adoptées.

Les choix, qui peuvent être faits par les pays en voie de développement, seront plus particulièrement discutés.

2. INTERACTION DES RAYONNEMENTS AVEC LA MATIERE

Les rayonnements ionisants ont des énergies de l'ordre de plusieurs millions d'électron-volts. Lorsqu'ils traversent la matière, ils provoquent des ionisations, c'est-à-dire l'arrachement d'électrons au cortège électronique des atomes.

Les électrons accélérés provoquent une ionisation directe, déplaçant des électrons tout au long de leur parcours qui peut être, dans l'eau, de quelques centimètres. L'électron incident est peu dévié, il perd progressivement son énergie et disparaît après un certain parcours qui est toujours le même pour un milieu et pour une énergie initiale donnés. On peut donc définir une profondeur de pénétration maximum qui, d'ailleurs, est inversement proportionnelle au nombre d'électrons présents dans le milieu, c'est-à-dire finalement à sa densité.

Les rayons gamma peuvent être considérés comme des particules neutres chargées d'énergie. Leur probabilité d'interaction avec les électrons du milieu est faible. Par contre, l'énergie cédée au cours d'un choc est une fraction importante de l'énergie initiale. Le photon disparaîtra par effet photoélectrique après un petit nombre de chocs, mais dans chacun d'eux, il aura communiqué à un électron une énergie de plusieurs centaines de kiloélectronvolts. Au cours d'un parcours de plusieurs millimètres, cet électron provoquera des ionisations "secondaires". Le plus souvent, après un choc, le photon est largement dévié de sa direction initiale. Les photons ont donc des parcours anguleux et extrêmement aléatoires (fig.1). Il n'est pas possible de définir une profondeur de pénétration maximum, et l'atténuation d'un faisceau de photons est grossièrement exponentielle avec un coefficient d'atténuation, à peu près proportionnel à la densité.

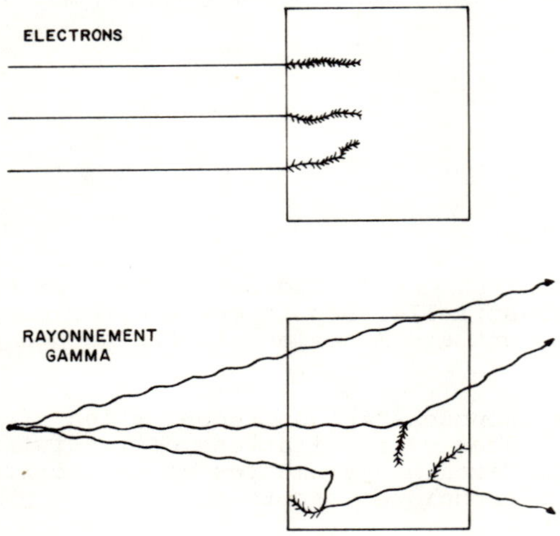

FIG.1. Interaction des électrons et du rayonnement gamma avec la matière.

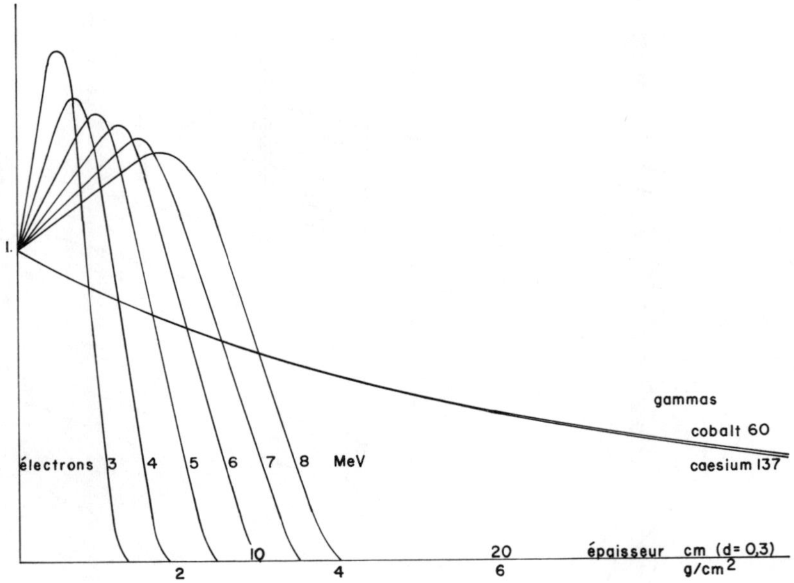

FIG.2. Courbes d'atténuation.

La figure 2 représente l'atténuation de faisceaux parallèles d'électrons et de photons du Cobalt 60 et du Caesium 137.

Il n'est peut-être pas inutile de rappeler, enfin, que les rayons gamma et les électrons de moins de 5 MeV n'induisent aucune radioactivité dans les matériaux irradiés.

3. DISPOSITIONS GEOMETRIQUES POUR L'IRRADIATION GAMMA

La source de rayonnement.

Le Cobalt 60 et le Caesium 137 sont les seuls radioéléments utilisables en radiostérilisation. Pour l'irradiation, on place généralement un ensemble de paquets tout autour d'un porte-source rectangulaire plan (fig.3).

La source est constituée de sources élémentaires parallèles. Le dessin de la source doit tenir compte de rechargements ultérieurs. En effet, au bout de quelques années, de nouvelles sources élémentaires seront mises en place. Les différentes sources élémentaires n'auront pas toutes la même activité. Leur arrangement ne sera pas indifférent, si on veut garantir une certaine uniformité de dose dans les paquets.

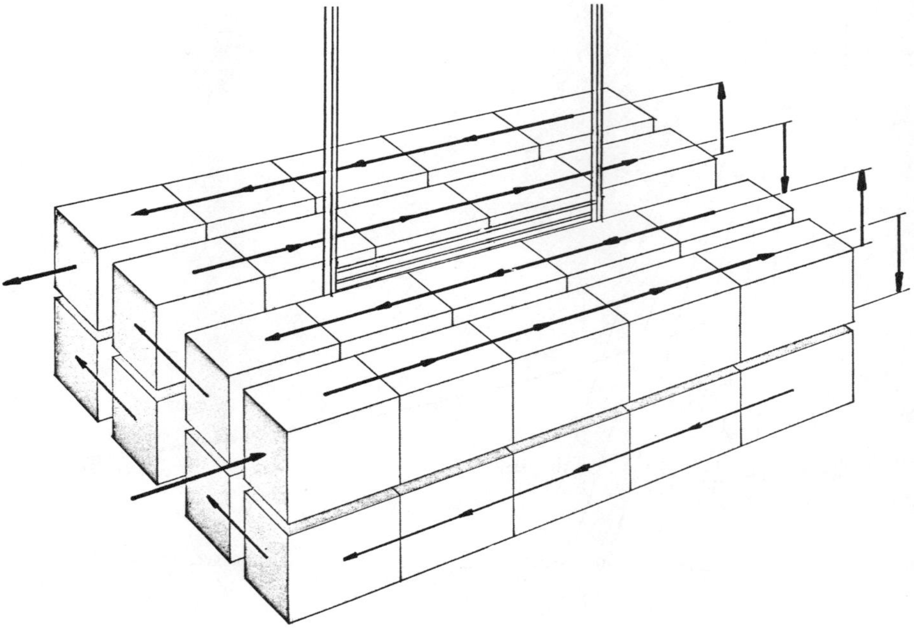

FIG.3. Disposition des paquets dans un irradiateur gamma.

Le dessin de la source doit aussi minimiser l'autoabsorption du rayonnement par la source elle-même. Avec une source plane, seule la fraction du rayonnement émise latéralement est utilisée, c'est donc dans les directions latérales qu'il faut minimiser l'autoabsorption. Les sources planes minces, de quelques millimètres d'épaisseur, présentent de ce point de vue un certain intérêt. Les sources cylindriques de environ 8 mm de diamètre sont également très répandues, leur autoabsorption est inférieure à 10 %. Par contre, les grosses sources de plusieurs centimètres de diamètre présentent une autoabsorption élevée, parfois supérieure à 20 %.

Bien que les vendeurs de cobalt tiennent compte de l'autoabsorption dans l'établissement des prix, il paraît souhaitable que les installations modernes soient prévues pour utiliser des sources à faible autoabsorption.

Géométrie des produits à irradier.

Dans un irradiateur à source plane (fig.3), les paquets sont rangés, de part et d'autre de la source, sur plusieurs lignes et plusieurs niveaux. En principe, pour assurer une bonne uniformité, chaque paquet passera successivement dans toutes les positions possibles. Dans l'établissement du projet, on rencontre des exigences contradictoires : le rendement de l'installation sera d'autant plus grand que l'épaisseur traversée par le rayonnement sera plus grande. Mais il n'est pas possible d'augmenter l'épaisseur des paquets au-delà de certaines limites, sinon le rapport de la dose maximum à la dose minimum deviendrait excessif. Souvent, il n'est pas non plus souhaitable de trop augmenter le nombre de rangées d'irradiation ou le nombre de niveaux, car le convoyeur deviendrait complexe, donc coûteux et sujet à des incidents de fonctionnement. Enfin, si on irradie d'une manière discontinue, la capacité du convoyeur devra être proportionnée à l'importance moyenne des lots à traiter.

Les rayonnements gamma sont d'autant mieux absorbés par un matériau quelconque que sa densité est plus forte. Par conséquent, pour une même disposition géométrique des produits à irradier, le rendement sera d'autant plus élevé que la densité est plus grande.

Pour chaque densité, il y a une épaisseur de paquet optimum, qui permet d'obtenir le meilleur rendement tout en respectant la limite imposée pour le surdosage.

On voit donc à quel point la définition des besoins de l'utilisateur est importante pour la conception de l'ensemble de l'installation. Pour une usine ne fabriquant qu'un petit nombre de produits, il est relativement facile de définir des conditionnements présentant la densité la plus forte possible et dont les dimensions correspondront à un rendement et à une uniformité de dose satisfaisants.

Dans une installation de traitement à façon et dans les pays où la production d'articles à usage unique est peu importante mais variée, il est beaucoup plus difficile de définir les besoins des utilisateurs.

C'est encore plus compliqué lorsque l'installation doit être utilisée pour des traitements divers, par exemple pour l'irradiation des aliments ou pour des recherches. Dans ce cas, on choisira un type de convoyeur facilement transformable. Les conteneurs, ou les balancelles, devront pouvoir transporter des paquets d'épaisseur variable.

4. TECHNOLOGIE DES IRRADIATEURS GAMMA

Les sources de Cobalt 60 sont obtenues par irradiation neutronique dans un réacteur nucléaire de cobalt naturel ou Cobalt 59. Les éléments de Cobalt 59 sont enveloppés d'une première enveloppe d'acier inoxydable ou d'aluminium. Par la suite, et avant leur livraison, elles recevront une deuxième enveloppe d'acier inoxydable parfaitement étanche.

Les sources commercialisées pour les irradiateurs industriels sont dites "sous forme spéciale". Cette expression, admise internationalement, signifie que, même après un accident de transport, la probabilité de dispersion de matière radioactive par fracture, écrasement, fusion, sublimation ou combustion est improbable.

Malgré les garanties très sérieuses offertes par la réglementation internationale et les réglementations nationales, il est bon que les responsables d'installations d'irradiation testent périodiquement l'intégrité de leur source.

Par ailleurs, il serait raisonnable de se fixer une certaine durée d'utilisation. Les sources très anciennes, ayant plus de vingt ans par exemple, ne présentent plus guère d'activité. Pour l'utilisateur, le rayonnement émis est négligeable, mais le danger de contamination ne l'est pas. La contamination d'une piscine, dans un pays mal équipé pour le traitement d'effluents radioactifs, peut poser des problèmes.

L'activité spécifique des sources industrielles au moment de leur conditionnement varie de 10 à 80 Ci/g. Au cours des dix dernières années, on a pu noter une tendance générale à l'augmentation de l'activité spécifique. Cette tendance correspond à la réalisation d'installations d'irradiation assez compactes et présentant néanmoins une puissance de source élevée : jusqu'à 1 et même 2 mégacuries.

La décroissance du Cobalt est de 12,5 % par an. Périodiquement, il y a lieu de recharger la source. Le plus souvent, on est amené, à cette occasion, à augmenter la puissance de la source, pour suivre l'augmentation de production des articles radiostérilisés.

Les porte-sources sont généralement prévus pour un seul type de sources, et ceci est regrettable car, pour les augmentations de capacité, le client n'a pas la possibilité de mettre en concurrence divers fournisseurs.

Il existe un autre radioisotope présentant des caractéristiques intéressantes pour l'irradiation : le Caesium 137.

Contrairement au Cobalt 60, c'est un sous-produit obligé de la fission de l'Uranium 235.

Dans une proportion de 6 %, la fission d'un noyau d'U 235 donne naissance à un noyau de Cs 137. Des milliers de mégacuries seront donc produits, gratuitement, au cours des prochaines décennies. Il serait d'autant plus intéressant de les utiliser qu'il s'agit d'un produit dont le stockage pose des problèmes.

Malheureusement, sa séparation des autres produits de fission est très coûteuse, son conditionnement est difficile. La Commission de l'Energie Atomique des Etats-Unis met en vente, depuis quelques mois, des lots de Caesium 137 à un prix compétitif avec celui du Cobalt 60. Toutefois, ces sources, de gros diamètre, ne sont peut-être pas directement utilisables sous cette forme.

Rien ne permet de dire si le Caesium 137 et le Caesium 134 qui lui est associé auront un jour une place dans la radiostérilisation industrielle.

Le convoyeur d'irradiation.

L'ensemble des paquets à irradier, placés sur plusieurs rangées et plusieurs niveaux, entoure la source.

Le convoyeur d'irradiation assure les mouvements des paquets, en sorte qu'ils soient irradiés uniformément. Ce résultat est obtenu en amenant les paquets à occuper successivement toutes les positions possibles.

On peut donc imaginer deux modes de fonctionnement.
Dans l'un, irradiation discontinue, l'ensemble des paquets est placé sur le convoyeur qui peut sortir ou ne pas sortir de la chambre d'irradiation. Le convoyeur est mis en marche et la source placée en position d'irradiation. Les paquets circulent en circuit fermé, passant successivement dans toutes les positions possibles. Si la durée totale d'irradiation est égale à un nombre entier de cycles du convoyeur, les paquets auront tous reçu la même irradiation.

Dans l'autre mode de fonctionnement, irradiation continue, le convoyeur sort obligatoirement de la chambre d'irradiation. Il fonctionne à une vitesse bien déterminée, périodiquement un paquet irradié est retiré du circuit et remplacé par un paquet non irradié. Il n'y a pas de cycle d'irradiation.

Chacun de ces deux modes de fonctionnement présente ses avantages et ses inconvénients. Ils peuvent être plus ou moins bien adaptés aux besoins de l'utilisateur.

L'irradiation continue convient bien à une production constante, unique et bien standardisée. Elle suppose que les postes de chargement et déchargement soient plus ou moins automatisés, car le fonctionnement est le même la nuit que le jour.

Elle implique obligatoirement que la dose soit toujours la même, et que la densité des produits varie peu.

L'irradiation discontinue convient bien au traitement de lots de produits. Chaque lot présentera un chargement homogène, surtout du point de vue de la dose demandée et de la densité. Mais d'un lot à l'autre ces paramètres pourront varier dans une large mesure.

Les opérations de chargement et déchargement pourront être partiellement automatisées, mais elles nécessiteront toujours la présence de main-d'oeuvre et d'un surveillant. Par contre, elles ne dureront que quelques dizaines de minutes ou quelques heures, à des intervalles de temps bien définis (toutes les 24 heures par exemple).

Enfin, l'irradiation discontinue peut être pratiquée avec des convoyeurs très simples ne sortant pas de la chambre d'irradiation, celle-ci peut alors comporter une porte au lieu d'un labyrinthe d'accès.

Par contre, les convoyeurs susceptibles de fonctionner en continu, avec sortie à l'extérieur, peuvent en général, sans inconvénients, être utilisés en fonctionnement discontinu.

Ici encore, on voit qu'il ne saurait y avoir un seul modèle "standard" d'irradiateur convenant à tous les utilisateurs. Beaucoup d'industriels européens et américains ont adopté un mode de fonctionnement continu. Ce système n'est pas forcément le meilleur pour le traitement à façon et pour le traitement de produits variés. Le choix ne peut être fait qu'après une réflexion approfondie sur le programme de l'utilisateur.

Toutefois, on peut penser que, pour les installations polyvalentes, le mode de fonctionnement "discontinu" présente de nombreux avantages.

Différents types de convoyeurs.

a) Convoyeurs à déplacement principal vertical.

Dans ce type de convoyeur, les conteneurs d'irradiation sont suspendus à des chaînes verticales et se déplacent eux-mêmes verticalement. Le principal avantage de cette disposition est qu'elle s'accommode bien de sources élémentaires placées horizontalement. Or, cette disposition est la seule qu'il soit réaliste d'envisager si on souhaite adopter un stockage sec de la source. Plusieurs irradiateurs de ce type ont été réalisés vers les années 64-65, mais, à notre connaissance, il n'a pas été employé depuis pour la radiostérilisation.

b) Convoyeurs à déplacement principal horizontal, à plusieurs niveaux (fig.4).

Dans cette disposition, les paquets ou les conteneurs effectuent un circuit autour de la source, d'abord à un niveau, puis refont un circuit identique à un autre niveau, et même

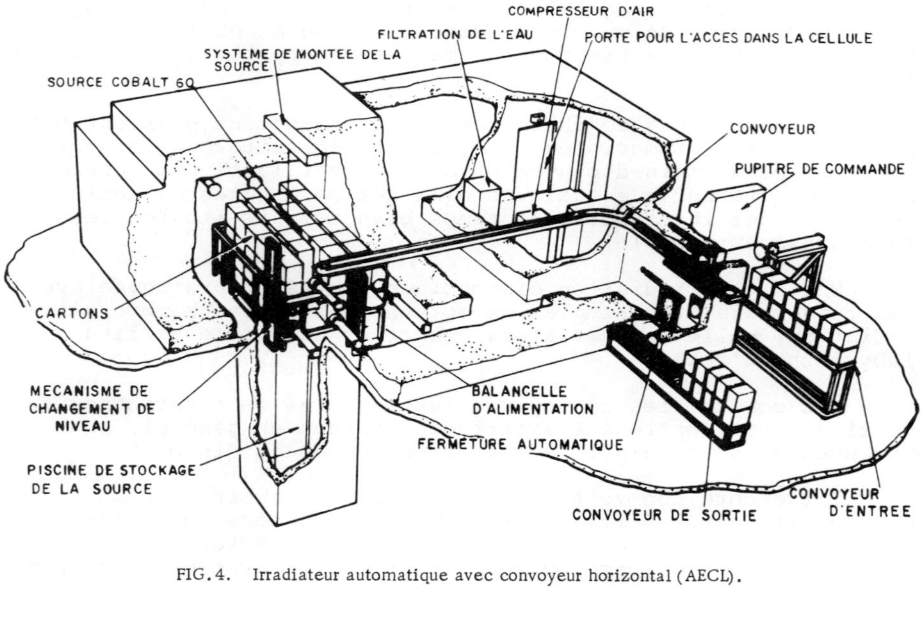

FIG.4. Irradiateur automatique avec convoyeur horizontal (AECL).

éventuellement à un troisième niveau. La circulation des paquets se fait sur des chemins de roulement horizontaux. Le plus souvent, les paquets ou les conteneurs sont poussés par des vérins pneumatiques. Le système de changement de niveau est également mû par des vérins pneumatiques. Ce type d'irradiateur est très répandu. Les conteneurs ne sont pas indispensables mais, si on traite directement les cartons, il faudra une standardisation de ceux-ci. Le transfert des paquets de l'extérieur de la cellule à l'intérieur est fait par un convoyeur auxiliaire.

Dans un autre modèle, les déplacements sont assurés par des vérins électriques mobiles sur des monorails. Toutes les opérations peuvent être effectuées par deux vérins seulement qui se déplacent dans les différentes positions d'utilisation (fig.5). L'alimentation électrique est assurée par des trolleys. L'ensemble ne comporte que des isolants minéraux et ne nécessite donc aucun remplacement périodique des pièces.

c) Convoyeur à déplacement principal horizontal, à un seul niveau.

Le dispositif à balancelles suspendues à un monorail est un moyen de transport bien connu et éprouvé. Son emploi dans une installation d'irradiation apporte une grande simplicité et une grande sécurité de fonctionnement. Les balancelles comportent plusieurs niveaux, le transfert d'un niveau à l'autre peut être fait à l'extérieur de la chambre d'irradiation au moyen d'une machine automatique. Si ce transfert est fait à la main, ce type de convoyeur permet le traitement des produits dans des conditionnements variés. Il est donc particulièrement intéressant pour les installations à but multiple.

Dans la même catégorie doivent être placés les petits convoyeurs discontinus comportant des conteneurs métalliques poussés par des vérins pneumatiques ou électriques (fig.6). Ce type de convoyeur, particulièrement économique et pouvant s'adapter à des cellules simplifiées, est intéressant pour le démarrage d'une production.

Quel que soit le type de convoyeur, la première qualité à exiger de ces appareils est la fiabilité. Cette exigence, déjà importante dans les pays développés, doit encore être renforcée dans les pays en voie de développement, où un incident mécanique, même mineur, peut provoquer un arrêt de longue durée.

Une des conditions d'une bonne fiabilité est l'emploi de matériel standard longuement éprouvé dans d'autres branches industrielles.

Les problèmes de protection.

L'émission du rayonnement ne pouvant être arrêtée, il est indispensable de prendre des dispositions pour qu'un écran absorbant suffisant se trouve toujours entre la source et le personnel. L'accès à la chambre d'irradiation ne sera permis que lorsque la source aura été placée en position "de stockage".

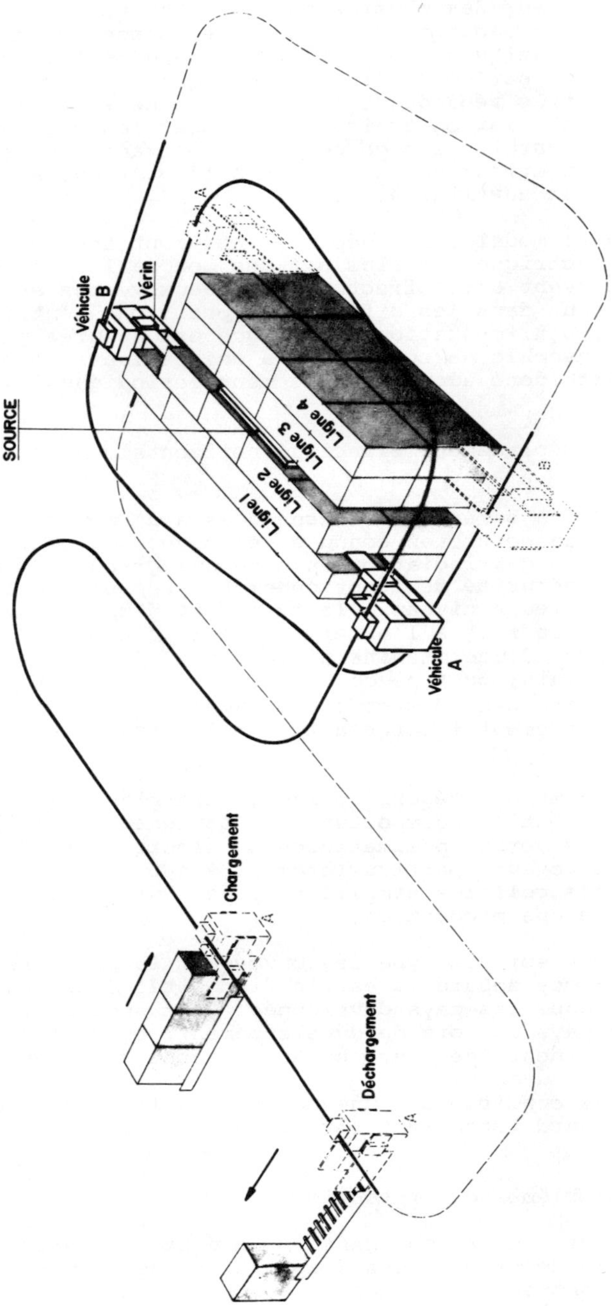

FIG. 5. Convoyeur à vérins électriques mobiles.

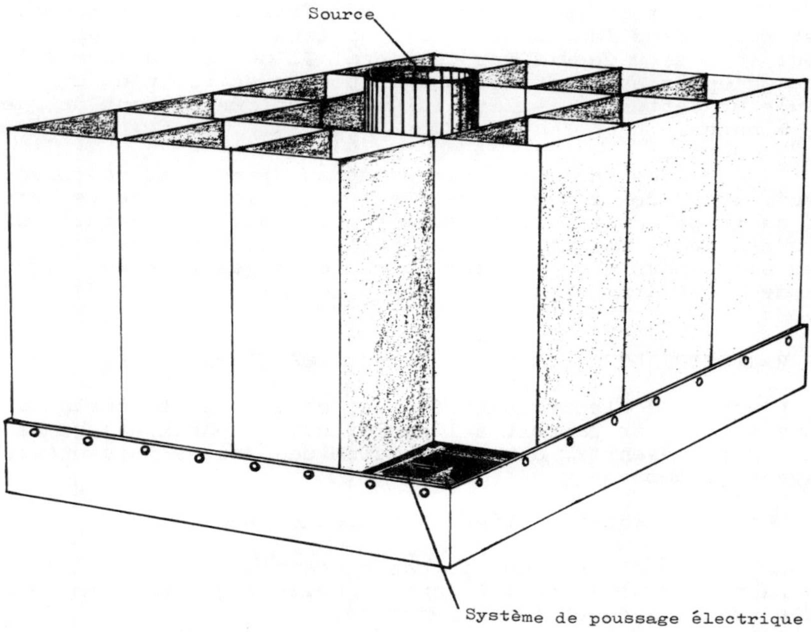

FIG. 6. Convoyeur d'irradiation discontinue (Conservatome-CEA).

Le stockage est généralement effectué au fond d'une piscine, sous 5 m d'eau. Cette disposition permet un rechargement facile des sources. Moyennant quelques précautions élémentaires, tout risque d'incident peut être éliminé.

Le stockage peut aussi s'effectuer à sec, au fond d'un puits. Cette disposition, commode et peu coûteuse pour le stockage, n'est pas très pratique pour le chargement des sources. Il semble qu'au cours des dernières années, le stockage sec soit de moins en moins utilisé pour les grosses installations.

La protection du personnel est complétée par l'existence d'un réseau de sécurité, basé sur les principes suivants :

a) L'accès de la cellule est interdit si la source est en position d'irradiation.

b) On ne peut mettre la source en irradiation si l'accès à la cellule est permis.

c) L'opérateur est tenu de s'assurer que personne n'est à l'intérieur de la cellule avant d'en interdire l'accès.

Le respect des deux premiers principes est obtenu par un ensemble de sécurités mécaniques, doublé d'un réseau d'alarme actionné par des détecteurs de rayonnement en cas de situation anormale.

Le troisième principe : vérification que personne n'est resté dans la chambre d'irradiation, implique que l'opérateur soit tenu de visiter celle-ci. Il ne pourra fermer la porte qu'après avoir appuyé sur un bouton se trouvant dans la cellule et cette possibilité lui sera retirée au bout de quelques minutes.

Si des incidents survenant au dispositif de convoyage peuvent avoir des conséquences fâcheuses, les incidents survenant au réseau de sécurité peuvent avoir des conséquences dramatiques, aussi est-il indispensable d'avoir recours à du matériel éprouvé et d'assurer un entretien extrêmement rigoureux de ces équipements.

5. RADIOSTERILISATION AVEC DES ACCELERATEURS D'ELECTRONS

L'emploi d'électrons accélérés, s'il présente quelques inconvénients par rapport à l'emploi des rayonnements gamma, peut aussi présenter, dans le domaine des fortes puissances, un avantage économique.

Les différentes parties d'un accélérateur sont :

La source d'électrons de basse énergie.
La zone d'accélération dans laquelle règne un champ électrique statique ou à haute fréquence.
La zone de mise en forme du faisceau (fig. 7).

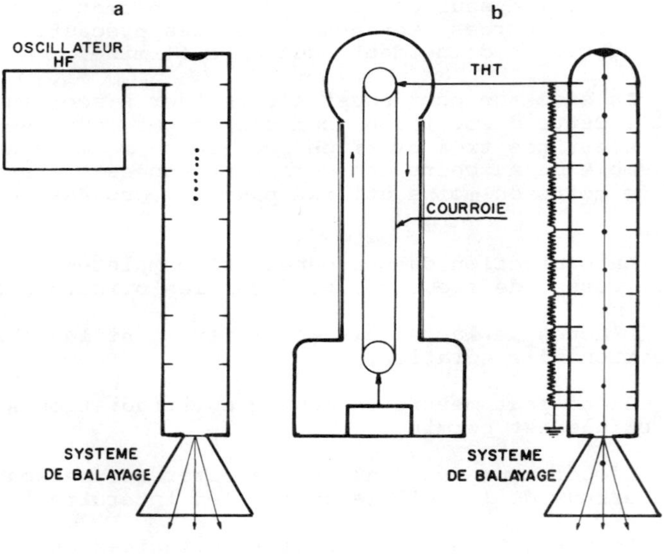

FIG. 7. Disposition d'accélérateurs utilisés pour la radiostérilisation (a – accélerateur linéaire; b – Van de Graaff)

L'ensemble est dans une enceinte fermée, dans laquelle on maintient le vide en permanence. Le faisceau d'électrons, mis en forme de rideau par déviation latérale alternative, franchit la paroi de cette enceinte à travers une feuille métallique de faible épaisseur appelée "fenêtre". Sous cette fenêtre, le faisceau est disponible pour le traitement.

Deux types de machines peuvent être employées en radiostérilisation :

Les machines électrostatiques : une haute tension assure l'accélération des électrons. Elle peut être entretenue par un dispositif mécanique porteur de charges (type Van de Graaff), ou par une cascade de transformateurs et de redresseurs.

Les accélérateurs linéaires ont un espace d'accélération cylindrique, dans lequel des électrodes circulaires sont soumises à un potentiel de très haute fréquence. Des bouffées d'électrons se succèdent le long de l'axe du cylindre.

Quel que soit le type de l'accélérateur, il se termine par une zone de balayage, dans laquelle le faisceau est "étiré" jusqu'à atteindre une largeur de 40 à 80 cm.

Après avoir traversé la "fenêtre" de la machine, le faisceau rencontre les paquets à irradier qui défilent sur un convoyeur extrêmement simple.

Pour des paquets d'épaisseur donnée et de densité homogène, il est possible de déterminer l'énergie qui assure une irradiation convenable dans toute l'épaisseur ; mais ceci n'est possible que si l'épaisseur, équivalente en matériau de densité 1, n'excède pas une certaine limite correspondant à l'énergie maximale de la machine.

La valeur des paramètres principaux du fonctionnement de la machine, en particulier l'énergie et l'intensité du faisceau, est contrôlée en permanence.

Un accélérateur nécessite une chambre d'irradiation à peu près aussi grande et une épaisseur de béton du même ordre que celles d'un irradiateur gamma. Lorsqu'on veut entrer dans la cellule d'irradiation, on peut arrêter l'appareil. Un dispositif de sécurité interdit l'accès quand la machine marche.

6. COMPARAISON DES DIFFERENTS SYSTEMES - ASPECTS PROPRES AUX PAYS EN VOIE DE DEVELOPPEMENT

Si l'intérêt de l'article à usage unique est considérable dans les pays développés, il semble qu'il soit encore plus important dans les pays en voie de développement. En effet, la stérilisation classique à l'autoclave est souvent pratiquée sans beaucoup de rigueur, le matériel n'est pas toujours en parfait état, le personnel n'est pas toujours suffisamment entraîné, en particulier dans les campagnes.

La radiostérilisation, qui apporte des garanties qu'aucune autre méthode ne peut fournir, qui permet d'approvisionner des centres hospitaliers peu importants et isolés avec un matériel de grande qualité, devrait connaître un très grand développement.

Aussi de nombreux pays envisagent-ils de se doter d'installations de radiostérilisation.

La difficulté de recruter du personnel compétent et de s'approvisionner en pièces détachées, et la faible quantité de matériel à traiter au cours des premières années suggèrent de s'orienter plutôt vers les irradiateurs gamma que vers les accélérateurs d'électrons.

Il nous semble également souhaitable de rejeter les solutions à stockage sec, un incident mécanique même mineur pouvant provoquer un arrêt de longue durée, faute de pouvoir disposer d'engins téléguidés d'intervention comme il en existe dans les pays qui ont développé une importante industrie nucléaire.

Il n'est peut-être pas non plus conseillé d'adopter un système de convoyeur entièrement automatique. Le chargement-déchargement à la main d'un convoyeur fonctionnant en discontinu peut être préférable et donner toute satisfaction, du moins dans une première phase. Par exemple, pour un débit de 1 m^3/jour, il est préférable d'effectuer, toutes les 24 heures, le déchargement et le chargement d'un mètre cube que d'intervenir fréquemment pour charger quelques décimètres cubes.

Le choix du type d'appareil ne peut être fait qu'après une étude détaillée des besoins de l'utilisateur, des conditions locales, des moyens disponibles, en hommes, en matériel, en argent.

Les différents types d'installations devront être jugés, nous semble-t-il, suivant les critères techniques suivants, par ordre d'importance :

Sécurité
Fiabilité
Uniformité de dose obtenue
Rendement et souplesse.

Ce n'est qu'après une étude approfondie que peut être effectué un choix donnant le maximum de satisfaction.

DISCUSSION

K.H. MORGANSTERN: In terms of the quantity of ^{60}Co used, at what point would you expect the cross-over between batch and continuous irradiation from the economic standpoint?

R. EYMERY: I have not made a detailed study, but at a rough guess I would say around 200 - 300 kCi.

J. LAIZIER: Mr. Morganstern tells us in his paper[1] that the cost of irradiation with electrons varies considerably with the power (capacity) of the facility. What is the cost in the case of a cobalt facility?

[1] IAEA-SM-192/8, these Proceedings.

R. EYMERY: I haven't any exact figures. There is of course such variation for cobalt, too, though it is much smaller than in the case of accelerators.

R.A. VAUGHAN: Could you comment further on your objections to dry storage of the source.

R. EYMERY: The first point is that the probability of mechanical hitches is greater and that the possibilities of taking corrective action are limited. I know one facility in which the loading of some of the tubes has become impossible because of the mechanical problems involved.

The second point is that, in my opinion, dry storage is possible only with horizontal source rods. If you use a conveyor moving vertically, the inhomogeneities of dose are integrated and you have no real problem. If you want to use a horizontal conveyor, it will be difficult to attain good vertical uniformity, especially when some lines have high activity, which can be the case after reloading of the facility.

K.H. CHADWICK: I cannot agree with your comments on the problems of dry storage. We have about seven years' experience with it and I am sure there are some among us with as much as 20 years' experience in this field. We have had no problems at all with dry storage, although I think there are advantages and disadvantages in both wet and dry storage systems. In a batch irradiator, to have a uniform dose distribution vertically one needs 'source overlap' — a longer source than product — and this produces no more difficulty with horizontal than with vertical source rods.

K.S. AGGARWAL: I share Mr. Chadwick's view that both systems have comparative merits and demerits. In India we have been using a dry storage facility for batch irradiation for the last three years. The system is simple to operate and fairly compact. In this particular case, a wet facility would not be economical.

R. EYMERY: For this type of irradiation facility dry storage may well be the best in view of the considerable economic advantage.

K.H. CHADWICK: I would like to comment on the uniformity ratio ($D_{max}:D_{min}$) in gamma radiation plants and in an electron irradiation system. At 3 MeV Mr. Morganstern showed a uniformity ratio of 1.7 giving, at 2.5 Mrad minimum, a maximum dose of 4.2 Mrad. In a gamma plant ratios of 1.3 - 1.4 are normally considered maximal, giving overdosing of 3.2 - 3.5 Mrad. Does this overdose create any material problems and, if so, are these problems different for electrons, as compared with gamma rays?

R. EYMERY: Whoever is in charge of carrying out the radio-sterilization operations or designing the facility has to comply with the overdosing ratio established by the product manufacturer. This overdosing ratio may vary from one product to another. Usually it lies between 1.2 and 1.4. There should not be any problems involved.

B.M. TERENT'EV: According to what you say, the ^{137}Cs source can now compete successfully with the ^{60}Co source in terms of cost. Can you tell us the cost of a ^{137}Cs source, and also its specific activity?

R. EYMERY: The basic cost of caesium-137 sold by USAEC is $0.10/Ci, and its specific activity is around 20 Ci/g. It should be recalled that the power emitted by 1 Ci of ^{60}Co is equivalent to the power emitted by 4 Ci of ^{137}Cs.

RADIATION DOSIMETRY PROBLEMS WHEN STERILIZING MEDICAL PRODUCTS IN GAMMA-IRRADIATION PLANTS

K. KRISHNAMURTHY
Isotope Division,
Bhabha Atomic Research Centre,
Trombay, India

Abstract

RADIATION DOSIMETRY PROBLEMS WHEN STERILIZING MEDICAL PRODUCTS IN GAMMA-IRRADIATION PLANTS.

This report briefly presents a discussion on all pertinent aspects of dosimetry in commercial gamma sterilization plants with reference to those arising out of design and prevailing conditions of operation of the plant. A brief description of ISOMED, India's first large-scale radiation plant for sterilization of medical products and the results of dosimetry obtained therein are given in order to provide a realistic basis for focusing the attention on various aspects discussed.

1. INTRODUCTION

Radiation dosimetry plays a vital role in commissioning and operating large-scale processing plants for sterilization of medical products, pharmaceuticals, etc. Many research workers have shown that a process which ensures absorption of 2.5 Mrad of gamma radiation energy (equivalent to ~25 J of power) ensures a high degree of sterility in the products so treated [1-3]. Even tests for sterility are considered superfluous for products that are manufactured under hygienic conditions and treated to a minimum specified dose. As such the results of dosimetry often dictate whether or not the products could be released to the consumer whose well-being, in some cases his life, may depend on the sterility of the products so treated. Therefore, it is of utmost importance that enough care is taken in the selection and use of dosimeters and also in the standardization of procedures employed for the characterization of radiation-sterilization plants both at the time of commissioning and also during routine operations.

ISOMED, India's first radiation-sterilization plant was commissioned in December 1973. Ever since, the plant has been used as a service irradiation facility for sterilizing a variety of medical products which include such diverse items as cotton, infusion sets, surgical gloves, eye ointment tubes, sutures, etc. Being a service facility, there is no choice or control on the amount of material of each type received for processing in the plant. These materials vary both in density and in chemical composition and may have different radiation absorption characteristics. They have to be irradiated in the best possible manner to ensure the absorption of 2.5 Mrad in the product. Control measures adopted to ensure this absorbed dose in all products have revealed a number of problems in adopting conventional dosimetric systems under plant-operating conditions.

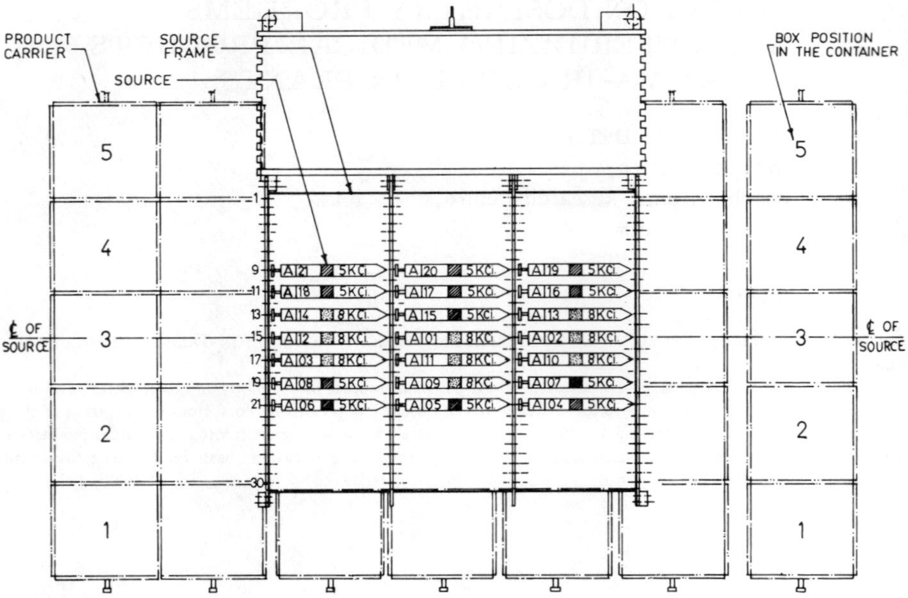

FIG. 1. Source-loading pattern in the source frame of ISOMED (1.12.1973).

2. ISOMED – BRIEF DESCRIPTION

The plant is designed to irradiate medical products of packing density ranging from 0.1 to 0.2 g/cm^3. The product conveyor system incorporated in the plant allows continuous processing of packages and helps to achieve a high degree of uniformity of dose absorbed in the product treated.

Products to be treated are accommodated in a standard cardboard carton of dimensions 59 cm long × 34 cm wide × 43 cm height (volume approx. 90 litres). The empty carton weighs about 1.1 kg. The maximum weight of the loaded carton is specified according to the nature of the products to be treated.

The source rack is designed to house 90 composite source units of Wantage design [4]. The source rack was loaded with 21 composite source units in a predetermined pattern as shown in Fig. 1. A total of 127 kCi of ^{60}Co was loaded in the plant at the time of commissioning (1.12.1973).

The product conveyor system consists of the following (See Fig. 2):

(1) Incell conveyor or source pass conveyor
(2) Labyrinth conveyor
(3) Feed conveyor and load-unload station.

Both the source pass conveyor and the labyrinth conveyor are monorail in type. These conveyors together hold 63 vertical product carriers, suspended from individual trolleys on the monorail (See Fig. 2, item 7).

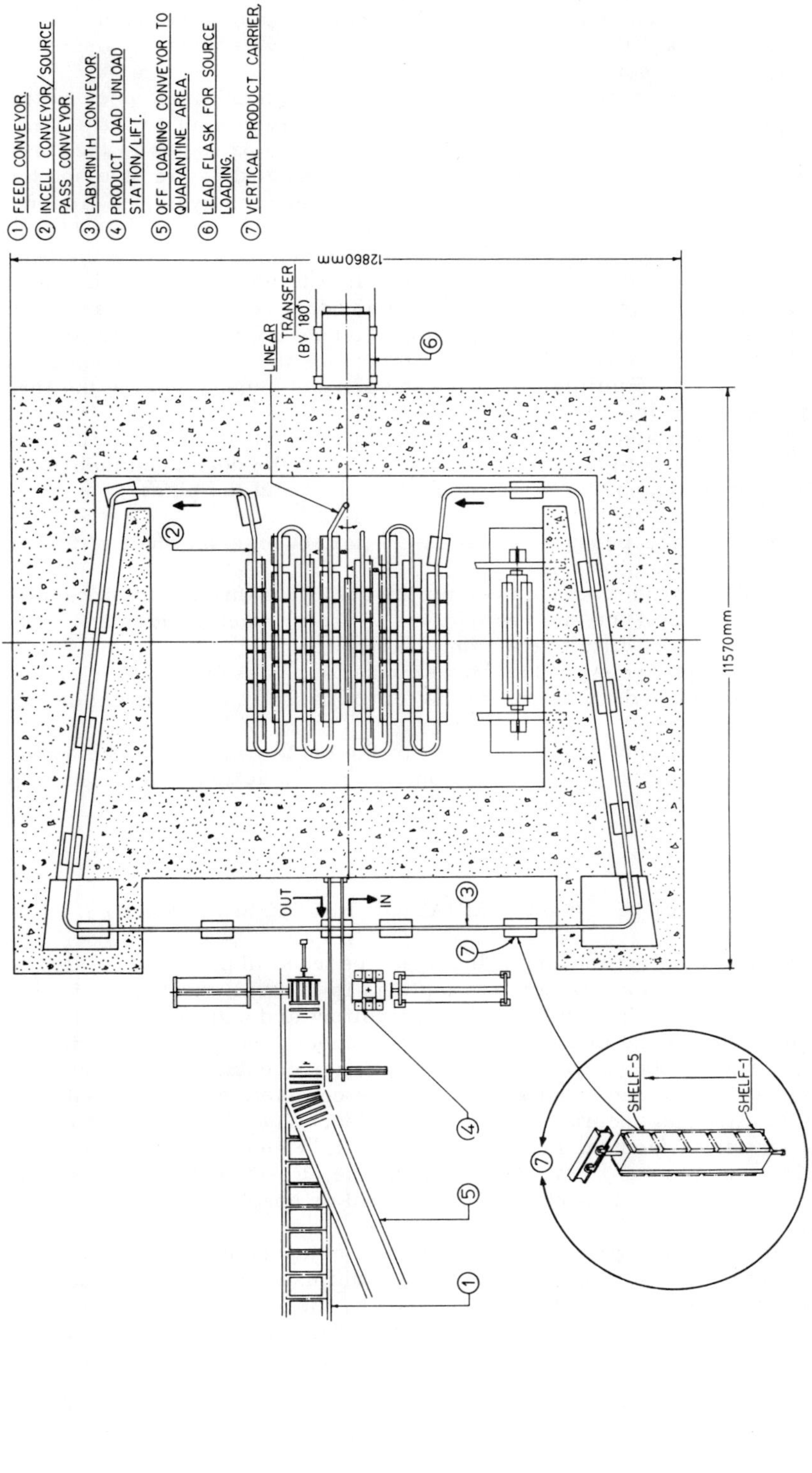

FIG.2. Product flow line in ISOMED.

Forty-eight of these carriers are within the cell at any time and arranged on the source pass conveyor in eight rows and six colums on either side of the source rack. Each vertical carrier holds five standard cartons one over the other in five separate shelves or tiers.

The product carriers continuously move with uniform speed in the cell. The speed of the conveyor is controlled to a precision of within ± 1%. This uniform motion ensures uniformity of dose in the packages in the direction of travel. The vertical dose uniformity in the packages is obtained by shifting the cartons progressively through each shelf in each cycle. A linear transfer mechanism located inside the cell reverses the carrier through 180° at the end of the fourth pass on one side of the source, thus enabling the two opposite faces of the box (normal to the source plane) to be exposed to the source.

The labyrinth conveyor carries the product between the loading station and the incell conveyor. The speed of the conveyor is adjusted to four times that of the in-cell conveyor in order to reduce the product inventory in the plant and also to minimize the time the products spend outside the irradiation field in each cycle.

The salient aspects of irradiation of a product box in routine processing is summarized here to highlight the conditions of the dose accumulation or time-mode of dose absorption in a product processed in the plant.

(1) Each new product carton enters the lower-most shelf of the product carrier.
(2) Each product carton completes one cycle through the irradiation cell in each of the shelves or tiers in the product carrier and exits at the end of the fifth cycle from the top tier.
(3) In each cycle product boxes make four passes on either side of the source and at the end of the fourth pass a 180° reversal of the product occurs.
(4) The product box spends 80% of the cycle time in the radiation field and about 20% of the time outside the radiation field. Thus irradiation is discontinuous.

3. DOSIMETRY OF ^{60}Co IRRADIATION PROCESSING PLANTS

An accurate estimation of the dose absorbed in products treated in the ^{60}Co irradiation plant is not as simple as is generally understood. It is considered that ^{60}Co dosimetry is straightforward and well established. This may be true of dose measurements made under laboratory conditions. However, the situation prevailing in a radiation processing plant is not all that congenial for the use of most of the known dosimetric systems with a degree of reliability and accuracy that is demanded. The IAEA Code of Practice for Sterilization of Medical Products states that: "Recognised dosimetric systems should be employed", e. g. ferrous sulphate, Fe-Cu sulphate, oxalic acid and film systems (polyvinyl chloride, coloured or uncoloured polymethyl methacrylate) [5].

This implies that all dosimeters have comparable properties or response characteristics and are suitable for in-plant use. In fact these dosimeters are not comparable and doses measured by these systems may differ widely in many cases. An optimum dosimetric system for use in irradiation processing plants should meet, among others, the requirements listed below:

(1) The dosimetric system should have a range of measurements between 1-5 Mrad and should be able to achieve an accuracy of ± 5% in absorbed dose measurements.

(2) Dosimetric material should have comparable radiation absorption characteristics to that of medical products processed.

(3) The system should be insensitive to dose-rate and temperature variation during irradiation. The anticipated dose-rate variation ranges from 10 krad/h to a few Mrads/h for a typical plant loading of 100 kCi of ^{60}Co (Fig. 3 shows the time-mode variation of dose-rate during irradiation process in ISOMED). Temperatures up to a maximum of 50°C may be expected during irradiation within the cell where no special cooling arrangement is provided.

(4) The dosimetric response should be independent of the variation of gamma energies. The gamma spectrum varies from point to point in the plant and depends on the location and nature of products interposed between the source and box position.

(5) Most important of all is that the dosimetric system should have a very good stability before, during and after irradiation. The irradiation period may extend well over 100 h in some cases and usually ranges from about 10 to 100 h in ^{60}Co plants. Besides, irradiation is not continuous and is done in cycles, each cycle contributing to 10 to 15% of the total dose absorbed. The dose accumulation in a product box during irradiation in ISOMED is presented in Fig. 4. Stability of radiation-induced changes during and after irradiation is important to deduce actual dose absorbed from the dosimetric response at the end of irradiation. Fast fading or fast growth in the response characteristics of dosimeters immediately following the cessation of irradiation could lead to erroneous estimation of absorbed dose values. (See Fig. 5). This is all the more important as the irradiation itself is not continuous and is carried out in a number of cycles.

(6) As a routine many dosimeters may have to be prepared for dose monitoring. Simple dosimetric systems which can be easily prepared and stored for some time should be preferred. The physical size of the dosimeters should be small to enable dose evaluation at a point within the volume of the product box.

4. CHOICE OF DOSIMETERS FOR INPLANT USE

The foregoing considerations limit the choice of dosimeters for use in a radiation sterilization plant under actual process conditions. Though calorimetric and ionization methods are ideal for absorbed dose measurements and are generally used as primary standards, they cannot be used for in-target dose measurements in radiation-sterilization plants for obvious reasons. To date the only dosimetry procedure generally accepted as "standard" is the ASTM procedure for the Fricke dosimeter [6]. The dose range involved here rules out the use of the Fricke dosimeter directly. However, it may be used indirectly for measuring doses in each cycle or each pass of the conveyor by adjusting the irradiation time suitably. The ferrous cupric dosimeter [7] and extended super Fricke dosimeter [8] not only do not satisfy the dose

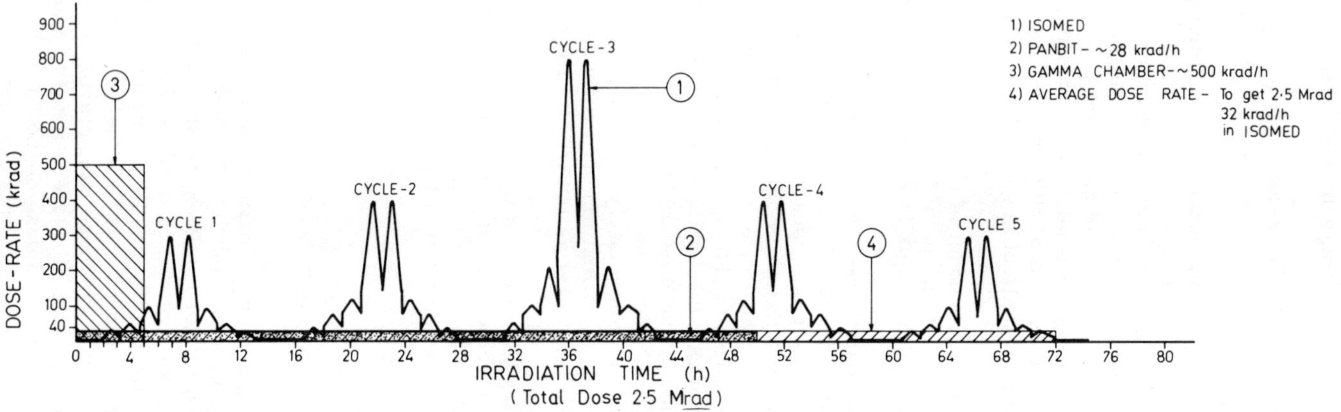

FIG. 3. Dose-rate conditions during irradiation of dosimeters.

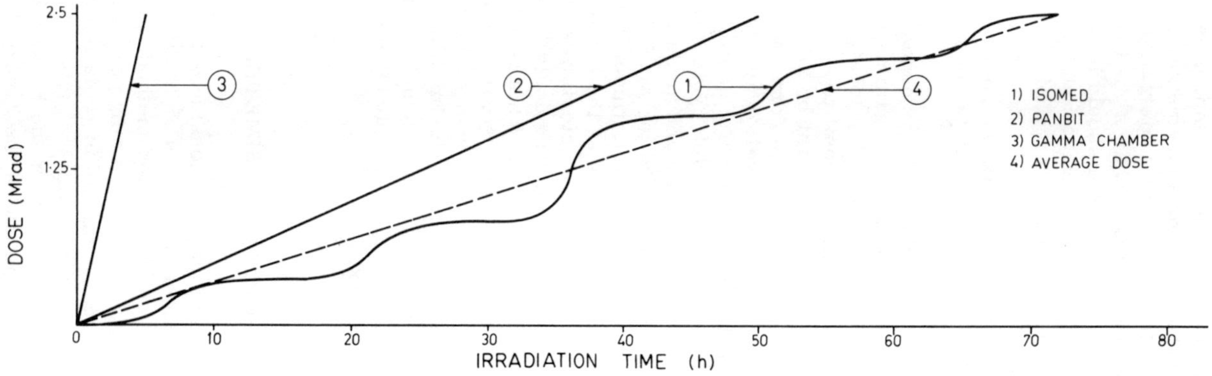

FIG.4. Time mode of dose accumulation.

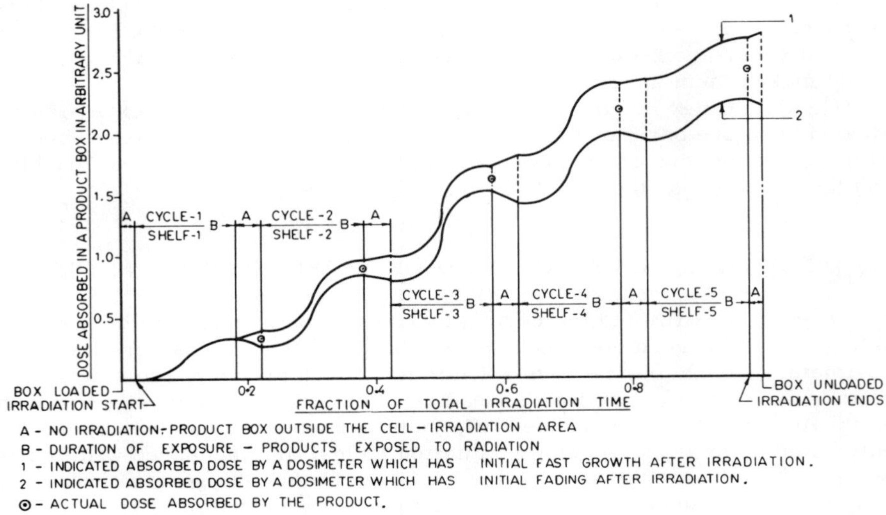

FIG. 5. ISOMED – indicated dose accumulation in a product box by a dosimeter. (Effect of post-irradiation stability of dosimeters)

range of interest but also have poor pre-storage and post-irradiation stability [9]. They are particularly unsuitable for prolonged irradiation [8]. Of the chemical dosimeters the oxalic acid dosimeter and the ceric sulphate dosimeter may be used in the dose range of interest. Although these dosimetric systems have been well studied and understood, difficulties have been experienced in getting the systems to work reliably [10]. The oxalic acid system [11], though useful in the range of interest and is water equivalent, its measurement could be tedious and troublesome because of the problem involved in the analytical measurement techniques employed. The ceric dosimetric system is very sensitive to chemical impurities. Recently methods have been suggested to overcome this effect and also to simplify measurement by potentiometric techniques [12]. The ceric system [13] is also known for its energy dependence at high cerium concentrations. However, at the level of concentration used for measuring doses in the 1 to 5 Mrad range, dose-rate dependency does not pose a serious problem.

Variation in response due to temperature dependence of the ceric system is about 5 to 10% in the temperature range of about 15 to 50°C and this factor should be taken into consideration during actual use. Despite these limitations, it may be said that the only chemical dosimeter suitable for use in the measurement of doses in the range of interest, under plant conditions, is the ceric sulphate system. The ceric system, if carefully prepared, is extremely stable before and after irradiation and can be stored for a few months. Reproducibility of the order of ± 4% can be easily achieved using the ceric system [14].

Solid-state dosimeters such as PVC films [15], lucite [16], clear perspex [17, 18] special glasses [19, 20], and red perspex [21], have been suggested for routine dosimetry in irradiation plants. PCV and film dosimeters are highly energy dependent and cannot be used for accurate dose measurement. Glass dosimeters such as silver-activated phosphate glass or cobalt-activated borosilicate glasses, though they cover the range of our interest here, are known to have a fast-fading characteristic after irradiation, which makes them unsuitable for use.

Clear perspex HX and red perspex dosimeters have been extensively studied, and are being routinely used in many irradiation plants. Their radiation absorption characteristics match those of many medical products and therefore make them suitable for in-plant use.

5. DOSIMETERS FOR ABSORBED DOSE MEASUREMENTS IN ISOMED

The perspex HX and the ceric sulphate dosimetric system have been selected for absorbed dose measurements in ISOMED. The perspex HX dosimeters (batch 3) size 4 cm × 1 mm have been procured from M/s Gillette (450 Basingstoke Road, Reading, Berkshire, UK) together with the calibration curve. The perspex HX, being a secondary dosimeter, has to be calibrated before use, in known radiation fields, and measured with the help of a primary dosimeter such as the Fricke. Though the general behaviour and response of the perspex HX dosimeters are well known [22] no detailed information is available on their response to intermittent irradiation. In view of this, some studies have been carried out in our laboratory on the suitability of perspex dosimeters under actual processing conditions in ISOMED.

Dosimeters of uniform thickness from 0.98 to 1.02 mm were selected, cleaned and wrapped in lens tissue paper and aluminium foil. All these dosimeters were further sealed in polythene. Measurements of optical density were carried out at 305 nm on a Pye Unicam SP-500 spectrophotometer. These measurements were further compared with the values obtained on a Beckman DB spectrophotometer for their consistency and reproducibility.

6. CALIBRATION OF THE DOSIMETER

The perspex dosimeters were calibrated at different dose-rates to understand the effect of the time mode of dose absorption on the induced optical density-dose relationship. Three dose-rates were used; 520 krad/h, 52 krad/h, and 28 krad/h. The last one represented the average dose-rate obtained in ISOMED (temperature during irradiation was found to be 38°C at 520 krad/h in the gamma chamber, and about 32 - 34°C at 52 and 28 krad/h) in the Panoramic Batch Irradiator, Trombay (PANBIT) [23]. The graphs presented in Fig. 6 clearly indicate that there is a small effect due to dose-rate, the optical density being lower at small dose-rates and long duration of irradiation compared with high dose-rates and a short irradiation period. Probably during irradiation there is slight fading in the radiation-induced optical density.

Post-irradiation stability of the induced optical density was therefore studied at higher and lower dose-rates. The graphs in Fig. 7 show that, following irradiation at low dose-rates, the optical density steadily fades at the rate 6 to 7% a day whereas the induced optical density at higher dose-rates, initially grows for a period of about 24 h and steadily decreases at a slower rate of 2 to 3% a day. In ISOMED, irradiation conditions alternate

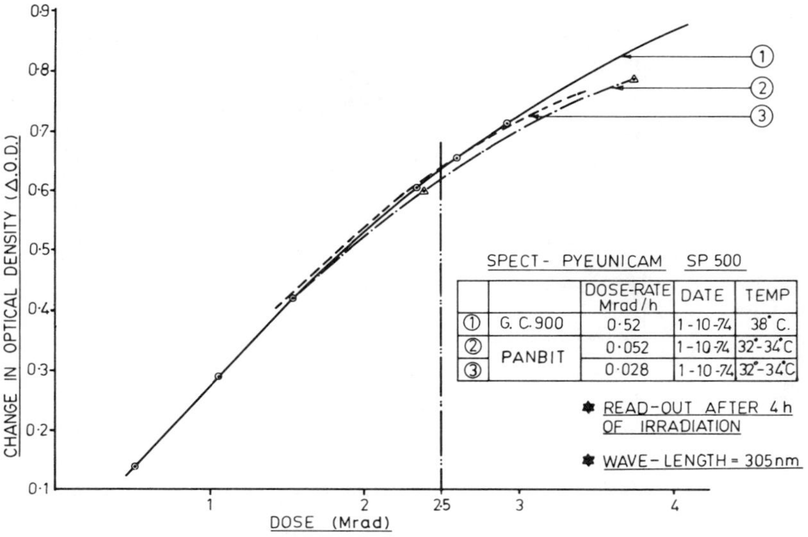

FIG.6. Calibration curve for perspex-HX dosimeter. (Batch 3, 1 mm).

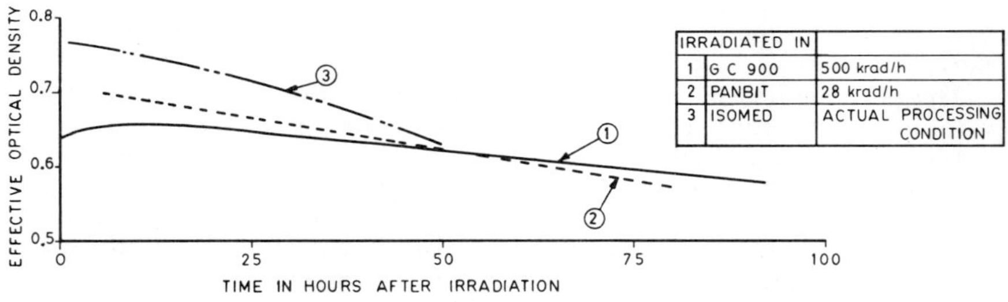

FIG.7. Fade curves of perspex-HX (Batch 3).

between smaller and higher dose-rates and extend over 70 h (See Figs 3 and 4). Such a condition could result in complex decay characteristics of the induced optical density. Therefore, these dosimeters were exposed to different doses in ISOMED under actual processing conditions and the fading characteristics were then followed for 3 d (See Fig. 8).

It is clear from the results presented in Figs 7 and 8 that the post-irradiation fading in optical density depends on the time mode of delivery of doses and also on the final dose absorbed. One startling fact, obvious from Fig. 8 was that very little difference in optical density was observed between dosimeters which have completed four and five cycles in the irradiation plant, though it was known (Fig. 9) that the last cycle contributed as much as 10% of the total dose. This behaviour of the dosimeter was further checked and

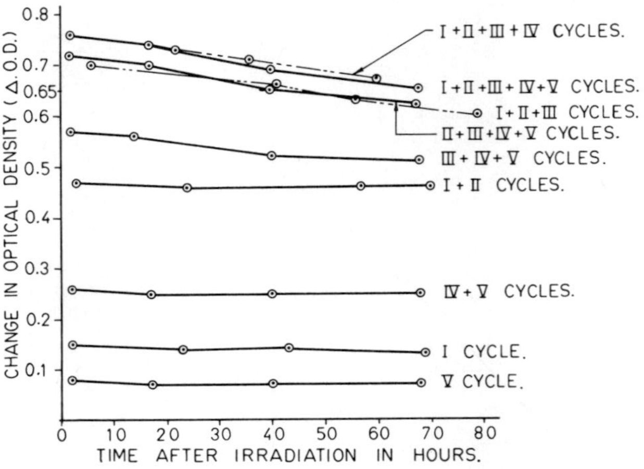

FIG.8. Decay curve for perspex-HX dosimeter (Batch 3, 1 mm). (Irradiations in ISOMED).

confirmed. The time mode of dose accummulation in ISOMED, as can be seen in Figs 3 and 4, is complex. Nearly 50% of the dose is delivered in the third cycle alone. Under such conditions the effect of intermittent irradiation, shutdown of the plant etc., will seriously affect the radiation-induced optical density in the perspex and hence the total dose deduced from these measurements.

The perspex dosimeters were also tested for their response characteristics to intermittent radiation. Our studies reveal that the perspex dosimeters could be used safely if the interval between two successive irradiations does not exceed 4 h. For a higher interval there was a definite fading or growth in the induced optical density depending on the dose-rate at which the irradiation was terminated. In view of this, all our measurements of optical densities were carried out within 4 to 5 h after the end of irradiation.

Besides dose-rate conditions we find that temperature during irradiation is also an important factor. In view of this our calibration graphs were made at temperatures 37 and 33-35°C, matching the temperature conditions closely during irradiation in ISOMED.

The odd behaviour of radiation-induced optical density during and after irradiation is very difficult to explain. Especially the changes occurring in the fading characteristics of the dosimeters with time mode of delivery of doses and total dose are not simple to determine without further studies.

The ceric sulphate dosimetric system is also used in conjunction with perspex dosimeters. Though this system is well understood there is no uniform agreement on the values of G to be used for ceric-cerrous conversion at different concentrations and at various temperatures. However, this being the only chemical dosimeter satisfying the range of our interest and of extremely good stability, this system was used. The ceric sulphate system consisted of 10 m$\underline{M}$ ceric ammonium sulphate in 0.8$\underline{N}$ H_2SO_4. This dosimeter was calibrated in a known radiation field measured by Fricke dosimeters. Temperature during irradiation was 38°C, closely matching

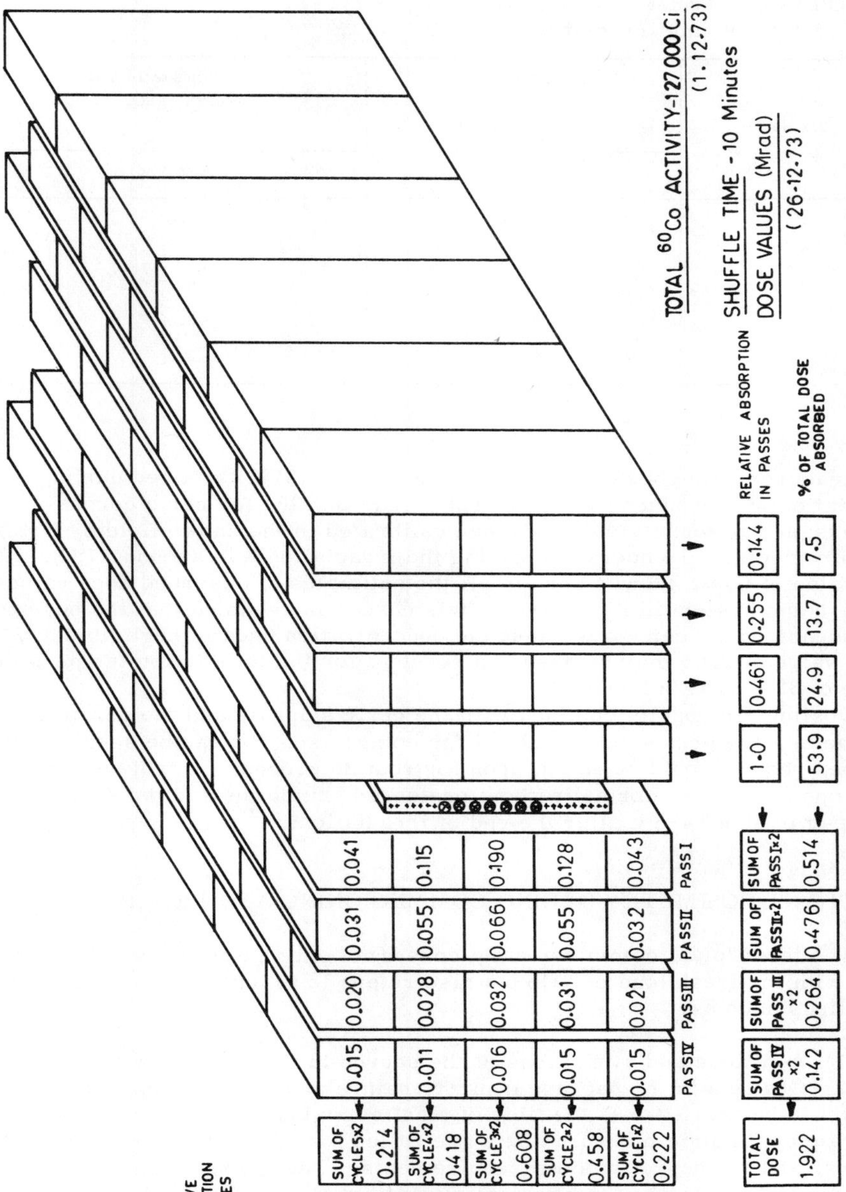

FIG. 9. ISOMED – Dose accumulation in a product box during irradiation.

TABLE I. COMPARISON OF DOSES MEASURED BY PERSPEX AND CERIC DOSIMETER IN ISOMED
Temperature during irradiation 33-37°C

Box No.	Perspex (1 mm) measured 4 h after irradiation	Ceric sulphate 10 m$\underline{M}$ in 0.8$\underline{N}$ H_2SO_4 G = 2.26
	Dose (Mrad)	Dose (Mrad)
1	3.50	3.52
2	3.54	3.48
3	2.42	2.47
4	2.37	2.52
5	2.40	2.58

the measured value of the maximum temperature (37°C) attained during irradiation of a product box in ISOMED. The G value for ceric cerous conversion was found to be 2.26 when calibrated in the known field by Fricke measurements. (Temperature during the irradiation was around 37°C). The value of G = 2.26 was used in all the estimates of absorbed dose values by the ceric system in ISOMED. In this case it is assumed that the energy dependence of the ceric system at the concentration used is negligible and the G value of 2.26 may be assumed constant for the irradiation temperature range of 32 - 37°C in ISOMED.

Dosimetric results obtained with the ceric sulphate and the perspex dosimeters are presented in Table I for comparison. It can be seen that the perspex results could be relied upon for routine process control only if the plant operations are normal without prolonged shutdowns and the dosimeters are read in about 4-5 h after the end of irradiation.

7. PLANT DOSIMETRY DURING COMMISSIONING AND PROCESS CONTROL

The term 'plant commissioning' covers not only the initial startup of the plant with the fresh load of ^{60}Co but also refers to all changes in irradiation conditions such as:

(1) Additions and alterations of the source in the rack; and

(2) Changes in the package density, nature and composition of the product to be treated. At the time of plant commissioning extensive and accurate dosimetric measurements are called for to set the process parameters, and to ensure the dose requirements. The International Code of Practice for Radiation Sterilization of Medical Products clearly defines the objectives of the radiation dosimetry in gamma-irradiation plants [5]. It reads, "The aim of dosimetry is to make sure that every unit receives at least the required minimum dose and that excessive doses are not given". The fulfillment of this requirement should be ensured on commissioning the plant and whenever the process parameters are altered, and should be

controlled during routine operations. Further, it recommends that all dosimetric measurements should be carried out in a product package under actual processing conditions. In a ^{60}Co plant the process parameters which define or alter the absorbed dose in a product are:

(a) Source activity and source position
(b) Products irradiated
(c) Conveyor speed or time setting

Source activity is a known factor at any time and hence can be considered fixed once the loading of the source is complete in a predetermined distribution. The source position is usually set to a great degree of accuracy in ^{60}Co plants.

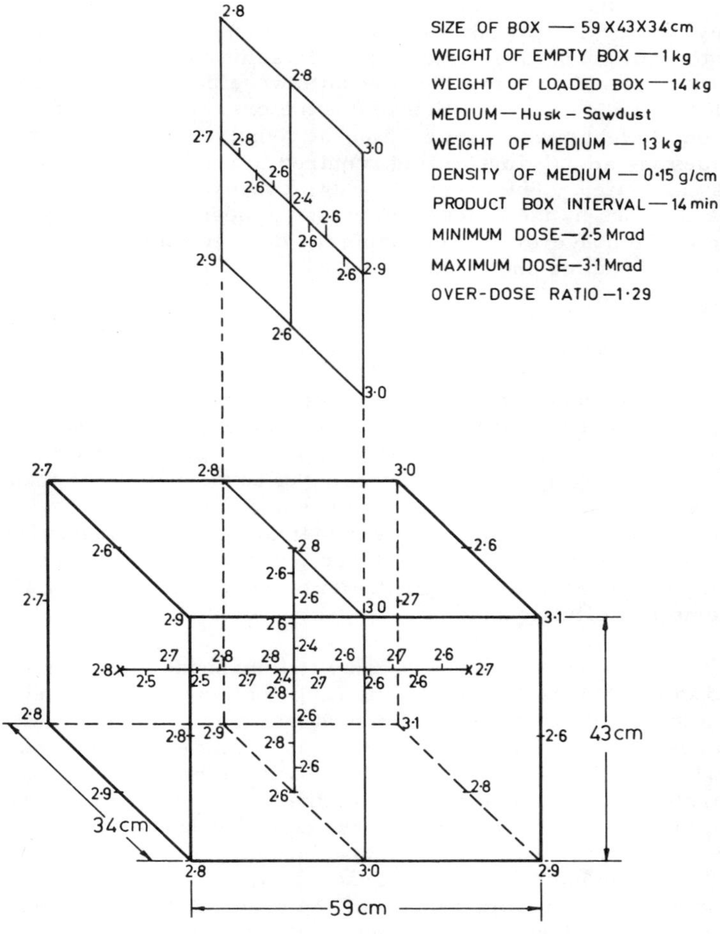

FIG.10. ISOMED dosimetry. Dose (Mrad) in an irradiation box.

Products of varying nature will be received for irradiation in service irradiation plants and therefore dosimetry at the time of commissioning will have to be carried out on dummy products. These are boxes filled with homogenous phantom materials such as husk, waste papers or plastic moulding that have a similar density and radiation absorption characteristics to the production packages received for processing. A set of such boxes of varying densities may be used to arrive at the relation between the conveyor speed and the desired absorbed dose in products of different densities. As this is usually difficult and time-consuming, a fixed density dummy product may have to be used.

The exact time setting is arrived at after studying the distribution of absorbed dose in the dummy box and from the known values of the minimum absorbed dose value in the distribution. (See Fig. 10).

In ISOMED the ceric sulphate dosimetry system has been employed in conjunction with the perspex HX dosimeters. All perspex dosimeters were used to evaluate the comparative values of the absorbed doses within the dosimetry box rather than for estimating the actual dose absorbed. To arrive at the exact minimum dose value, the ceric sulphate dosimetric measurements were initially relied upon and these results were then cross-checked by the Fricke dosimetry. Using the Fricke dosimeters, the values of absorbed dose due to each of eight passes and each of the five cycles were estimated. From these values the actual time setting required to provide the minimum absorbed dose was calculated. The lower of the two conveyor speeds corresponding to the times obtained by the Fricke and the ceric dosimeters was then fixed as the speed of the conveyor for imparting the desired minimum dose to the product during routine operations.

8. PROCESS CONTROL

The products received in the plant, especially in a service plant like ISOMED, may not all be of the same density or chemical composition and this could cause serious concern in maintaining the minimum absorbed dose in the product. Ideally speaking, processing of boxes containing these products constitutes alteration in process conditions and the speed of the conveyor should be changed accordingly. It is rather difficult to translate this into practice as the amount of material of each type processed at any one time may not be sufficient to fill the plant. Two methods are usually advocated to overcome this situation:

(1) The conveyor speed is set so as to impart the minimum dose of 2.5 Mrad to the densest product in the range of densities expected [24]. The acceptable range of densities will be such that the products of lower density in this range do not receive excessive doses. This effectively means that the total weight of the product to be irradiated will be controlled by suitably mixing lighter and denser materials and the weight of the individual product box shall not exceed the limits set by the conveyor speed for the specified density maximum.

(2) The other method often suggested [25] is to set the conveyor speed to correspond to the average density of the products to be processed. The range of densities allowed will be such as to limit the variation in absorbed dose between 2.5 to 2.9 Mrad. It must be borne in mind that in this practice

the specified minimum dose of 2.5 Mrad is not usually adhered to. However, in our practice we find it is desirable to get the lower limit to 2.5 Mrad, so that the average minimum dose will be somewhere between 2.7 to 3.0 Mrad and the lowest value of 2.5 will be reached only with the densest product. In view of the problem of product mix in actual production boxes, it is often desirable and essential that the dose absorbed be checked regularly by placing dosimeters in boxes which are filled with materials differing in density or composition. An effective and accurate dosimetric system must be employed to obtain reliable dosimetric results. In our experience we have found that it is not always possible to predict or to identify the points of maximum and minimum dose in a product box based on the dose distribution studies in a dummy box. The perspex HX dosimeters could be relied upon for routine process control dosimetry of the plant if there is no shutdown extending more than 4 h and the dosimeters are read at a fixed time interval of say 4-5 h after irradiation. It is always desirable to use the ceric dosimetric system for process control as these dosimeters can be prepared well in advance and can be stored for a long period. An alternate method of measurement such as a potentiometric system can be easily adopted in routine practice [12].

9. MICROBIOLOGICAL MONITORS

The use of microbiological dosimetry at the time of plant commissioning and routine process control is recommended [5]. These dosimeters during routine use merely act as go-no-go-type indicators and no quantitative information is obtained. This often restricts the delivery of the products treated to the customers if a positive count is obtained. In many cases these positive counts may later turn out to be false. Besides, the accuracy and reproducibility of the microbiological system is very much dependent on the

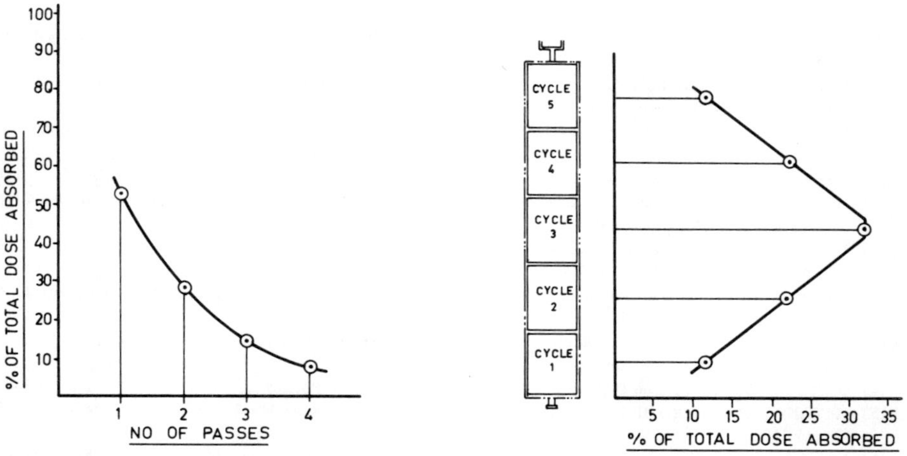

FIG. 11. Dose absorption in a product box.

skill of the personnel employed and also on the availability of proper equipment. Application of these microbiological monitors are rather expensive and in fact may not be necessary during routine process control of ^{60}Co radiation plants.

10. RADIATION UTILIZATION EFFICIENCY

The pattern of dose distribution obtained as shown in Fig. 9 has enabled us to calculate the percentage contribution of each row or pass and each of the cycles to the total dose. This is schematically represented in Fig. 11. It can be seen from Figs 9 and 11 that the contribution of the fourth row or pass (7.5%) is relatively lower than that of the first or fifth (12%) cycle even though the source activity is concentrated in the middle of the plaque. Any increase in the actual size of the source is likely to cause reduction in the radiation utilization efficiency.

ACKNOWLEDGEMENT

The author wishes to acknowledge with thanks the help rendered by Shri S.V. Navada, Shri Deodutt and Shri R.N. Deshpande during the dosimetry studies in ISOMED.

REFERENCES

[1] VAN WINKLE, W., Radiosterilization of Medical Products, Pharmaceuticals and Bioproducts, Tech. Rep. Series No. 72, IAEA, Vienna (1967) 24.

[2] ARTANDI, C., Large-Scale Application of Radioisotopes to Sterilization, Proc. Amer. Nucl. Soc., Topical Meeting 1, Part IV, (1966) 33.

[3] JEFFERSON, S., LEY, F.J., ROGERS, F., Radiation Sterilization of Medical Supplies, Nucl. Engng, August (1964).

[4] Radiation Sources for Industry and Research, Amersham Catalogue, Radiochemical Centre, Amersham (1971) 54.

[5] IAEA, Radiation Sterilization of Medical Products (Proc. Symp. Budapest, 1967), IAEA, Vienna (1967) 427, Section 4.3, Note 1.

[6] Tentative Method of Measuring Absorbed Gamma Radiation Dose by Fricke Dosimetry, ASTM Designation: D 1671-59T (1959).

[7] JARRETT, R.D., HALLIDAY, J.W., Dosimetry in Food Preservation by Ionising Radiation, US Army Natick Lab. Activities Rep. No. 15 (1963).

[8] HASSBROEK, F.J., Extending the range of Fricke dosimeter, Nat. Conf. Nuclear Energy, Application of Isotope and Radiation, Pretoria, Republic of South Africa (1961).

[9] KRISHNAMURTHY, K., NAVADA, S.V., Use of Ferrous Cupric Dosimeters, (To be published).

[10] HOLM, N.W., SEHESTED, K., Influence of Dose Rate or Large Absorbed Doses on the Oxalic Acid Dosimeter, Risø Rep. No. 112 (1968).

[11] HOLM, N.W., BJERGBAKKE, E., SEHESTED, K., An Investigation of the Oxalic Acid System for Co-60 Dosimetry, Risø Rep. No. 111 (1969).

[12] MATHEWS, R.W., Potentiometric estimation of megarad dose with the ceric-cerous system, Int. J. App. Radiat. Isotopes 23 (1972) 179.

[13] BJERGBAKKE, E., "The ceric sulphate dosimeter", Manual on Radiation Dosimetry (HOLM, N.W., BERRY, R.J., Eds) Marcel Dekkar Inc., New York (1970) 323.

[14] BRYNJOLFSSON, A., The Production Test Irradiation of 30 000 Pounds of Bacon, BNL Rep. No. 10652, (1966

[15] ARTANDI, C., STONEHILL, A.A., Polyvinyl chloride, A new high level dosimeter, Nucleonics 16 (5) (1958) 118.

[16] MULLER, A.C., RIZZO, F.X., MOGK, R.B., The Measurement of Megarad Doses of Cobalt-60 Gamma Rays with Ultraviolet Transmitting Lucite, BNL Rep. No. 985 (1966).

[17] BOAG, J.W., DOLPHIN, C.W., ROTBLAT, J., Radiation dosimetry by transparent plastics, Radiat. Res. 9 (1958) 589.
[18] ORTON, C.G., Clear Perspex dosimetry, Phys. Med. Biol. 11 (1970) 377.
[19] CHEKA, J.S., BECKER, K., High level glass dosimeters with low dependence on energy, Nucl. Appl. 6 (Feb. 1969).
[20] SCHULMAN, J.H., et al., Measuring high doses by absorption changes in glass, Nucleonics 13 (2) (1955) 30.
[21] WHITTEKER, B., Red Perspex Dosimetry, Manual on Radiation Dosimetry, (HOLM, N.W., BERRY, R.J., Eds), Marcel Dekkar Inc., New York (1970) 363.
[22] CHADWICK, K.H., Radiation Effects and After-Effects in the Clear Polymethyl Methacrylate Dosimeter, Ph. D. Thesis (1971) (Published by the Centre for Agricultural Publishing and Documentation, Wageningen).
[23] KRISHNAMURTHY, K., et al., The Panoramic Batch Irradiator, Trombay (PANBIT) BARC Rep. No. I-145 (1971).
[24] Bulletin published by Gillette, Surgical Gamma Irradiation Plant, England (May 1967).
[25] VAN WINKLE, W., et al., "Destruction of radiation resistant microorganisms on surgical sutures by Co-60 irradiation under manufacturing conditions", Radiation Sterilization of Medical Products (Proc. Symp. Budapest, 1967), IAEA, Vienna (1967) 169; and Code of Practice, p.423.

DISCUSSION

K.H. CHADWICK: You have given us a very good example of a problem that I have been discussing for some time now – the question of accuracy in calibration and the accuracy of dosimeters in practice, when the irradiation times and conditions are very different. This is a problem that arises with very many different dosimeters and which needs further investigation.

The effect of different fading in the perspex after a short or long irradiation is very simply explained by the rotation of the OD spectrum, which occurs following the short irradiation, causing the OD at 305 nm to increase for ± 24 h and fade more slowly thereafter. At 314-315 nm, the cross-over wavelength, the OD does not increase but fades at a rate of ± 2% per day, due to a balance between O_2 diffusion and the spectrum build-up. After a long irradiation the OD spectrum has completely rotated, and this can be simulated by heating heavily irradiated perspex sample for half an hour at 50°C, so that there is no longer a build-up of OD at 305 nm, but only a fading caused by O_2 diffusion, which leads at 305 and 314 nm to a fading of 5-6% per day. This effect is illustrated in the Figs A and B. The use of measurements at 314-315 nm would preclude the turning effect at 305 nm that would occur in the plant irradiation, but the fading effects cannot be avoided. Personally, I would like to see use made of a 2-mm-thick PMMA dosimeter since it would at least make this fading less critical.

K. KRISHNAMURTHY: I am interested to hear that you have also had this problem. Whether or not one adopts 305 or 314 nm for optical density measurements, the fact remains that the induced OD at these two wavelengths fades with time and at a considerable rate to boot. Since the OD induced could also fade while the dosimeter was inside the plant, for example, if the plant were for any reason shut down for a certain time, say for a day or more, or if the dosimeter were subjected to a very low dose-rate while in transit from the cell, one could not measure the actual dose absorbed by the dosimeter, to say nothing of the actual dose absorbed by the product. Since the fading characteristics of the perspex are inherent in the material, unless something is done to stabilize them the dosimeter cannot be recommended for use as an universal instrument for plant operation or monitoring. For plant operation, the problem is how to find out the actual dose absorbed by the

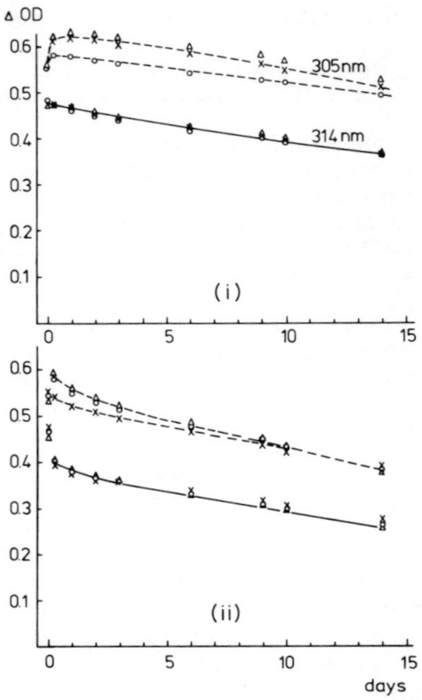

FIG. A (i). Normal untreated, 1-mm HX, 2.56 Mrad at 900 krad/h.
(ii). Treated 4 h at 55°C, 1-mm HX, 2.56 Mrad at 900 krad/h.

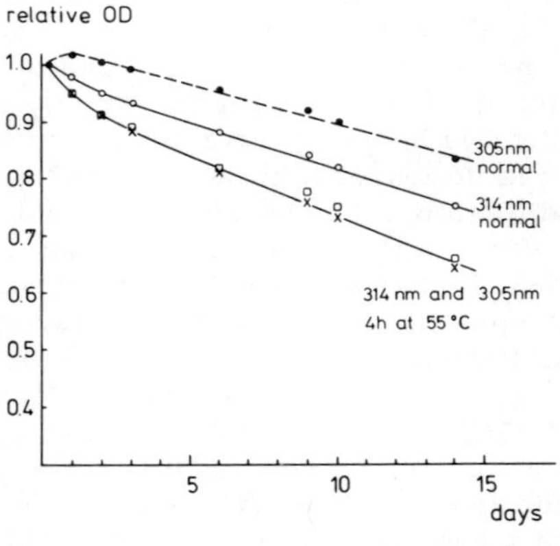

FIG. B. 1-mm perspex fading in air in dark.

product, and not how or why fading occurs in the perspex, or how it can be corrected once the dosimeter is outside the plant. Use of thicker perspex and selection of 315 nm as the measurement wavelength might bring about a slight improvement, but the nature of the perspex itself cannot be changed very much in this way. What we really need is a co-operative calibration effort in this field.

PLANNING OF GAMMA-FIELDS: FORMING AND CHECKING DOSE-RATE HOMOGENEITY IN IRRADIATION FACILITIES

V. STENGER, G. FÖLDIÁK, Zs. HORVÁTH, L. NASZÓDI
Institute of Isotopes of the Hungarian Academy of Sciences,
Budapest, Hungary

Abstract

PLANNING OF GAMMA-FIELDS: FORMING AND CHECKING DOSE-RATE HOMOGENEITY IN IRRADIATION FACILITIES.

The optimal geometry of the sources of an 80 000-Ci ^{60}Co irradiation facility was calculated. The array of the sources is suitable for fundamental research and pilot-plant radiosterilization simultaneously. A method was developed to compensate the inhomogeneity of the dose-rate field: it is no worse than that of the continuous large-scale facilities. In five years the activity of the sources decreased by about half; therefore, this recharge became inevitable. Experience proved that with the new source geometry optimalized by calculations a dose-rate of 1.2 ± 10% became available within the packages.

To accelerate the practical use of results achieved in radiation technology research — especially in developing countries, with their relatively modest resources — a pilot-scale production with irradiation sources originally built only for research purposes may frequently be ncessary. This paper outlines some of the problems that had to be overcome under such circumstances at the Institute of Isotopes of the Hungarian Academy of Sciences, in a pilot-scale radiosterilization process.

For the determination of the optimal field the dose-rate was calculated for more than 30 000 points in the irradiation chamber. The dose-rate distribution of point-like, cylindrical, rod-like and plate sources were compared and the possibilities of their utilization were established [1,2]. It was found that the irradiation field is strongly influenced by the desired dose, the geometry, the activity and the specific activity of the sources, the size of the irradiation chamber, the density of items to be sterilized, the capacity of the facility and, indirectly, by the dimensions of the box and the technique of material transport.

Our 80 000-Ci nominal activity ^{60}Co γ-irradiation facility was commissioned in 1969. The sources are sited at the centre of an iron and concrete shielded, $4 \times 4 \times 5.5\text{-m}^3$ chamber [3,4]. The system is made up of a total of 20 cassettes with four 1000-Ci source elements in each of them (Fig.1(a),(b)). In our facility a cylindrical array of sources is generally used [4-6]: the $cc_{max.}$ 7-10 $Mrad \cdot h^{-1}$ dose-rate inside the cylinder is used for fundamental research, while outside it pilot-plant irradiation is carried out. The dose-rate in the chamber drops with the distance from the sources, at the same time, however, the absolute homogeneity of the irradiation field increases (Fig.2). At a distance of 1100 mm from the centre a vertical dose-rate difference of up to 23-60 $krad \cdot h^{-1}$ may appear within a package (Fig.3). Therefore, an irradiation technology had to be developed which would, nevertheless, guarantee the dose homogeneity achievable with continuously operating large-scale commercial facilities [7].

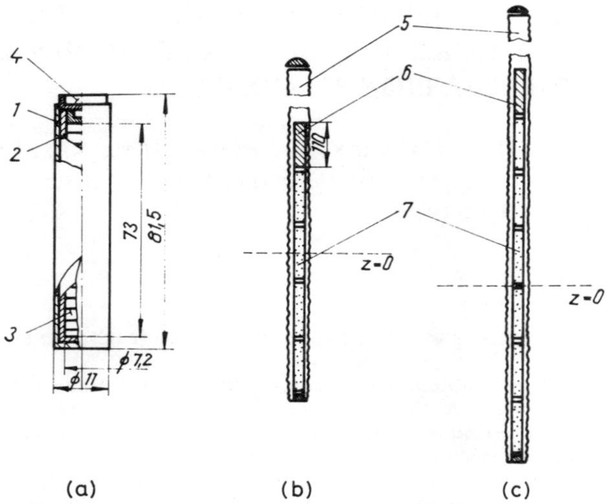

FIG. 1. (a) Source element; (b) Original irradiation cassette; (c) Cassette after recharge.

1. Outer stainless steel case
2. Inner stainless steel case
3. Radioactive cobalt and inactive aluminium discs
4. Welded cap
5. Flexible tube
6. Iron core
7. Active charge

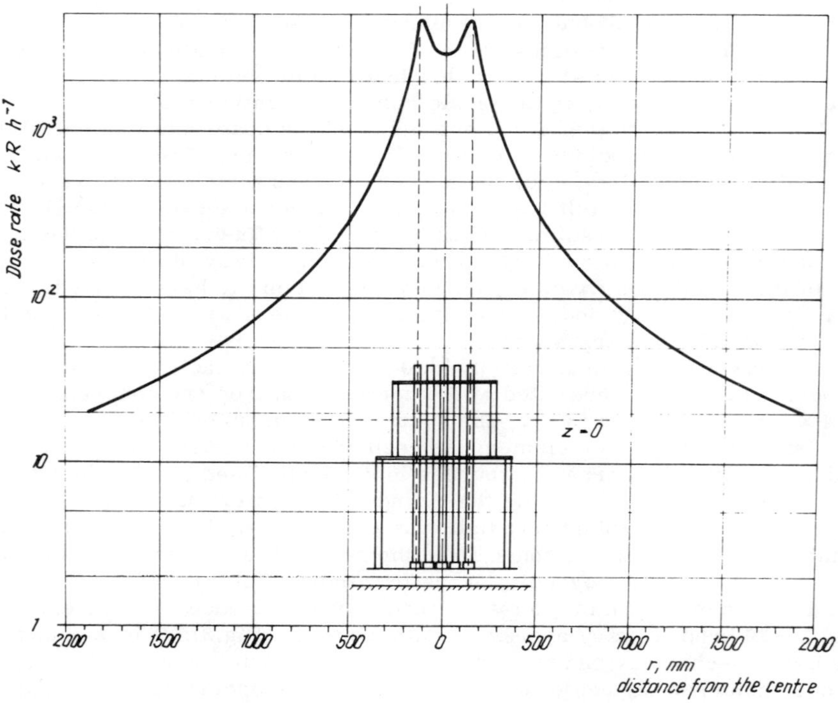

FIG. 2. Dose-rate as a function of the distance from the sources (r) in the z = 0 plane (z = 0, see Fig. 5).

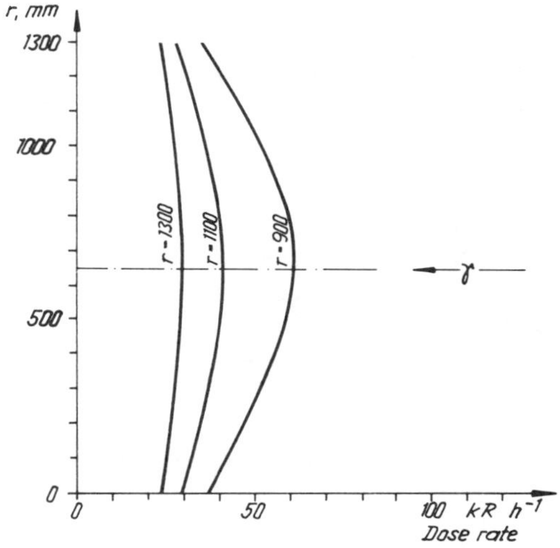

FIG. 3. The dose-rate as a function of the height (z) of the irradiation chamber. (z, see Fig. 5.)

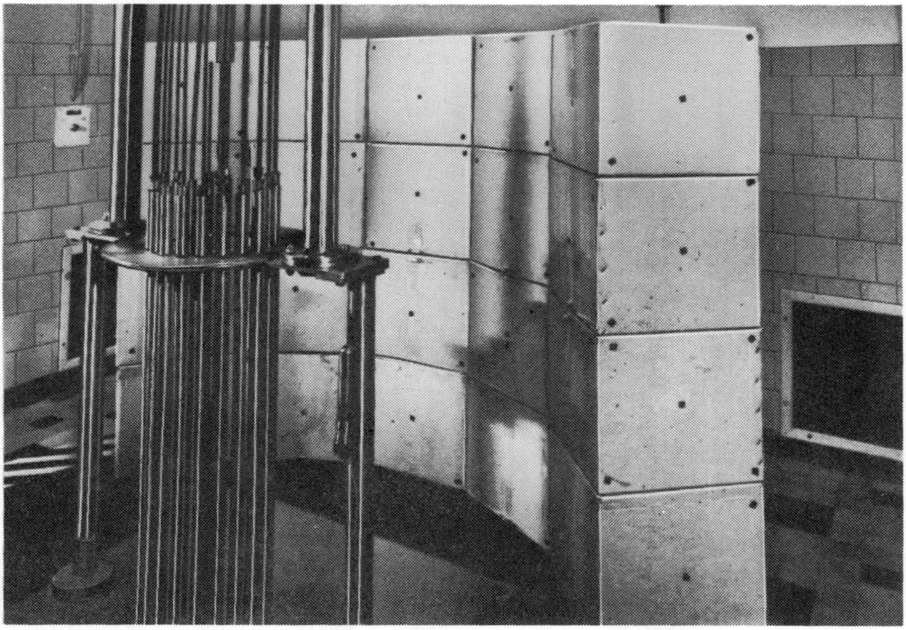

FIG. 4. Boxes to be sterilized in the irradiation chamber.

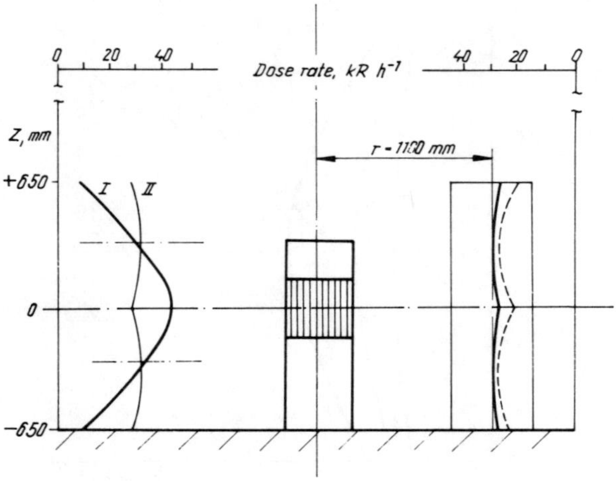

FIG. 5. Dose-rate distribution in a box under sterilization.

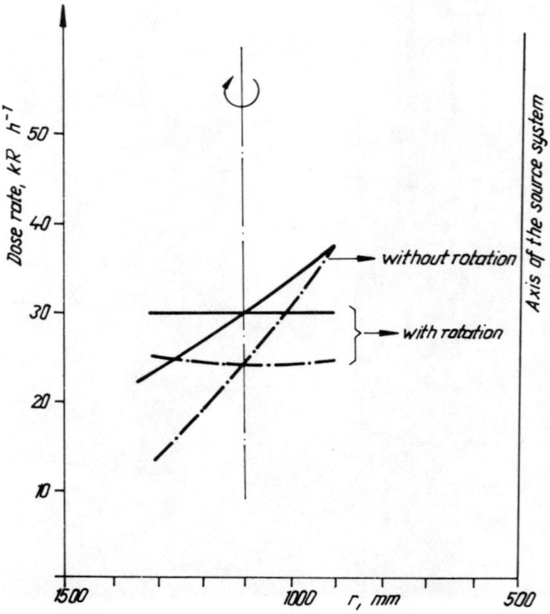

FIG. 6. Dose-rate distribution produced by turning the boxes during irradiation (z = | 650|).

In a commercial irradiation plant the boxes are exposed to radiation on their parallel sides for equal times and those moving over and under the source centre are also continuously interchanged [8]. However, such a continuous operation is economical for radiosterilization only if the activity of the ^{60}Co charge is higher than 100 000 Ci.

In the case of a pilot-plant sterilization, in order to compensate for the large inhomogeneity of irradiation field, several possibilities of box transport were tested. It was established that the field not farther than 1200 mm and not higher than 1600 mm from the centre of the sources can be most effectively utilized. In this particular case the 400 × 400 × 350-mm^3 boxes contained polyethylene eye droppers with an average bulk density of 0.13 g·cm^{-3} (Fig.4). According to our technology the boxes are rotated twice by 180° around their vertical axis at a quarter and three-quarters of the irradiation time, replaced at half the irradiation time; or turned by 180° around their vertical axis and the same time the boxes above and under the source horizontal symmetry plane are interchanged at half the irradiation time.

The dose-rate distributions in the irradiation chamber and in the boxes at a distance of r = 1100 mm from the centre of the source and at distances above and below the horizontal symmetry plane of the dose field between z = 650 and z = +650 are shown in Figs 5 and 6.

The curves on the left of Fig.5 represent the distributions obtained without (I) and with (II) vertical field compensation. A significant improvement in homogeneity of the dose-rate field can be achieved by compensation. The curves on the right of the figure refer to the dose-rate distribution along the vertical central axis of a box under sterilization in the case of no absorption (full line), and when the dose-rate deformation due to the packaged material was taken into account (dashed line).

In Fig.6 the dose-rate field measured by rotation of the box between z = -650 and z = +650 and between r = 900 and r = 1300 mm can be seen. The full line traces the distribution in air, while the dash-dotted one the distribution in absorbent with a bulk density of 0.3 g·cm^{-3}.

This way the surface irradiation dose fraction at the boxes (i.e., the ratio of the highest to the lowest dose-rate) could be brought down to 1.16. The dose values were evaluated on the basis of data determined by alcoholic chlorobenzene dosimetry [6,9,10]. From an evaluation of about 800 measurements it was calculated that the prescribed dose of 2.8 Mrad could be guaranteed with an accuracy of ±4%.

In five years the activity of source elements dropped to about 500 Ci, i.e. by half. So, in 1973 it became necessary to recharge them. To achieve the original 4000-Ci summary activity the four 500-Ci activity elements were completed with two 1100-Ci nominal activity elements. We therefore had to construct longer cassettes, capable of containing six, instead of the originally four source elements (Fig.1(c)).

First of all the optimal array of 500 and 1100-Ci source elements had to be determined. The different dose curves belonging to the possible arrangements of these elements were calculated. An algorithm was worked out for an electronic computer to establish quickly and economically the distribution of the relative dose-rate in the irradiation field. The dose values on the surface of a cassette are given in Fig.7(a), while those at a distance of 160 mm from the cassette can be seen in Fig.7(b).

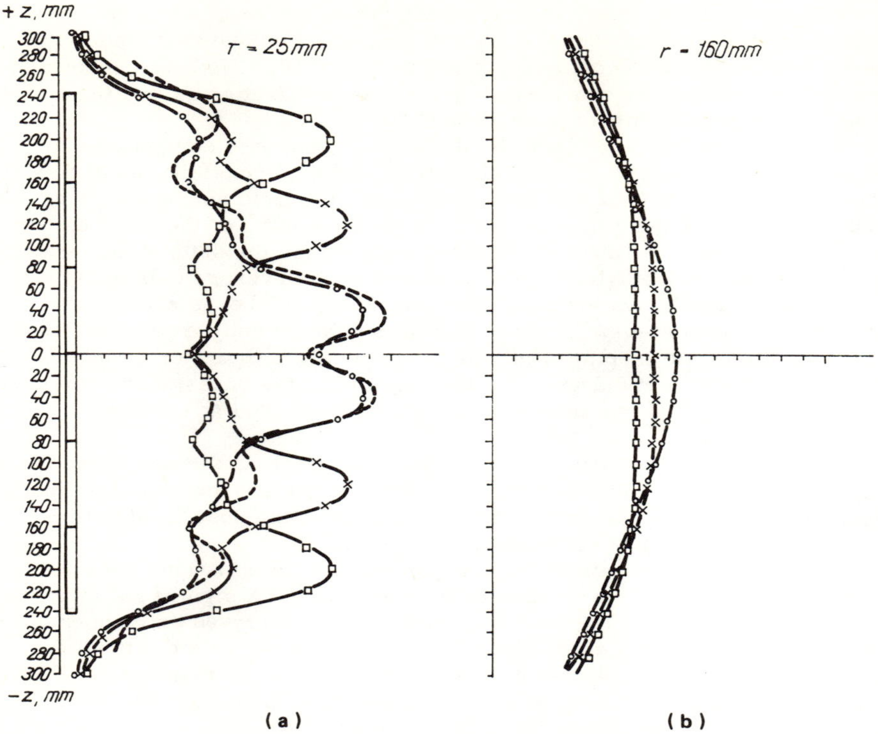

FIG. 7. Calculated and measured dose-rate curves of a six-element (Fig. 1(c)) source rod.

——— calculated	o	500, 500, 1100, 1100, 500, 500 Ci
	×	500, 1100, 500, 500, 1100, 500 Ci
------ measured	□	1100, 500, 500, 500, 500, 1100 Ci
		500, 500, 1100, 1100, 500, 500 Ci

Control measurements were carried out after the recharge of the radiation sources. The computer data were checked by physico-chemical dosimetry carried out in the irradiation chamber [11]. Semiconductor dose-rate detectors were developed [12-14] for the rapid, simple and low-cost determination of high dose-rates (10^3 - 10^7 R$\cdot$h^{-1}). The results of these measurements are given in Fig.7. These semiconductors are suitable also for the dose-level control of intermittently or continuously operated radiation-sterilizing facilities.

The capacity of the 80 000-Ci irradiation facility amounts to about 600 m^3 yearly (at a dose of 2.5 Mrad and material density of 0.20 g$\cdot$cm^{-3}), which corresponds to an efficiency of about 12%. Using two rows of the boxes around the sources the efficiency rises to a maximum of 18%, about 50% of the efficiency of large-scale irradiators.

REFERENCES

[1] FEJES, P., HORVÁTH, Zs., STENGER, V., Isotopenpraxis 6 (1970) 98.

[2] HIRLING, J., STENGER, V., FEKETE, Z., "Development of high-efficiency γ-irradiation plants", Izotóptechnikai kutatások, MTA Izotóp Intézete, Budapest (1972) 269 (in Hungarian).

[3] HIRLING, J., STENGER, V., Energia és Atomtechnika 22 (1969) 466 (in Hungarian).

[4] STENGER, V., Izotóptechnika 15 (1972) 485 (in Hungarian).

[5] HORVÁTH, Zs., STENGER, V., FÖLDIÁK, G., FEJES, P., "Determination of the dose rate field of high-intensity gamma-sources", Izotóptechnikai kutatások, MTA Izotóp Intézete, Budapest (1972) 41 (in Hungarian).

[6] FÖLDIÁK, G., HORVÁTH, Zs., STENGER, V., "Routine dosimetry for high-activity gamma-irradiation facilities", Proc. Symp. Dosimetry in Agriculture, Industry, Biology and Medicine, IAEA, Vienna (1973) 367.

[7] HORVÁTH, Zs., "Dosimetric inspection system of γ-irradiation facilities", Proc. Johnson and Johnson Meeting, Sterilization by Ionizing Radiation, Multiscience Publication (1974) (in press).

[8] FÖLDIÁK, G., STENGER, V., Atomtechnikai Tájékoztató 12 (1969) 255 (in Hungarian).

[9] HORVÁTH, Zs., BÁNYAI, É., FÖLDIÁK, G., Radiochim. Acta 13 (1970) 150.

[10] HORVÁTH, Zs., FÖLDIÁK, G., Izotóptechnika 15 (1972) 531 (in Hungarian).

[11] HORVÁTH, Zs., FÖLDIÁK, G., Atomtechnikai Tájékoztató 13 (1970) 332 (in Hungarian).

[12] STENGER, V., "Using semiconductors for measurement of high-energy gamma radiation", Proc. Johnson and Johnson Meeting "Sterilization by Ionizing Radiation", Multiscience Publication (1974) (in press).

[13] STENGER, V., PAVLICSEK, I., Atomtechnikai Tájékoztató 13 (1970) 339 (in Hungarian).

[14] OSVAY, M., TÁRCZY, K., Izotóptechnika 16 (1973) 307 (in Hungarian).

A LOW-COST IRRADIATION FACILITY FOR PILOT-SCALE PROCESS IRRADIATION STUDIES

K. KRISHNAMURTHY, K.S. AGGARWAL
Isotope Division,
Bhabha Atomic Research Centre,
Trombay, India

Abstract

A LOW-COST IRRADIATION FACILITY FOR PILOT-SCALE PROCESS IRRADIATION STUDIES.

In this paper a few salient aspects of a low-cost irradiation facility designed and constructed by the Isotope Division of BARC, India, are described. A batch-type shuffle-dwell conveyor system, which has been developed to optimize the dose conformity and irradiation efficiency for the medical products, is briefly detailed. Experimental observations of parameters such as radiation heating in the storage container, variation of irradiation efficiency with products of different densities and dose uniformity such as products of different types, are also indicated.

1. INTRODUCTION

During the last two decades impressive advances have been reported in the applications of intense radiation sources in the field of industrial processing. These include radiation sterilization of medical products, preservation of foods, synthesis of bio-degradable detergents and chemicals, production of new consumer products such as wood plastics, concrete polymer composites, etc. Despite the significant achievements in the laboratory research in radiation applications, a wide gap exists between these findings and commercial requirements. The process industry today is very competitive and is slow to take up any new venture unless the process is proved to be economical. The laboratory findings have to be transferred to production plant operations through stages to establish the technological feasibility and commercial viability of any process. This transition from laboratory to industrial-scale operation is normally achieved through pilot-scale facilities. Pilot plants are product oriented and custom designed. In view of these, they are expensive and cannot serve diverse applications. Most of the commercially available small-scale gamma irradiators are small in volume (4 litres) and have fixed radiation intensity over this irradiation volume. There is a great need for a compact and general-purpose irradiation facility with maximum versatility and minimum cost.

As part of the radiation technology programme of the Bhabha Atomic Research Centre to develop competence for designing, manufacturing and operating radiation units, a multi-purpose radiation facility, PANBIT, has been designed, built and commissioned at Trombay and has been in operation for the last two years.

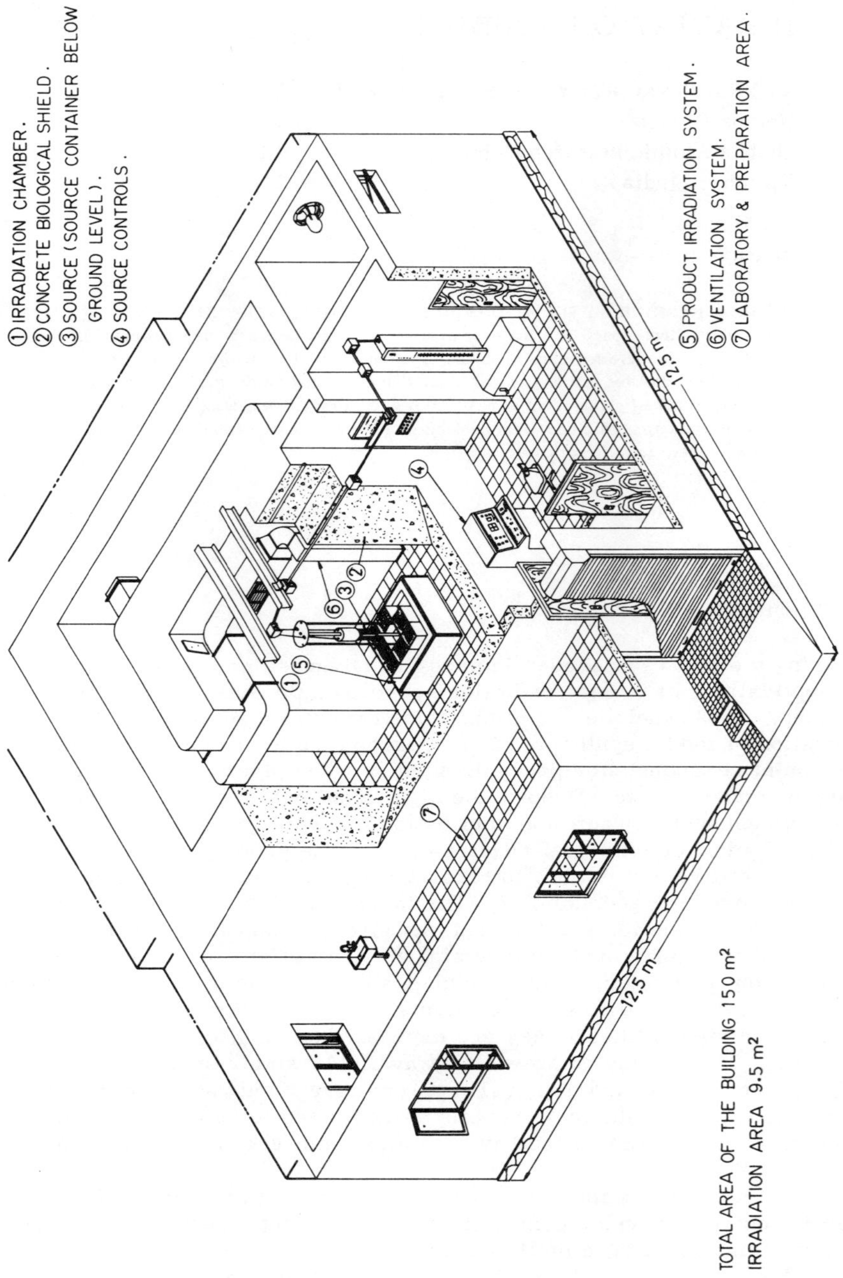

FIG. 1. Isometric view of the PANBIT facility.

The design was evolved to provide the following:

(1) A large-volume irradiation space sufficient to meet most of the radiation process requirements;
(2) A wide range of dose-rate variations to study the effect of dose-rate on process parameters and product mix;
(3) A batch-type shuffle-dwell conveyor to obtain uniform irradiation dose over a large volume of products;
(4) Experimental service lines to establish the technological and commercial feasibility of a particular process; and
(5) Training for personnel in the operation and maintenance of irradiation facilities.

2. DESCRIPTION OF THE FACILITY

An isometric view of the PANBIT facility is shown in Fig.1. The total area of the building inclusive of the laboratory and preparation area is 150 m^2. The irradiator occupies 95 m^2 of floor space. Basically the facility consists of:

(1) Irradiation chamber
(2) Concrete biological shield and labyrinth access
(3) Source container
(4) Source controls
(5) Product irradiation system
(6) Ventilation and safety system
(7) Laboratory and preparation area

The irradiator shielding is designed to provide a safe exposure of over 100 000 Ci. The labyrinth passage, 120 cm wide, is designed to take the source storage container inside the irradiation chamber with ease and to reduce the scattered radiation level near the control-room entry to an acceptable level (0.25 mR/h) at the maximum activity loading. The shielding walls are built from prefabricated concrete blocks of 2.35 g/cm^3 density and are arranged in an array to minimize radiation leakage through interfaces. All services such as electrical, source-raise/lower wire ropes, air lines, auxiliary lines etc., are taken through the embedments in the shielding. In the control room there are embedment sleeves at the ceiling level and also at 100 cm above ground level. Provisions have been made for sleeves at ceiling level in the laboratory area. These experimental holes are normally kept closed by lead plugs.

The irradiation chamber is 300 cm × 300 cm in area and 300 cm high. In the centre of the irradiation chamber a pit 100 cm × 100 cm × 100 cm has been provided to house the radiation source.

A shuffle-dwell batch conveyor occupies the central position in the irradiation chamber as seen in Fig.1. Materials can be irradiated inside the conveyor and also can be placed anywhere in the irradiation chamber depending on the desired dose-rate and the doses to be delivered to the product.

The dimensions of the irradiation chamber chosen are expected to meet most of the irradiation requirements in terms of the size of the products to be irradiated and the range of dose-rate variation.

Access holes have been provided at the ceiling level in the laboratory area for passing pipes and gas lines.

3. RADIATION SOURCE AND THE CONTAINER

The designers of the radiation source and the container for the PANBIT facility envisaged the container to serve both as an irradiator-cum-storage device and also as a transport flask. Evidently the flask design has to meet a number of unusual and conflicting requirements. For example, one would like to have a radiation source large in size to obtain good uniformity of dose distribution within a product box of reasonable dimensions. This would inevitably make the container very large, heavy and expensive for transport. Again the source in an irradiation device should be held free enough to allow easy in-and-out movement of the source, whereas in a transport container it should be held rigid enough to meet the transport conditions and requirements. Taking such factors into consideration a cylindrical source design which would allow a compact source of large capacity was arrived at. The cylindrical source cage accommodates 24 standard source pencils in a 15-cm diam. annular ring. The design takes into consideration the use of ^{60}Co of a specific activity ranging from 30 to 40 Ci/g.

Shielding calculations for the design of a lead flask are involved and complex when the sources and a container of finite sizes are considered. Simplified approaches based on actual multiple paths of radiation from the source through containers of cylindrical geometry and of varying thicknesses were attempted. As a result, an equivalent thickness of lead shield, which would reduce the dose on the surface to a level allowed for transport and permanent storage, was arrived at.

Radiation heating is another important factor to be considered when ^{60}Co sources of 100 kCi and above are to be stored and transported. Regulations for transport dictate that the surface temperature on the flask should not exceed 80°C. This condition imposed an additional limitation on the activity that could be accommodated in the container. Both computational and experimental results showed that additional cooling fins may have to be fixed on the container if the stored activity should exceed 80 000 Ci of ^{60}Co. Based on the above considerations a flask was designed as shown in Fig.2.

Experimental values of the temperature on the surface of the flask and at various points within the flask obtained with 47 000 Ci of ^{60}Co loading are shown in Fig.3.

4. SOURCE CONTROL

The source drive system consisting of source hoist, the driving unit and associated electrical controls is located in the control room. The source is raised or lowered by an external electric motor-gear unit system coupled to the source plug through the wire ropes passing over a series of pulleys. All the fixtures such as pulleys brackets etc., are easy to

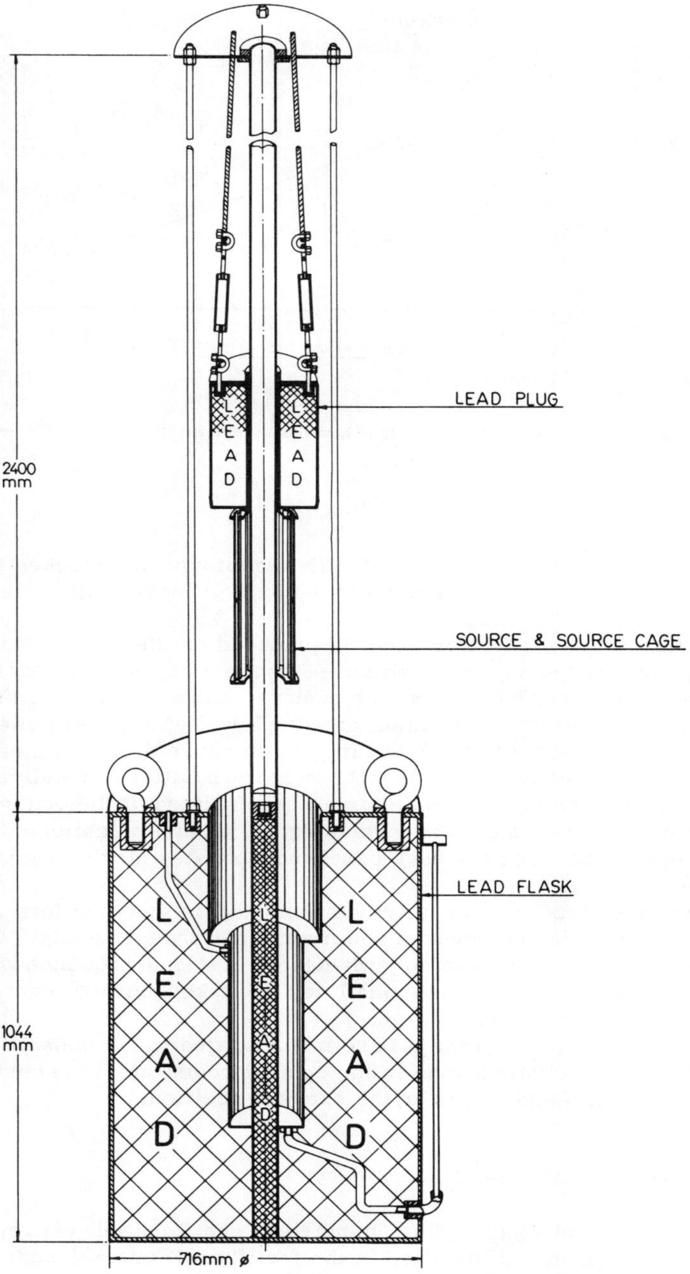

FIG. 2. Irradiation container (source in exposed position).

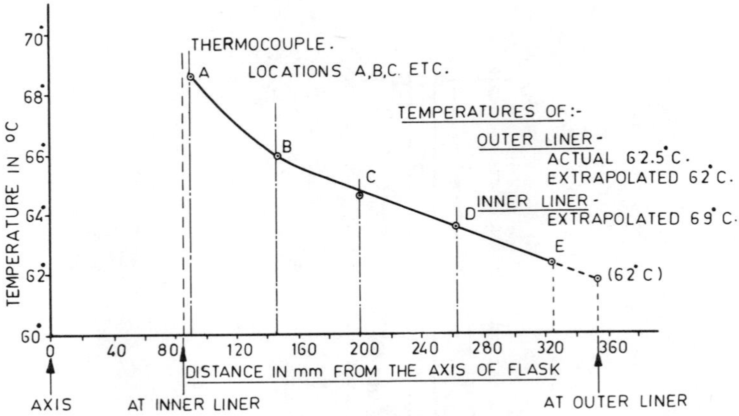

FIG. 3. Steady-state temperature distribution in lead flask. (Activity 47 000 Ci of ^{60}Co — April 1973).

dismantle and modular in type. The drive motor is interlocked through a series of switches sequentially operated by a process called "search operation" inside the cell.

An illuminated indicator panel is provided on the wall of the control room to identify the source position — whether exposed or shielded. Indications of emergency situations, such as failure of electric power, the operation of trip-wire mechanism, are displayed on a panel powered by a battery. In the event of power failure the entrance door remains securely locked and the source can be brought to store position manually.

All operations of the irradiator are centralized at the control console. To the top panel of the console are assigned all the indications of the system operations, and the base panel incorporates all the operational switches.

Complete electrical circuitry is designed in modular form. Each block or module can be checked and maintained independently. Power to different stages is available only when the operational sequence is meticulously followed. Higher current-carrying lines are taken through contactors to provide operational safety.

An area gamma monitor is also incorporated in the console to check the background radiation level in the control room, and to ascertain whether the source is in the shielded or the exposed position.

5. PRODUCT IRRADIATION SYSTEM

A new concept of an endless batch-type shuffle-dwell conveyor system has been developed and installed in the facility. There are eight box positions on the conveyor. However, at any time only seven positions are occupied by the product boxes. One box position is left empty for shuffle operations. Each box dwells for a predetermined irradiation time in one

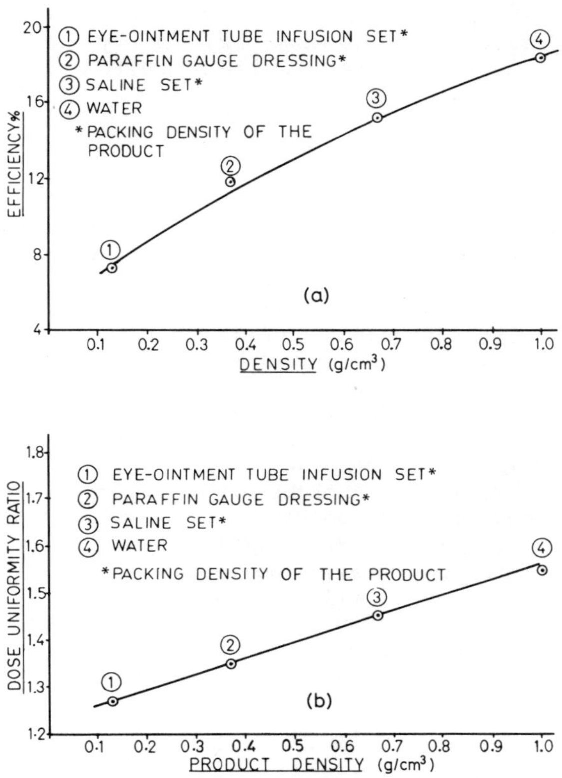

FIG. 4. (a) Radiation utilization efficiency with different product densities in PANBIT.
(b) Dose uniformity ratio within a product box. (Box size: 33 × 33 × 33 cm).

position and is then moved over to the next position by a set of pneumatic actuators. After 28 such shuffles all the four sides of each of the seven boxes get equally exposed to the radiations. This method of four-sided irradiation ensures good dose uniformity in the product.

Independent controls are available for adjusting the number of cycles and the time for product shuffling.

The conveyor design also incorporates a modular concept. The pneumatic actuators, cylinders and electrical components are mounted independently on removable plates. In the case of the failure of any component the defective module can be replaced by a spare one and the service of the defective one can be done later.

Solenoid actuating valves are placed in a pit in the irradiation chamber to avoid excessive gamma dose to these valves.

The external dimensions of the product box are 40 × 40 × 40 cm and about 60 litres of the product can be accommodated in each box.

The dose-distribution patterns in product boxes filled with materials of different density have been studied.

The graphs in Fig.4 show the uniformity ratios within a product box and the radiation utilization efficiency of the facility when medical products of various densities and compositions are irradiated.

6. VENTILATION AND SAFETY

When exposed, multi-kilocuries of ^{60}Co sources may produce a considerable amount of ozone and oxides of nitrogen in the cell. Therefore, an effective ventilation system is required to remove these obnoxious and corrosive gases. Special ducts have been installed in the irradiation chamber for a continuous change of air in the cell.

Flow pattern and exchange rate of air were evaluated by using SF_6 gas as a tracer to rate the efficacy of the installed system. The results indicated that the duct design adopted ensures adequate air changes and stagnant air columns are not present in the irradiation cell.

Maximum precautions have been taken in the design of the irradiator for the safety of operators. All crucial operations such as the opening and closing of the cell door, raising of the source etc., are interlocked so as to minimize dependence on the operators' vigilance and to make the facility work on a fail-safe principle. Provision has been made for automatic change-over of all the operations by battery power in the event of mains power failure.

As an additional feature, a closed-circuit television system has been incorporated to observe and check the operations in the irradiation chamber. The television camera is located inside the cell and can be withdrawn by remote control to a safe region of low radiation level within the cell.

CONCLUSIONS

The facility has been built to operate around the clock with one trained operator each shift and with complete fail-safe mechanisms. The cost of the construction was kept to a minimum by building shields of pre-fabricated concrete blocks arranged in a specified manner and using modular-type controls and fixtures. The entire cost of the facility excluding the cost of ^{60}Co was less than $ 40 000.

With the installed capacity of 80 000 Ci of ^{60}Co, medical products can be sterilized in about 7 h per batch and hence there will not be any product hold-up which is one of important economic factors to be considered. In addition, areas adjacent to the conveyor can be used simultaneously for other purposes — for processes such as wood plastic composites, radiation-induced chemical reactions etc. Thus, it can be seen that the facility represents an optimized system in its class for pilot-scale studies in radiation processing.

RADIATION STERILIZATION OF PHARMACEUTICAL SUBSTANCES

(Session VI)

Chairman: G.O. PHILLIPS (United Kingdom)

ПРОМЫШЛЕННЫЕ УСТАНОВКИ ДЛЯ РАДИАЦИОННОЙ СТЕРИЛИЗАЦИИ МЕДИЦИНСКИХ ИЗДЕЛИЙ

В.Б.ОСИПОВ, С.В.МАМИКОНЯН,
Г.Д.СТЕПАНОВ, Ю.С.ГОРБУНОВ,
А.А.КУДРЯВЦЕВ, Б.М.ТЕРЕНТЬЕВ,
И.И.САРАПКИН, Ю.И.САФОНОВ, Н.Г.КОНЬКОВ,
Б.М.ВАНЮШКИН, С.Ю.КРЫЛОВ,
Э.С.КОРЖЕНЕВСКИЙ, В.М.ЛЕВИН,
В.А.ГЛУХИХ, В.И.МУНТЯН
Всесоюзный научно-исследовательский институт
радиационной техники,
Москва,
Союз Советских Социалистических Республик

Abstract–Аннотация

INDUSTRIAL FACILITIES FOR THE RADIATION STERILIZATION OF MEDICAL PRODUCTS.

The authors explain the reasons for setting up industrial radiation facilities for sterilizing medical products in the Soviet Union and describe the principle of operation, the design features and the main technical and operating characteristics of these facilities, which use the LUE-8/5B electron accelerator or ^{60}Co as their radiation source.

ПРОМЫШЛЕННЫЕ УСТАНОВКИ ДЛЯ РАДИАЦИОННОЙ СТЕРИЛИЗАЦИИ МЕДИЦИНСКИХ ИЗДЕЛИЙ.

В докладе приводятся предпосылки, послужившие созданию отечественных промышленных радиационных установок для стерилизации медицинских изделий. Описывается принцип действия, конструкция и основные технические и эксплуатационные характеристики разработанных установок с использованием в качестве мощного источника излучения как электронного ускорителя ЛУЭ-8/5В, так и изотопа кобальт-60.

Радиационная стерилизация медицинских изделий (в частности, из полимерных материалов) является одной из основных областей промышленного освоения радиационной технологии [1]. Показательны данные Э.Э.Фаулера, приведенные им в обзорном докладе на IV Женевской конференции [2]. Как следует из состояния работ в данной области, радиационная стерилизация в целом перестала быть научно-технической проблемой и переживает ныне период становления в индустриальную отрасль медицинской промышленности.

Решение и развитие данной проблемы в нашей стране началось несколько лет назад. Основанием к этому послужили результаты отечественных исследований, показавшие, что радиационная стерилизация обеспечивает:

– гарантированную высокую степень бактерицидности различных стерилизуемых изделий (в том числе и объемную);

– возможность стерилизации термолабильных пластических материалов и изделий из них различной конфигурации;

– безопасность применения простерилизованных изделий и материалов;

– сохранение достигнутой степени стерильности за счет обработки изделий в герметичных установках того или иного вида, являющихся "прозрачными" для ионизирующего излучения при стерилизации и "непрозрачными" для проникновения через них микроорганизмов после процесса стерилизации и при последующем хранении;

– возможность организации непрерывно-поточной (автоматизированной) линии радиационной обработки в условиях заводского производства соответствующих изделий в упакованном виде;

– конкурентоспособность метода по сравнению с другими методами стерилизации при его внедрении в промышленное производство некоторых изделий медицинского назначения.

Имевшая место за рубежом и в СССР дискуссия по вопросу применения типа источников ионизирующего излучения (электронные ускорители и радиоизотопные источники гамма-излучения) не выявила доминирующего положения какого-либо источника. Поэтому в СССР разработки установок данного назначения выполняются с использованием как источников гамма-излучения (кобальт-60), так и электронных ускорителей. Это согласуется с современными тенденциями, вытекающими также из анализа материалов IV Женевской конференции, а именно тенденциями в области создания установок для радиационной стерилизации с массовым применением для обработки изделий гамма-излучения радиоизотопных источников (большей частью кобальта-60) и линейных ускорителей электронов [2].

1. УСТАНОВКА ДЛЯ РАДИАЦИОННОЙ СТЕРИЛИЗАЦИИ ИЗДЕЛИЙ МЕДИЦИНСКОГО НАЗНАЧЕНИЯ – СТЕРИЛИЗАЦИЯ-III

Установка с облучателем из кобальта-60, показанная на рис.1, предназначена для стерилизации пластмассовых изделий одноразового пользования и металлических игл. Установка (разработчик – ВНИИРТ) включена в технологическую линию завода по производству указанных изделий.

Принцип действия. Медицинские изделия стерилизуются гамма-излучением изотопа кобальта-60. Из совокупности стандартных источников кобальта-60 формируется плоскостной облучатель. Объекты в таре с помощью механизма загрузки помещаются в подвесы конвейера и транспортируются по лабиринту в зону облучения. Облучение объекта происходит при движении его с одной и другой стороны облучателя. Далее объекты по лабиринту выводятся из зоны облучения на позицию вертикального перемещения тары в подвесе или выгрузки.

Конструкция. Установка, размещающаяся в двухэтажном здании с площадью застройки 300 м^2 и высотой 7 м, состоит из камеры облучения с лабиринтом, облучателя, хранилища облучателя, транспортного конвейера типа ТП-80, камеры сборки, рабочего стола, контейнера-накопителя, манипулятора и пульта управления.

Камера облучения имеет площадь приблизительно 100 м^2 и высоту 3,5 м. Толщина биологической защиты (бетон с $\rho = 2,1 \div 2,3$ т/м^3) – 2 м. Мощность экспозиционной дозы за пределами биологической защиты и на выходе из лабиринта не превышает 1,4 мР/ч. Лабиринт предназначен для прохода подающей и разгрузочной ветвей конвейера и снабжен дверью с электромеханическим замком. В камере располагаются транспортный конвейер, облучатель и хранилище облучателя.

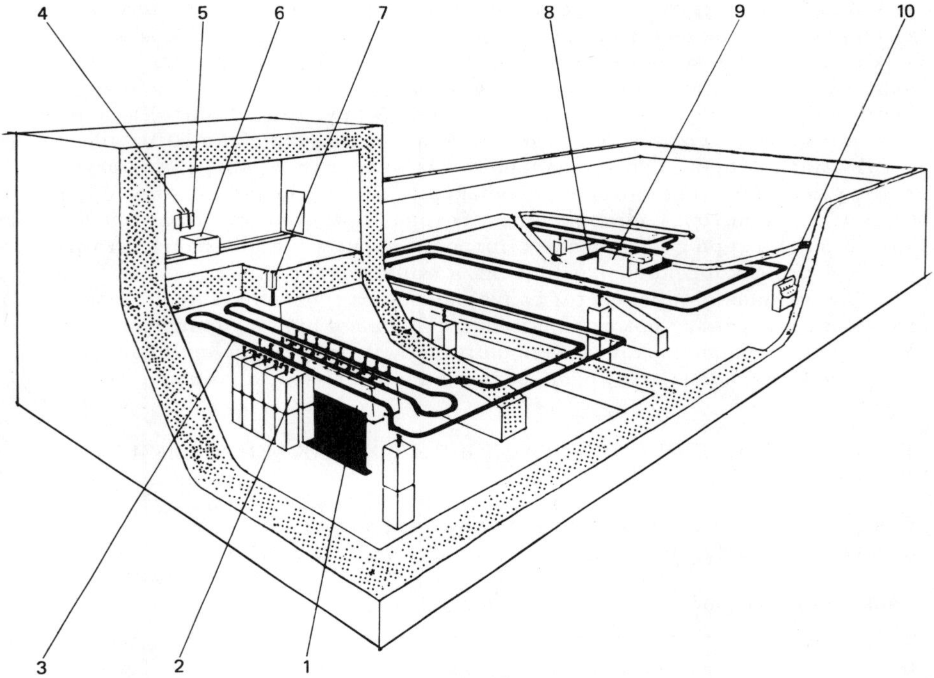

Рис.1. Промышленная гамма-установка для радиационной стерилизации медицинских изделий. 1– облучатель, 2– облучаемые изделия, 3– конвейер, 4– манипулятор, 5– смотровое окно, 6– накопитель источников, 7– загрузочный канал, 8– рольганги, 9– механизм загрузки-выгрузки, 10– пульт управления.

Облучатель – вертикальная плоскость с габаритами: длина – 2,4 м, высота – 1,0 м, которая образована 103 стержнями. Каждый стержень состоит из стальной трубки с источниками излучения и ступенчатой пробки. Верхняя часть облучателя выполнена в виде ступенчатой плиты. С боковых сторон облучателя укреплены ползуны, скользящие по направляющим при перемещении облучателя. Облучатель оборудован системой аварийного сброса источников в хранилище с гидравлическим тормозом. В конструкции облучателя предусмотрена возможность периодического пополнения облучателя источниками кобальта-60 с целью поддержания требуемой производительности в течение всего срока амортизации установки ($T_{ам}$ = 10 лет).

Хранилище облучателя выполнено в виде прямоугольной ступенчатой щели в полу камеры облучения и оборудовано защитным тепловым экраном в виде стального блока со змеевиковым теплообменником для охлаждения источников и бетона хранилища.

Для транспортировки продукции используется конвейер с автоматическим адресованием груза ТП-80, состоящий из прямых и поворотных секций. Конвейерная линия расположена в 4 ряда относительно плоскости облучателя (по 2 ряда с каждой стороны). Тара с медицинскими объектами облучения расположена в 3 яруса (по высоте) в подвесах, которые перемещаются толкающей цепью со скоростью 10 м/мин.

Камера сборки, предназначенная для загрузки и разгрузки облучателя, размещена над камерой облучения и имеет размеры 4 × 5 × 3,5 м. Стены, пол и потолок выполнены из бетона, являющегося биологической защитой. Камера оборудована смотровым окном и в ней располагаются рабочий стол, контейнер-накопитель и манипулятор М-22.

Наблюдения за процессом радиационной стерилизации производятся из операторского помещения, где установлен пульт управления. Управление обеспечивается наличием систем: перемещения объекта облучения, сигнализации, блокировки и дозиметрического контроля.

ТЕХНИЧЕСКИЕ И ЭКСПЛУАТАЦИОННЫЕ ХАРАКТЕРИСТИКИ ГАММА-УСТАНОВКИ

Размеры отдельного стандартного кобальтового источника, Ø × Н, мм	11 × 80,5 (МРТУ 10-62-68)
Начальная активность указанного источника, кюри	$1{,}91 \cdot 10^3$
Интегральная поглощенная доза, Дж/кг	$2{,}5 \cdot 10^4$
(Мрад)	(2,5)
Коэффициент равномерности облучения, %	20
Коэффициент использования излучения, %	более 30
Размеры отдельной тары для объектов облучения, мм	570×570×760
Схема облучения	непрерывная, двухсторонняя и многорядная.
Напряжение питания, В	380/220
Потребляемая мощность, кВт	20
Количество потребляемой воды, $м^3$/ч	5
Расход воздуха на вентиляцию и охлаждение, $м^3$/ч	10^4
Режим работы	круглосуточный, непрерывный.

2. ПРОМЫШЛЕННАЯ УСТАНОВКА С ЛИНЕЙНЫМ УСКОРИТЕЛЕМ ЭЛЕКТРОНОВ ДЛЯ СТЕРИЛИЗАЦИИ МЕДИЦИНСКИХ ИЗДЕЛИЙ

При создании подобных мощных радиационных установок должны учитываться некоторые основные требования, предъявляемые к ним, а именно:

– источник излучения (линейный ускоритель электронов) должен быть простым по конструкции, надежным в эксплуатации, а энергия падающих электронов должна быть достаточной для полного проникновения электронов на всю толщину облучаемого объекта ;

– должна быть обеспечена непрерывная подача стерилизуемой продукции в зону облучения и стабилизация скорости транспортного устройства.

Предлагаемая установка (рис.2) предназначена для стерилизации пучком ускоренных электронов медицинских изделий из полимерных материалов [3].

В зависимости от производительности и требований к надежности установка может поставляться с одним или двумя линейными ускорителями и, как правило, по технологической цепочке производства размещается после окончательной упаковки медицинских изделий.

Принцип действия основан на облучении с помощью линейного ускорителя электронов ЛУЭ-8/5В упакованных медицинских изделий, перемещающихся под пучком ускоренных электронов, развернутых в полосу поперек движения конвейера.

Конструкция. Установка имеет две независимые технологические линии по радиационной обработке и размещается в двухэтажном здании с площадью застроек 700 $м^2$ и высотой 9,5 м. В технологическую линию входят: линейный ускоритель электронов, камера облучения, транспортное устройство, система автоматики и защитной блокировки, пульт управления и контроля.

Для осуществления радиационной стерилизации в промышленном масштабе разработан односекционный линейный ускоритель ЛУЭ-8/5В, обладающий более узкими возможностями в части изменения параметров пучка ускоренных электронов, но зато более простой по конструкции, более надежный в эксплуатации и менее дорогой [4,5]. Вертикальная компоновка ускоряющей системы в блоке излучателя позволяет обходиться без поворотного магнита при установке ускорителя над горизонтальной конвейерной линией промышленного предприятия (рис.3).

Устойчивость работы ускорителя обеспечивается стабилизацией питающего напряжения (с точностью ± 1 %) и температуры ускоряющей системы и СВЧ-генератора (с точностью ±1°С). Режимы работы от включения к включению воспроизводятся с точностью до ±2,5 % без каких-либо дополнительных подстроек.

Пучок ускоренных электронов выводится наружу через алюминиевую или титановую фольгу после рассеивания его в полосу путем сканирования с частотой от 0,5 до 2 Гц отклоняющим магнитом. Размер поля облучения на расстоянии 200 мм от фольги представляет собой полосу длиной 500 мм и шириной 30 мм. Неравномерность плотности потока электронов по длине полосы при сканировании пучка не превышает ±5 %.

Пуск и управление технологической линией осуществляется дистанционно с пульта управления.

Камера облучения площадью 120 $м^2$ и высотой 9,5 м представляет собой защитный бетонный блок с лабиринтными устройствами для подачи и выгрузки продукции. Она спроектирована таким образом, что поступающие на стерилизацию потоки продукции изолированы друг от друга. Толщина биологической защиты (бетон с $\rho = 2{,}1 \div 2{,}3$ т/$м^3$) – 2,8 м. Мощность экспозиционной дозы за пределами биологической защиты и на выходе из лабиринтных устройств не превышает 1,4 мР/ч. В камере

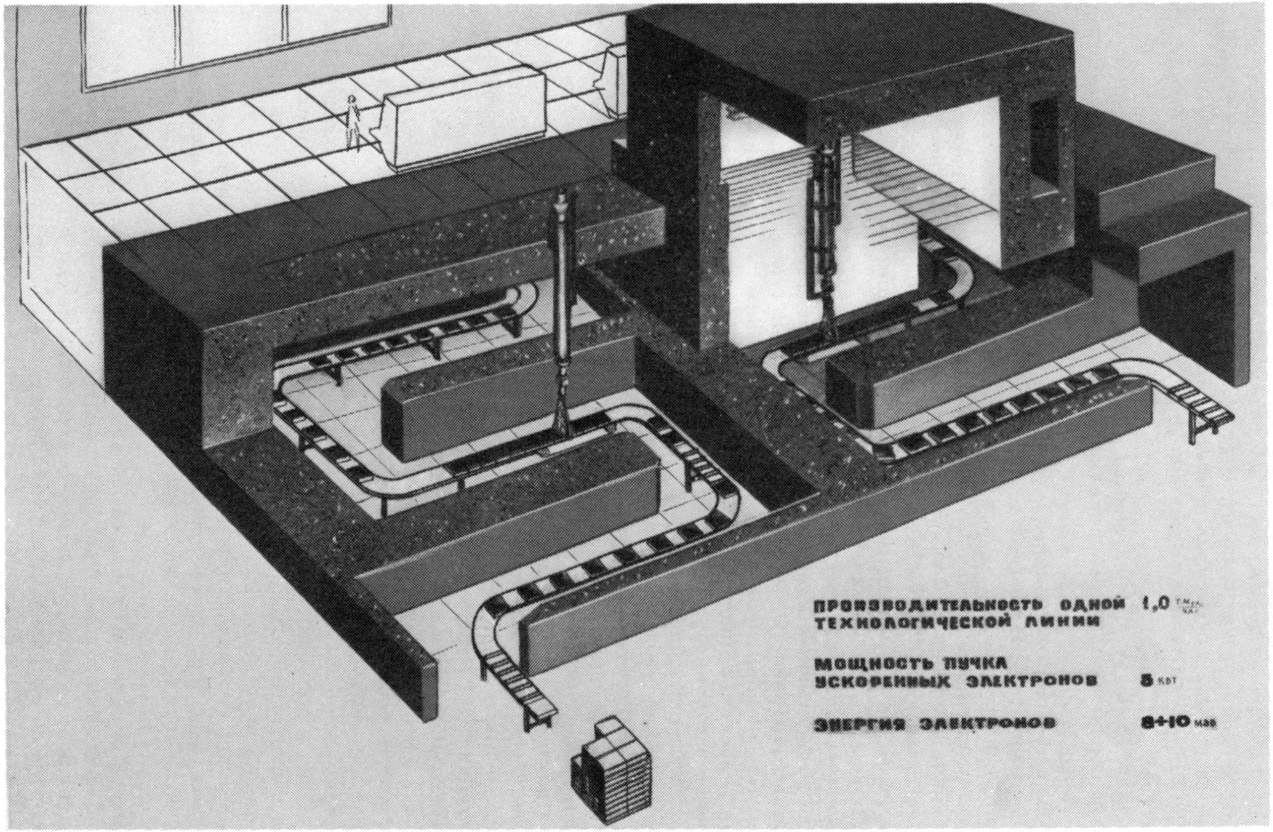

Рис.2. Промышленная установка с двумя линейными ускорителями электронов для радиационной стерилизации медицинских изделий. Производительность одной технологической линии – 1,0 т·Мрад/ч; мощность пучка ускоренных электронов – 5 кВт; энергия электронов – 8 ÷ 10 МэВ.

Рис.3. Линейный ускоритель электронов ЛУЭ-8/5В.

располагаются линейный ускоритель электронов, транспортное устройство и площадки обслуживания. Вход в камеру и обслуживание ускорителя и транспортного устройства осуществляется с первого и второго этажей по лабиринтам. На входе в лабиринтные устройства установлены двери, имеющие электромеханический замок и блокировку по питанию ускорителя.

Транспортное устройство предназначено для перемещения стерилизуемой продукции с позиции загрузки на облучение и далее на позицию разгрузки. Оно выполнено в виде четырех конвейеров — подающего, собирающего, лучевого, разгрузочного — последовательно расположенных в горизонтальной плоскости на высоте 0,8 м от пола. Ширина движущихся частей конвейеров равна минимальной ширине развертывающего устройства ускорителя. Подающий и разгрузочный конвейеры представляют собой роликовый транспортер, состоящий из прямых и поворотных секций. Роликовый транспортер позволяет надежно перемещать блочные объекты как по прямолинейным, так и криволинейным участкам при проходе через лабиринтные устройства.

Собирающий конвейер выполняет роль накопителя продукции и предназначен для непрерывного питания "лучевого конвейера". Основным звеном всего транспортного устройства является "лучевой" конвейер (находящийся в зоне облучения), который представляет собой транспортер с металлической сеткой. Все движущиеся части транспортного устройства выполнены из нержавеющей стали, а лучевой конвейер полностью изготовлен из стали Х18Н10Т, включая и транспортную ленту. Такое конструктивное решение удорожает стоимость изделия, но позволяет обеспечить надежную работу в условиях, где имеется мощный источник ионизирующего излучения, и поэтому предъявляются повышенные требования к радиационной стойкости материалов. Все конвейеры кинематически связаны между собой и приводятся в движение от одного привода, установленного за пределами камеры облучения. Мощность привода – 1,3 кВт.

Скорость движения сетки ленточного транспортера стабилизирована с точностью не хуже ±2% и может регулироваться в пределах от 0,03 ÷ 3 см/с. Номинальная скорость перемещения продукции при минимальной стерилизующей дозе 2,5 Мрад составляет 0,8 см/с.

Установка оборудована контрольно-измерительной, защитной, сигнализирующей и дозиметрической аппаратурой, позволяющей осуществлять контроль основных параметров установки и обеспечить безопасность работы на ней. Для визуального наблюдения за всем процессом радиационной стерилизации установлена телевизионная камера. Контроль и управление производятся из операторского помещения, где установлен пульт управления.

ТЕХНИЧЕСКИЕ И ЭКСПЛУАТАЦИОННЫЕ ХАРАКТЕРИСТИКИ

Источник излучения – линейный ускоритель электронов ЛУЭ-8/5В.

Номинальная энергия ускоренных электронов, МэВ 8-10

Мощность пучка ускоренных электронов, кВт 5

Производительность одной линии при дозе 1 Мрад и плотности изделий 0,15 ÷ 0,2 г/см^3, т/ч. 0,9

Интегральная поглощенная доза, Мрад 2,5

Неравномерность поглощенной дозы, % 25

Коэффициент использования излучения, % . . . . более 30

Габариты облучаемой продукции, см . . . . 60 × 50 × 20

Ширина транспортного устройства, см. 50

Длительность импульса тока ускоренных электронов, мкс 2,8

Максимальная частота следования импульсов, 1/с 500

Промышленный КПД ускорителя, % . . . . около 10

Время ввода в режим, мин. . . . . не более 30

Потребляемая мощность установки (на 1 линию), кВт 100

Напряжение питания, В 3/380

Количество потребляемой воды (на 1 линию), $м^3/ч$ 10

Расход сжатого воздуха (на 1 линию), $м^3/ч$. 50

Следует отметить, что разработчики отечественных радиационных установок для стерилизации исходили из более жестких исходных данных, чем это практикуется за рубежом. Эти отличия – вследствие того, что в нашей стране создаются специализированные радиационно-технологические установки, встраиваемые в технологические линии конкретных промышленных предприятий (а не центры по стерилизации медицинских и других изделий, существующие независимо от соответствующей промышленности). Именно в случае первых разработок (т.е. специализированных радиационно-технологических установок) потенциально достижимым является наиболее высокий уровень технико-экономических показателей.

ЛИТЕРАТУРА

[1] Мощная Радиационная Техника, Под ред. С.Джефферсона, Перев. с англ., М., Атомиздат, 1967.

[2] FOWLER, E.E., 4th Int. Conf. Peaceful Uses At. Energy (Proc. Conf. Geneva, 1971) 14, UN, New York, and IAEA, Vienna (1972) 29.

[3] КОНЬКОВ, Н.Г. и др., "Стерилизация медицинских изделий из полимерных материалов с помощью линейного ускорителя электронов", В сб. Радиационная Техника, М., Атомиздат, 6 (1971) 186.

[4] НИКОЛАЕВ, В.М., Линейные ускорители электронов для стерилизации и радиационной химии, Доклад на Всесоюзном научно-техническом совещании по использованию ускорителей в народном хозяйстве, Ленинград, февраль 1971 г.

[5] "Liner Electron Accelerators for Sterilization and Radiation Chemistry", International Nuclear Industries Fair Nuclex 69, Oct. 1969.

DISCUSSION

K.H. CHADWICK: I agree entirely with your comment on the importance of siting an irradiation plant near the industry which is manufacturing the products to be irradiated. I think this point should be emphasized. The questions I have are as follows: first could you say how the observation of the sterilization process in the ^{60}Co plant is achieved from the control room; second, what is the energy of the linac electrons; and third, can you enlarge on the dosimetry techniques used in both types of plant?

B.M. TERENT'EV: Thank you for your comment, Mr. Chadwick. As regards the questions, we observe the sterilization process from the control room by means of closed-circuit television, or in the case of a cobalt plant through an observation window. This is to check on the loading of the irradiator.

The rated electron beam energy of the linac (model 147-8/5) when adjusted for radiation sterilization is 8 MeV. The dosimetric techniques applied are those described in the paper of Mr. Balitskij[1].

[1] IAEA-SM-192/41, these Proceedings.

RADIATION STERILIZATION OF PHARMACEUTICALS AND BIOMEDICAL PRODUCTS

R. BLACKBURN, B. IDDON, J.S. MOORE,
G.O. PHILLIPS, D.M. POWER, T.W. WOODWARD
Department of Chemistry and
Applied Chemistry,
University of Salford, Salford,
United Kingdom

Abstract

RADIATION STERILIZATION OF PHARMACEUTICALS AND BIOMEDICAL PRODUCTS.

Sterilization of pharmaceuticals by radiation is accompanied by chemical degradation which must be eliminated or minimised if the method is to be successfully applied. In order to devise ways in which the pharmaceutical can be protected it is necessary to know the yield and nature of the decomposition products, the mechanisms by which degradation occurs, and the rate constants for the reactions involved. We have obtained such data for a variety of pharmaceutical compounds, viz. vitamin B12, benzyl penicillin, sulphonamides, indoles, heparin, alginates and phenylmercurics, both in the solid state and in aqueous solution. The scope and limitations of radiation sterilisation are discussed in the light of these results.

The successful use of radiation for the sterilisation of medical equipment has led to the possibility of sterilising pharmaceutical preparations in the same way (1). The method is particularly attractive for those systems where conventional sterilisation methods have proved difficult or inadequate and has advantages in that it can be used after packaging in bulk. In principle, the successful sterilisation of any material, whether surgical hardware or drug, is subject to only one criterion, viz. removal of the microbe population or, at the least, its reduction to a non-viable level. In practice the nature of the material or mixture to be treated and its reaction to the sterilising agent is important. Effects of ionising γ and electron sources vary greatly from one material to another and there are many important differences between the inherently radiation-resistant metallic and vitreous materials for which radiation sterilisation has been successfully used and the chemical compounds used as pharmaceuticals. The latter cover a wide range of chemical types and their behaviour towards radiation also varies greatly. The basic practical difficulty to be overcome in applying the sterilising dose of 2.5 Mrads, say, is the prevention of gross decomposition of the biologically-active agent and any compounds which may be associated with it as a vehicle or excipient.

Earlier studies have been over-concerned with the superficial changes induced by radiation, for example, of colour, potency, acid content, gas formation, and spectral properties. Insufficient attention was given to practically important parameters such as the presence of oxygen and other radical-active species. These ad hoc studies have demonstrated many of the disadvantages of radiation-sterilisation without suggesting any procedures for its improvement and have possibly had a detrimental effect on the development of the technique. The objective of our research group at the University of Salford has therefore been to apply the principles of radiation chemistry to the study of pure "unformulated" pharmaceuticals and, from a detailed knowledge of radiation chemical pathways, find methods by which degradation mechanisms can be eliminated or turned to advantage. We have also studied the behaviour of excipients such as alginates since in many pharmaceutical formulations the biologically active ingredient is present in only small concentration, the major component being the excipient or vehicle.

Protection of sensitive compounds against radiation damage can be afforded in three ways:

A) Transfer of absorbed energy from the species it is desired to protect to a suitable additive. Energy-transfer protection is of special application to aromatic and carbohydrate systems.

B) In aqueous solution, free radicals derived from radiolysis of the water are usually responsible for solute breakdown. The effect of high energy radiations on neutral deaerated water (2) may be summarised as:

$$H_2O \rightsquigarrow 2.7(e^-_{aq}) + 2.7(\cdot OH) + 0.55(H\cdot) + 0.45\ (H_2) + 0.71(H_2O_2) + 2.7(H^+)$$

the numbers before the chemical symbols representing G-values, i.e. the number of molecules of each species formed per 100 eV absorbed. Addition of suitable radical scavengers can provide protection by removing the harmful primary radicals and converting them to non-reacting species. In this conncection the reactivity of a wide range of radicals is relevant to the design of a suitable sterilisation procedure.

C) In a few instances it is possible to have present in the system a reagent which will react with the eventual radiolysis products in such a way as to reform the original pharmaceutical or a compound with comparable biological activity.

Our purpose here is to present our radiation data for a variety of compounds, either pharmaceuticals themselves or parent substances for a class of pharmaceuticals, and to show how radiation chemical techniques can be applied. In addition we discuss some of the factors which our experiments have shown to be relevant.

1. CYANOCOBALAMIN (VITAMIN B12) AND HYDROXOCOBALAMIN (VITAMIN B12a)

These Co(III) compounds are used in the treatment of pernicious anaemia and administered as a sterile injection from a unit-dose solution. Our main objective here was to understand the behaviour of a complex structure containing a metal atom capable of exhibiting several oxidation states and the system illustrates many points of importance. Of the radical species produced by radiolysis of water only e^-_{aq} and ·OH, and to a lesser extent H·, are immediately significant in the attack of cobalamins although the formation of hydrogen peroxide could have long-term storage relevance. Both cobalamins react with e^-_{aq} to form Vitamin B12r, a Co(II) compound. This species is readily oxidised by oxygen to hydroxocobalamin and the destructive effect of e^-_{aq} can thus be reversed by post-irradiation exposure to oxygen. If the oxygen is present during radiolysis the protective effect is somewhat greater since H· is scavenged to give HO_2·. The latter is a non-abstracting radical and reacts with the cobalamin much less effectively. The reduction and re-oxidation of hydroxocobalamin can be represented:

$$\{\text{Co(III) hydroxocobalamin}\} \xrightarrow{e^-_{aq}} \{\text{Co(II) Vitamin B12r}\} + OH^-$$

$$O_2/H_2O$$

For cyanocobalamin the reactions are more complex, with elimination of CN^- occurring on reduction:

$$\{\text{CO(III) cyanocobalamin}\} + e^-_{aq} \rightarrow \{\text{Co(II) Vitamin B12r}\} + CN^-$$

$$\downarrow O_2/H_2O$$

$$\{\text{Co(III) hydroxocobalamin}\}$$

The stronger CN^- ligand then reforms cyanocobalamin by the reaction:

$$\{\text{Co(III) hydroxocobalamin}\} + CN^- \rightarrow \{\text{Co(III) cyanocobalamin}\} + OH^-$$

In earlier sterilisation studies no mention is made of Vitamin B12r formation and this is undoubtedly the reason for the great discrepancy in reported product yields, since its production renders the system very sensitive to the presence of oxygen both during and after irradiation (3).

Reaction with ·OH radicals ($k_{II} = 6.5 \times 10^9\ dm^3\ mol^{-1}\ s^{-1}$) produces permanent degradation to yield a brown organocobalt compound, and to prevent loss of Vitamin in this way it is necessary to remove this radical. This can be done by adding t-butanol or glucose, both of which are effective ·OH scavengers at concentrations of $\sim 10^{-2}$M. An alternative ·OH scavenger, perhaps not wholly acceptable in a pharmaceutical system, is formate ion. At concentrations of $10^{-2} - 10^{-1}$M the hydroxyl radicals react virtually completely with formate in preference to the more dilute cyanocobalamin:

$$\cdot OH + HCOO^- \rightarrow H_2O + CO_2^-$$

$$\cdot H + HCOO^- \rightarrow H_2 + CO_2^-$$

The strongly reducing radical CO_2^- ($E_{\frac{1}{2}}$ against standard calomel electrode is about -1.3v (4)) might be expected to reduce cobalamins to Vitamin B12r, thus making the following protection sequence possible:

$$\text{Co(III)cobalamin} + \left.\begin{matrix} H\cdot \\ \cdot OH \\ e_{aq}^- \end{matrix}\right\} \xrightarrow{HCOO^-} \begin{matrix} CO_2^- + \text{Co(III)cobalamin} \\ e_{aq}^- \end{matrix}$$

$$\text{Co(III)hydroxocobalamin} \xleftarrow{O_2/H_2O} \text{Co(II) Vitamin B12r}$$

These possible processes were supported by the increased yield of Vitamin B12r in deaerated cyanocobalamin solution containing formate and the considerable decrease (x 6) in cobalamin destruction in oxygenated-formate solution. In the latter case the following alternative is possible:

$$O_2 + e_{aq}^- \rightarrow O_2^-$$

$$CO_2^- + O_2 \rightarrow O_2^- + CO_2$$

Since O_2^- does not reduce cobalamins, the latter would be protected. These continuous irradiation experiments were complemented with pulsed-electron radiolysis studies designed to investigate reactions on a 10 - 4000 μs time-scale, and these showed conclusively that neither of the above hypotheses is wholly correct, or even possible. While hydroxocobalamin is reduced by CO_2^-, cyanocobalamin is not. The increased yield of Vitamin B12r in deaerated formate solution and the strong protective action of oxygen and formate involve the sequence:

$$\text{Co(III)cyanocobalamin} + e^- \rightarrow \text{Co(II)Vitamin B12r}$$

$$k_{II} = 3.8 \times 10^{10}\ dm^3\ mol^{-1}\ s^{-1}$$

$$\text{Co(II)Vitamin B12r} + CO_2^- \rightarrow \text{Co(I)Vitamin B12s}$$

$$k_{II} = 8.2 \times 10^{8}\ dm^3\ mol^{-1}\ s^{-1}$$

$$\text{Co(I)Vitamin B12s} + \text{Co(III)cyanocobalamin} \rightarrow 2\ \text{Co(II) Vitamin B12r}$$

A full discussion of kinetic data is given elsewhere (4).

The marked effect on reaction pathway due to the simple ligand change in going from cyano- to hydroxo-cobalamin demonstrates the danger of extrapolating results obtained on one compound to use with one even very closely similar and suggests that radiation sterilisation demands a very detailed investigation of each individual system. In no way is the technique a panacea.

Oxygen-saturated formate solution ($\{HCOO^-\} = 10^{-1}\underline{M}$), while giving a greatly increased measure of protection to cyanocobalamin ($\{cobalamin\} = 3.7 \times 10^{-5}\underline{M}$) still results in a loss of 36% at a dose less than 0.1 Mrad, considerably lower than the minimum sterilising dose of 2.5 Mrad. Most of this loss could be eliminated if side reactions due to e^-_{aq}, one-third of which react with cobalamins in such a way that Vitamin B12r is not formed, could be avoided. This might be possible if e^-_{aq}, ·OH and ·H were quantitatively converted to a radical capable of specifically reducing the cobalamin to Vitamin B12r. The "reversal" procedure shows promise and is certainly to be preferred to simply scavenging unwanted attacking radicals. In this connection it is worth noting that t-butanol, a widely used ·OH scavenger, is not suitable for use as a protective agent for cobalamins since the radical products of the reaction

$$(CH_3)_3\ C(OH) + {\cdot}OH \rightarrow (CH_3)_2\ C(OH)\ CH_2{\cdot} + H_2O$$

adds readily to Vitamin B12r to form a coenzyme-like compound (5). The latter is photosensitive and could give rise to unpredictable side reactions. We are, however, currently investigating the possibility of turning these reactions to advantage in a sterilisation sequence beginning with Vitamin B12-coenzyme, whose formula can be taken as (B12 - CH_2 - Ribose - Adenine):

$$ {\cdot}OH + (CH_3)_3\ C(OH) \rightarrow (CH_3)_2\ C(OH)\ CH_2{\cdot} + H_2O$$

$$e^-_{aq} + (B12 - CH_2 - Ribose - Adenine) \rightarrow Vitamin\ B12r + (CH_2 - Ribose - Adenine)^-$$

$$k_{II} = 3 \times 10^{10}\ dm^3\ mol^{-1}\ s^{-1}$$

$$Vitamin\ B12r + (CH_3)_2\ C(OH)\ CH_2{\cdot} \rightarrow B12 - CH_2\ C(OH)\ (CH_3)_2$$

$$k_{II} = 2.4 \times 10^8\ dm^3\ mol^{-1}\ s^{-1}$$

The coenzyme-like compound so formed is probably acceptable for direct use since it is in the form of a coenzyme that the cobalamin is utilised metabolically. Should this prove unacceptable a further step, post-irradiation U.V. photolysis, rapidly converts the cobalamin to Vitamin B12r. On injection into an oxidised fluid this is converted to hydroxocobalamin. Alcohols other than t-butanol can be used but the redox potential of the resulting radical is highly critical since reduction of Co(II) → Co(I) can take place in preference to addition (5).

When solid Vitamin B12 is irradiated G(-cyanocobalamin is 0.6. Little inorganic cobalt is produced, most of the product being Vitamin B12r. Thus after a sterilising dose no major composition changes would occur. The stability of the solid is presumably related to the aromatic nature of parts of the molecule.

TABLE I. YIELDS FROM GAMMA RADIOLYSIS OF A 10^{-3} M AQUEOUS SOLUTION OF BENZYLPENICILLIN.

Product	G Value Degassed	N_2O saturated
Benzylpenilloic acid	1.5	1.2
Benzylpenillic acid	0.5	< 0.05
Benzylpenicilloic acid	0.47	0.97
Hydroxybenzyl penicillin (ortho)	0.24	0.46
" " (meta)	0.13	0.23
" " (para)	0.12	0.23
Carbon dioxide	1.0	0.6

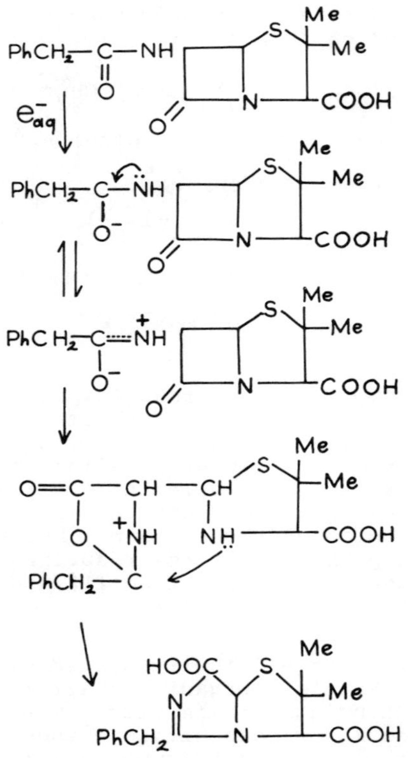

Fig. 1. Formation of benzylpenillic acid by reaction with e^-_{aq}, probably by the processes given here.

2. BENZYLPENICILLIN

In dilute aqueous solution (10^{-4} - 10^{-2}M) there is extensive degradation of sodium benzylpenicillin. At 10^{-4}M a G(-) value of 3.8 suggests that both e^-_{aq} and ·OH radicals participate in attacking the solute, yields of major products being given in Table I. At concentrations above 10^{-3}M G(-) exceeds the total radical yield from water radiolysis indicating some kind of limited chain reaction possibly involving benzylpenilloic acid. Using appropriate radiation chemical techniques (6) it was found that reaction with ·OH radicals leads to hydroxylation of the benzene ring in the penicillin side-chain. Yields of o-, m-, and p-hydroxybenzylpenicillins are in the ratio 2:1:1, the rate constant for reaction being 3.4 x 10^9 dm^3 mol^{-1} s^{-1}. Benzylpenicilloic acid is formed exclusively by ·OH attack.

Benzylpenillic acid is formed only by reaction with e^-_{aq}, probably by the processes given in Figure 1. A major product, benzylpenilloic acid, is formed by reaction with either ·OH or e^-_{aq}

Benzylpenicillin $\xrightarrow{e^-_{aq} \text{ attack of C=O group and elimination of CO}}$

Benzylpenicillin $\xrightarrow[\text{of C-N bond, and elimination of CO}]{\text{·OH abstraction of a hydrogen at C-6, cleavage}}$

There is no doubt that in aqueous solution, benzylpenicillin is extremely sensitive both to the oxidising and reducing radicals produced by radiolysis and it is surprising that claims have been made for its sterilisation by γ-irradiation. In contrast, its radiation stability in the solid state is extremely good. Decomposition below a dose of 3 Mrad is negligible and sterilisation of the solid should present no problem. Other penicillins may, however, show different behaviour. The high stability may be due to the conjugated nature of the molecule but is surprising since radiolytic decarboxylation usually takes place rather readily in the solid state.

3. SULPHONAMIDES AND RELATED COMPOUNDS

Five typical sulphonamides were studied, viz. sulphanilamide, sulphaguanidine, sulphathiazole, thalamyd and sulphasuccidine. In aqueous solution all these compounds were found to be extensively degraded, G(-sulphonamide) values varying between 3.5 and 5.1 under anoxic conditions. These yields indicate that both e^-_{aq} and ·OH are involved. A factor of potential importance for radio-sterilisation is the breakdown of some of these compounds to give others which are also biologically active. For example, when the N_4 substituted compounds thalamyd and sulphasuccidine are irradiated scission of the CO-NH bond occurs, giving sulphacetamide (G = 1.4) and sulphathiazole

(G = 0.7), respectively. The inhibitory action of nitrous oxide on the formation of these products proves that e^-_{aq} attack is responsible for them. The effect of ·OH radicals is to produce phenolic compounds; sulphanilamide has been identified as a product from sulphathiazole. Rates of e^-_{aq} and ·OH attack vary only slightly from one sulphonamide to another, being approximately 10^{10} and 5×10^9 dm^3 mol^{-1} s^{-1}, respectively. The site of ·OH reaction was shown from pulse radiolysis transient spectra to be the S-N sulphonamido bond. Experimental details are available elsewhere (7).

Sodium sulphacetamide also reacts with both e^-_{aq} and ·OH, rate constants being respectively 4.1 and 0.47×10^{10} dm^3 mol^{-1} s^{-1}. Attack by e^-_{aq} yields sulphanilic acid, a major irradiation product. A very similar reaction is observed with sulphathiazole where sulphanilic acid is formed by scission of the S-N bond. G(-sulphacetamide) is 3.6 at 10^{-3}M and reaches a maximum of 4.9 at 1.2M (i.e. 30%). This represents a decomposition of only 5% at a dose of 2.5 Mrad and, providing that the products of radiolysis are not harmful, radio-sterilisation could be viable for such concentrated solutions. For dilute solutions suitable non-toxic protective agents would be needed, calling for a thorough knowledge or relevant mechanisms and kinetics.

In the solid state sulphonamides are quite resistant to decomposition, G(-) lying in the range 0.15 - 0.6. Sulphanilamide is produced from both sulphathiazole and sulphaguanidine, sulphathiazole from sulphasuccidine, and sulphacetamide from thalamyd. Sulphacetamide is very resistant with G(-) < 0.01 (8). All these solids therefore are perfectly able to withstand a sterilising dose of 2.5 Mrad with only slight decomposition.

4. INDOLE

A number of pharmaceuticals are based on indole, and as a preliminary to detailed investigation of some of these we made a study of the irradiation behaviour of the parent compound. In aqueous solution the major gamma radiolysis products are 2,2-bis-(3-indoloyl)indol-3(2H)-one and polymer. Low yields of indoxyl, indoxyl red, indigo and indirubin are also formed. G(-indole) values are high varying from about 8 in deaerated solution to 14 in the presence of oxygen. The main processes involved when oxygen is present are complex and have been discussed elsewhere (9). Pulse radiolysis shows that ·OH radical attack is characterised by rate constants of 3.18×10^{10} dm^3 mol^{-1} s^{-1} for indole and 3.83×10^{10}, 3.41×10^{10}, and 3.34×10^{10} dm^3 mol^{-1} s^{-1} for indoline, 2-methylindole and 3-methylindole. The high rate constants and G(-) values do not augur well for radiation sterilisation of this class of compounds.

5. EXCIPIENTS

In many pharmaceutical preparations additives are frequently used as preservatives, gelling agents, or as flavours and colouring. Most such additives are not essential to the efficiency of the preparation and should one of them prove unacceptable it could be omitted or another substituted. Since flavours and colourings have received attention in the related field of radiation food preservation we have restricted our investigation to an alginate gelling agent and phenylmercuric compounds. The latter, as well as being spermicides and fungicides in their own right, have been used in pharmacy as bacteriostatic agents. In dilute neutral solution both phenylmercuric acetate and chloride undergo considerable degradation to give mainly phenol, inorganic mercury salts, and a polymer together with small amounts of benzene and diphenyl. Other phenylmercurics, both ionic and covalent, behave in a similar way. The use of phenylmercuric compounds in preparations to be sterilised by radiation therefore results in unacceptably high concentrations of toxic materials. Irradiation in the solid state gives essentially the same products, G(-) for the acetate being very much larger than that for the chloride, reflecting the radiation-sensitive nature of the carboxyl group.

Several alginate preparations are available, a common one being "Manucol DM", a polysaccharide (Mol. Wt. = 150,000) containing a preponderance of manuronic groups. When calcium ions are added to a 2% solution of the alginate a stiff gel is formed. Deaerated gels are rather sensitive to radiation, the viscosity decreasing until a fluid syrup is produced after a dose of 1 Mrad; the effect of oxygen is to slightly increase the sensitivity. The presence of good ·OH scavengers such as propylene glycol, benzoate, and salicylate greatly reduces radiation damage indicating that it is this radical which is responsible for breakdown. A concentration of 10^{-2}M sodium benzoate gives good protection against a dose of 2.5 Mrad, the gel retaining most of its rigidity. Damage appears to be of two types, viz. scission and the production of reducing groups. If the ·OH radical attacks at the C_1 carbon atom of the repeating unit only scission occurs, but attack at C_4 and C_5 leads to the formation of a reducing group as well as scission. In deaerated gels G(scission) is 3.1 suggesting that H· as well as ·OH attacks the molecule; G(reducing groups) = 2.2. The presence of nitrous oxide raises the G(scission) value to 5.6. Pulse radiolysis gives a second order rate constant for ·OH attack of 4×10^8 dm^3 mol^{-1} s^{-1}.

6. HEPARIN

Heparin is a glycosaminoglycan found in the presence of mast cells and has found medical application because of its anti-coagulant properties. Irradiation in the solid state gives a G(radical) yield of 3.6, decomposition taking place by depolymerisation, acid formation (G = 10), formation of reducing products (G = 2) and release of sulphate (G = 14). Using reflectance spectroscopy (10) a new absorption at ∿ 275 nm infers the production of αβ unsaturated carbonyls. If suitable cationic counter-ions are complexed with heparin in the solid state, significant reduction in damage can be achieved. Indeed, complexing with the detergent cetyl pyridinium chloride entirely prevents depolymerisation, sulphate and acid formation, and reducing products. Formation of αβ unsaturated carbonyls is greatly decreased but not eliminated.

In solution, in common with all native polysaccharides, heparin is unreactive towards the hydrated electron (11) but it reacts rapidly with hydroxyl radicals ($k_{II} = 2.2 \times 10^8 \ dm^3 \ mol^{-1} \ s^{-1}$). Thus the addition of nitrous oxide, which converts e^-_{aq} to ·OH, leads to marked increases in the destruction of the solute with formation of acids and reducing products. Loss of dye binding capacity (12) and anticoagulant properties (13) accompany these chemical changes.

DISCUSSION

Provided that the degradation pathways of a solute are known, considerable protection can be afforded by elimination or conversion of the species responsible for breakdown. This can be achieved by addition of suitable radical scavengers and from a kinetic competition viewpoint the protective effect of these will be greatest when the pharmaceutical is present in low concentration. However, since it is never possible to achieve complete protection because of side reactions, and since small losses are proportionately more serious in dilute solutions, radiation sterilisation may have very limited application to such systems. There is some scope, however, for using radiation to effect partial reductions in the microbe population. For concentrated aqueous solutions percentage losses can be kept small, as in the case of sodium sulphacetamide. In such cases attention centres mainly on the nature of the products, for example, as to whether they are toxic or otherwise undesirable. Thus applicability of radiation sterilisation to unit dose preparations will depend to some degree on whether these are normally used in concentrated or dilute solution.

Even if the requirement is for sterilisation of a concentrated solution, the mechanisms of radiolytic breakdown are best investigated by standard radiation chemical techniques using dilute solutions and it is in this respect that future investigations will differ from the ad hoc phenomenonological approach of past studies. Not only will the detailed data so obtained be of use in deciding whether the method is technically viable, or in suggesting ways in which it can be made so, but they will be necessary to obtain legal acceptance for the product. At present the minimum requirement for this is that the irradiated material should be scrutinised as a completely new product (14).

An aspect which we have largely ignored in our mechanistic studies is that of long-term post-irradiation storage. Apart from free radicals, which react rapidly, radiolysis of water produces hydrogen peroxide and this can induce slow oxidation reactions. Deleterious effects on the original pharmaceutical or on radiation protective additives can be conveniently tested by chemical means but the behaviour of possible organic radiolysis products could be a problem. A further point which seems to have been generally disregarded concerns the role of free radicals in inactivating the microbe population of a solution. Consideration must be given to whether radical scavengers added to protect the pharmaceutical may not simultaneously diminish the efficiency of sterilisation. In this connection much will depend on whether the products of radical scavenging are still capable of de-activating microbes and also on the degree to which radical, as opposed to direct, action is responsible for de-activation.

In the pharmaceuticals we have irradiated in the solid state relatively little breakdown has been induced by the minimum sterilising dose of 2.5 Mrad. In general little trouble need be anticipated provided that the compound does not possess an inherently radiation-sensitive structure. In this connection it would perhaps save useless repetition of work if a shortlist of such compounds were to be prepared based on data already available in the literature of radiation chemistry. In the case of compounds able to form intermolecular networks, for instance by extensive hydrogen-bonding, the state of physical sub-division can also be important (1). The high stability of solids compared with that of solutions suggests sterilisation procedures in which the solid and water are irradiated in a container designed geometrically so as to keep the two phases separate during irradiation but allowing them to mix afterwards. This potentially fruitful subject, however, is beyond the scope of the present discussion.

REFERENCES

(1) PHILLIPS, G. O., Manual on Radiation Sterilisation of Medical and Biological Materials (Technical Reports Series No. 149) IAEA, Vienna (1973) Ch. 19.

(2) SWALLOW, A. J., "Radiation Chemistry", Longman, London (1973).

(3) BLACKBURN, R., COX, D. L., PHILLIPS, G. O., J. Chem. Soc. (Faraday I) 68 (1972) 1687.

(4) BLACKBURN, R., ERKOL, A. Y., PHILLIPS, G. O., SWALLOW, A. J., J. Chem. Soc. (Faraday I) 70 (1974) 1693.

(5) BLACKBURN, R., KYAW, M., PHILLIPS, G. O., SWALLOW, A. J., to be published.

(6) PHILLIPS, G. O., POWER, D. M., ROBINSON, C., J. Chem. Soc. (Perkin II) (1973) 575.

(7) PHILLIPS, G. O., POWER, D. M., SEWART, M. C. G., Radiation Research 53 (1973) 204.

(8) PHILLIPS, G. O., POWER, D. M., SEWART, M. C. G., Radiation Research 46 (1971) 236.

(9) IDDON, B., PHILLIPS, G. O., ROBBINS, K. E., DAVIES, J. V., J. Chem. Soc. (B) (1971) 1887.

(10) MOORE, J. S., PHILLIPS, G. O., RHYS, D., Radiation Research 50 (1972) 479.

(11) BALAZS, E. A., DAVIES, J. V., PHILLIPS, G. O., SCHEUFELE, D. S., J. Chem. Soc., (C) (1968) 1420.

(12) JOOYANDEH, F., MOORE, J. S., MORGAN, R. E., PHILLIPS, G. O., Radiation Research 45 (1971) 455.

(13) SUNDBLAD, L., BALAZS, E. A., "The Amino Sugars" Vol. IIB, Ch. 47, Oxidation-reduction systems and radiation. (BALAZS, E. A., JEANLOZ, R. W., Eds.), Academic Press, New York (1966).

(14) CRAWFORD, C. G., Manual on Radiation Sterilisation of Medical and Biological Materials. (Technical Reports Series No. 149) IAEA, Vienna (1973) Ch. 12.

DISCUSSION

N.G.S. GOPAL: Is there a steric factor involved in the disproportionation reaction between vitamin B_{12S} and vitamin B_{12}?

R. BLACKBURN: In actual practice apparently not, but I agree that you would expect it.

N.G.S. GOPAL: Can tertiary butanol be used as a preservative (i.e. as a scavenger) in radioactive vitamin B_{12} preparations?

R. BLACKBURN: Yes, it can, though under different circumstances. When preparing radioisotope-labelled pharmaceuticals, or in fact labelled chemicals of any description, one encounters the problem of decomposition by self-irradiation, i.e. self-radiolysis. One of the standard ways of prolonging the shelf-life of such products is to add alcohol, notably tertiary butanol. It doesn't afford full protection, but at least it doubles what would otherwise be the lifetime.

A.K. KAPADIA: Have I understood correctly that you are referring in your paper to simple solutions of sodium sulphacetamide in water. I ask this because manufacturers of sulphacetamide eye drops add preservatives and stabilizers during manufacture and these might produce complex reactions.

R. BLACKBURN: The compound used by us was the normal chemically pure grade with no additives.

K.H. CHADWICK: Could you tell me how fast your pulse radiolysis is, and what dose-rate is given per pulse. Also, could you comment, in view of the practical application of either electron facilities with high dose-rates and gamma facilities with relatively lower dose-rates, on the likelihood of any dose-rate effects on the concentrations of the final products.

R. BLACKBURN: The pulse length is 0.1 μs, the dose per pulse being 700-800 rad. I do not believe that there will be any serious radical/radical interaction at the high dose-rates delivered by electron irradiation facilities. The mathematics of dose-rate effects has been extensively discussed by Burns of Harwell.

РАДИАЦИОННАЯ СТЕРИЛИЗАЦИЯ РАДИОФАРМАЦЕВТИЧЕСКИХ ПРЕПАРАТОВ

В.В.БОЧКАРЕВ, В.Т.ХАРЛАМОВ,
А.К.ПИКАЕВ, Л.П.ШУБНЯКОВА,
З.М.ПОТАПОВА, Е.П.ПАВЛОВ
Институт биофизики
Министерства здравоохранения СССР,
Москва,
Союз Советских Социалистических Республик

Abstract–Аннотация

RADIATION STERILIZATION OF RADIOPHARMACEUTICAL PREPARATIONS.

The majority of radiopharmaceuticals are solutions for injection and have to be made available in sterile form. However, some of them do not withstand thermal sterilization, and autoclaving is technologically impracticable for mass production owing to the high level of radiation from the solutions produced. In view of this, radiation sterilization may well be the most effective method; it should be possible to sterilize radiopharmaceuticals in the package, by remote control, and in a continuous process. Our investigations show that gamma irradiation of a number of radiopharmaceutical preparations labelled with ^{131}I results in their radiochemical decomposition, manifested by a drop in pH, reduction in concentration of the basic substance and the formation of labelled and non-radioactive radiolysis products. The inclusion of certain acceptors of radicals formed during the radiolysis of water in some cases prevents these preparations from decomposing and enables them to retain the basic properties called for by pharmacopoeial specifications. At the same time it has been established that, given the bactericidal characteristics of certain radiopharmaceutical preparations, the relatively low microbe contamination and the high radiosensitivity of the microflora present in radiopharmaceutical preparations produced in the USSR, they can be reliably sterilized with a dose of around 1 Mrad. On the basis of the results obtained the radiation method can be considered promising for the sterilization of radiopharmaceutical preparations.

РАДИАЦИОННАЯ СТЕРИЛИЗАЦИЯ РАДИОФАРМАЦЕВТИЧЕСКИХ ПРЕПАРАТОВ.

Большинство радиофармацевтических препаратов (РФП) являются инъекционными растворами и должны выпускаться в стерильном виде. В то же время некоторые из них не выдерживают термической стерилизации; кроме того, автоклавирование при массовом производстве технологически неудобно из-за наличия высокого уровня радиации от производственных растворов. В связи с этим для РФП может представить интерес применение метода радиационной стерилизации, в принципе позволяющего осуществлять стерилизацию продукции в упакованном виде, дистанционно, в непрерывном технологическом потоке. Наши исследования показали, что при γ-облучении ряда РФП, меченных йодом-131, происходит их радиационно-химическое разложение, проявляющееся в снижении pH, уменьшении концентрации основного вещества, образовании меченых и нерадиоактивных продуктов радиолиза. Введение в препараты некоторых акцепторов радикалов, образующихся при радиолизе воды, в ряде случаев позволяет предохранить эти препараты от разложения и сохранить соответствие их основных показателей требованиям Фармакопеи. Одновременно установлено, что бактерицидные свойства некоторых РФП, относительно малая микробная обсемененность и высокая радиочувствительность микрофлоры, присутствующей на выпускаемых в СССР радиофармацевтических препаратах, обеспечивает надежную их стерилизацию облучением в дозе около 1 Мрад. Полученные данные позволяют считать радиационный метод перспективным для стерилизации РФП.

Радиофармацевтические препараты (РФП) играют важную роль в диагностике и терапии различных заболеваний [1-3]. Так как большинство этих препаратов вводят в организм парентерально, при производственном выпуске должна быть обеспечена их стерильность. В последнее время значительное внимание уделяется применению радиационного метода для стерилизации медицинских материалов и лекарственных средств [4]. Изучение возможности радиационной стерилизации применительно к РФП представляет особый интерес, поскольку обычные методы обеспечения их стерильности иногда вообще не могут использоваться или технологически неудобны из-за высокого уровня радиации от производственных растворов. Радиационная стерилизация в этой связи привлекает возможностью проводить процесс непосредственно в упаковке, непрерывно и, что особенно важно для РФП, дистанционно.

Наличие "собственного" излучения от РФП является обстоятельством, которое обусловливает их неустойчивость и ограничивает срок годности, что приводит к необходимости предпринимать специальные меры предосторожности, позволяющие стабилизировать состав и фармакологические свойства РФП. Это излучение, вместе с тем, может в принципе обеспечить стерильность соответствующих препаратов. Большинство РФП, однако, имеет такие уровни удельных активностей, а входящие в их состав радиоактивные изотопы — такие периоды полураспада и энергии излучения, что доза от "собственного" излучения, получаемая за сравнительно короткое время, прошедшее от изготовления до выпуска, не позволяет достичь стерильности. В связи с этим для стерилизации необходимо воздействовать на РФП излучением от внешнего источника, имеющего достаточно высокую мощность дозы, что вызывает дополнительные трудности в сохранении кондиционности препаратов.

Практическое применение радиационного метода стерилизации связано с необходимостью изучения действия излучений высокой энергии на РФП. В результате воздействия возможно значительное повреждающее действие излучения на растворенное вещество, проявляющееся в особенности при облучении водных растворов, каковыми являются все РФП.

Среди РФП, применяемых в медицине, важное место занимают препараты, меченные радиоизотопами йода. Эти препараты, весьма разнообразные по химическому составу, имеют одно общее свойство — они представляют собой водные растворы относительно низких концентраций, в основном, йодорганических веществ, в молекулы которых, наряду со стабильным йодом, входит какой-либо из его радиоактивных изотопов (главным образом йод-131 или йод-125). При изучении возможности радиационной стерилизации йодорганических РФП важно установить, в частности, общие закономерности радиационно-химического разложения, обусловленные наличием йода в молекулах, идентифицировать и количественно оценить образующиеся продукты. В настоящее время уже установлено, что при облучении этих препаратов в стерилизующих дозах происходит их радиационно-химическое разложение, степень которого зависит от состава и структуры молекул растворенных веществ, концентрации последних в растворах, дозы облучения и т.д.

Возможность применения радиационной стерилизации РФП была исследована на группе йодорганических препаратов, меченных йодом-131: натриевых солей о-йодбензоиламиноуксусной кислоты (о-йодгиппурат натрия, гиппуран), бис-(2,4,6-трийод-2-карбоксианилид) адипиновой кислоты (билигност), β-(4-окси-3,5 дийодфенил) α-фенилпропионовой кислоты (билитраст), о-, м-, п-йодбензойных кислот, а также на препаратах йодида натрия — ^{131}J с концентрацией йодид-ионов $1 \cdot 10^{-4}$ моль/л и менее.

Проведенное изучение показало, что при облучении водных растворов йодорганических веществ наблюдается значительное снижение рН

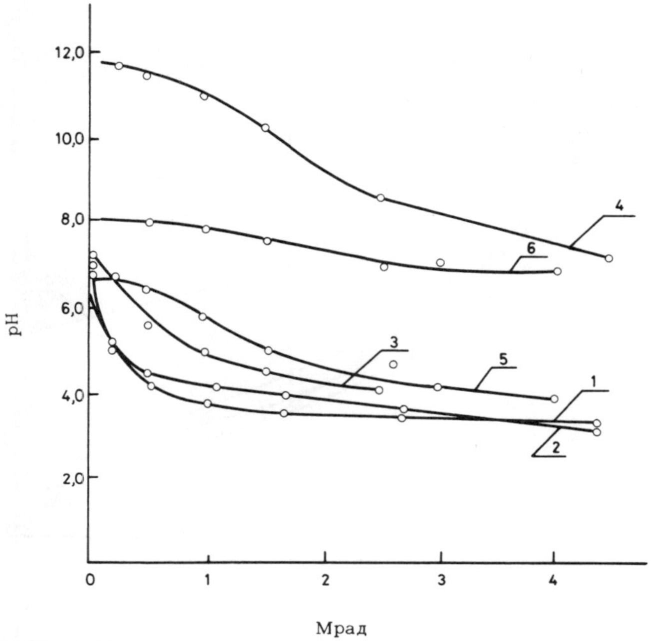

Рис.1. Изменение pH растворов некоторых йодорганических – ^{131}J веществ в зависимости от дозы облучения: 1 – 0,01 М о-йодгиппурат натрия, 2 – 0,014 М о-йодбензоат натрия, 3 – 0,014 М м-йодбензоат натрия, 4 – 0,014 М п-йодбензоат натрия, 5 – 0,005 М билигност, 6 – 0,004 М билитраст.

растворов и уменьшение растворимости некоторых веществ, которое в ряде случаев обусловливает выпадение осадков. На рис.1 представлена зависимость pH водных растворов указанных йодорганических препаратов от дозы облучения (источник гамма-излучения-^{60}Co, мощность дозы 52 рад/с). Можно сделать заключение, что в ряду изученных йодорганических веществ уменьшение pH их водных растворов происходит независимо от природы заместителя и положения йода в бензольном кольце относительно этого заместителя. Хотя механизм радиационно-химического процесса, приводящего к снижению pH растворов, остается неясным, очевидно, что к этому приводят реакции растворенных веществ с продуктами радиолиза воды. Несмотря на то, что в принципе стабилизации pH водных растворов можно добиться применением подходящего буфера, состав последнего должен быть установлен, исходя из фармакологических свойств и химического состава каждого препарата.

Установлено, что одним из основных процессов, обусловливающих радиационно-химическую неустойчивость йодорганических препаратов рассматриваемой группы, является отщепление йодид-ионов, степень которого зависит от структуры йодорганического вещества и которое,

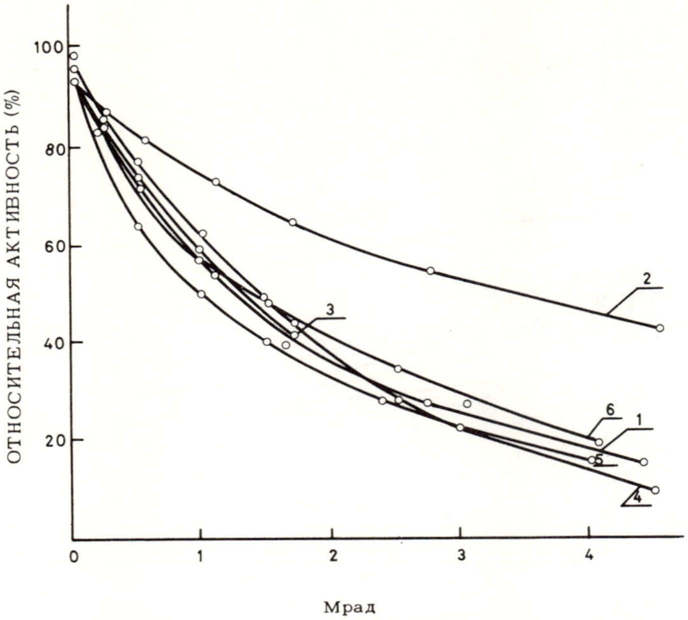

Рис.2. Накопление йодид-ионов при радиолизе растворов некоторых йодорганических – ^{131}J веществ в зависимости от дозы облучения: 1 – 0,01 М о-йодгиппурат натрия, рН 7,3; 2 – 0,014 М о-йодбензоат натрия, рН 7,6; 3 – 0,014 М м-йодбензоат натрия, рН 6,8; 4 – 0,014 М п-йодбензоат натрия, рН 11,5; 5 – 0,005 М билигност, рН 8,8; 6 – 0,004 М билитраст, рН 8,1.

независимо от его природы, тем больше, чем выше доза облучения и меньше концентрация вещества в растворе. На рис.2 представлена зависимость относительной активности йодид-ионов в препаратах от дозы облучения, а на рис.3 – от концентрации в них рассматриваемых веществ при дозе 2,5 Мрад[1].

Представленные зависимости наглядно иллюстрируют вышеуказанное заключение. Обнаружено, что при облучении образуются также другие йодорганические вещества, меченные по йоду, и нерадиоактивные продукты, в том числе содержащие в бензольном кольце гидроксильную группу. Найдены продукты димеризации, декарбоксилирования и т.д. Установление природы образующихся веществ требует специальных исследований с применением различных физико-химических методов. На рис.4 можно видеть долю основного вещества для тех же препаратов, остающуюся без изменения в растворе при его облучении в различных дозах.

При радиолизе разбавленных растворов йодида натрия – ^{131}J значение рН не изменяется; из посторонних веществ преимущественно образуются JO_3^- и J_2. На рис.5 приведены зависимости относительных

---

[1] Величина дозы, обычно применяемая в мировой практике для радиационной обработки медицинской продукции [5-7].

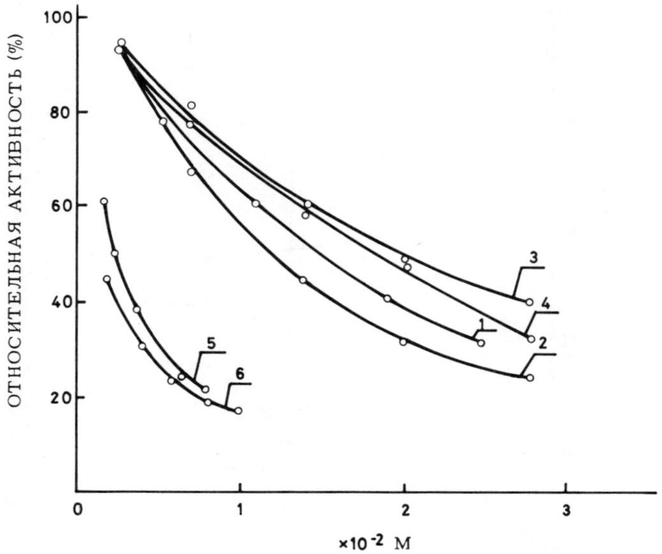

Рис.3. Изменение относительной активности йодид-ионов в растворах некоторых йодорганических – ^{131}J веществ в зависимости от концентрации последних при облучении в дозе 2,5 Мрад: 1 – о-йодгиппурат натрия, 2 – о-йодбензоат натрия, 3 – п-йодбензоат натрия, 4 – п-йодбензоат натрия, 5 – билигност, 6 – билитраст.

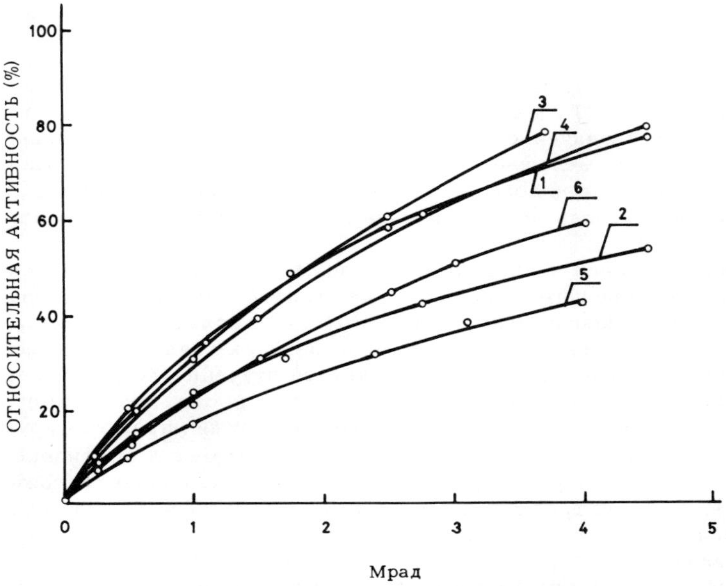

Рис.4. Изменение относительной активности некоторых йодорганических – ^{131}J веществ от дозы облучения: 1 – 0,01 М о-йодгиппурат натрия, pH 7,3; 2 – 0,014 М о-йодбензоат натрия, pH 7,3; 3 – 0,014 М м-йодбензоат натрия, pH 6,8; 4 – 0,014 М п-йодбензоат натрия, pH 11,5; 5 – 0,005 М билигност, pH 8,8; 6 – 0,004 М билитраст, pH 8,1.

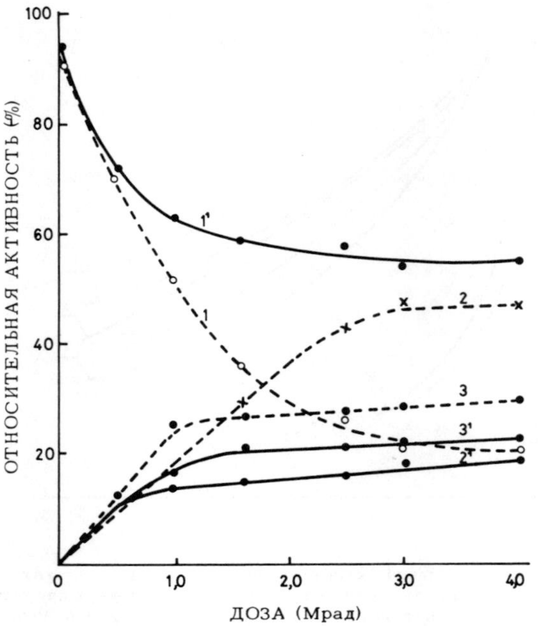

Рис.5. Изменение радиохимического состава водного (1-3) и изотонического (1^1 - 3^1) растворов $Na^{131}J$ ($\sim 8 \cdot 10^{-7}$ М, рН 7,0) в зависимости от дозы: 1,1^1 - J^-; 2,2^1 - JO_3^-; 3,3^1 - J_2.

активностей различных компонентов йода-131 от дозы облучения для $8 \cdot 10^{-7}$М раствора J^-, насыщенного воздухом при рН 7,0. Можно видеть, что с увеличением дозы облучения относительная активность J^- существенно уменьшается и в зависимости от состава раствора падает до 25-60% при дозе 2,5 Мрад. Степень радиационно-химического разложения тем больше, чем меньше концентрация йодид-ионов в растворе (рис.6).

Наблюдающееся значительное разложение препаратов обусловлено протеканием первичных реакций между растворенными веществами и продуктами радиолиза воды – свободными радикалами. Характер радиационно-химических процессов, приводящих к разложению вещества, может быть установлен лишь в результате изучения указанных первичных реакций. Для этого используется, в частности, метод импульсного радиолиза, позволяющий наблюдать короткоживущие продукты, существующие в течение нескольких микросекунд. Этим методом найдено, например, что ответственными за радиационно-химическое разложение о-йодгиппурата натрия в водных растворах являются образующиеся при радиолизе воды гидратированные электроны и радикалы ОН. Установлено, также, что скорости реакций этих частиц с указанным веществом довольно высоки и практически одинаковы (константы скорости реакций составляют, примерно, $5 \cdot 10^9$ моль$^{-1}$ с$^{-1}$)[2]. Исходя из известных свойств

[2] Данные получены при участии Л.И.Карташевой.

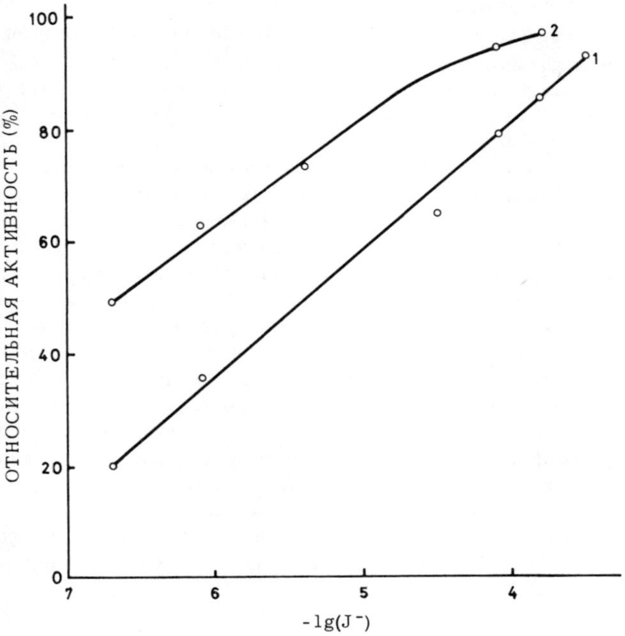

Рис.6. Изменение относительной активности йодид-ионов при γ-облучении (доза 2,5 Мрад) водного (1) и изотонического [3,6 мг/мл фосфора в виде ортофосфата натрия] (2) растворов $Na^{131}J$ в зависимости от исходного содержания йодид-ионов.

указанных первичных частиц [8], следует признать, что радиационно-химическое разложение о-йодгиппурата обусловлено происходящими при облучении его водных растворов окислительно-восстановительными реакциями. В случае йодида натрия ответственными за радиационно-химическое разложение являются преимущественно радикалы ОН.

Для подавления нежелательных радиационно-химических процессов, приводящих к разрушению молекул растворенных веществ, эффективным может оказаться введение в растворы препаратов специальных добавок-акцепторов радикалов, обладающих повышенной реакционной способностью по отношению к продуктам радиолиза воды в сравнении с "защищаемым" веществом. Связывая свободно радикалы, акцепторы препятствуют реакции взаимодействия их с растворенными веществами, в результате чего основная масса препарата остается без изменения. В качестве примера можно привести результаты экспериментов по "защите" одного из наиболее важных йодорганических препаратов — о-йодгиппурата натрия — ^{131}J. Из рис.4 видно, что при гамма-облучении в дозе 2,5 Мрад в препарате остается неизменным лишь около 30% исходного вещества. Введение в этот препарат $8 \cdot 10^{-2}$ моль/л цистамина позволяет сохранить до 85% о-йодгиппурата при облучении в той же дозе.

Более эффективной оказалась "защита" йодида натрия — ^{131}J. Введение в препарат с концентрацией йодида натрия $(1,3\div8)\cdot10^{-7}$ М (pH 6-10) тиосульфата натрия, цистеина, меркамина или тиомочевины $5 \cdot 10^{-3}$ моль/л и выше полностью предотвращает его радиационно-химическое разложение.

Несмотря на принципиальную возможность "защиты" РФП от радиационно-химического разложения, поиск эффективных акцепторов радикалов, пригодных для определенной системы, требует проведения дополнительных исследований.

При радиационной стерилизации РФП должны быть преодолены и другие трудности, обусловленные медицинским назначением препаратов. В связи с инъекционным характером ряда РФП должна быть обеспечена изотоничность их растворов. Это означает, что при "защите" акцепторы радикалов не могут вводиться в препараты в концентрациях, приводящих к получению гипертонических растворов. Между тем, "защитное" действие акцепторов зависит от их концентрации в растворе. Мало того, акцепторы радикалов должны быть биологически совместимыми с организмом, не должны быть токсичными и не должны образовывать при облучении токсичных продуктов. Эти ограничения значительно сужают круг возможных акцепторов радикалов и требуют обстоятельных всесторонних исследований облученных растворов. Наконец, следует упомянуть о том, что стерилизованные радиационным способом препараты должны сохранять свои фармакологические свойства.

В свете рассматриваемого вопроса целесообразно указать на некоторые возможности снижения радиационно-химического разложения путем уменьшения дозы облучения при условии, конечно, сохранения стерилизующего эффекта. Имеется несколько обстоятельств, позволяющих говорить об этом. Одним из них является низкая микробная обсемененность выпускаемых в СССР РФП и высокая радиочувствительность обычно присутствующей в них микрофлоры. Так, при изучении обсемененности жизнеспособными микроорганизмами РФП до стерилизации оказалось, что в среднем в препаратах содержалось не более 10^1 микроорганизмов, а показатели $Д_{10}$ для них не превышали 150 крад [9].

На основе проведенных исследований, результаты которых подробно изложены в отдельном докладе, представленном на данный симпозиум, были разработаны рекомендации по величинам доз облучения для стерилизации РФП. При инициальной контаминации до 10^2 микробных клеток и коэффициенте безопасности 10^6 эти дозы не превышают 1 Мрад. Если учесть вышесказанное соображение и провести облучение о-йодгиппурата в сопоставимых условиях в дозе 1 Мрад в присутствии $8 \cdot 10^{-2}$ моль/л цистамина, то при этом останется неизменным вещества значительно больше.

Другим обстоятельством, которое следует учитывать при выборе дозы облучения для радиационной стерилизации РФП, является то, что некоторые из них, в том числе бенгальская роза – ^{131}J, йодогност – ^{131}J и другие, обладают отчетливо выраженным бактерицидным действием [10].

Несмотря на ряд трудностей, преодоление которых необходимо при переходе на радиационный способ стерилизации РФП, полученные предварительные данные являются обнадеживающими и свидетельствуют о перспективности этого метода.

ВЫВОДЫ

1. Изучена радиационно-химическая устойчивость ряда РФП, меченных йодом-131, при внешнем гамма-облучении стерилизующими дозами.

2. Найдено, что эффект от облучения группы йодорганических препаратов – ^{131}J, наряду со стерильностью, проявляется в снижении pH растворов, уменьшении концентрации действующих веществ в растворе, отщеплении йодид-ионов, образовании ряда продуктов радиационно-химического разложения.

3. Рассмотрена возможность снижения радиационно-химического разложения препаратов путем введения в РФП перед облучением акцепторов радикалов, образующихся при облучении воды, и обсуждены возможности снижения дозы облучения при радиационной стерилизации РФП.

ЛИТЕРАТУРА

[1] ЗЕДГЕНИДЗЕ, Г.А., ЗУБОВСКИЙ, Г.А., Клиническая Радиоизотопная Диагностика, М., Изд-во "Медицина", 1968, стр.19.

[2] КИМБИ, Э., ФЕЙТЕЛЬБЕРГ, С., СИЛЬВЕР, С., Радиоактивные Изотопы в Клинической Практике, М., Медгиз, 1963, стр.116.

[3] КОЗЛОВА, А.В., Основы Радиевой Терапии, М., Медгиз, 1956, стр.33.

[4] PHILLIPS, G.O., Manual on Radiation Sterilization of Medical and Biological Materials, IAEA, Technical Report, series N 149, Vienna, 1970, p.207.

[5] БОЧКАРЕВ, В.В., ВАШКОВ, В.И., ПАВЛОВ, Е.П., ХРУЩЕВ, В.Г., РАМКОВА, Н.В., ЖМЭИ, 8 (1973) 56.

[6] БОЧКАРЕВ, В.В., ПАВЛОВ, Е.П., СЕДОВ, В.В., ХРУЩЕВ, В.Г., Хим.-фарм.Ж., 7 (1972) 31.

[7] АНТОНИ, Ф., КОЗИНЕЦ, Г., КОТЕЛЕШ, Г., Atomic Energy Review 4 3 (1966) 39.

[8] ПИКАЕВ, А.К., ЕРШОВ, Б.Г., Успехи Химии, 36 (1967) 1427.

[9] БОЧКАРЕВ, В.В., ТУШОВ, Э.Г., ПАВЛОВ, Е.П., ХРУЩЕВ, В.Г., СЕДОВ, В.В., КОТЛЯРОВ, Л.М., Мед. Радиол., 12 (1973) 34.

[10] КОТЛЯРОВ, Л.М., КОНЯЕВ, Г.А., ЖМЭИ, 8 (1971) 5.

DISCUSSION

J. FLEURETTE: If you include protectors in your medical products as protection against radiation, do they not also protect the microorganisms present?

E.P. PAVLOV: Yes, unfortunately, they do to some extent. One of the problems is to find stabilizing substances that protect the pharmaceuticals and not the microorganisms. We are still working on this.

N.G.S. GOPAL: I would like to take the opportunity, in connection with your paper, of mentioning some results on the effects of radiation on pharmaceuticals that Mr. K.A. Sundaram and I have obtained in our work. The effect of external irradiation of certain radiopharmaceutical injections, such as ^{131}I-serum albumin, ^{131}I-rose bengal, ^{131}I-hippuran, ^{51}Cr-EDTA and ^{169}Yb-DTPA have been examined at three dose levels – 0.5, 1 and 2 Mrad, the dose-rate constituting 0.5 Mrad/h. The solutions were irradiated in penicillin-type vials and the radiochemical purity determined by paper electrophoresis and/or paper chromatography.

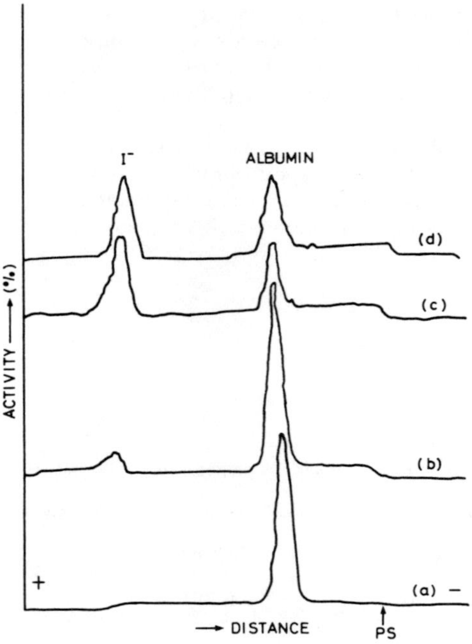

FIG.A. Paper electrophoresis of radio-iodinated (^{131}I) serum albumin. (a): control (b): 0.5 Mrad (c): 1 Mrad (d): 2 Mrad

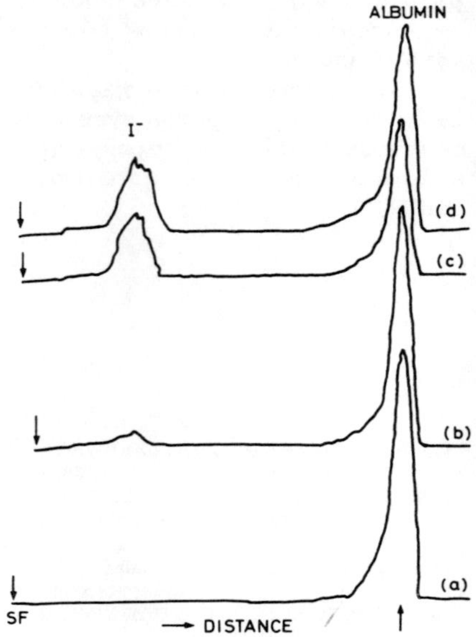

FIG.B. Paper chromatography of radio-iodinated (^{131}I) serum albumin.

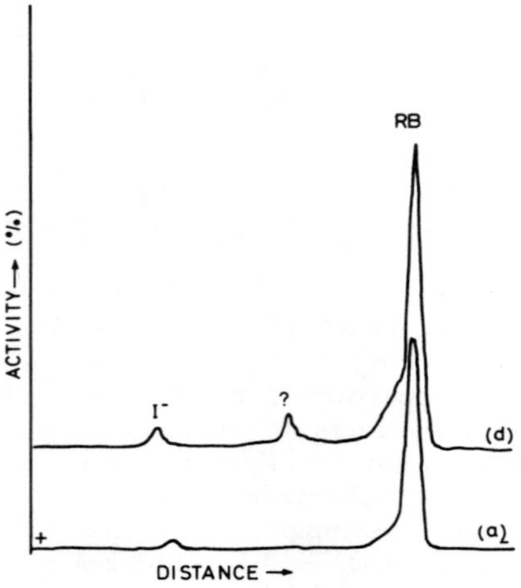

FIG.C. Paper electrophoresis of rose bengal (^{131}I).

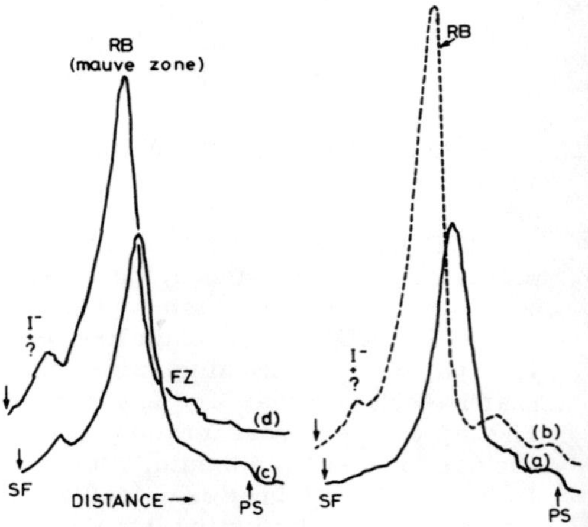

FIG.D. Paper chromatography of rose bengal (^{131}I). (FZ = fluorescent zone.)

^{131}I-human serum albumin

Paper electrophoresis: Whatman 540, 0.05M Na_2HPO_4, 8 V/cm, 2 h (Fig.A)

Dose (Mrad)	0	0.5	1	2
		% activity		
Albumin	98	86	63	64
I-	2	14	36	34

Paper chromatography: Whatman No. 1, methanol: water (85:15) (Fig.B)

Dose (Mrad)	0	0.5	1	2
		% activity		
Albumin Rf = 0.05	95	84	62	64
I- Rf = 0.75	2	16	37	34

This preparation undergoes extensive radiolysis, the free iodide concentration increasing with the dose. Sterilization of this injectable by radiation is therefore not advisable.

Rose bengal (^{131}I)

(a) Paper electrophoresis: details as above (Fig.C)

Dose (Mrad)	0	0.5	1	2
		% activity		
Rose bengal	97	94	88	83
Unidentified species	2	1	4	13
I^-	1	3	7	5

(b) Paper chromatography: Whatman No.1, 0.15M ammonia (Fig.D)

Dose (Mrad)	0	0.5	1	2
		% activity		
Rf = 0	6	5	7	6
Fluorescent zone Rf = 0.25	10	8	9	9
Rose bengal Rf = 0.50	77	76	76	72
I^- + ? Rf = 0.75	5	9	8	11

(c) Paper chromatography: Whatman No.1, 0.1M acetic acid

Dose (Mrad)	0	0.5	1	2
		% activity		
Rose bengal Rf = 0.06	99	97	96	94
Intermediate zone	-	-	2	3
I^- Rf = 0.81	1	1	2	3

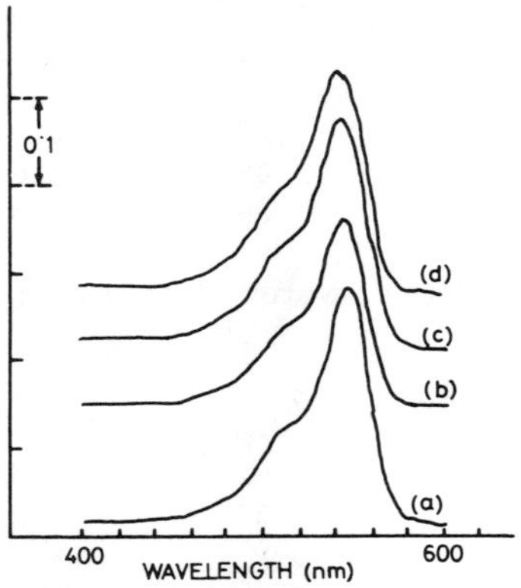

FIG.E. Absorption spectra of rose bengal (^{131}I).

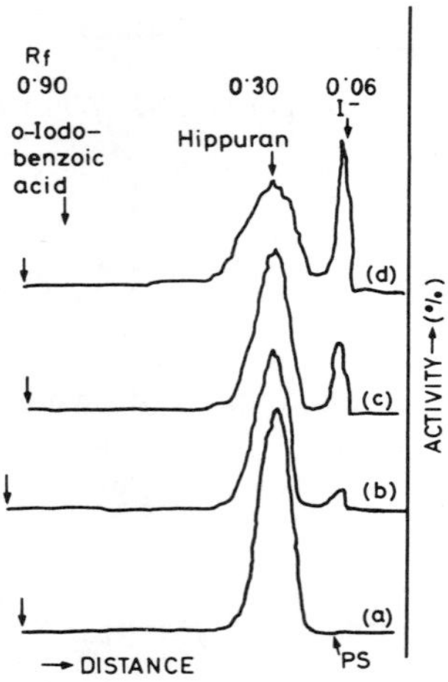

FIG.F. Paper chromatography of hippuran (^{131}I).

(a) shows that the activity distribution in the iodine zone increases at doses beyond 0.5 Mrad; (b) shows that part of the activity in the iodide zone may be due to an unidentified species; and (c) shows that the bulk of the activity is still precipitable by acid and that the unidentified component may therefore be an iodo-organic acid.

The absorption spectra of control and irradiated solutions in 0.1M sodium acetate were similar (Fig.E), but there is a slight shift of the absorption maximum towards the shorter wavelength. While the absorbance (normalized to a finite dilution) at 515 nm was almost constant, the absorbance at 546 nm showed a slight decrease with increasing dose.

Dose (Mrad)	0	0.5	1	2
Å 515 nm	0.12	0.11	0.12	0.12
Å 546 nm	0.28	0.26	0.25	0.23

It is apparent that irradiation of this injectable causes a perceptible change in some physio-chemical characteristics.

It would, however, be interesting to see whether the irradiated solutions show liver uptake and clearance rates similar to, or different from, those of the controls.

Hippuran (^{131}I)

Paper chromatography: Whatman No.1, C_6H_6: AcOH: H_2O = 2:2:1 (Fig.F)

Dose (Mrad)	0	0.5	1	2
		% activity		
I^- Rf = 0.06	1	8	15	23
Hippuran Rf = 0.30	98	98	85	76
o-iodo-benzoic acid Rf = 0.90	< 1	< 1	< 1	< 1

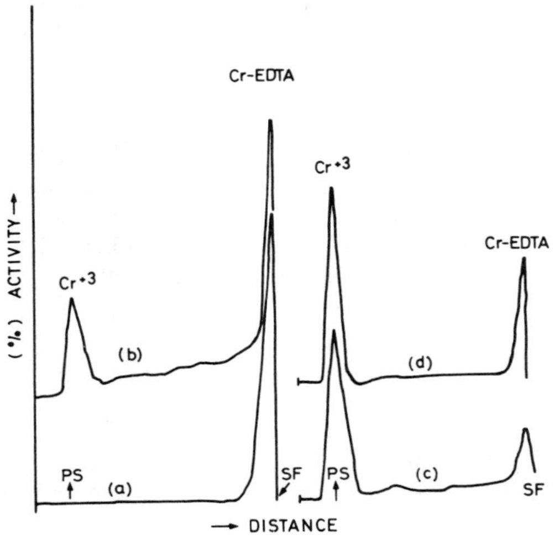

FIG.G. Paper chromatography of chromium (^{51}Cr) EDTA.

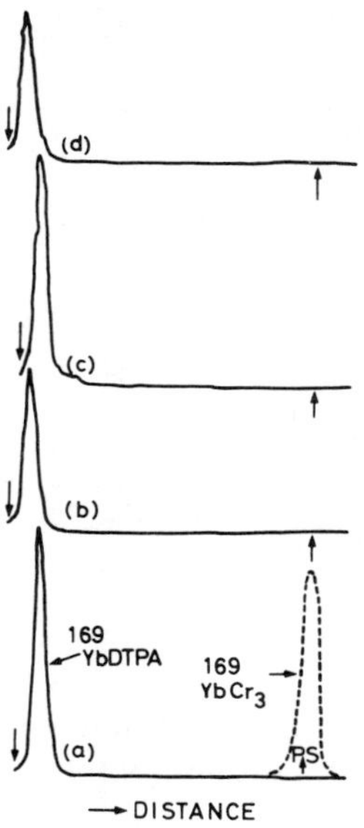

FIG.H. Paper chromatography of ^{169}Yb DPTA.

While the ortho-iodo-benzoic acid component shows no increase, there is a steady increase in the free iodide content. Irradiation would therefore not be a suitable sterilization method.

Chromium (^{51}Cr) EDTA

Paper chromatography: Whatman No.1, 0.05M ammonium acetate: acetone (20:1) (Fig.G)

Dose (Mrad)	0	0.5	1	2
		% activity		
Cr^{+3} Rf = 0.06	1	20	56	53
Cr-EDTA Rf = 0.94	99	60	29	39

When 0.01M EDTA Na_2 was irradiated at 0.25, 0.5, 1, 1.5, 2 and 2.5 Mrad, there was an increase in pH as well as a loss of EDTA, as seen below:

Dose (Mrad)	0	0.25	0.5	1	1.5	2	2.5
pH	4.7	5.1	5.45	5.60	5.95	6.35	6.20
% loss	-	3.3	7.1	16.9	20.8	29.0	37.1
(G) (-EDTA)	-	1.26	1.37	1.63	1.33	1.39	1.42

Radiation sterilization is therefore not suitable, since there is considerable decomposition and the activity in the Cr^{+3} zone increases rapidly with dose.

Ytterbium (^{169}Yb) DTPA

Paper chromatography: Whatman No.1, 0.05M ammonium acetate: acetone (20:1) (Fig.H)

Dose (Mrad)	0	0.5	1	2
		% activity		
Yb^{+3} Rf = 0.06	< 1	< 1	< 1	< 1
Yb DTPA Rf = 1	> 99	> 99	> 99	> 99

When 0.01M DTPA solution was irradiated at the doses employed for 0.01M EDTA Na_2, the pH change, % loss and G(-) values were as follows:

Dose (Mrad)	0	0.25	0.5	1	1.5	2	2.5
pH	7.75	7.80	7.75	7.80	7.90	7.85	7.90
% loss	-	3.6	7.6	15.3	31.6	35.2	37.2
G (-ETPA)	-	1.47	1.58	1.58	2.17	1.81	1.53

Though irradiation does not destroy the complex, radiation sterilization does not appear to offer any advantage over autoclaving, which is reported to be satisfactory.

For a solution with a geometry of 1 cm height and 1 cm radius, the total absorbed dose (D_∞) and the dose absorbed at the end of 32 days (D_{32}) for ^{131}I and D_∞ for ^{51}Cr are as below:

		^{131}I	^{51}Cr
D_∞	(Mrad/mCi)	0.12	0.03
D_{32}	(Mrad/mCi)	0.11	-

When compared with the externally delivered doses, the absorbed dose due to internal irradiation would be too low to cause a comparable degree of decomposition of the above-mentioned radiopharmaceuticals before their expiry date.

RADIOPASTEURIZATION IN THE PROCESSING OF NON-STERILE PHARMACEUTICAL PREPARATIONS AND BASIC MATERIALS

R. F. ARMBRUST
Laren N. H.,
Vinkebaan,
Netherlands

Abstract

RADIOPASTEURIZATION IN THE PROCESSING OF NON-STERILE PHARMACEUTICAL PREPARATIONS AND BASIC MATERIALS.

It is evident that in the near future international and national pharmacopoeia will have requirements on the microbial contamination of non-sterile basic materials and pharmaceutical preparations. As radiopasteurization is a useful method for obtaining the prescribed microbial levels, the necessity for a code of practice for this method is discussed.

The practice of radiation sterilization for a wide range of medical supplies has reached routine industrial operation in many countries. However, with regard to medicaments, the applicability of radiation sterilization has so far not met with satisfactory technical success or with acceptance by the Public Health Authorities for safe use for human patients. The reason is, as has been discussed in detail in this Symposium, the possible occurrence of undefined degradation product(s) following the radiation chemical actions on some components and/or the additives of the medicament being toxic or having other undesirable effects on health.

Besides the pharmaceutical substances which are desired for their "sterile use" as they come in contact with the interior of the human body and possibly with the blood stream, there are numerous medicaments in use which are not specified for "sterile use" in the Pharmacopoeia of various countries.

In such cases of medicaments no information could be easily obtained by the user with regard to their microbial content. Such a factor is not always without risk.

The unfortunate incidents in 1963 in Sweden involving heavy bacterial contamination in a pharmaceutical product can serve as a typical example. These examples concern eye ointments containing cortisone, a steroid hormone and two broad-spectrum antibiotics. Owing to low water content and the presence of bacteriostatic agents the chance of bacterial growth is low. Despite these precautions and its formulation, this ointment happened to cause a large number of severe cases of eye infection, in one case even with loss of an eye. An inspection of the infected eyes and of many tubes from different batches of the ointment indicated the presence of the same contaminating bacteria. These bacteria were also found at several places on the production premises. Reports of a similar incident have been made regarding thyroid tablets infected with high counts of Salmonella (some tablets even had more than 25 million germs per gramme).

After some of these tragic occurrences were revealed (many reports) a number of countries, including Scandinavia, the USA, Czechoslovakia, became aware of the importance of the question of microbial purity of medicaments and of the basic pharmaceutical ingredients. They insisted on a much improved implementation of the steps in the Good Manufacturing Practices. The microbial requirements for pharmaceutical preparations have either appeared in their Pharmacopoeia, or will appear in the near future.

In most cases these Pharmacopoeia requirements suggest limiting (as opposed to total removal or destruction of microorganisms implicit in the sterilization) the amount of microbes to a certain level, and the complete elimination of specific pathogenic ones. The report of the Federation Internationale Pharmaceutique (FIP), and the Group of Industry Pharmacists (April, 1972) divided the pharmaceuticals into four classes:

(1) Injectables: (sterile regarding the requirement of sterility prescribed in the Pharmacopoeia)

(2) Eye preparations: Microbial absence in 1 g

(3) Topical preparations: Ear and nasal drops, ointments for damaged skin etc.
$< 10^2$ germs/g/ml
No _enterobacteriaceae_, _Pseudomonas aeruginosa_, or _Staphylococcus aureus_.

(4) All other preparations: Oral
$< 10^{3-4}$ germs/g/ml
No _Escherichia coli_ (salmonella)
or $< 10^2$ moulds

At the moment it is not quite clear to what extent the national and international Pharmacopoeia will undertake the matter of microbial requirements. Some countries prefer to exclude oral preparations; except for those preparations containing organic, herbal or some other specific highly contaminated types of raw materials and additives.

The most simple way to get low microbial levels is to use dry or wet heat. Many pharmaceutical preparations do not withstand this method. A number of authors are pleading for a Good Manufacturing Process (called G.M.P.), starting with low contaminated basic materials and additives, thus obtaining stable pharmaceutical preparations with a low contamination level. G.M.P. means production under optimal hygienic conditions, preventing microbial growth.

To obtain these low contaminated raw materials, gamma irradiation has many advantages over the gas method. For example, the enclosed germs (in cells or crystals) are destroyed.

Many results concerning irradiation of pharmaceutical products have been published. Unfortunately it has emerged that the formulation of composed pharmaceutical preparations mostly has an influence on the resistance of germs and materials against irradiation. Relatively small amounts of additives can have a large effect in this respect. The free-radical inducers, like water, oxygen, glycerol, D.M.S.O., ethanol, nitrates, halogens etc., decrease the resistance of germs and materials to irradiation, while radical scavengers like sulphydryl groups (sulphites-proteins), carbohydrates etc., increase the resistance.

To change the formulation of pharmaceutical preparations will necessitate complete new tests, since they may then be considered to be new products. The effect of additives can hardly be predicted. In my opinion the best way will be to decontaminate the raw materials; the G. M. P. must then provide stable pharmaceutical preparations, according to the requirements of the Pharmacopoeia. Radiopasteurization is a favourite for this decontamination:

(a) If one is working with dry substances, contrary to composed preparations, the results are predictable because there is no influence of other materials.
(b) The group of highly or dangerously infected raw materials is relatively small, compared with the immense group of composed pharmaceutical preparations.

Radiopasteurization is in principle different from radiosterilization, but let us first examine the term sterile.

Sterility is statistically not absolute, maybe it is better to define it as the negative logarithm of the chance of contamination.

Normally we accept a sterility grade of 6 (according to steam of 120°C during 20 min), which means that there is a chance of one germ on 1 million g/ml or item.

To obtain this very low chance of contamination the system of the over-kill is used.

On the contrary pasteurization means reducing the microbial level to an acceptable one; it is a selective killing of the germs.

The results can be controlled by counting and identifying the survivors. The radiopasteurization dose depends on:

(a) The pre-contamination and the desired post-pasteurization microbial level; and
(b) The sensitivity of the microbes to the materials to be radiopasteurized.

From our own experiments we have learned that in almost all cases 0.4 - 1 Mrad is sufficient (the D_{10} value is mostly rather low — 0.125 - 0.175).

Concerning the absence of specific pathogenic germs, most of them are rather sensitive to γ-rays.

Highly infected materials like organic and herbal ones, with contaminations above 10^6, sometimes need more than 1 Mrad to obtain an acceptable microbial level.

Though I admit that sterilizing with a dose lower than 2.5 Mrad means playing the devil's advocate, for pasteurization with counting the number of survivors there is no devil at all.

The question remains why radiopasteurization is not widely used. As mentioned in the beginning, the Public Health Authorities in practically all countries see irradiated materials as something new, so they request new investigations concerning toxicity etc.

This is not the place to discuss this matter, but it is interesting that heat and gas sterilization, where new products are also formed, never command this devoted attention.

Therefore, in several countries gamma irradiation of consumable products and medicaments is forbidden. Considering how the IAEA with its

Code of Practice for the Radiosterilization of Medical Products has overcome the resistance and objections to this method, in practically all countries, and noting how the IAEA has organized by international co-operation a well-developed centre to discuss all the theoretical and practical problems (as demonstrated at this meeting), it is quite clear to me that the IAEA must become the same neutral international centre for gamma pasteurization as it is now for gamma sterilization. I should like to make the following suggestions:

(a) To prevent duplication, research in gamma pasteurization must be co-ordinated and centralized. A centre is needed where results from different research institutes can be registered. (Some years ago the Federal German Pharmaceutical Industries already started a centralized research programme on gamma irradiation of medicaments. The way it was organized can be an example of how it should be done.)

(b) A code of practice must be worked out to obtain uniformity in methods and controls.

(c) In the future there must be a unity in the national regulations for this new form of pasteurization.

In such a way the IAEA would expand the peaceful uses of nuclear energy for the care of health throughout the world.

With this end in mind I sent out a questionnaire to all the research centres I know (see Annex I).

On this questionnaire are the raw materials, additives and preservation agents, which will probably appear in the European Pharmacopoeia, with a microbial limitation. The Netherlands has a large production and export of starches, used for tabletting. Starches can be a source for contamination. In co-operation with the pharmaceutical technological departments of the University of Leiden (Professor Polderman) and Groningen (Professor Lerk) we started a research project to find out if irradiation with 0.5 and 0.8 Mrad has any influence on the physical properties needed for tabletting. The results show that there is no influence, since no difference was found between irradiated and non-irradiated tablets.

I sincerely hope that discussions between experts, such as those present at this Symposium, will result in setting up international standards of the discussed method as soon as possible.

When within some years the pharmacopoeia establish microbial limitations for raw materials as well as for preparations, an international code of practice for irradiation pasteurization must also be included.

ANNEX I

QUESTIONNAIRE

1. Do you have any experience with gamma-irradiation of subjointed substances ?
2. If yes, did you publish your results and where ? (see question no. 7)
3. If you did not publish are you willing to give a summary of the obtained results or a photocopy of your report?
4. If yes, are these summaries or reports confidential?
5. Do you like to have a summary of the results obtained from this enquiry as far as they are not confidential?
6. Will you please indicate at what min. radiation dose significant changes of activity in the listed pharmaceutical substances you have found.
7. Will you please indicate where and when you published your results or please send us a copy or reprint.

SUBSTANCES (pure) (not in preparation or solution)	1		2		3		4		5		6	7
	yes	no	yes	no	yes	no	yes	no	yes	no		
Acaciae gummi												
Acidum citricum												
Amyla												
Barii sulfas												
Belladonnae folium												
Benzalkonii chloridum												
Benzocainum												
Bismuthi subcarbonas												
Bismuthi subnitras												
Calcii gluconas												
Cascarae cortex												
Cetrimidum												
Cinchonae cortex												
Corticosteroidi (general)												
Dextrosum												
Digitalis folium												
Frangulae radix												
Gentianae radix												
Hyoscyami folium												
Ipecacuanhae radix												
Krameria radix												

SUBSTANCES (pure) (not in preparation or solution)	1		2		3		4		5		6	7
	yes	no	yes	no	yes	no	yes	no	yes	no		
Lactosum												
Laevulosum												
Liquiritae radix												
Magnesii carbonas												
Magnesii oxidum												
Matricariae flos												
Menthae folia												
Polyoxyethyleneglycols												
Riboflavinum												
Sennae folium												
Sennae fructus												
Sorbitan oleate and related substances												
Stramonii folium												
Sucrosum												
Talc												
Thiaminum												
Thyroid												
Zinci oxidum												

SUBSTANCES used for preservation, such as :	1		2		3		4		5		6	7
	yes	no	yes	no	yes	no	yes	no	yes	no		
Methylis oxybenzoas												
Propylis oxybenzoas												
Benzalkonium chloride												
Phenyl Mercuri borati												
Phenyl Mercuri nitrati												

DISCUSSION

A. TALLENTIRE: While I am in sympathy with the principles that you have outlined in your presentation, I must challenge the general statement that the protection of microorganisms against radiation inactivation does not occur in the dry condition. There is an appreciable amount of published evidence showing that such effects do exist. For example, cysteamine, thiourea, broth, serum and glucose protect dried cells against both O_2-dependent and O_2-independent damage. Each situation should therefore be examined individually and extrapolation from one to another should be avoided.

R.F. ARMBRUST: I agree with you, but in the case I am discussing dry substances means well-defined dry chemical products, not organic or herbal ones.

J.O. DAWSON: I am glad to see you have used the term "radiation pasteurization" for the cleaning operation. I think it is much more appropriate than "radiation sterilization at reduced doses", which is ambiguous in such a context and can lead to confusion in international working groups.

K.H. WALLHÄUSSER: I think that the point which emerges from your presentation is that in the European Pharmacopoeia the purpose of the techniques described is to lower the contamination of basic materials and thereby that of the final products. In this sense the word "sterilization" is not really applicable as it is too absolute. In the Federal Republic of Germany we are now interested in finding very simple materials and reducing their germ content. If we take starch or sugar in powder form, for example, we can reduce the germ count with 2-2.5 Mrad. Or we can take an enzyme powder and with radiation sterilization reduce the count from 10^8 to 10^4. But these figures are still too high. Eventually, we ought not to have more than about 10^3 or 10^4 germs in an enzyme powder. In those cases where irradiation is not a suitable technique we are looking for a method to replace ethylene oxide sterilization.

It is very necessary that proper collaboration is maintained between the people representing the industry and the Agency and those of the Pharmacopoeia Commission. They should jointly try to develop suitable methods for each substance. I think Mr. Dawson's paper[1] is relevant to this point.

In the Federal Republic of Germany some tests have now been started by a number of groups. Each group consists of five experts, one dealing with the radiation side, another specializing in microbiology, another in toxicology, and so on. This, in my opinion, is the proper approach, although, as I have said, we are dealing only with very simple substances for the moment.

K.H. CHADWICK: The principal idea you are putting forward seems to involve a reduction of the radiation dose to the basic products. Could this reduction in dose be attained by a combined heat and radiation treatment, for example, as discussed earlier by Mr. Trauth[2]? I am not thinking so much of a simultaneous treatment, but rather of heating before or after

---

[1] IAEA-SM-192/86, these Proceedings.

[2] IAEA-SM-192/38, these Proceedings.

irradiation, which can also lead to a synergistic effect. This is being studied also with the aim of reducing the dose level in food irradiation research.

R.F. ARMBRUST: The heating of basic materials in bulk is rather difficult. Apart from that, if we can simply irradiate at doses between 0.4 and 1 Mrad, then I see no point in making the process more complicated by introducing heat.

FEASIBILITY STUDIES ON RADIATION STERILIZATION OF SOME PHARMACEUTICAL PRODUCTS

N.G.S. GOPAL, S. RAJAGOPALAN,
G. SHARMA
Isotope Division,
Bhabha Atomic Research Centre,
Trombay, India

Abstract

FEASIBILITY STUDIES ON RADIATION STERILIZATION OF SOME PHARMACEUTICAL PRODUCTS.

The paper deals with some studies carried out to evaluate the feasibility of radiation sterilization or treatment of some medical products and pharmaceuticals of immediate importance to their respective industries. The products include penicillin G sodium, ampicillin sodium, tetracyclin hydrochloride ointment, hydrocortisone acetate and its ointment, aqueous sodium chloride solutions (0.9 and 20%), fluorescein sodium strips, urea, ethylmorphine hydrochloride, aqueous solution of chlorobutanol and one of its commercial preparations, phenylmercuric nitrate and its aqueous solutions, aqueous solutions of methyl and propyl paraben, lactose, gum karaya, absorbent cotton and poly-(vinyl chloride) based medical products. The irradiated products have been examined for pharmacopoeial specifications wherever available. In general the products have been examined for changes in colour, pH, ultra-violet and infra-red absorption spectra. Thin-layer chromatographic analyses have been carried out to establish the purity of some of the irradiated products. The feasibility or otherwise of radiation sterilization or treatment of the various products from the physico-chemical and microbiological (pharmacological) clinical considerations is also described.

1. INTRODUCTION

In pharmaceutical parlance, sterilization is a process designed to kill the microorganisms present. The application of the various methods of sterilization, as recommended by the British Pharmaceutical Codex 1973, is given in Table I [1].

The usefulness of radiation as a unique means of sterilization preferred over other methods can be seen from the comparison given in Table II. Sterilization by radiation is particularly suitable for products that are heat sensitive and enclosed in air-, moisture- and microbe-proof packaging. It is reported, for instance in Canada, that the increase in radiation sterilization is related to the recent comments of the Food and Drug Administration and of industry on the potential hazards of residual ethylene oxide gas now used for sterilization, as well as to the deeper penetration by radiation, making it possible to sterilize products in their final packaged form [2].

TABLE I. METHODS OF STERILIZATION RECOMMENDED BY BRITISH PHARMACEUTICAL CODEX, 1973

Product	Method					
	1	2	3	4	5	6
Glass & metal apparatus	+	+				+
Rubber - Natural		+				
- Silicone	+	+				
Plastic containers					+	+
Aq. solutions		+	+	+		+
Aq. suspensions		+	+			+
Non-aqueous liquids & products	+					
Powders	+				+	+
Surgical dressings	+[a]		+		+	+

1. Dry heat
2. Autoclaving
3. Heating with a bactericide
4. Filtration
5. Ethylene oxide
6. Ionizing radiation

a For paraffin gauze.

TABLE II. COMPARISON OF STERILIZATION PROCESSES

Factor	Autoclave	Filtration	Ethylene oxide	Radiation
Temperature	+	-	+	-
Time	+	+	+	+
Pressure	+	+	+	-
Vacuum	+	+	+	-
Concentration	-	Area	+	-
Wrapping	+	Asepsis	+	- (a)
Humidity	+	-	+	-
Post-sterilization drying/degassing	+	-	+	-
Residual toxicity	- (b)	-	- (b)	- (b)
Type of process	Batch	Batch	Batch	Continuous

(+) affects the process;
(-) does not affect the process;
(a) in the final impermeable packaging;
(b) depends on the type of substance.

1.1. Medical and pharmaceutical products

Products which require irradiation, either for the purpose of sterilization or for radurization, may be classified into three categories:

Medical products;
Pharmaceutical products required in a sterile form;
Pharmaceuticals, health-care items, cosmetics of satisfactory microbiological quality, etc.

The medical products include a large class of products such as:

(i) Surgical dressings and sutures, employed in minor and major surgery and for application to wounds;
(ii) Transfusion and infusion assemblies;
(iii) Devices such as catheters employed for draining physiological fluids;
(iv) Surgical gloves, etc.

Pharmaceutical products include (i) antibiotics, alkaloids, anaesthetics, steroids, carbohydrates, etc. which are required to be sterile; (ii) enzymes, opthalmic preparations such as ointments, eye drops, etc.; and (iii) some raw materials which need not be sterile but may contain some non-pathogenic microorganisms not exceeding a certain limit imposed by the national health and drug authorities. In some countries ophthalmic products are required, like injections, to be sterile.

Health care includes items such as sanitary pads; and cosmetic materials include talc, fatty acid esters, proteins, etc., which may be the source of microbial contamination.

Earlier investigations on the effect of radiation on some medical products and pharmaceuticals have been reviewed elsewhere [3 - 7]. The main problem of applying radiation sterilization to drugs lies in establishing the efficacy, toxicity and stability of these irradiated drugs. While one cannot deny the possible formation of some radiolytic products, in very low concentrations, it must be recognized that they may not be detected even by the most sensitive analytical techniques available today. A possible solution to this problem would lie in utilizing the radio-tracer technique or isolating the suspected impurity by techniques such as preparative high-performance liquid chromatography for a subsequent evaluation. However, if an irradiated product conforms to pharmacopoeial standards (except for a slight change in colour) and is free from any toxicity and its bioavailability is unaltered, then it may be considered suitable for radiation sterilization or treatment.

Studies carried out in the authors' laboratory on some pharmaceuticals and medical products are described in this paper.

2. EXPERIMENTAL

2.1. Materials and methods

The chemicals employed in the course of these studies were of reagent grade quality. The medical products and pharmaceuticals employed in these investigations were obtained from commercial sources. Irradiation with ^{60}Co gamma rays was carried out in various types of containers: I. penicillin-type vials; II. polyethylene bags; III. medium density polyethylene bottles (500-ml capacity); IV. PVC bags (500-ml capacity); V. glass ampules (evacuated); VI. collapsible aluminium tubes.

Spectrophotometric measurements were made with the double-beam recording instrument, Shimadzu-UV 200, using 1-cm quartz cells.

Infra-red spectra were taken with a Perkin Elmer instrument using KBr dispersion, Nujol mull and/or hexachlorobutadiene mull.

Optical rotation measurements were made with a Lippich Polarimeter. Paper chromatography was carried out by the ascending technique on Whatman No.1 strips.

For thin-layer chromatography (TLC) silica gel C (E. Merck) and cellulose-TLC (Serva-Entwicklungslabor) were employed in 0.25-mm-thick layers on glass plates.

The solvent systems employed for paper and thin-layer chromatography are given below:

(1) Acetone:acetic acid (95:5)
(2) Chloroform:methanol:aq.ammonia (17%) (2:1:1) – upper layer.
(3) Iso-amyl alcohol:methanol:formic acid:water (65:20:5:10)
(4) Propanol:pyridine:acetic acid:water (15:10:3:12)
(5) Aq.sodium chloride (0.1M)
(6) Aq. ammonia (0.56%)
(7) Aq. disodium hydrogen phosphate (0.5M), pH 8.7
(8) Amyl alcohol:aq. ammonia (25%) (99:1)
(9) n-butanol:water:aq.ammonia (25%) : ethanol (200:88:2:40)
(10) Methylenedichloride
(11) Chloroform:acetone (4:1)
(12) Benzene:acetone (1:1)
(13) Chloroform:methanol:ammonia (7:2.5:0.5) containing 0.5% phenothiazine.
(14) Methanol:n-butanol:benzene:water (12:3:2:1)
(15) Ethanol:acetic acid:water (6:3:1)

The spray reagents employed for paper and thin-layer chromatography are given below:

(1) Ninhydrin (0.3%) in acetone (for ampicillin)
(2) Aq. potassium permanganate (10%)
(3) A 1:1 mixture of (i) 1.5 g cupric chloride + 3 g ammonium chloride in 50 ml water, and (ii) aq. hydroxyammonium chloride (20%)
(4) Resorcinol-zinc chloride-sulphuric acid:
    (i) Some zinc chloride is added to a 20% solution of resorcinol in ethanol
    (ii) 4N sulphuric acid
    (iii) Aq. potassium hydroxide (40%)

The plate is sprayed with (i), heated 10 min at 150°C, sprayed with (ii), heated 20 min at 120°C and finally sprayed with (iii).

(5) Methanol or ethanol:sulphuric acid (1:1).
Develops colour on heating at 120°C and fluorescences in u.v.

(6) Anisaldehyde: sulphuric acid: acetic acid (0.5:1:50) Develops colour on heating at 90-100°C.
(7) 0.1 g of picric acid in 36 ml acetic acid + 6 ml perchloric acid. Develops colour on heating at 70-80°C.
(8) A 1:1 mixture of (i) triphenyltetrazolium chloride (4%) in methanol and (ii) N NaOH. Develops colour on heating at 100°C.
(9) Iodine vapour; ammonia vapour
(10) Bromine vapour; ammonia vapour

Fluorescent substances on paper and thin-layer chromatograms were detected by the u.v. detector (2537 and 3560 Å) of Baird and Tatlock (London) Ltd.

2.2. Benzyl penicillin and ampicillin (sodium salt)

It is reported that irradiation has little effect on the potency of sodium benzyl penicillin in the dry state [8 - 10]. Benzyl penicillin (I), potassium salt of I(II), procaine salt of I(III), phenoxymethyl penicillin (IV), potassium salt of IV(V), methyl ester of IV(VI), syntapen or cloxacillin (VII), and sodium salt of VII(VIII) have been studied by Dziegielewski et al. [11] by EPR and ORD measurements, and they found that the free acids are more sensitive to gamma radiation than their salts, the order of sensitivity of the two groups changing in the following order:

$$IV < VII < I \text{ and } III < VI < V < VIII < II, \text{ respectively.}$$

Sodium salts of benzyl penicillin and ampicillin were irradiated at 0.5 Mrad/h to a dose of 2.5 Mrad in type I container. They were examined and found to conform to pharmacopoeial specifications except that the powders had a slight yellow tinge. This colour change was more pronounced in the case of ampicillin. (The colour change of penicillin when irradiated in a type-V container was less than when irradiated in a type-I container.) The purity of the control and irradiated samples of both antibiotics was determined by TLC on silica gel G using solvent systems (1) to (4). The TLC plates were examined under u.v. light (3560 Å) and then sprayed with reagents (1) to (3). The Rf values of the various zones in solvent systems (1) to (4) were as follows:

Solvent system	Benzyl penicillin sodium	Ampicillin sodium
1	0 (Traces)	0 (Traces
	0.83 (main zone)	0.21 (main zone)
		0.76 (faint)
2	0.93	0.93 - 1.0
3	0.68	0.03 (faint)
		0.07 (main zone)
		0.32 (faint)
4	0.93	0.95

The chromatographic patterns of the control and irradiated samples were alike.

The purity of control and irradiated ampicillin sodium was also determined by TLC on cellulose with solvent system (5). The TLC plate was sprayed with reagent (1). The Rf values of the control and irradiated samples were 0.69 and 0.68, respectively. No other zones were present.

The infra-red absorption spectra of the irradiated samples exhibited maxima which were at the same wavelengths and had similar relative intensities to those in the control samples.

Commercial ampicillin is reported to contain a small percentage of di- and polymerized products [12]. Whether or not irradiation augments them, must be carefully examined. Bioavailability studies on ampicillin are in progress.

2.3. Tetracyclin hydrochloride ointment

Tetracyclin has been reported in literature as quite stable to radiation (0.5 - 10 Mrad) in solid state and in the ointment bases lanolin and petrolatum [13, 14]. A commercial tetracyclin hydrochloride ointment (1% in soft paraffin base) was irradiated to 2.5 Mrad and assayed for tetracyclin hydrochloride spectrophotometrically at 380 nm. It was found that $E^{1\%}_{1cm}$ value for the irradiated ointment was about 10% more than that of the control which may mean that there could be a slight change in the irradiated ointment. Further studies on its chromatographic purity, in particular for the presence of anhydro-, 4epi-, and 4-epi-anhydrotetracyclins, are in progress. Pharmacological, clinical efficacy and tolerance studies show that they retain their full efficacy and are free from toxicity [15].

2.4. Hydrocortisone acetate

The effect of gamma radiation on hydrocortisone acetate has been investigated by Hortobagyi et al. [16] and Hangay et al. [17]; they have observed that the sample irradiated in air retains its full activity though there is a slight colour change. In the present investigation we have not only examined the irradiated hydrocortisone acetate vis-à-vis the various pharmacopoeial specifications, but also carried out extended TLC studies to ascertain the absence of any radiation decomposition products.

The product irradiated to 2.5 Mrad was found to conform to the pharmacopoeial specifications except for a slight yellow tinge. The absorption spectra in methanol and ethanol, and the ratio of absorbances at the absorption maxima and minima were nearly identical for the control and irradiated samples. The assay for the steroid was carried out by u.v. absorption measurements at 242 nm (due to the 3-keto group), and by using blue tetrazolium at 252 nm (due to the substituent at C-17). TLC was carried out on silica gel G-plates with the solvent systems (11) to (13). The quantity of hydrocortisone acetate spotted was about 100 μg. The TLC plates were sprayed with reagents (5) to (10). The Rf values of hydrocortisone acetate (control and irradiated) in the three solvent systems were as below:

Solvent system	(11)	(12)	(13)
Rf	0.46 - 0.53 (main zone) 0.08 * 0.0 *	0.91 - 0.94	0.92 - 0.93

(*Intensity comparable with that of 0.01 - 0.05 μg of hydrocortisone acetate.)

The nearly identical TLC pattern of control and irradiated hydrocortisone acetate and the absence of any extraneous bands amply go to show that hydrocortisone acetate is very stable to a dose of 2.5 Mrad in solid state.

Similar experiments were carried out on hydrocortisone acetate isolated from control and irradiated ointments (in soft paraffin base) and creams (in an aqua-soft paraffin base). Here also the TLC patterns of the control and irradiated materials were alike and no other decomposition products were observable.

2.5. Aqueous sodium chloride solutions

Normal saline in polystyrene syringe in leak-proof double-polyethylene package or in polyethylene droppers or squeeze-type dispensers have been gamma irradiated and employed as ready-to-use sterile solution in hospitals and eye-camps without any unfavourable effect [18]. Bachman reports that radiolysis of aqueous sodium chloride solution (0.9%) does not result in change in pH or Cl^- concentration but produces H_2O_2 [19]. Berry reports that irradiated saline solutions are not toxic to mammalian cells in tissue culture [20]. Ogg has also reported that intradermal injections of saline, irradiated to 5 Mrad, into human volunteers did not produce any untoward reaction [18].

We have examined here radiation sterilization by 2.5 Mrad, of 0.9% sodium chloride in type-I,-III and -IV containers and of 20% sodium chloride solution in type-I container. The solutions were examined for oxychlorides, pH and peroxide content soon after irradiation and after a few months of storage at room temperature (25°C). Tests for oxychlorides proved to be negative. The other data are in Table III and we may conclude that:

(i) The pH of freshly irradiated solution in either glass or polyethylene container remains well within the pharmacopoeial limits (4.5 - 7); however, there is a significant fall after prolonged storage.

(ii) There is practically no change in pH in 20% sodium chloride solution.

(iii) The peroxide content (expressed as H_2O_2) of 0.9% solution in glass containers decreases markedly, but in the polyethylene containers there is very little change on storage.

(iv) The peroxide content in 20% solution is very low even in freshly irradiated solution.

(v) In a PVC bag there is a steep fall in pH but this can be prevented by adding suitable additives.

Pharmacological tests on irradiated solutions in polyethylene bottles indicate that there is no manifestation of toxicity. Clinical efficacy tests show that they are well tolerated in human subjects [15].

TABLE III. EFFECT OF ^{60}Co GAMMA RADIATION ON SODIUM CHLORIDE SOLUTION

Container	NaCl (%)	Dose (Mrad)	Age	pH		H_2O_2 (ppm)
				Control	irradiated	
Glass	0.9	1.0	1 W	5.80	5.45	3.4
			28 m	-	4.70	<< 1
		2.5	1 W	5.80	5.35	3.7
			20 m	-	4.40	≪ 1
		3.5	1 W	5.80	5.00	3.9
	20.0	2.5	3 d	5.70	5.75	< 1
			6 m	5.75	5.80	< 1
Polyethylene	0.9	2.5	2 m	5.80	5.30	4.0
			7 m	5.90	4.95	3.0
PVC	0.9	2.5	4 m	5.45	3.60	≪1

(d = day; W = week; m = month)

2.6. Fluorescein sodium

Fluorescein sodium is employed in the form of a 2% aqueous solution as an ophthalmic diagnostic aid. The successful sterilization of aqueous solutions of fluorescein sodium has already been reported in the literature [5b].

There is considerable interest in the use of radiation sterilization of fluorescein sodium impregnated paper strips for ophthalmic diagnosis. One such commercial product has been investigated here for its radiation stability. The strips (I) were irradiated at 0.5 Mrad/h to a dose of 2.5 Mrad and for purposes of comparison fluorescein sodium (II) and its 2.2% aqueous solution (III) were also irradiated under identical conditions and were examined by spectrophotometry and by paper chromatography.

The absorption spectra of the control and irradiated samples of I, II and III in water and in ethanol, were nearly identical. The absorption maxima in water and ethanol were at 490 and 499 nm, respectively. Paper chromatography of aqueous solutions of II and of aqueous extracts of I were carried out with solvent systems (6) to (9) and the zones were examined under u.v. light (3560 Å). The chromatograms showed a single zone only and the Rf values were as below:

Solvent system	6	7	8	9
Rf	0.80	0.25	0.0	0.25

The chromatogram of III, however, showed an additional pink zone whose Rf values in the solvent systems (6) and (9) were 0.62 and 0.14, respectively, and in system (7) there was a pink streak from the point of spotting right up to the main spot.

It may be seen from the above results that fluorescein sodium in the impregnated strips is quite stable to a dose of 2.5 Mrad and relatively more stable than the aqueous solution to irradiation.

2.7. Urea

Urea is used as a diuretic in large volume infusions. Aqueous solutions of urea decompose on standing or on heating. Autoclaving at about 120°C for 12 min is reported to cause 2.5% destruction [28]. Urea is also unstable to dry heat as it decomposes at about 150°C into biuret and ammonia. Therefore, for injectable purposes urea has to be made available as a sterile solid. There appears to be no report on irradiation studies on urea.

Urea from two different sources (A and B), irradiated at 0.5 Mrad/h to a dose of 2.5 Mrad in type-I container did not undergo any colour change. Their aqueous solutions (4%) were examined for pH and biuret content. The pH of the solutions of the irradiated samples was 0.2 to 0.4 unit more than that of the controls. The biuret content [21] of control as well as irradiated samples of urea (A) was about 2.5% and of urea (B) about 1.5%, thus indicating that there is no increase in the content of biuret owing to irradiation.

2.8. Ethylmorphine HCl

Ethylmorphine HCl which is used in some ophthalmic ointments, has been examined for its stability towards irradiation (2.5 Mrad at 0.5 Mrad/h). It was observed that, except for a slight change in colour, the irradiated product conformed to the specifications of the Indian Pharmacopoeia 1966 and NF XII. The absorption maxima and minima (283 and 260 nm, respectively) and the ratio of absorbences at the maxima and minima (2.4) were nearly identical for the control and irradiated samples. TLC was carried out on silica gel G [22a] with solvent systems [14] and [15]. Spray reagent [5] and Droggendorff reagent [22b] were employed to detect the zones on the TLC plates. The chromatograms of the control and irradiated samples showed a single zone only and the Rf values of the alkaloid in the above two solvent systems were 0.40 and 0.31, respectively.

2.9. Phenylmercuric nitrate

Phenylmercuric nitrate is employed as an antiseptic and also as a preservative in injections, ophthalmic solutions and ointments at concentrations of about 0.002 to 0.005%. There appears to be no report on irradiation studies on this compound and its aqueous solutions.

Solid phenylmercuric nitrate was irradiated at 0.5 Mrad/h to doses of 0.25, 0.5, 1, 2 and 2.5 Mrad in type-I container. Its stability was studied spectrophotometrically by estimating it with KI at pH 4.7 [23]. The solid turned distinctly off-white after about a 1-Mrad dose. Solutions (0.005%) of the control and irradiated samples were prepared, and the assay showed that the compound irradiated at various doses did not significantly differ from the control.

(Mrad)	0	0.25	0.5	1	2	2.5
Relative absorbence in units	1	1	1.04	0.91	1	1

Irradiation of its aqueous solutions (0.005%) resulted in a 70 - 90% loss of the compound from the solution in the dose range 0.25 to 2.5 Mrad. Irradiation of its aqueous solutions (0.05%) resulted in a degradation of about 40% at 0.25 and 90% in the dose range of 0.5 to 2.5 Mrad.

2.10. Methyl and propylesters of p-hydroxy benzoic acid

These two compounds, more familiarly known as methyl and propylparaben are employed as preservatives in parenteral and ophthalmic preparations in the concentration range of 0.05 to 0.2%. Aqueous solutions of the two esters were prepared separately and irradiated at 0.5 Mrad/h to 0.5, 1, 1.5, 2 and 2.5 Mrad. While the control solutions were colourless, the irradiated solutions turned faintly brown. The change in pH and absorbence at 252 nm (for the methylester) and 254.5 nm (for the propylester) were as in Table IV. It is evident from the Table and from Fig.1 that methylparaben is less sensitive than propylparaben to radiolysis and there is a significant degradation of methylparaben even at 0.5 Mrad.

2.11. Chlorobutanol

Chlorobutanol is used as a preservative in ophthalmic preparations at a concentration of about 0.6% (nearly saturation concentration at about 25°C). Its aqueous solutions are unstable to heat owing to hydrolysis and fall in pH. When such an aqueous solution was irradiated at 0.5 Mrad/h to a dose of 3.1 Mrad, apart from a fall in pH from 4.25 to 1.70, there was formation of a white precipitate instead of a yellow precipitate [24]. Work with regard to identification of the white precipitate is in progress. The conditions under which aqueous chlorobutanol solution is prepared also influences

TABLE IV. EFFECT OF ^{60}Co GAMMA RADIATION ON AQUEOUS SOLUTIONS OF METHYL AND PROPYLPARABEN

Dose (Mrad)	Methylparaben (0.02327%)			Propylparaben (0.01045%)		
	pH	Absorbence[a]	% degradation	pH	Absorbence[b]	% degradation
0	8.40	0.45	-	7.80	0.42	-
0.5	8.00	0.41	8	6.70	0.29	43
1	7.35	0.38	17	6.30	0.21	58
1.5	7.05	0.36	20	6.10	0.18	64
2	6.80	0.33	27	5.90	0.15	69
2.5	6.80	0.33	27	5.75	0.15	69

[a] 50-fold dilution; [b] 25-fold dilution

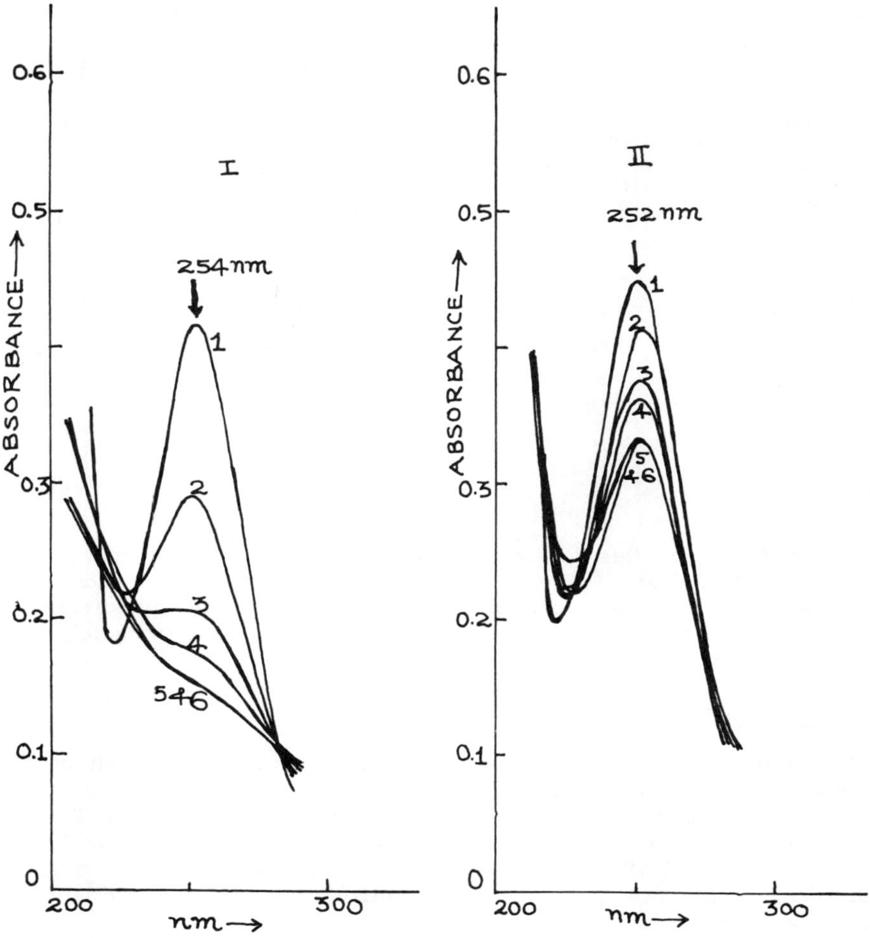

FIG.1. Ultra-violet absorption spectra of (I) propyl- and (II) methylparaben. 1. Control, 2. 0.5 Mrad, 3. 1.0 Mrad, 4. 1.5 Mrad, 5. 2.0 Mrad, 6. 2.5 Mrad.

its behaviour towards radiation as can be seen from Table V. As the tertiary carbon in $-C(CH_3)_2OH$ is radiation sensitive, the loss of a hydroxyl radical and dimerization of the organic radical could lead to the formation of insoluble $Cl_3CC(CH_3)_2-(CH_3)_2CCCl_3$.

It is employed in combination with a number of other ingredients such as sulphacetamide sodium, boric acid, $ZnSO_4$ and the methyl and propyl-esters of para-hydroxy benzoic acid. When one such commercial preparation was irradiated to different doses in type-I container, it was found that the pH of the irradiated preparation was 6.5, nearly the same as that of the control and the solutions showed very little change in their colour up to

TABLE V. EFFECT OF ^{60}Co GAMMA RADIATION ON AQUEOUS SOLUTIONS OF CHLOROBUTANOL

Dose (Mrad)	Soln. prepared at 25°C			Soln. prepared by heating in BWB for 15 min	
	pH	% loss	Clarity	pH	% loss
0	5.75	-	Clear	3.95	-
0.5	2.85	13	Clear	2.55	18
1.0	2.45	13	Clear	2.25	22
1.5	2.35	17	Opalescent	2.10	26
2.0	2.05	33	White ppt.	1.95	30
2.5	1.90	33	White ppt.	1.90	47

1.5 Mrad. Beyond this dose the discolouration was more pronounced, becoming deeply brown at 2.5 Mrad. This shows that such a preparation cannot be treated with radiation even at low doses, especially as constituents like chlorobutanol, methyl and propylparaben are prone to radiolytic decomposition (see section 2.10).

2.12. Lactose

Lactose is employed in pharmaceutical industry as a diluent and for tablet making. It was irradiated at 0.5 Mrad/h to doses of 0.2, 0.4 and 0.8 Mrad and examined for its pharmacopoeial characteristics. The changes induced by radiation were the straw-yellow tint and increased acidity. The acidity increased with increasing dose, as can be seen from the quantity of alkali required (0.1N NaOH) to neutralize a solution of 5 g of lactose in 50 ml water:

Mrad	0	0.2	0.4	0.8
ml	0.14	0.5	1.0	1.7

The u.v. absorption spectra of the saturated solutions did not show any absorption peaks in the region 400 to 200 nm. The infra-red spectra of control and irradiated samples in KBr dispersion exhibited maxima at the same wavelengths and had similar relative intensities. However, the infra-red spectra in Nujol mull showed that the 0.2- and 0.4-Mrad-treated samples exhibited the same maxima and relative intensities as that of control; in the case of the 0.8-Mrad-treated sample, the relative intensities of the peaks in the region 6 - 12 μm became weaker than in the case of the control, and the peaks corresponding to 6.1, 7.75 and 9.5 μm disappeared.

Whether the slight increase in acidity would impede the utilization of the irradiated lactose, as a diluent or as an ingredient in tablet making, should be examined.

2.13. Gum karaya

Natural gums like gum karaya, which are used for their demulcent property in pharmaceutical preparations, are likely to have high presterilization counts. They can be upgraded from the microbiological point of view with relatively low radiation doses without seriously affecting their swelling ability.

Dose (Mrad)	0	0.25	0.5	0.75	1	2.5
Microbial (counts/g)	6500	350	100	20	10	nil
Swelling ability (ml water absorbed per g)	40	29	17	13	7.5	3

It is evident from these data that doses as low as 0.25 Mrad would be sufficient to upgrade the natural gum, which may have a few thousand microorganisms per gramme in its natural state.

2.14. Absorbent cotton

This constitutes an important medical product employed in the preparation of surgical dressings. Though the bleaching process may remove most of the soil- and water-borne microorganisms from raw cotton, the subsequent washing, drying and handling processes result in contamination of the absorbent cotton with microorganisms found on manufacturing premises. The level of such contamination in some of the commercially available absorbent cotton has been found to be a couple of hundred microorganisms per gramme of cotton. Absorbent cotton from different sources, sealed in polyethylene bags, and irradiated to 2.5 Mrad, has been examined for the following characteristics: effect of heat, oxidizing substances, water soluble extractives and absorbency.

The tests were conducted on freshly irradiated cotton and also after different time intervals on the samples stored at about 25°C.

The two characteristics, effect of heat and oxidizing substances, were not affected by irradiation. The water soluble extractive is distinctly found to be enhanced by irradiation.

	Source A					Source B		Source C		
Control	0.16	0.07	0.13	0.15	0.25	0.27	0.20	0.39	0.65	0.47%
Irradiated	0.33	0.15	0.24	0.26	0.42	0.52	0.37	0.58	0.87	0.57%

The water absorbencies of absorbent cotton as determined on control and irradiated samples, at different intervals of time are given below:

	Source A	Source B			Source C	
Age	1 d	1 d	6 month	15 month	1 d	12 month
Control	4 s	4 s	3 s	3 s	12 s	30 s
Irradiated	4 s	4 s	3 s	3 s	15 s	60 s

In freshly irradiated samples the absorbency does not differ appreciably from that of control. On prolonged storage, however, there is a marked decrease in the absorbency of the control and irradiated samples of source C but not in those of B. This would therefore reduce the storage time of such batches.

Although irradiation may affect the water soluble extractive and absorbency to a slight extent, absorbent cotton formulations, which are used once only, can be conveniently radiation sterilized [3].

2.15. PVC-based medical products

In some countries radiation-resistant grades of PVC formulations are available. Local manufacturers are trying to produce PVC compositions of such a grade. Granules, tubings, bags, catheters, infusion and transfusion sets locally fabricated from PVC have been irradiated to a sterilization dose of 2.5 Mrad and examined for the following physico-chemical characteristics:

(i) Non-volatile residue, residue on ignition, heavy metals and buffering capacity (see USP XVIII);
(ii) Heavy metals, tin and barium (see BP 1973); and
(iii) Extinction between 360 to 220 nm (see Pharmacopoeia Helvetica, 1971).

All these products were found to show a noticeable change in colour and odour. It was noticed that the degree of discolouration varied from product to product. It is interesting to note that when one variety of PVC tubes was irradiated to 2.5, 3, 3.5 and 4 Mrad (at a dose-rate of 0.5 Mrad/h) the colour change as compared with control was negligible, but when irradiated at a lower dose-rate to about 3 Mrad (duration about 90 - 100 h) the tube became distinctly discoloured.

All the irradiated products conformed to the specifications given in USP. While most of them conformed to BP requirements, some did not satisfy the requirements in respect of tin and barium. All these products (including the controls) showed a higher extinction than that allowed in Pharmacopoeia Helvetica.

There have been quite a number of reports about the possibility of ingredients such as plasticizers, catalysts, filters, stabilizers and unpolymerized monomers being leached out by solutions either during storage or during infusion [25 - 27].

The u.v. absorption spectrum of aqueous extracts of most of the samples of infusion sets showed absorption maxima which were comparable with that of commercially available di-2-ethyl hexyl phthalate (DOP). The tubes of control and irradiated infusion sets were refluxed with methanol for 2 h and the methanol extract was evaporated on water bath and the residue was examined by infra-red spectra in Nujol mull and by TLC on silica gel G with solvent system (10). The peak position in the infra-red spectra of the extracts were similar to those in reference DOP. The TLC plates were sprayed with reagent (4). The zones on the TLC plates of the control and irradiated samples not only corresponded to each other but also to that of reference DOP. No other zones could be detected, which would indicate that irradiation to 2.5 Mrad caused no detectable change in the nature of the methanol leachable ingredients.

When the control and irradiated PVC tubes of an infusion set were equilibrated with normal saline (1 ml/cm^2) at about 28°C, the following quantities of u.v. absorbing species were leached out at the end of 4 and 11 days equilibration. As the u.v. spectra of the extracts were similar to that of DOP, the u.v. absorbing species are expressed in terms of DOP.

	4 days	11 days
Control (ppm)	6	10
Irradiated (ppm)	18	24

This would indicate that, though there is a higher leach-out from irradiated sets, only a negligible quantity of DOP may be leached out from the sets during transfusion operation which may last a few hours. But this leach-out could be of greater significance in control or irradiated PVC containers used for storing infusions, and therefore it requires careful consideration. Clinical studies show that discoloration does not affect the safe use of the irradiated infusion sets [15].

The pharmacopoeias of the United States of America, the United Kingdom, the Scandinavian countries, Switzerland and France, and the Department of Health and Social Security of the United Kingdom, have laid down general specifications for the plastic-based medical products and also for the PVC products in particular. An examination of these specifications reveals that there is a wide disparity in the method of preparation of extracts for various chemical tests, the limits of impurities and specifications for biological tests.

Therefore, we feel that it is of paramount necessity to draw up specifications for the various materials, infusion sets, catheters, containers, implantable materials etc., separately, depending on their intended use.

3. CONCLUSIONS

Though some workers have felt that attention has been mainly devoted to colour changes, loss of potency, pH change, u.v. absorption spectra etc., it should be noted that such studies coupled with an examination of the toxicity, bioavailability and/or clinical efficacy of the irradiated product, should considerably help in deciding whether or not a product is suitable for irradiation. Products manufactured under good manufacturing conditions, and therefore with either almost no detectable contamination or with very few microorganisms, could be treated with radiation doses lower than 2.5 Mrad, to render them microbiologically satisfactory, either from the sterility aspect, or to decrease the level and nature of contamination to a satisfactory level. Such lower doses would ensure that any radiation-induced changes would also be minimal. Concomitant studies on the radiation chemistry of these compounds should, however, be carried out in order to understand the mechanism of radiation degradation processes.

The feasibility of radiation sterilization or treatment of the various products considered here may be summarized as follows:

ACKNOWLEDGEMENT

The authors wish to thank Dr. V.K. Iya for his keen interest in the work and Mrs. K.M. Patel for carrying out the microbiological work.

REFERENCES

[1] British Pharmaceutical Codex 1973, The Pharmaceutical Press, London (1973) 924.
[2] Isotope Radiat. Tech. 8 (1970) 128.
[3] BRIDGES, B.A., POWELL, D.B., in Massive Radiation Technique (JEFFERSON, S., Ed.), George Newnes Ltd. (1964).
[4] Chemist Druggist 189 (1968) 563.
[5] PHILLIPS, G.O., in Manual on Radiation Sterilization of Medical and Biological Materials, Tech. Rep. Ser. No.149, IAEA, Vienna (1973), (a) pp.207-28 (b) p.225.
[6] IAEA, Radiosterilization of Medical Products (Proc. Symp. Budapest, 1967), IAEA, Vienna (1967).
[7] Use of Gamma Radiation Sources for the Sterilization of Pharmaceutical Products, Rep. Assoc. Brit. Pharmaceutical Industry (1960).
[8] CONTROULIS, J., et al., J. Am. Pharm. Assoc. 43 (1954) 65.
[9] COLOVOS, G.C., CHURCHILL, B.W., J. Am. Pharm. Assoc. 46 (1957) 580.
[10] GRAINGER, H.S., HUTCHINSON, W.P., J. Pharm. Pharmacol. 9 (1957) 343.
[11] DZIEGIELEWSKI, J., et al., Chem. Abstr. 80 (1974) 151057 b.
[12] BUNDGAARD, H., J. Pharm. Pharmacol. 26 (1974) 385.
[13] POCHAPINSKII, V.I., et al., Chem. Abstr. 58 (1963) 3272.
[14] RADOVIC, G., et al., Chem. Abstr. 76 (1972) 103730 m.
[15] KULKARNI, R.D., GOPAL, N.G.S., IAEA-SM-192/87, these Proceedings.
[16] HORTOBAGYI, G., et al., in Ref.[6], p.25.
[17] HANGAY, G., et al., in Ref.[6], p.55.
[18] OGG, A.J., in Ref.[6], p.49.
[19] BACHMAN, S., Chem. Abstr. 80 (1974) 63832 V.
[20] BERRY, R.J., in Ref.[6], p.54.
[21] Official Methods of Analysis of the Association of Official Analytical Chemists, (WILLIAM HORWITZ, Ed.), Association of Official Analytical Chemists, Washington (1970) 20.
[22] EGON STAHL, Thin-Layer Chromatography, George Allen and Unwin, Ltd., London (1969) (a) p.529, (b) p.873 (reagent No.96).
[23] GOPAL, N.G.S., Unpublished results.
[24] RASERO, L.J., Jr., SKAUEN, D.N., J. Pharm. Sci. 56 (1967) 724.
[25] NEMATOLLAHI, J., et al., J. Pharm. Sci. 56 (1967) 1447.
[26] ATKINS, R.C., et al., Lancet II (1968) 1014.
[27] JAEGER, R.J., RUBIN, R.J., Lancet II (1970) 151,778.
[28] KALA, H., et al., Chem. Abstr. 80 (1974) 87452 X.

SAFETY AND CLINICAL EFFICACY OF SOME RADIATION-STERILIZED MEDICAL PRODUCTS AND PHARMACEUTICALS

R.D. KULKARNI
Department of Pharmacology,
Grant Medical College, Bombay

N.G.S. GOPAL
Isotope Division,
Bhabha Atomic Research Centre,
Trombay, India

Abstract

SAFETY AND CLINICAL EFFICACY OF SOME RADIATION-STERILIZED MEDICAL PRODUCTS AND PHARMACEUTICALS.

Medical products and pharmaceuticals must conform to certain minimum physico-chemical microbiological and biological requirements. The biological requirements comprise principally testing for toxicity, safety, and pyrogens. Besides the above-mentioned criteria, there are two other important characteristics, viz. clinical efficacy and tolerance in animal and/or human beings. These latter requirements, expected from the final product released for general human use, are not carried out routinely. In the present-day pharmaceutical and medical technology, numerous new products are appearing, many of them requiring radiation sterilization as they are not stable towards the other conventional methods of sterilization. While the post-irradiation physico-chemical changes in a product may not be significant, the bio-availability or biological activity of the product may be affected to a more significant extent. Some evidence of this has recently been reported. Hence, it is desirable to carry out studies on safety as well as clinical efficacy on irradiated products. This paper describes some studies on plastic-based medical products, a pharmaceutical raw material, a typical infusion fluid such as normal saline, antibiotics and their ointments.

1. INTRODUCTION

Radiation sterilization is a technology, which has rapidly come into the field of pharmaceutics and medicine, to provide a means for efficient sterilization of heat-sensitive products in their final packaging. Today, physicians and surgeons employ many types of medical products and devices in routine use (Table I). The fact that radiation sterilization can be carried out on the final packaged medical products and devices is highly reassuring to physicians and surgeons as they are relieved of the tedium and responsibility of having to carry out routine sterilization.

Many of these medical products are made of plastics such as polyethylene, polyvinyl chloride, polypropylene, polystyrene, polyethylene-terephthalate and nylons.

In pharmaceutics, plastic materials are employed as containers for oral or topically applied drugs, injectibles and large-volume infusions.

The plastics are often not pure polymers but compounded with a variety of additives such as plasticizers, stabilizers, anti-oxidants, etc. The possibility of release of these additives either into a medicament contained by them, or into the body tissues, strongly influences their choice for intended use.

TABLE I. LIST OF MEDICAL PRODUCTS

Absorbent cotton-wool and its formulations
Polyurethane foam burn dressings
Rubber and plastic sheets
Medicated tapes
Examination and surgical gloves
Absorbent powder for surgical dressings
Absorbable surgical sutures (catgut — chromated and plain)
Non-absorbable surgical sutures (silk, nylon, polystyrene, metallic wires, etc.)
Irrigation sets
Mucous extractors
Medical tubings (duodenal tubes, Karr's tubes, neonatal tubes, resuscitation tubes, etc.)
Arterial and venous sets for haemodialysis
Blood administration set
Electrolyte infusion set
Catheters (oxygen, intravenous, urethral, cardiac, etc.)
Drainage bags
Pump units
Probes
Heart valve and other prostheses
Surgical instruments like spatula, razor blades, needles, scalpel, scissors, saw, etc.
Disposable syringes, etc.

The USP XVIII, National Formulary XIII, British Pharmacopoeia 1973, and several national pharmacopoeias have formulated physico-chemical, biological, pyrogen and safety tests to serve as guidelines in the evaluation of their acceptability for safe use. Many of the plastic products are thermolabile. The gas-sterilization process requires careful control of a number of process parameters, an some concern is expressed about possible toxicity of residual ethylene oxide. In the light of those considerations plastic products are now considered patently suitable for radiation sterilization. The thermolability and the not-too-easy and safe means of gas sterilization single the products out as patently suitable for radiation sterilization.

Radiation appears also to be the only means of effecting sterilization of a number of pharmaceutically important heat-sensitive powders, ointments and solutions in their final containers. Most of the pharmacopoeias recommend a dose of 2.5 Mrad of ionizing radiation for the purpose of sterilization. Some pharmaceutical raw materials, e.g., reserpina powder, may be radiation pasteurized to upgrade them from a microbiological aspect for export purposes.

As it is well known that radiation interacts with matter by ionization and excitation, it is likely that some degree of change in the quality may result from irradiation. These changes may be insignificant physico-chemically or biologically, or may be observable only by either of the two techniques. Over the last two decades studies in other countries have shown that products such as benzyl penicillin, chloromycetin, streptomycin etc. in dry state, are stable to radiation from the pharmacopoeial aspect [1, 2].

Pochapinskii et al.[3] have studied irradiation effects (2 Mrad) on streptomycin sulphate, its $CaCl_2$ complex, dihydro streptomycin sulphate, PAS-mycin, streptopenicillin and paromycin, and observe that the antibiotic activity, therapeutical index, toxicity etc., of all except lyophilized streptomycin and dihydrostreptomycin sulphates remain unchanged. They note that slight discoloration of dry antibiotics has no effect. They have also irradiated tetracycline, oxy- and chlortetracyclin to 2 Mrad and report that the chemical composition and biological activity remained unchanged and ointments of these antibiotics were stable for 14 months at room temperature.

Radovic [4] has shown that oxy-tetracyclin is quite resistant to chemical and biological changes when irradiation is carried out in dry state from 0.5 - 10 Mrad. Crippa [5] has irradiated rifampicin at 2.5 Mrad and finds that the ESR results show a negligible effect and confirm the possibility of using radiation sterilization of rifampicin.

Polyenic antibiotics like levorin, nystatin, amphotericin B, Na-levorin, Na-nystatin have been studied by Tsygonov et al. [6], by subjecting them to 1 Mrad and they report that irradiation decreases the potency by 10% and that there was no difference in toxicity between irradiated and non-irradiated samples. Similar results were observed on 1-year-old irradiated samples and they expressed the view that gamma radiation can be used to sterilize these antibiotics without loss of biological activity [7].

Bacitracin-Mn and Bacitracin-Zn have been studied by Baldwin [7] at 1.25 Mrad, the control and irradiated samples assaying 59 and 56 units/g.

However, the off-white shade of some irradiated products appears to be a crucial fact inhibiting utilization of this technology for terminal sterilization of products, which cannot be sterilized by any other means. If irradiated products are still of pharmacopoeial quality and if they are non-toxic and maintain their biological characteristics and bio-availability, there ought to be no hesitation in using radiation as a means of sterilization.

This paper deals with some studies on the alteration in safety or efficacy of (a) eye ointments, (b) infusion fluids and infusion sets, and (c) crude vegetable drugs after radiation sterilization.

2. MATERIALS AND METHOD

Materials in the study include:

(a) (1) Penicillin eye ointment
 (2) Tetracyclin eye ointment
 (3) Chloramphenicol eye ointment

(b) Infusion sets made by three different manufacturers A, B and C, including infusion fluid (sodium chloride) in the case of A.

(c) Serpina powder from a commercial supplier.
(The materials in their primary containers were irradiated with ^{60}Co gamma radiation to a dose of 2.5 Mrad)

The following animal and human tests were performed:

(i) Local irritation test on rabbit eye, using six rabbits for each test;
Microbiological assay for antibiotic content;
Clinical efficacy and tolerance study in patients with conjunctivitis due to susceptible organisms, using 10 patients for each study.
(ii) Animal safety test according to BP, 1973;
Human volunteer infusion study on six healthy volunteers for each study.
(iii) Chemical estimation of total alkaloids and hypotensive effect on anaesthetized dog in comparison with reserpine.

3. RESULTS

Table II gives the results obtained with antibiotic eye ointment with regard to local irritation, antibiotic content and clinical efficacy. It is seen that irradiated penicillin, tetracyclin and chloramphenicol eye ointments retain their full efficacy and are free from toxicity.

Table III gives the results of a study on irradiated disposable infusion sets and infusion fluids. It is seen that irradiated infusion of normal saline is well tolerated and hence is suitable for sterilization by this method.

Table IV shows the total alkaloid content and hypotensive potency of irradiated serpina powder.

TABLE II. LOCAL TOLERANCE AND CLINICAL EFFICACY OF EYE OINTMENTS AFTER RADIATION STERILIZATION, COMPARED WITH CONVENTIONAL PRODUCTS

	Local tolerance: rabbit eye			Antibiotic content of stated potency (%)	Average time required for cure (d)	Local tolerance
	Conjunctival congestion	Cornea	Corneal reflex			
Penicillin eye ointment	Absent / Absent	normal / normal	present / present	105 / 110	4 / 3-2	good / good
Tetracyclin eye ointment	Absent / Absent	normal / normal	present / present	98 / 115	2-6 / 3	good / good
Chloramphenicol eye ointment	± / ±	normal / normal	present / present	109 / 110	3-1 / 3-5	good / good

NB. The denominator represents the conventional product.

TABLE III. ANIMAL TOXICITY AND HUMAN SAFETY STUDIES ON IRRADIATED INFUSION ADMINISTERED THROUGH IRRADIATED PLASTIC INFUSION SETS OF THREE MANUFACTURERS

Manufacturers	Animal toxicity test		Human safety test			Any other adverse reaction
	Change in B.P.[a]	Alteration in carotid reflex	Change in B.P.	Change in pulse rate	Rigours	
A	+ 2 mm	nil	nil	nil	Absent	nil
B	nil	nil	nil	nil	Absent	nil
C	nil	nil	nil	nil	Absent	nil

[a] B.P. = blood pressure.

TABLE IV. MEAN ALKALOIDAL CONTENT AND HYPOTENSIVE ACTIVITY OF POWDERED RAUVOLFIA ROOT AFTER IRRADIATION

	Alkaloidal content (mg/g)	Peakfall in B.P.[a] (mm)	Duration of fall in B.P. (min)	Potency in comparison to reserpine (%)
Non-irradiated	6.2	18	36	70
Irradiated	6.4	16	31	70

[a] B.P. = blood pressure

4. DISCUSSION

Irradiation of an organic substance may cause certain alterations in their structures and properties and hence it is necessary to test the irradiated products for their pharmacopoeial quality, toxicity, and biological characteristics/bioavailability. In the present study we selected three groups, which are products most likely to lend themselves to radiation sterilization. Amongst the drugs we tested were eye ointments containing antibiotics like penicillin, tetracyclin and chloramphenicol. These were tested for local tolerance in rabbit eye and as can be seen from Table II they are well tolerated and appear free from local toxicity. Clinically, these antibiotic ointments were used in patients suffering from conjunctivitis. Sensitivity of the causative organisms to the antibiotic in question was simultaneously tested, and failures in cases where organisms were not sensitive to the antibiotic were eliminated from consideration. The average period for cure and local tolerance were tested by examining the patients every day. It can be seen that there is also no difference in the conventional and irradiated ointments in this respect.

Similarly from Table III, it can be seen that radiation-sterilized infusion sets did not reveal undue toxicity when tested according to BP 1973. When the irradiated infusions were given through these sets to normal volunteers, these were also well tolerated. It is necessary to note here that some of the sets sterilized thus had undergone discoloration, but this did not affect their safety.

Table IV shows the effects of radiation sterilization on a crude preparation of the rauvolfia root powder. This preparation was also stable to radiation as seen by the unchanged alkaloidal content and hypotensive potency in dogs vis-à-vis reserpine.

Thus it appears that radiation sterilization may be used for a variety of medical products, but in each case the safety and efficacy will have to be ascertained. Also in the present study, the evaluation was done soon after irradiation. There is a possibility that the effects of irradiation may be slow in becoming apparent and hence it is necessary to study the post-irradiation shelf life of such products.

REFERENCES

[1] PHILLIPS, G.O., in Manual on Radiation Sterilization of Medical and Biological Materials, Tech.Rep.Ser. Ser.No.149, IAEA, Vienna (1973) 208.

[2] BRIDGES, B.A., POWELL, D.B., in Massive Radiation Techniques, (JEFFERSON, S., Ed.), George Newnes Ltd. (1964).

[3] POCHAPINSKII, V.I., et al., Chem.Abstr. 56 (1962) 10296e; 58 (1963) 3272.

[4] RADOVIC, G., et al., Chem.Abstr. 76 (1972) 103730 m.

[5] CRIPPA, P.R., Nucl.Sci.Abstr. 28 (1973) 586.

[6] TSYGANOV, V.A., et al., Chem.Abstr. 74 (1971) 795554.

[7] BALDWIN, R.S., Chem.Abstr. 80 (1974) 112688.

DISCUSSION

on papers IAEA-SM-192/6 and IAEA-SM-192/87

A. KEALL: Mr. Kulkarni, how many animals were used in your toxicity test and in the clinical trial was a double-blind exercise undertaken?

R.D. KULKARNI: A minimum of six and maximum of ten animals and human beings were used. We did not make the double-blind test.

R.F. ARMBRUST: In view of the European Pharmacopoeia Commission's requirement that all substances used for preservation must be sterile, I was very glad to hear from Mr. Gopal's presentation that they can be sterilized in the dry solid state with 2.5 Mrad.

K.H. CHADWICK: Mr. Gopal, how pure are the chemicals which you are irradiating and how reproducible are these products from batch to batch? Could you give a few more details on the toxicity testing which you plan for these irradiated products.

N.G.S. GOPAL: We assessed the purity of chemicals from the point of view of pharmacopoeial requirements. We took samples in duplicate, triplicate and quadruplicate for the toxicity testing. The results were statistically analysed to deduce the associated confidence level. As regards toxicity tests, I think Mr. Kulkarni can answer your question.

R.D. KULKARNI: Apart from the specific tests mentioned in the Pharmacopoeia, we are planning to carry out acute and chronic toxicity tests. However, since they involve so much expense, we may not make them on a routine basis.

Nazly HILMY: Mr. Gopal, I would like to know whether you have done any work on the microbiological aspects, such as isolation and identification, radiosensitivity and pyrogen sensitivity, of the microorganisms present in the pharmaceuticals/medical devices. Such studies might be helpful in selecting the minimum sterilization dose for tropical countries.

N.G.S. GOPAL: We normally isolate the microorganisms, but we try to identify only the species, not the strain. We are thinking however of pursuing more detailed studies, of the type you mention. Perhaps Mrs. Patel, who is working in this field, could add something?

Kikiben PATEL: We have isolated and identified microorganisms present in pharmaceuticals such as penicillin, eye ointment, lactose and gum karaya. They include *Bacillus subtilis*, *Staphylococcus citreus*, *Staphylococcus albus*, flavobacterium species, as well as mould and yeast. These are all radiosensitive, except for *B. subtilis*, which exhibits the same radioresistance as *Bacillus pumilus 601*.

P.V. IYER: Mr. Gopal, you indicated that the aqueous sodium chloride solution meant for infusion was found clinically satisfactory after irradiation. I would like to caution that in the case of intravenous infusion solutions of large volume, in addition to sterility, non-pyrogenicity should also be ensured. I am wondering what the effect of gamma rays in on pyrogens in the case of a dose of 2.5 Mrad.

N.G.S. GOPAL: It is stated in the literature, that a dose of 2.5 Mrad is required to inactivate pyrogens. If pyrogens appear after the solution has been prepared, then, of course, 2.5 Mrad would not be enough to remove them. Clearly, the sooner the solution is irradiated after preparation, the better.

General discussion

V.K. IYA: Mr. Chairman, many of us here would be interested to hear about the general situation regarding the radiation sterilization of pharmaceuticals in present-day European industrial practice.

G.O. PHILLIPS (Chairman): Perhaps Mr. Blackburn could make a few remarks on this.

R. BLACKBURN: I would like to say that the pharmacopoeial requirements seem to place too much stress on spectroscopic measurements. While these might be acceptable in the case of routine processing, I don't think they are suitable as analytical tools in radiation chemical breakdown. With substances like phenol mercurics, for instance, it is easy to register absence of change, since the bulk of any change would be due to the π-aromatic ring, which is present before and after irradiation. I would imagine that more extensive application of, for example, thin-layer chromatography, actual separation of the products, and the use of isotope-labelled compounds would be a better analytical approach in the laboratory for the basic work.

G.O. PHILLIPS: Mr. Gaughran, what is the situation in the United States of America?

E.R.L. GAUGHRAN: In the United States the irradiation of drugs is strongly discouraged by the Food and Drug Administration. Unofficially, however, certain members of the Administration have expressed the opinion that, in the case of a drug which must be sterile, but can now only be sterilized by filtration and aseptically packaged, irradiation in the final container would offer a more desirable alternative. To my knowledge, only limited use has been made of irradiation for the sterilization of ophthalmic ointments.

G.O. PHILLIPS: One of the present problems in testing irradiated pharmaceuticals is to find a very small quantity of material in a 100% solid or 100% solution. The method that has revolutionized this field is gas-liquid chromatography linked with mass spectrometry. The detection of extremely small amounts of material is now possible by this technique, together with identification, and I know that many of the drug companies have purchased the requisite equipment.

I would like to urge, however, that when expressing decomposition we should not use percentage since the latter does not mean very much in radiation chemical terms. The quantity we really want to know is the number of molecules changed and the amount of energy input — these are what are actually indicative of the true state of affairs. It would therefore be helpful if we could switch from percentage decomposition to -G value, even if only within broad limits.

K.H. WALLHÄUSSER: The first supplement to the European Pharmacopoeia contains some data on the purity of tetracyclines, including the limits for epi-tetracycline anhydro-tetracycline and epi-anhydro-tetracycline. This is, I think, the first time these figures have been published. With regard to the stability of ointments, we have done experiments in the Federal Republic of Germany showing that the stability of tetracycline ointments is very good over a long period. It would seem that there is no need for the radiation sterilization of ointments.

On the other hand, when working with emulsions or creams one encounters certain problems. For example, directly after sterilization there is 90% potency, but this drops to 40% three months later, and to less than 10% after six months. Consequently, this type of sterilization is unsuitable for pharmaceuticals of this kind and a new approach, such as working in aseptic conditions, will be required if we are to have sterile products.

E.P. PAVLOV: When discussing the subject of radiation sterilization of medical products, I think we ought to give a little more attention to such important aspects of the matter as changes in the organoleptic properties of the irradiated materials. The point is that a number of pharmacopoeias, including the Soviet one, contain rigid requirements as to colour, hence difficulties arise when the products change in appearance through the action of radiation. For example, certain semi-synthetic penicillins when sterilized undergo this change. Although these antibiotics are highly radiation-resistant, the question of what recommendations should be made for that radiation sterilization in industry has not yet been settled; a deterioration in the appearance of the commercial product would seem to be the main difficulty.

A second point is the study of the ageing of irradiated drugs. The effect of radiation as an artificial ageing factor in a variety of materials is well known. I think this fact should be given attention and careful study made of the properties of irradiated products during storage.

REPORTS ON CURRENT STATE OF RADIATION STERILIZATION OF MEDICAL PRODUCTS IN MEMBER STATES

(Sessions VII and VIII)

Chairman (Session VII): Pamela A. WILLS (Australia)
Chairman (Session VIII): J.O. DAWSON (United Kingdom)

FACTORS INVOLVED IN PLANNING RADIATION-STERILIZATION PRACTICES AND TECHNOLOGY IN THE DEVELOPING COUNTRIES, AND THE AGENCY'S PROMOTIONAL ROLE

R. N. MUKHERJEE, H. C. YUAN
International Atomic Energy Agency,
Vienna, Austria

Abstract

FACTORS INVOLVED IN PLANNING RADIATION-STERILIZATION PRACTICES AND TECHNOLOGY IN THE DEVELOPING COUNTRIES, AND THE AGENCY'S PROMOTIONAL ROLE.

The application of ionizing radiation for sterilizing ready-to-use medical supplies, sutures and grafts provides a broad scope for the up-grading of public health care and family planning programmes in the developing countries. Sterile ready-to-use medical supplies become particularly important for improving the standard of those services given through the improvised camp-hospitals and mobile medical units for the remote areas of such countries, if needed. The practices generated in the technologically advanced countries will form the basis of the planning, but the necessary adjustments should be made in their implementation to suit best the local conditions and needs and to promote utilization of local raw materials. Necessary research and development and an effective infrastructure should be emphasized. Plastic materials are among the major pollutants of the environment. Timely parallel practical steps need be adopted and an action programme planned to preserve the quality of the human environment.

INTRODUCTION

In a number of technologically advanced countries the sterilization of medical products by ionizing radiation is a well-established industrial process. However, considering that this technology was initiated in the United States of America and the United Kingdom during the late 'fifties and early 'sixties, it can in no way be considered as old. Over a period of fifteen years the development of as many as about sixty industrial-scale radiation-sterilization facilities throughout the world is indeed encouraging for the great success of this technology and the public acceptance of the processed medical supplies for "sterile" use.

The advantages of radiation-sterilization practices for pre-packed hermetically sealed medical supplies in terms of products' safety as well as process control have been amply covered at this Symposium. Many advantages of this "cold" sterilization process undoubtedly qualify this as the best sterilization method. Furthermore, it is the only method of choice for a wide range of medical supplies in to-day's market which are made of materials sensitive to heat and chemicals and hence cannot be sterilized by conventional methods.

During the last five years there has been a growing desire in a large number of developing countries for the introduction of radiation-sterilization practices in their medical supply manufacturing industries. Some of those countries already have operating industrial-scale or pilot-scale demonstration facilities on production, while others are in the process of implementation, and still many others are at the planning and preparatory stages.

TABLE I. WORLD INVENTORY OF GAMMA IRRADIATORS FOR STERILIZATION OF MEDICAL PRODUCTS

Location	Number	Maximum capacity (MCi)	Commissioning date
Argentina	1	1.0	1970
Australia	3	4.0	1960-72
Belgium	1	0.25	1973
Brazil	1	0.5	1972
Canada	4	2.0	1964-74
Czechoslovakia	1	1.0	1972
Denmark	2	2.0	1969-72
* Egypt	1	0.5	
France	4	3.3	1960-73
Germany, Dem. Rep. of	1	0.5	1967
Germany, Fed. Rep. of	3	3.0	1966-68
Greece	1	0.5	1973
* Hungary	1	0.5	1975
India	1	1.0	1974
Ireland	1	1.0	
Israel	1	1.0	1972
Italy	5	3.5	1967-74
Japan	3	1.5	1970-73
* Korea (South)	1	0.3	1975
Netherlands	2	1.5	1970-73
New Zealand	1	1.0	1966
Norway	1	0.12	1970
Poland	1	0.25	1972
South Africa	1	1.0	1971
Spain	1	0.330	1971
Sweden	2	2.0	1968-71
Switzerland	1	0.5	1972
United Kingdom	10	6.5	1960-73
United States of America	8	6.0	1964-74
USSR	1	1.0	1974
	65	47.5	

Two sessions of this Symposium are devoted to the presentation of current status reports by the participants from the host country, India, and from Israel, Philippines, Korea, Indonesia, Arab Republic of Egypt, Argentina and Bangladesh, and followed by general discussions. The major consideration underlying the introduction of radiation sterilization in these countries is to up-grade the standard of existing public health and medical care systems through the availability of ready-to-use disposable medical supplies in the countries' hospitals, clinics, village health centres and camps for the campaign for family planning programme.

This paper, therefore, is primarily limited to the enumeration of some general factors and the preparatory steps which need to be considered to introduce radiation-sterilization practices and technology into developing countries. This is followed by reports from several developing countries on their practices of radiation sterilization.

RADIATION STERILIZATION IN THE PIONEERING COUNTRIES

For the analysis of factors significant in the promotion of radiation sterilization of medical supplies in the developing countries, a retrospective survey of the similar developments in the technologically advanced countries seems to be pertinent. Broadly speaking, during the 'fifties a rapid development of nuclear technology and electronics engineering led to the appearance, respectively, of high-energy gamma-emitting ^{60}Co-radioisotope sources, and electron accelerators. These radiation sources were utilized extensively to investigate the effects of penetrating high-energy ionizing radiations on the microorganisms as well as on other biological systems and organic macromolecules of biological importance.

Around the same time the advent of synthetic plastic polymer technology played an extremely important role by showing considerable promise for the mass production of health care items suitable for replacing many of the then existing expensive, repeated-use medical supplies. The availability of a penetrating radiation source with its efficient microbicidal properties as well as the disposable plastic medical supplies pre-packaged with plastic materials caught the imagination for mass industrial production of radio-sterilized ready-to-use items. This new technology in the medical manufacturing industry heralded a new era for the history of health care.

The traditional practices of hospital sterilization of reuse items thus found a better, safer, and labour-saving substitute among the provision of single-use disposables sterilized by ionizing radiation in a hermetically sealed condition. The practice significantly contributed towards reducing the risk of health hazard through cross-contamination (nosocomial infection) which frequently accompanied the reuse of medical supplies. The manufacturing concerns promptly responded to these needs through the large-scale production of single-use items at reasonable cost in good quality and employing reliable and controlled sterilization processes. The practical outcome of these is illustrated by the existence to-date of over sixty industrial-scale ^{60}Co-facilities (Table I) and about ten electron accelerators (Table II) round the world engaged in the sterilization of a rapidly growing list of single-use medical supplies, sutures, implants and tissue allografts for therapeutic aid and surgery to restore human health and welfare.

TABLE II. ACCELERATOR PLANTS FOR RADIOSTERILIZATION OF MEDICAL PRODUCTS

Location	Number	Type of accelerator	Maximum energy of electrons (MeV)	Power output (kW)	Date of commissioning
Denmark	2				
Risö		LINAC (Varian Ass.)	10	5	1960
Glostrup nr. Copenhagen		LINAC (Varian Ass.)	10	10 - 15	1967
France	1				
Corbeville nr. Saclay		LINAC CIRCE 10	6 - 10	7 - 10	1967
Germany, Fed. Rep. of	1				
Cologne		Van de Graaff	3	6	1957
Poland	1				
Warsaw		LUE 13-9	13	9 - 13	1972
United Kingdom	1				
Birmingham		Van de Graaff	4		1963
United States of America	4				
Radiation Dynamics Inc.					
High Voltage Eng. Corp.					
Haimson Res. Corp.					
Natic Labs., Mass.					
USSR	1				
Kurgan		LUE 8-5B	8 - 10	5	1974
	10				

MAJOR CONSIDERATIONS FOR RADIATION STERILIZATION IN THE DEVELOPING COUNTRIES

As in any sector of industry there are problems to be solved relating to the feasibility of implementing the radiation-sterilization methods in the manufacture of medical supplies in a developing country. A satisfactory answer to that aspect relies largely upon the determination of the types of medical supplies used by the public health services and the medical profession; the current and projected size of the average annual consumption of the different products; the attitude of the local manufacturers of medical products towards the adoption of a new sterilization technique; legal and public health clearance aspects of the radiosterilized products and, above all, the cost of the final sterile products. These in turn will dictate the size, design and the operational policy of the irradiation facility as well as the requirement of the various categories of trained technical and maintenance staff to run the project. The results of these assessments will also help defining the extent and nature of the supporting research for further process control and development. Although the situation may vary widely from one country to the other, the following considerations may be generally applicable:

A. Prospective aspects

(1) Market survey of medical supplies in the national health care system
(2) Market survey of customers for the irradiator facilities
(3) Survey of the plastics and other locally available raw material resources
(4) Market promotion strategies including relevant educational and publicity services
(5) Steps pertinent to the improvement of hygienic conditions of the manufacturing sites
(6) Cost and quality estimates for the products
(7) National code of practice and inclusion in the pharmacopeia
(8) Active liaison with the National Public Health Authorities

B. Technical and engineering aspects

(1) Planning and design and site selection
(2) Radiation source and capacity
(3) Conveyor parameters
(4) Sterilizing dose
(5) Dosimetry calibration and efficiency control of the facility on commissioning and in routine operation
(6) Safety devices and measures

C. Auxiliary activities leading to process development

(1) Monitoring of microorganisms in the premises and on products prior to sterilization
(2) Relevant radiobiological and radiation microbiology investigations
(3) Research on radiation chemistry of component materials
(4) Inclusion of topics in the academic curriculum of the technical and medical universities

In the rest of this paper only selected important factors relevant for the developing countries' interest will be discussed in detail while others will be briefly touched upon.

Market survey

A market survey is among the most essential initial pre-investment steps. Information is vital on the various items of medical supplies including devices, sutures, implants and pharmaceuticals in use and to be introduced in the country's health care services. The survey should include the names of the manufacturing concerns and the types and quantities of the different products, suitable for radiation sterilization, which they manufacture annually. The existing number of hospitals and the patient beds as well as their projected expansion and demand for the supplies are all the components of the estimated size of the product throughput to be processed by radiation sterilization.

The location of the radiation sterilization facility should be selected in the region having a maximum concentration of the manufacturing medical supplies industries. Ease and the low cost of transport of the products to the irradiator facility adds to the economic feasibility of the process.

Maximum efforts to utilize indigenous raw materials

Local availability of some or most of the raw materials including the various types of plastics for manufacturing and packaging materials should play a major role towards the feasibility of implementing this technology at a cheaper cost. However, the introduction of a radiation-sterilization facility in a country with plastic manufacturing capability should boost the growth of the latter industry.

The plastics industry might suffer owing to the current world petroleum crisis both in terms of availability and cost. This aspect should be taken into consideration and evaluated during the planning stages. The authorities responsible for the distribution of industrial raw materials should adjust policies to safeguard an uninterrupted supply of plastics to the manufacturers of disposable medical devices.

Cost considerations of radiosterilized medical supplies

Cost of the finished products is one of the most important considerations for the feasibility of the involved manufacturing technology. However, with regard to a superior quality medical supply ensuring safe public health care, such feasibility aspects involve some additional considerations. The potential savings from the waste of the valuable time of doctors and nurses through the application of ready-to-use supplies, and of the patients through the added safety against nosocomial infection and undesirable occupancy of hospital beds, cannot be evaluated against any small surcharge of the product cost. The experiences of a number of countries using radiosterilized single-use medical supplies leave little doubt that the overall economics of the process is satisfactory.

Microbiological problems

Since the purpose of a sterilization operation is the destruction of all forms of contaminating microorganisms, it is essential to gain a thorough knowledge of the composition and characteristics of the local microflora. Before a facility is commissioned such microbiological parameters, with regard to the production sites as well as of the pre-sterilization medical supplies and the raw materials, should be established. A number of microbiological surveys with the resulting data have been reported (Christensen et al. [1]; Kronenwett [2]; Tattersall [3] which can serve as the model system for organizing others.

The hot humid climate of the tropics is most favourable for extensive microbial growth. Furthermore, the composition of the flora may differ significantly from those countries where surveys have been undertaken. Some available literature reports many species of bacteria, viruses, a great preponderance of moulds and other parasitic protozoans.

Sterilizing dose

The question of sterilization dose and its scientific basis has been the subject of great controversy. Opinions and acceptance standards differ widely between countries. Generally speaking it may be stated that mainly at the pre-commissioning stage and also as routine the sterilization reliability may be established in a variety of ways. None of these alone may give the desired sterility assurance, but in combination they do give:

(1) Positive knowledge of plant hygiene conditions
(2) Measurement of contamination of products prior to sterilization by numbers and types
(3) Testing of the various contaminating microorganisms from pre-sterilization products for their radiosensitivity
(4) Use of biological indicators to challenge the process routinely.

The generally accepted index of sterilization is the D_{10} value defined as the radiation dose required to destroy 90% of the population. This is a measure of the radiation response of a microorganism species. Standard microbial preparations (biological indicators) using spores and vegetative forms of radioresistant microorganisms are used to determine the minimum sterilizing dose.

In the calculation of the processing cycle for a product to be radiation sterilized, the total concentration of microorganisms in the batch to be sterilized determines the number of D-values (the total radiation dose) needed. If on the average the items have a contamination level of 100 organisms per unit (established through the hygienic standard of the production site) and there are 10^4 items, it requires only six D-values to reduce this level of contamination to one survivor. The usually accepted probability of a survivor in sterile medical supplies is less than one in a million. This means that an additional six D-values should be added to enhance the safety assurance of the processing cycle.

If one assumes that the radiation resistance of Bacillus pumilis (indicator microorganism) is on the average greater than that of the general contaminating population, then the radiation D-value of 0.2 Mrad for this

TABLE III. THE D_{10}-VALUES (Mrad) OBTAINED WITH SPORES OF *Bacillus pumilus* E.601 DRIED FROM VARIOUS SUSPENDING MEDIA ON TO DIFFERENT SUPPORTING SURFACES AND IRRADIATED IN THE PRESENCE OF AIR (Burt and Ley, 1963 [4])

Supporting surface	Suspending medium				
	Buffer	Nutrient broth	Serum	PA solution	Greasy base
Glass	0.169 (0.159 - 0.180)	0.156 (0.143 - 0.170)	0.153 (0.147 - 0.158)	0.282 (0.266 - 0.300)	0.342 (0.321 - 0.366)
Perspex	0.159 (0.152 - 0.167)	0.151 (0.141 - 0.161)	- -	0.321 (0.299 - 0.346)	- -
Polystyrene	0.178 (0.171 - 0.185)	- -	- -	- -	- -
Aluminium	0.165 (0.158 - 0.171)	0.144 (0.136 - 0.151)	0.154 (0.149 - 0.160)	0.283 (0.272 - 0.295)	- -
Stainless steel	0.155 (0.148 - 0.161)	- -	- -	- -	- -
PA film	0.211 (0.196 - 0.228)	0.324 (0.293 - 0.361)	0.329 (0.296 - 0.370)	0.363 (0.348 - 0.380)	- -

TABLE IV. SURVIVAL OF SPORES DRIED ON TO PAPER DISCS OR GUT SAMPLES (BARTHA et al. [5])

	Per cent samples giving positive cultures after irradiation with						
	(control)	0.5 Mrad	1.0 Mrad	1.5 Mrad	2.0 Mrad	2.5 Mrad	3.0 Mrad
Paper discs							
B. pumilus E 601	100	100	90	-	-	-	-
B. subtilis subsp. niger NCTC 10073	100	100	-	-	-	-	-
B. stearothermophilus NCIB 8919	100	100	70	-	-	-	-
Gut							
B. pumilus E 601	100	100	100	70	30	10	-
B. subtilis subsp. niger NCTC 10073	100	100	100	40	10	-	-
B. stearothermophilus NCIB 8919	100	100	100	60	10	-	-

TABLE V. INFLUENCE OF ENVIRONMENT DURING IRRADIATION ON THE DOSE REQUIRED FOR REDUCTION BY A FACTOR OF 10^5 OF POPULATIONS OF PERTINENT ORGANISMS (LEY, F.J. [6])

Organism	Dose (Mrad)	Conditions during irradiation
S. typhimurium	0.1	Buffer suspension in air
	0.3	Buffer suspension anoxic
	0.1	Dried on to kaolin powder in air
	0.5	Dried in bone meal in air
Strept. faecium $A_2 1$	0.6	Buffer suspension in air
	1.9	Buffer suspension anoxic
	0.7	Dried on to kaolin powder in air
	3.0	Dried in serum in air
B. pumilus spores	0.9	Buffer suspension in air
	1.5	Buffer suspension anoxic
	0.7	Dried on to kaolin powder in air
	1.7	In grease in air

organism, when multiplied by 12 D-values gives rise to a total dose of 2.4 Mrad. Most medical supplies in the United States of America, United Kingdom and many other countries are processed at a dose level of 2.5 Mrad. On the contrary, the Scandinavian countries recommend a total dose of 4.5 Mrad for radiation-sterilized medical supplies, on the basis of a different and more radioresistant indicator microorganism, Streptococcus faecium with a D-value of 0.3-0.4 Mrad.

The striking modifying effects of irradiation environments on the radiation response of microorganisms have been reported (Tables III, IV, V). Much further research is needed for the development of some suitable biological indicators representative of the local contaminants and the dose level to be determined with due regard to their radiation response at the environmental conditions of the industrial operations. This approach might in future result in a recommended minimum sterilizing radiation dose even lower than the currently prescribed 2.5 Mrad and at the same time maintain the safety assurance level required by the public health authorities.

TECHNOLOGICAL ASPECTS

Radiation source and conveyor parameters

In the industrial and pilot-plant-scale operations of radiation sterilization of medical products, gamma rays from ^{60}Co have been most frequently utilized both in the developing and developed countries although this sterilization technology began with the electrical machines such as the

Van de Graaff linear accelerators, set up by the Ethicon Inc. in the early 1950s. The merits and demerits of the different radiation sources are discussed in the paper of Mr. K.H. Morganstern[1] and by others. For the conditions of the many developing countries the experts generally recommend ^{60}Co gamma-irradiation sources in either dry or wet storage system.

So far, experiences in many countries have indicated that the dependence on the large-scale availability of skilled engineers and technicians is less for the radioisotope sources than for the electron accelerators in the course of its routine operation, maintenance and servicing in the event of breakdown.

Furthermore, one must consider the fact that the facilities in the developing countries most likely have to provide service sterilization to a number of different manufacturers and are therefore expected to deal with a number of products specifications. Under such circumstances the source geometry and the deep penetrating nature of the high-energy emitted radiation should allow a greater flexibility in volumes, shapes and sizes of the carton boxes containing the prepacked medical supplies. Therefore, the choice of a ^{60}Co source would be preferable.

The conveyor system may be preferable to move the hanging packaged cartons through the irradiation cell containing the ^{60}Co source plaque in several cycles in a multipass. The conveyor parameters should include a provision for the variation of its speed. The utility of a number of physical chemical dosimeters in the routine monitoring of the source efficiency and the delivery of the prescribed minimum sterilizing dose has been extensively reviewed at this Symposium in a number of papers.

The output storage conveyor should be designed to lead the finished goods in the storage area well isolated from that for the input items to avoid accidental admixture and consequent health hazards.

Generally recommended steps of radiation health safety with regard to the source operation as well as with the surrounding environment should be adopted. Automated interlocking devices, as well as the shutdown of the source in the event of any irregularity, should be installed. Irradiated sterile product boxes must bear appropriate labels with distinctive colour for easy identification.

AGENCY'S PROMOTIONAL ROLE

During the last ten years the Agency's Life Sciences programme has emphasized the beneficial applications of radiobiological effects in the developing and developed Member States. The subject of sterilization of medical products and biological tissues has had top priority in those programmes. Through the organization of symposia, conferences, panels and working group meetings held at appropriate intervals and geographical locations, the latest information accumulated at the leading institutes in the Member States was discussed and attempts for further promotion were made in other areas. Table VI lists those meetings held to date.

The occasion of the present symposium is indeed an excellent illustration of the role that such international meetings can play in fostering the dissemination of latest technical information and generation of new experience

---

[1] IAEA-SM-192/8, these Proceedings.

TABLE VI. IAEA SCIENTIFIC MEETINGS AND PUBLICATIONS

Title of Meeting	Date and place	Proceedings published
Application of Large Radiation Sources in Industry and Especially to Chemical Processes	8-12 September, 1959 Warsaw	1960
Application of Large Radiation Sources in Industry	27-31 May, 1963 Salzburg	1963
Radiosterilization of Medical Products, Pharmaceuticals and Bioproducts	17-19 January, 1966 Vienna	1967
Code of Practice for the Radio-sterilization of Medical Products	5-9 December, 1966 Vienna	1967
Radiosterilization of Medical Products	5-9 June, 1967 Budapest	1967
Radiation Sterilization of Biological Tissues for Transplantation	16-20 June, 1969 Budapest	1970
Utilization of Large Radiation Sources and Accelerators in Industrial Processing	18-22 August, 1969 Munich	1969
Radiation Sterilization of Medical Products, Pharmaceuticals and Biological Tissues	22-23 November, 1971 Risö	-
Revision of the IAEA Recommended Code of Practice for the Radiation Sterilization of Medical Products	5-9 June, 1972 Risö	1974
Radiation Sterilization of Medical Products, Pharmaceuticals and Biological Tissues	14-16 February, 1973 Budapest	-
Manual on Radiation Sterilization of Medical Products		1973
Ionizing Radiation for Sterilization of Medical Products and Biological Tissues	9-13 December, 1974 Bombay	1975

among the various Member States. A total of about 150 participants and observers from 32 Member States (including 19 developing ones) are participating to discuss all aspects of the processes concerned. Many of the specific problems encountered by some of the experts are being jointly reviewed for possible solution. In this way the experts are complementing one another's expertise, experiences and resources.

The rapid development in a small number of advanced countries of radiation-sterilization technology and related research has raised the field to a high level of technical sophistication. This status of development often imposes as prerequisite the need for highly specialized and trained personnel prior to its introduction in many of the technologically less advanced countries. This limiting aspect has been attempted for recovery through the timely selection of relevant topics for the award of Agency fellowships and organizing advanced training courses for the scientists of the developing countries. During the last five years two such training courses have been held in India and Argentina, respectively, while still others have been proposed. Published proceedings of the Agency meetings on this subject as well as a comprehensive Manual has provided (Table VI) valuable reference material for research guidance and as texts in the academic curriculum. The Agency's Nuclear Information Services (INIS) adds to this objective goal as an effective adjunct.

Additional effective machinery can be identified in the co-ordinated Research Programme on Radiation Sterilization of Medical Products and Biological Tissues, participated in by the scientists from the developed and developing countries. All the participating experts undertake Agency approved research projects convergent on one or a few well-defined closely related relevant goal(s). Their approaches and techniques, however, pay due regard to the priority needs for their own countries or the regions concerned for the promotion and early implementation of the practices in the existing health care systems.

The Agency's Technical Assistance Department as well as the United Nations Development Programme (UNDP) jointly sponsor a limited number of selected Country Projects aimed at providing early major economic and welfare returns to the recipient developing countries. The subject of radiation sterilization has been the nucleus of a successfully completed project (ISOMED) in India, the venue of the symposium. Currently, similar other projects are underway for the introduction of industrial-scale radiation-sterilization facilities in Hungary and Korea. In addition, assistance through experts and market surveys for radiation-sterilized medical supplies has been provided in the Philippines, Israel and Argentina. The ISOMED project in India has already provided on-site training to several microbiologists, chemists and engineers from a number of countries in the Far East, Asia and the Middle East, interested in these practices.

The widespread recognition of the many advantages associated with the single-use radiosterilized medical supplies have made the finished goods reach the expanding consumer markets beyond the national boundary of the producer country. This trend is further increasing. This brings forth a timely question of the standardization of the manufacturing and sterilizing processes and the quality control of products in accordance with the internationally agreed standards and specifications. The Agency's role in collaboration with the WHO has resulted in the draft formulation of the Recommendations for Radiation Sterilization of Medical Products. The latest version of this draft document is presented for joint up-dating and revision by all the Member States' representatives and participating experts at the final session of this symposium.[2] Many of the recommendations have already been incorporated in the National Code of Practice of a number of Member States.

---

[2] IAEA-SM-192/90, these Proceedings.

In the Agency's programme activities the emphasis for the promotion of radiation-sterilization practices in the developing countries will continue. Preparatory to that objective, relevant research topics to solve specific local problems and to develop locally suited techniques and processes using nuclear methods will be encouraged. For this purpose individual or regionally co-ordinated efforts towards the identification of appropriate topics will be most welcome.

CONCLUSIONS

The application of ionizing radiation for the sterilization of ready-to-use medical supplies, sutures and grafts provides a broad scope for the upgrading of public health care and family planning programmes in the developing countries. Sterile ready-to-use medical supplies become particularly important for improving the standard of those services given through improvised camp-hospitals and mobile medical units for the remote areas of such countries, if needed. The practices generated in the technologically advanced countries will form the basis of the planning, but the necessary adjustments should be made in their implementation to suit best the local conditions and needs and to promote utilization of local raw materials. Necessary research and development and an effective infrastructure should be emphasized. Plastic materials are among the major pollutants of the environment. Timely parallel practical steps need be adopted and an action programme planned to preserve the quality of the human environment.

REFERENCES

[1] CHRISTENSEN, E.A., MUKHERJEE, S., HOLM, N.W., "Microbiological control of radiation sterilization of medical supplies." I. Total Count on Medical Products (Disposale Syringes and Donor Sets) Prior to Radiation Sterilization, Risö Rep. No. 122, Danish Atomic Energy Commission, Risö, Denmark (1968).

[2] KRONENWETT, F.R., American Biological Control Laboratories, Unpublished Rep. (20 Dec. 1968).

[3] TATTERSALL, K., "Problems of microbial contamination in prepackaged preparations", Ionizing Radiation for the Sterilization of Medical Products, Proc. Panel on Gamma and Electron Irradiation, London (1965) 15.

[4] BURT, M.M., LEY, F.J., J. appl. Bacteriol. 26 (1963) 484.

[5] BARTHA, T., MÉRŐ, E., SZITA, J., Bacteriological model experiments of radiosterilization of catgut, Acta. microbiol. Acad. sci. hung. 16 (1969) 31.

[6] LEY, F.J., Radiation sterilization — An industrial process, Proc. Fifth Int. Congr. Radiat. Res., Seattle, USA (1974).

DISCUSSION

V.K. IYA: You have described at length some of the services provided by the Agency for the benefit of the developing countries. I would like to stress the great value of holding symposia and other meetings in those countries. As we all see, the present symposium has enabled a large number of scientists and other persons working in the pharmaceutical industries in India to attend the discussions and hear about the latest developments in the field. Similarly, a number of scientists from Asia and Africa have been able to attend the meeting here in India and likewise benefit therefrom.

R.N. MUKHERJEE: In reply to your comment I would like to emphasize that it is not the policy of the Agency to hold its meetings exclusively in Europe or America. In fact, the Director General actively supports the idea that every year a number of the planned symposia and advisory group meetings should be held, as far as is practicable, in the developing countries so as to facilitate their participation. This is particularly the case when the topic for such a meeting has a unique relevance to the current or scheduled development programmes of the country or region concerned.

F. FERNANDES: I fully concur with Mr. Iya's remarks on the desirability of the Agency arranging to hold more and more symposia and conferences in the developing regions of the world. It is true that the organization of such meetings often involves heavy expenditure, particularly as international experts have to be transported to these regions. But to me finance appears to be only a minor aspect of the problem when it is a question of transferring scientific knowledge and technology to the developing regions, which is one of the Agency's main responsibilities.

International experts attending these meetings will have an opportunity of making an on-the-spot study of the specific problems of the developing world, which they cannot otherwise do. It must also be realized that for economic reasons most of the young scientists from the developing nations cannot attend meetings held in developed countries and are thus denied the resulting benefits.

R.N. MUKHERJEE: I fully agree with you regarding the value of, and the pressing need for, large-scale dissemination of relevant information among scientists from the developing countries, and the establishment of contacts with the international experts, all of which are becoming more and more difficult these days due to world-wide monetary problems, among other things. I can only repeat what I said earlier, namely that the Agency continues to remain aware of the needs and interests of the developing Member States. As far as practicable, a limited number of grants are awarded to the participants from the developing countries to facilitate their participation at Agency meetings, and the same procedure has been followed for this symposium, too. Unfortunately, the small budget available limits the number of such grants.

K. KRISHNAMURTHY: I believe that the Agency is playing a most valuable role in bringing the modern technology of radiation sterilization to the developing countries. I should like at this point to raise the practical problem of measuring the absolute value of the absorbed sterilization dose in the radiation plant where there is an accepted minimum dose level of 2.5 Mrad. Does the Agency have any plans for the dose-intercalibration of operating facilities, or the promotion of dose-measurement systems useful for appropriately implementing this prescribed minimum dose of 2.5 Mrad in each facility?

R.N. MUKHERJEE: The Agency's Dosimetry Section (Life Sciences Division) undertakes intercomparison and calibration of therapeutic radiation sources located at various centres in Member States. But this, of course, relates to much lower dose levels. So far this programme has not included intercalibration services dealing with the megarad dose level used in industrial practice. This topic might be of interest for coverage in a future programme. I shall certainly pass on your comments to the Dosimetry Section.

M.A.R. MOLLA: Would you tell us whether the Agency is considering starting a co-ordinated research programme on the sterilization of medical products, appliances and biological tissues in Asia and the Far East, and if so, when would it be? I personally feel that a programme of this type would be extremely useful for the developing countries.

R.N. MUKHERJEE: Among the Agency-approved programmes, activities relating to the promotion of the radiation sterilization of medical products continue to receive strong support. In recent years there has been a trend in a number of countries of Asia, the Far East and the Middle East towards early introduction of this nuclear technique and technology so as to improve the standard of their existing public health services. Agency-sponsored programmes in this subject area have already played a significant role in those developments. If the countries of the regions concerned can identify some relevant research area in which they have a common or related goal, and where their individual research efforts can complement one another and provide new and useful information of practical interest for the region(s), then it should qualify as a topic potentially suitable for the initiation of a co-ordinated research programme. I am not able to say at this stage whether the Agency will be organizing a programme of the type you mention in the future. But in any event, if the interest shown in the Member States of the region concerned leads to such a proposal, it will be reviewed for its technical feasibility and programmatic relevance by the Director General before any decision is made.

REVIEW OF THE STERILIZATION OF SURGICAL SUTURES

J.O. DAWSON
Ethicon Limited,
Edinburgh,
United Kingdom

Abstract

REVIEW OF THE STERILIZATION OF SURGICAL SUTURES.

The sterilization of surgical sutures by ionizing radiation was the first commercial application of this process to the sterilizing of medical products. The history of the sterilization of sutures goes back 100 years to Lord Lister and the developments parallel those of other medical products. The evolution of such products is the result of ever improving technology in microbiological techniques, evaluation of physical properties.

The Sterilised Surgical Suture Industry had its origins just over 100 years ago with the work of Lord Joseph Lister. Although there were other workers in the field of microbiology it can be said that Lister pulled the ends together and produced and applied a system which considerably reduced the incidence of post-operative infection. Indeed the current symposium in its own way stems from the same philosophies and may be regarded as constituting another step in the fight against the problem, a problem the cause of which was not known until the middle of the last century.

Surgery has in fact been practised for many hundreds, even thousands of years. The ancient Egyptians were skilled in the art of trephining and there is evidence from excavation work and ancient literature of other forms of surgical intervention apart of course from repair work on accidental injuries. There is reference to the use in the Indian sub-continent several thousand years B.C. of the use of dried twisted intestines as a ligature. Examples of other materials used in surgery in ancient times include silk, hair, hemp and flax. These must be viewed, however, as being materials available to hand and not as threads having specific therapeutic properties. The use of what we nowadays call catgut was well established in many ancient civilisations for musical instrument strings showing its ability to withstand tension and abrasion and it must not be forgotten that even today some of the finest surgery, plastic and neurological, still makes use of human hair. Indeed it is only within the last 20 years that silkworm gut and kangaroo-tail tendon have been discontinued.

Despite the development of surgery over the centuries, and particularly after the Renaissance, the undergoing of a surgical operation was a major traumatic experience since there was no anaesthesia. The patient was doped with opium or hemp or various mixtures, gin, and the surgeon had to act fast. If the patient survived the operation he was likely to succumb to post-operative infection.

The nineteenth century saw the evolution and application of new knowledge in a number of fields. Anaesthesia was introduced firstly in the field of dentistry in 1844 with the use of nitrous oxide and of ether.

In 1847 in Edinburgh Simpson used chloroform as an anaesthetic for the very first time and this now gave the surgeon more time to carry out his work and to carry out more refined surgery, and have a longer time to infect the patient. P. S. Physik is credited with having re-introduced catgut as a ligature in 1816 and certainly Astley Cooper used it in 1817 for ligating the popliteal artery. Luigi Porta working in Flavia in 1845 carried out numerous animal experiments, some 600 in all, on various threads and claimed that of 80 catgut ligatures implanted, 33 disappeared. Some of them did not disappear for 2 - 3 years and would probably not satisfy the present-day surgeon since other catgut ligatures lasted even longer than this.

The century was dominated by Koch, Pasteur, and Lister - with help from a number of other people including Semmelweiss. The germ theory of disease was evolved, some surgeons actually cleaned their instruments and washed their hands before operating but it was Lister who drew the whole lot together and introduced ANTISEPTIC SURGERY. His system involved the liberal application of phenol to everything and everybody in the operating theatre by means of his carbolic spray. He treated his silk sutures with an aqueous solution of carbolic acid to kill the organisms lurking in the interstices of the thread and likewise turned his attention to catgut. Despite the physical properties of his treated product (i.e., with aqueous carbolic acid) he ligated the carotid artery of a calf and some three weeks later carried out a post mortem to find to his disappointment that the ligature had not disappeared; however, on closer examination he discovered that the ligature had been replaced by a band of living tissue. Lister's objective was to prevent infection of the dead or necrosed tissue resulting from ligation and in this he was successful. He worked further on the preparation of catgut and finally in 1869 he published a paper in the Lancet describing his method of preparation employing carbolic acid and olive oil which was the beginning of the Sterilised Absorbable Surgical Suture.

The Lister method of preparation utilising carbolic was employed with a greater or lesser degree of success by many surgeons for some 12 - 13 years. But by 1881 Koch was promoting the use of corrosive sublimate (mercuric chloride), and Kummel recommended its use for catgut in 1883. These were the days of rivalry and allegiances to individuals. There were still followers of the Lister method of preparation and indeed he devised some improvements notably the introduction of 'Carbol Chromic Catgut' and finally he finished his published work in 1908 with his method of preparing 'Sulphochrome Catgut' in which incidentally he used a mixture of mercuric chloride and chromium sulphate.

Between 1890 and 1910 a variety of treatments was proposed and indeed practised and indeed the first mention of anhydrous heat occurs. Many of these methods would be laughed off today and would not have been approved by the present Committee on Safety of Drugs or the F.D.A. Experimental procedures were often crude and the claims for the resultant superior qualities of the sutures so produced were somewhat extravagant.

The twentieth century began with a most significant advance, the publication by Claudius in 1902 of his method of Iodine Sterilisation, a process which with modifications was in use on a commercial scale for some 50 years and indeed is still employed today by a few companies. The first 50 years of this century saw not only the development of this method commercially but also the development of the anhydrous heat process. In general it could be said that heat was the main interest in the U.S.A. whilst the chemical iodine-type process was the mainstay of European manufacture. During and after the Second World War anhydrous heat assumed a growing importance in Europe.

The Sterilised Surgical Suture can be regarded as the first sterilised disposable Medical Product and therefore its commercial development is of interest to anyone engaged in the sterile medical disposable field.

At the beginning of this century catgut sutures were merely the catgut strings of commerce, for instance violin strings, chosen in a musical instrument shop by the surgeon personally and handed to his theatre sister for 'preparation'. Violin strings range in diameter from 0.65 mm for an 'E' string, 0.75 mm for an 'A' string, and 0.96 mm for a 'D' string. Harp strings can go up to 2.5 mm in diameter. Even as recently as 1947 some hospitals in the United Kingdom were still purchasing iodine-sterilised string in bulk for final surface sterilisation by iodine. However, as with most discoveries of this nature, specialist pharmaceutical firms were applying them commercially. The firm of MacFarlane of Edinburgh were manufacturing 'aseptic ligature silk' in 1905 prepared with aqueous carbolic acid. Subsequently firms began to specialise in this sort of work starting with the raw sheep intestine and producing strings specifically intended for surgical use which were subsequently sterilised.

With the growth of surgery it was obvious that theatres could not keep up with the demand for sterilised sutures and adequate control of the chemical processes of sterilisation which for security extended over a period of weeks could not be exercised. The development of anhydrous heat methods and more sophisticated chemical methods placed them outwith the scope of the normal equipment available in a hospital. An autoclave could sterilise silk but not catgut so over a period of time sterilised absorbable surgical sutures came within the province of the specialised company who could control the product's preparation all the way through.

In the United Kingdom a significant step was taken when as a result of Reports published by Bulloch, Lampitt, and Bushill in London and T. J. Mackie in Edinburgh in 1929 and 1928 respectively, sterilised surgical sutures of catgut and of other animal origins became subject to Regulations laid down under the Therapeutic Substances Act and for the first time their manufacture and control with especial regard to sterility became subject to legislation. Many then Commonwealth countries at some time or other have adopted similar legislation , and it is of interest that several years ago the then India Drugs Rules were virtually a reprint of the old U.K. Therapeutic Substances Regulations. The importance of this legislation is that for the first time in the U.K. we had legislation enabling Governmental control - albeit applied sensibly - of the four "P's", i.e., the Plant, the Process, the People, and the Packaging, the forerunner if you like of Good Manufacturing Practice.

Subsequently in the U.K. and the U.S.A. with an ever-increasing demand for sutures the tendency was to move towards the anhydrous heat system of sterilisation. This, however, still required the aseptic operations of filling the tubes with sterile reconditioning fluid and sealing.

Surgical catgut presents a challenge to a Code of Good Manufacturing Practice. The starting material, the sheep's intestine, is a potentially heavily contaminated material not necessarily per se but because of its method of collection and handling at the abbatoir. However, the methods of 'cleaning' employed by the suture manufacturer, the use of physical and chemical methods of removing unwanted tissues are such as to result in a remarkably pure raw catgut string. I would hazard the guess that the modern raw catgut string is probably less heavily contaminated than many of the so-called sterile strings employed by surgeons at the beginning of this century. The string is submitted to many handling operations such as gauging and polishing, the attaching of eyeless needles, the winding and packaging, all of which must be carried out under clean conditions, sterilisation possibly followed by an aseptic operation.

Suture manufacturers were always on the alert for new and improved methods of sterilisation adapted to large-scale manufacture.

In the 1920's suggestions were made that cathode rays might be used for sterilising but it must be remembered that ideas may not at the time of conception be capable of practical application. Serious studies were undertaken at M.I.T. around about 1948 mainly concerned with the preservation of foodstuffs. Out of this arose the commercial development of electron machines in the U.S.A. Experimental work on a small scale was done on the sterilisation of sutures but proved to be uneconomic. Dr. R. S. Hannan of the Low Temperature Research Station at Cambridge visited M.I.T. and studied the work being carried out there. Subsequently he reported to the pharmaceutical industry that he felt that the process could be applied to pharmaceuticals but in his opinion the first field of application would be sutures. In the meantime Dr. A. Charlesby irradiated some sutures for the author in B.E.P.O. at Harwell and a series of experiments was started on the 4-MeV Linear Accelerator under the control of Dr. Walter Miller at Metropolitan Vickers Radiation Laboratory at Trafford Park. This led to a revival of interest by Ethicon Inc. in Somerville who began work with a 1.5-MeV Van de Graaf machine subsequently going to a 3-MeV machine and then a linear accelerator. There was a considerable period of transatlantic co-operation in which initially the U.S.A. concentrated on fundamental studies and Edinburgh, because of its facility to spin in spores to catgut strings and also because of the higher energy available in Dr. Miller's laboratory and also thanks to Dr. H. L. Gray of the Hammersmith Hospital, concentrated on sterilising grossly contaminated suture strands in glass tubes. The outcome was that in 1956 Ethicon U.S.A. secured F.D.A. approval for the use of the process. The first commercially irradiated sutures were on the market by 1957. Ethicon Scotland continued its work but did not adopt the process commercially on the grounds of cost and capacity at that time.

However, during this period the Technological Irradiation Group was formed at Harwell later to move to Wantage. Originally intended to find an application for the radiations from the Caesium 137 of spent fuel rods, the group used Cobalt 60 as its source and eventually designed and built P.I.P. which came on stream in 1958. Virtually running neck and neck was the development in Australia of the Westminster Carpet company's Cobalt-60 irradiator for sterilising goat hair, and for the first time there were available what were for that time large gamma irradiation plants which could be employed for development work and, indeed on a commercial contract basis. These installations led to considerable activity in applying the process commercially to sutures. As early as June 1960 a firm called Lovell in Australia was using the Australian facility for γ-irradiating its sutures and by 1962 Johnson and Johnson Australia were marketing irradiated sutures.

In the U.K. Ethicon Edinburgh were working closely with the Ministry of Health and the T.S.A. Licensing authorities for approval for the process which we obtained in early 1962 - non-absorbable materials which were not subject to the same control were marketed earlier than this.

At the same time Ethicon Edinburgh was building the first fully automatic gamma irradiation plant intended solely for surgical sutures.

It must be remembered that little was known about the ability of machinery to withstand prolonged exposure to irradiation and that simple conveyor designs were intended to handle large bulk low cost materials. They were not suitable for small volume high cost items so that for that time a special loading and unloading system had to be evolved. Subsequent on the commissioning of the Edinburgh plant in 1962, Ethicon U.S.A. installed two gamma plants and abandoned their machines.

It is not intended to discuss the relative merits of machines and sources other than to say that in the surgical suture field it would appear that for the individual manufacturer wishing to control his whole process a gamma irradiation facility appears to offer greater opportunity for expanding capacity subject to original design. It needs fewer skills to operate and in remote areas is less dependant on outside help. The specialist contract or offering an irradiation service of high capacity who can support skilled technicians may on the other hand elect to use a machine.

Developments of this nature led to other improvements sometimes borne of necessity. Although hermetically sealed glass tubes are ideal for preserving sterility and have aesthetic appeal, particularly the large ones necessary to accommodate sutures with large needles are exceedingly difficult to open and are liable to shatter, damaging the suture and often the nurse. The adoption of a cold sterilisation method enables the use of plastics or foil packages, facilitating opening and saving weight and volume.

In this review emphasis has been placed on the work in the U.K. and the U.S.A. and Australia. Other countries have played their part as in Germany but the speed of application has depended on the availability of machines or sources. Sutures can be regarded as one of the success stories of the irradiation field. The application of the process has led to those responsible for the safety of the product. The method of sterilisation, however, still requires the application of Good Manufacturing Practice throughout the process of manufacture.

BIBLIOGRAPHY

BULLOCH, W., LAMPITT, L.H., BUSHILL, J.H., The Preparation of Catgut for Surgical Use (1929).

LISTER, J., Observations on ligatures of arteries on the antiseptic systems, Lancet, London, i. 451 (1869).

MACKIE, T.J., An Inquiry into Post-Operative Tetanus, a report to the Scottish Board of Health (1928).

CURRENT STATE OF RADIATION STERILIZATION OF MEDICAL PRODUCTS IN INDIA: PROBLEMS AND SPECIAL ADVANTAGES IN DEVELOPING COUNTRIES

V.K. IYA, R.G. DESHPANDE,
K. KRISHNAMURTHY, M.V. RAO
Isotope Division,
Bhabha Atomic Research Centre,
Trombay, India

Abstract

CURRENT STATE OF RADIATION STERILIZATION OF MEDICAL PRODUCTS IN INDIA: PROBLEMS AND SPECIAL ADVANTAGES IN DEVELOPING COUNTRIES.

Radiation sterilization is now firmly established as an industrial process for sterilizing a wide variety of medical products. The introduction of this technique has enabled the development of a number of inexpensive single-use products, offering a high degree of quality assurance for use in medical and surgical procedures. The need for introducing this technique in India was felt in the 'sixties. The developments leading to setting up ISOMED, a radiation plant for the sterilization of medical products at Trombay, and the specialized services offered to manufacturers of medical products, are described. The legal aspects of radiation sterilization in India are briefly discussed. Finally, the benefits of this technology to developing countries and some of the problems encountered in introducing it are also discussed.

1. INTRODUCTION

The commissioning of ISOMED, Radiation Plant for Sterilization of Medical Products, in January 1974, has been an important milestone in the industrial applications of large radiation sources in India. The setting up of this plant has been motivated by the efforts being made by the Government of India for improving health standards in the country, the growing awareness amongst manufacturers of medical products, and among hospital authorities, of the specific advantages of the radiation sterilization process, and the very active role being played by the Indian Department of Atomic Energy in promoting peaceful uses of atomic energy. The introduction of this technique has enabled the development of a number of ready-for-use medical devices, offering a high degree of quality assurance. This paper describes the various factors which have influenced the introduction of the radiation-sterilization technique on an industrial scale and the experience gained so far in this field. The special problems attached to the introduction of this technique in developing countries and also the benefits accruing are also described.

2. NEED FOR RADIATION-STERILIZATION FACILITY

The need for a radiation-sterilization facility in the country was recognized in the 'sixties. The main reasons which highlighted this need are:

(i) The growing demand for pre-sterilized, ready-for-use medical devices based on locally available plastic materials, e.g. infusion assemblies, blood donor and transfusion sets, catheters, etc.

(ii) The non-availability of reliable sterilization equipment and the difficulties experienced in its maintenance, particularly at hospitals located in remote areas and in surgical camps organized in different parts of the country.

(iii) The relatively high levels of microbial contamination in some medical products such as absorbent cotton wool, gauze, etc., which are often produced in small-scale industries and in somewhat unhygienic conditions.

(iv) Radiation sterilization is ideally suited for sterilization of products such as surgical sutures, containers and closures for pharmaceutical preparations (e.g. eye ointment). Alternative processes are either not convenient or cumbersome to adopt.

A preliminary market survey conducted in the Bombay area, which has a high percentage of the medical products manufacturing units and of pharmaceutical industries, indicated that a radiation-sterilization plant with an ultimate capacity of about 1×10^6 ft^3/yr would be required to cater to the needs of this area. The anticipated product-mix which would be necessary to be processed was designated as follows:

Characteristics of products	Product
High volume, low value, low density	Absorbent cotton wool, surgical dressings, sanitary pads, etc.
High value, low volume, low density	Surgical sutures, prostheses.
Medium value, low density	Disposable plastics including syringes, catheters, infusion assemblies, etc.
High density, high volume	Talc (surgical grade, and for infant use).

Following a decision to embark on such a programme, a joint Government of India-UNDP project was begun in January 1972. The construction of the plant building was completed by April 1973. The installation of the mechanical and electrical equipment of the plant was completed by November 1973. The ^{60}Co sources were loaded in the plant by the first week of December 1973. The radiation plant was thus commissioned approximately two months ahead of schedule. The mechanical component for the irradiator was obtained under UNDP assistance from H.S. Marsh Nuclear Energy Ltd. UK, while the radiation source of 125 000 Ci of ^{60}Co was made at the Bhabha Atomic Research Centre. The civil and electrical work of the plant building was executed by BARC.

3. PILOT-PLANT STUDIES

Pending the construction of the ISOMED plant, it was considered essential to offer irradiation services on a limited scale to enable BARC to carry out feasibility studies on various medical products, and also for test marketing such products. For this purpose a panoramic batch irradiator (PANBIT), loaded with 50 000 Ci ^{60}Co, was set up at Trombay. This has proved to be an important step in the rapid promotion of radiation sterilization as an industrial process. Further, it enabled the ISOMED staff to obtain practical experience in fields such as packaging, dosimetry and microbiology and to sort out the legal aspects of radiation-sterilization techniques.

4. PLANT DESCRIPTION AND OPERATION

The ISOMED plant has been initially loaded with 125 000 Ci ^{60}Co and has an ultimate capacity for 10^{12} Ci. The building housing the plant and the associated laboratories is spread over an area of 10 000 ft^2 (nearly 1000 m^2). The irradiation cell has an internal dimension of $7.5 \times 6.2 \times 3.7$ m high and is provided with approximately 6 ft of concrete shielding. The radiation source rack is a vertical frame in which 30 horizontal tubes of 140-mm diam. and 2200-mm length are spaced 45 mm apart. The rack can take a maximum of 90 composite source units each 38 mm in diam. and 695 mm long.

The product boxes are made of 5-ply cardboard and have outer dimensions of $59 \times 43 \times 34$ cm, corresponding to a volume of 0.09 m^3 (3 ft^3) and can be loaded to a maximum of 18 kg, corresponding to a density of 0.2 g/cm^3. A total of 63 vertical containers each holding five standard cartons are provided. The containers undergo four passes on either side of the source and the estimated efficiency of the system is 30%.

The initial charge of 125 000 Ci of ^{60}Co in the form of source rods was fabricated at the Isotope Division, Bhabha Atomic Research Centre, using low specific activity ^{60}Co slugs. A total of 21 source rods with ^{60}Co activity, varying from approximately 5000 - 8000 Ci, were used for the initial charge.

The dosimetry in the plant was carried out after commissioning using 'clear' perspex HX dosimeters. For this purpose all 315 cartons in the plant (including boxes containing dosimeters) were filled with material corresponding to density of 0.15 g/cm^3, and these were designated as dummy boxes. In the dosimetry box 53 perspex dosimeters were used, together with 13 ceric sulphate dosimeters for confirming the results obtained with the perspex dosimeters. The dosimeter boxes were introduced into the plant after every 10 dummy cartons. The conveyor speed was set approximately to the required value ($\sim$ 14 min/box) by a trial-and-error method. The optical density of the irradiated dosimeters was read on a Unicam SP-500 spectrophotometer at a temperature of 25°C and at wavelength of 305 mm. The calibration of perspex dosimeters was carried out using a standard radiation field (0.5 ± 0.02 Mrad/h). The average value of 53 dosimeters was 2.72 Mrad and only two dosimeters indicated a minimum dose of less than 2.5 Mrad (actual value: 2.4 Mrad). The maximum dose recorded in the dosimeters was 3.1 Mrad and the over-dose ratio worked out to 1.29.

During routine plant operation clear perspex HX is being used. However, it has been noticed that under the existing operating conditions, with large residence time in the irradiation cell (about 80 h), considerable post-irradiation fading is noticed in the dosimeters. The fading is of the order of 5 - 6% for 24 h after the completion of irradiation. This has necessitated the measurement of the dosimeters within 4 h of the completion of irradiation in order to minimize the fading effects. It appears from our observations that with long irradiation times encountered at low ^{60}Co loadings, the use of clear perspex HX dosimeters poses some problems in routine dosimetry work.

In addition to perspex dosimeters, biological indicators (Bacillus pumilus – 10^6 per disc) are placed in irradiation cartons and after irradiation are checked for any survivors. Biological indicators are used in one carton per week.

5. AUXILIARY SERVICES

To promote the radiation-sterilization technique rapidly it was recognized early in our programme that it would be necessary to offer auxiliary services and consultancy on microbiology, packaging, radiation effects on materials, and other related fields, to manufacturers of medical products. For this purpose laboratory facilities and trained personnel have been provided in the ISOMED plant to take up this peripheral work.

5.1. Microbiology

The microbiology group routinely carries out the determination of pre-sterilization counts on all products recieved for irradiation using standard microbiological procedures. The levels of pre-sterilization microbial counts determined on the products received vary from product to product. For example, for disposable plastic appliances, the counts range up to 300 per unit, whereas for absorbent cotton-wool they range from 50 to 300/g, and for surgical dressings 100 - 1000 per piece. Talc, however, shows high contamination levels usually ranging from 10^4 to 10^5/g. The information obtained is provided to the manufacturers to enable them to locate sources of contamination of the product, if any, during the manufacture and to help them to improve their manufacturing practices. Also, wherever necessary, the microbiology group carries out an environmental count on the manufacturers' premises to assess the level of manufacturing hygienics in the production area.

5.2. Chemical studies

To enable the manufacturers of medical products to assess the compatability of plastic material used in the medical devices to gamma irradiation ISOMED staff assists the manufacturers by carrying out detailed chemical investigations on such irradiated products. It is well known that most of the plastic materials contain various types of plasticizers, stabilizers, antioxidants, ultraviolet absorbers and fillers. The possibility of release of these additives either into the medicament contained in them or into the body tissues from an indwelling tube, strongly influences the choice of use. The acceptability of these devices for safe use is evaluated by a variety of physicochemical and biological tests including those prescribed in USP XVIII and NF XII.

ISOMED also assists manufacturers of pharmaceuticals in studies relating to chemical effects of irradiation on various pharmaceutical products amenable to radiation-sterilization process. The details of the studies carried out so far have been presented in another paper at this symposium.[1]

5.3. Packaging

It is recognized that adequate packaging is essential for ensuring a high level of sterility in medical products sterilized by radiation. Since packaging requirements are quite stringent and since some of the manufacturers of medical products have not gathered adequate expertise in this field, ISOMED has provided the required expertise to the manufacturers. In association with the local manufacturers of packaging material and packaging equipment, appropriate materials and sealing equipment have now been developed locally. In addition, ISOMED also offers packaging services in the initial stages to prospective users of radiation-sterilization techniques for packaging their medical products. In the case of surgical cotton and gauze dressings particularly, such assistance has helped in introducing the use of pre-packed sterile dressing kits in operating theatres in some of the leading hospitals in Bombay.

It is once again emphasized that a radiation-sterilization agency should be prepared to offer such auxiliary services in microbiology, packaging etc., in developing countries, since the medical products industry may not have acquired the necessary expertise. This also facilitates the early acceptance of the radiation-sterilization process.

6. PRODUCTS STERILIZED BY RADIATION/PRODUCT DEVELOPMENT

The products currently being sterilized by radiation in India include the following:

Plastic disposables	Infusion sets, blood-donor and bleeding sets, catheters, tubings, scalp vein sets, intra-uterine devices.
Cellulosic products	Absorbent cotton wool, surgical dressings, paraffin gauze, maternity pads.
Surgical products	Sutures, surgical gloves, blades, scalpels, etc.
Pharmaceutical containers and products	Ointment tubes, caps, droppers, bags, petri dishes, etc.

In addition, a number of pharmaceutical and medical products have been irradiated on a trial basis to evaluate the effects of a sterilizing dose on these products.

---

[1] IAEA-SM-192/6, these Proceedings.

7. KITS FOR FAMILY PLANNING AND CAMP SURGERY

An important contribution of ISOMED has been the development of kits for specific surgical operations. This is being done in consultation with medical experts and in collaboration with Public Health authorities. Potentially the most important of these kits is that developed for vasectomy operations.

Vasectomy is important for the governmental policy of tackling the population problem in the country. In rural areas, these operations must often be carried out at small centres and mobile units, which are inadequately equipped for ensuring thorough sterilization, particularly for achieving extensive results. Any infections or complications that may arise often cause considerable anxiety and slow down the family-planning programme in that area.

A disposable, ready-to-use radiation-sterilized kit has been developed. This consists of the following items: a disposable polystyrene syringe packed separately with needle, surgical steel disposable blade, five pieces of absorbent cotton and gauze dressing, linen suture, polyethylene drape of 1 m × 75 cm. The entire kit is assembled in a polyethylene bag and radio-sterilized. At present cost levels, it is estimated that a typical vasectomy kit would cost approximately Rs.5/- (or US ¢ 60).

The only additional items required by the surgeon are a local anaesthetic and a pair of steel scissors. These kits are now undergoing field trials in Bombay. The large-scale use of such sterile vasectomy kits would help to make vasectomy procedures safer and faster. Radiation sterilization could thus make an important contribution to family-planning programmes in developing countries. Similarly, large parts of rural India are covered medically by camp surgery for eye and ear operations. Here, too, sterile disposable kits can play an important part in reducing post-operative infections.

8. LEGAL ASPECTS OF RADIATION STERILIZATION

Radiation sterilization has been a legally accepted process in India for several years. The sterilizing dose specified is 2.5 Mrad. To enable manufacturers to carry out radiation sterilization of products included in the pharmacopoeia at ISOMED, the following requirements must be fulfilled.

(1) The radiation-sterilization agency has to obtain a licence from the Food and Drugs Administration, Maharashtra, to carry out radiation sterilization at ISOMED.

(2) The manufacturers of a medical product (products included in pharmacopoeia or those covered by drug rules) have to obtain a loan licence from the Food and Drugs Administration, for sterilization of their medical products at ISOMED.

At present, many products such as catheters, infusion sets, tubing etc., are not included in the pharmacopoeia or covered by drug rules.

The Department of Atomic Energy and the Ministry of Health, Government of India, jointly set up a Working Group on radiation sterilization of medical products in July 1973, with the following terms of reference:

(i) To recommend the dose required to be given to medical products for sterilization by radiation;
(ii) To evolve a national Code of Practice for the manufacture of medical products to be sterilized by radiation, and for the radiation sterilization of medical products.

The Working Group consists of representatives of Health Regulatory authorities, eminent microbiologists and officials of the Bhabha Atomic Research Centre. The Working Group, on the basis of the data collected at ISOMED, and published work, has recommended a minimum of 2.5 Mrad as the sterilizing dose for medical products. The Working Group in its report has also made specific recommendations on various points such as definition of medical products, packaging, pre-sterilization microbial counts, routine process control, test for sterility, biological indicators and division of responsibility between primary manufacturers of medical products and the radiation sterilization agency. The report of the Working Group is currently being considered by the Ministry of Health.

9. BENEFITS OF THE RADIATION-STERILIZATION PROCESS, AND PROBLEMS OF INTRODUCING THIS TECHNIQUE IN DEVELOPING COUNTRIES

While radiation sterilization is already established as an industrial process in many advanced countries, it is only in recent years that considerable attention has been paid to this process in developing countries. This is so because the many advantages offered by this process are not fully appreciated. It is necessary to emphasize these often to both manufacturers and users.

The radiation-sterilization process requires only one parameter to be controlled, namely irradiation time, while for other conventional sterilization processes, four or five parameters have to be simultaneously controlled. Further radiation sterilization does not require microbiological control after sterilization, provided the initial pre-sterile counts are kept at low levels. This is so because the safety factors obtained with this process are very high. Above all, the suitability of the process for heat-sensitive materials, and the fact that the process makes sterile products available which can be stored for months or years in ready-to-use packages, not only for remote or rural areas, but also for emergency and routine use should be explained and publicized widely to medical and public health authorities. In developing countries, where manufacturers are about to launch into the manufacture of disposable medical appliances, these considerations should favour the adoption of radiation as the sterilization process. Again, the need to establish a large number of ethylene oxide in-house plants to meet the requirements of individual industries and the provision of highly skilled manpower to operate them, makes gas sterilization far less practicable than a centralized radiation plant.

The benefits offered by single-use products are obvious. They are:

Confidence in the sterility of the product;
Safety by reducing chances of cross-infection;
Comfort to the patient in rapid treatment because of ready availability and less trauma compared with re-usable products.
Efficiency in treatment as no time is required for preparing sterile products;
Economy: Apart from direct benefits to the patient mentioned above, the hospitals benefit indirectly from the better utilization of hospital staff engaged preparing and sterilizing medical products for other critical work, and also from the larger turnover possible because of the reduction in cross-infection, and consequently the reduction in residence time of the patient in the hospital.

The introduction of the radiation-sterilization technique helps to promote the local design and manufacture of single-use medical products, and thus enables economies of scale to be carried out. Local industry should be encouraged to mass produce unit packages so that the maintenance of requisite quality control measures becomes feasible because of extensive production. Finally, industries should take over the preparation and sterilization of medical products from the hospitals, leaving hospital staff to discharge their primary responsibility, namely patient care. This is particularly relevant to situations in developing countries where there is a shortage of both medical and para-medical personnel.

In the initial stages of introducing this technology to developing countries, some of the following difficulties are likely to be faced:

(1) An absence of a ready market for single-use products;
(2) Resistance from hospitals and medical users to change from conventional methods and products;
(3) Lack of funds for the purchase of disposable medical supplies.

While no ready solutions are available for overcoming these problems, each country must find appropriate ways given the local prevailing conditions. Nevertheless, a good strategy would be to start with an area where there is an acute need for single-use products, e.g. camp surgery, hospitals in rural areas, and defence establishments located in inaccessible areas. Simultaneously, a well-planned publicity programme to highlight the benefits of single-use products and the advantages of the radiation-sterilization process, and aimed at disseminating information to the medical users and medical products manufacturers should be launched. This could be done by periodically dispatched information bulletins, or by seminars and symposia organized to cater to particular groups of industries, thus not relying only on personal contacts. The radiation-sterilization agency should also offer technical consultation in various aspects of radiation-sterilization processes, e.g. chemistry, microbiology, packaging, etc. Finally, if necessary, the radiation-sterilization agency should be prepared to develop and manufacture some of the important single-use products and market them until an adequate market is developed for such products, thus attracting the manufacturing industry to undertake their routine production.

The criteria for introducing radiation-sterilization technology on an industrial scale could be summarized as follows:

(1) Manufacturing level of various medical products suitable for radiation sterilization
(2) Sterilization methods currently adopted.
(3) Need for improving quality of medical products (particularly if they are manufactured in small-scale industry under sub-standard conditions).
(4) Status of the manufacture of plastic materials used in making medical devices, packaging materials, etc.
(5) Availability of trained scientific manpower to operate the plant and exercise adequate process control.

A market survey for identifying the total market for sterile goods based on existing available products, the need for new products, and their development, and the total anticipated volume of various products, should be undertaken before launching on a large-scale programme.

CONCLUSION

With the establishment of the ISOMED plant, the radiation-sterilization programme in India has gathered considerable momentum. The impact of the plant's commissioning is already being felt on the health care programme in the country by way of upgrading the quality of various medical products used. It has also provided an impetus for fostering the growth of ancillary industries, especially for manufacturing plastic medical appliances. We believe that, as a result of the R & D work being carried out and the auxiliary services offered by ISOMED, the scope for the application of radiation sterilization will increase for many medical and pharmaceutical products and that in the coming years a wider variety of these products will be sterilized by radiation.

In the future, we feel that radiation sterilization will be increasingly applied in India. This will necessitate setting up similar plants in other parts of the country where the medical products industries are located. A preliminary market survey is being conducted in the eastern region of the country and a decision to set up a second radiation-sterilization plant will then be taken. The ISOMED plant and its associated R & D facilities are available to scientists in this region for training and familiarization with radiation-sterilization processes.

DISCUSSION

H.C. YUAN: I would like to thank Mr. Deshpande for his account of the development of the ISOMED project and to compliment him, along with Drs Iya, Krishnamurthy and Rao, on their unstinted efforts to make the project successful. I am happy to report that, first, the plant is being loaded with a domestic ^{60}Co source, manufactured by the Isotope Division of BARC; second, that the building and facility have been completed ahead of schedule through the joint efforts of the Isotope Division and other engineering divisions at BARC; and third, that the plant has served as a training ground for international fellowships, even during the construction period.

Because of the success in the planning and construction of the plant, I would like to suggest that a timetable listing the major events in the project activities could serve as a useful reference for the planning of radiation-sterilization projects elsewhere.

RADIATION STERILIZATION IN ISRAEL – PAST, PRESENT AND FUTURE

M. ARONSON, E. EISENBERG*, M. LAPIDOT
Tel Aviv University, Tel Aviv,
Israel

Abstract

RADIATION STERILIZATION IN ISRAEL – PAST, PRESENT AND FUTURE.

The last decade has seen a considerable effort to introduce commercial radiation sterilization into Israel. Despite a highly developed and modern medical care system, this effort has been an uphill struggle. However, with the help of detailed and frequently updated economic feasibility studies, and very close co-operation with industry and medical care authorities, considerable progress has been achieved in the last few years. Recent development and promotional efforts should result in quicker implementation of additional products and processes aimed at doubling the radiation-sterilized products volume every two or three years for another decade.

1. INTRODUCTION

Preliminary surveys conducted during the first year of operation of the large Radiation Sources Section at the Soreq Nuclear Research Center (1964) revealed a significant potential market for radiation sterilizable products in Israel [1], and an insignificant amount of such products for sale or in use in local hospitals. Despite the relatively high standard of hospital practice and the modern equipment in many hospitals, radiation sterilization was either unknown, or frowned upon. Many officials and hospital staff considered the use of disposables too expensive.

At that time progress had already been achieved in the field of radiation sterilization in several countries, mainly in Australia, the United States of America, the United Kingdom and Scandinavia. The potential market in Israel seemed substantial, despite the relatively small population. Efforts were therefore initiated and they proceeded in several directions to justify a local radiation-sterilization facility and to implement the technology on an industrial scale. Technological and economic feasibility studies were undertaken, microbiological control techniques were developed, and local technical skill was expanded for the production of radiation sterilizable items.

This briefly reviews a decade of activities which led to the construction of a commercial radiation-sterilization plant now completing its second year of operation. Also, as a result, the development of a number of projects on radiation sterilizable medical and biological products was undertaken, many of which should be produced on a commercial scale in the next few years.

---

* Soreq Nuclear Research Centre, Yavne

2. PRELIMINARY ACTIVITIES

2.1. Economic feasibility studies

To create an initial interest in radiation sterilization, a brochure [2] describing this new technique and its potential was distributed to the medical, nursing and administrative staffs of hospitals and clinics. This was followed by meetings and lectures at hospitals and dispensaries.

As a result, permission was obtained to survey the consumption of items requiring sterilization in a small hospital [3], to assess the potential demand for radiation sterilizable products in Israel [4]. The survey was performed by engineers of SNRC who collected data over a period of several months from the supply stores and wards (where special forms were filled in daily by the nurses). This data concentrated on the breakage and wear of non-disposable items, as well as on the number of uses of such items; the number of disposable items already in use was insignificant.

This preliminary survey estimated a potential annual volume of 1000 m^3 of products (valued at $ 900 000) needing radiation sterilization, for nearly 8500 beds (general and surgical wards) in Israeli hospitals if replacement of certain non-disposable items by disposable items became feasible.

This volume could be more than doubled if pads, dressings and similar products were included. (The additional value, however, was only $ 300 000). Should all non-disposable items be radiation sterilized, the volume would increase to 20 000 m^3 (for 8500 beds). However, besides technological problems regarding the radiation sensitivity of diverse non-disposable medical devices and hospital supplies, the high cost of radiation sterilization (~ $100/$m^3$) relative to the product value, makes such a procedure economically prohibitive if a hospital irradiator for sterilization is conceived. One or two centrally located sterilization service plants could eventually become economically feasible if logistic problems were solved.

The above extrapolation of estimates from a 160 bed hospital to a total of 8500 beds (i.e. a scale-up factor of over 50:1) was considered sufficient to predict general trends, but a more detailed approach was essential to predict accurately the potential market for radiation sterilizable disposables. A number of surveys was therefore performed in several representative institutions [1].

The data were transmitted to an economic consultant group which prepared a detailed economic feasibility study for the production and sterilization of disposable hospital supplies [15, 16]. The study concluded that local consumption justified (from an economic viewpoint) the replacement of non-disposable by disposable items, where possible. The items to be considered were: syringes, petri dishes, test-tubes, dressings, gloves, infusion and transfusion sets. The savings to the country's economy were calculated. The most promising items were syringes and petri dishes.

---

[1] These included two large hospitals (~ 1000 beds each) [5, 6], two medium-sized hospitals (~ 500 beds) [7, 8], another small hospital (300 beds) [9], one psychiatric and nervous diseases hospital [10], several dispensaries of the local medicare system (Kupat Cholim) [11] and some microbiological and medical laboratories [12]; data from the stores of the Ministry of Health (controlling ~ 40% of the general hospitals) [13] and Kupat Cholim (controlling ~ 30% of the beds in the general wards and central dispensaries for ~ 75% of the population) [14] were also included. Data was gathered for items that were either disposable (dress pads etc.) or could be replaced by disposables (syringes and needles, petri dishes, test tubes, pipettes, gloves, catheters, infusion and transfusion sets).

Alternate employment for the redundant personnel was considered. The calculation showed a possible start at 400 m^3/yr to increase within five years to 4000 m^3.

This was followed by a series of more specific and detailed studies regarding the economic feasibility of a plant for the production of radiation sterilized disposable items [17-19]. At the same time, negotiations proceeded with a local firm to implement the plan. These negotiations failed since the local company, apprehensive of the high investment costs in the irradiation facility, decided to proceed with the production of ethylene oxide sterilized syringes and needles, based on a foreign license.

As a result of this failure, another attempt was made to negotiate the erection of a radiation-sterilization service facility, which would enable local manufacturers to consider the production of radiation sterilizable articles without the added risk of investment in an irradiation facility. For this purpose two more economic feasibility studies were prepared, aimed only at the sterilization part of the production process [20, 21], and based on five offers from leading radiation plant manufacturers received on the basis of detailed specifications [22].

2.2. Development of physical control techniques

With the purchase of a 30 000-Ci pilot-scale irradiation facility [23] by Isorad (Isotopes & Radiation Enterprises) Ltd., techniques had to be adapted to allow accurate control of the radiation-sterilization process. This was done by the dosimetry group of the SNRC Large Radiation Sources Section. The techniques developed were based on the use of cerium sulphate solutions [24] for the initial calibration of the plant and red perspex dosimeters for process control [25]. These techniques are now used in both the pilot and industrial scale facilities (cf. section 4.1).

2.3. Development of microbiological control techniques

Having had no previous experience in this field, a literature survey was initiated which immediately revealed a dilemma: the absence of an international code of practice. One had to decide between the Scandinavian school with its rigorous sterility requirements, and the Anglo-Saxon school with its pragmatic approach. From the purely scientific point of view, the Scandinavian approach was certainly preferable. It would undoubtedly be desirable to evaluate the (pre-irradiation) contaminating bacteria, to determine its resistance to radiation and to adjust the radiation dose accordingly. However, in practice, it involved an excessively expensive control apparatus for a relatively small industry. In addition, the bacterial strains recommended by the Scandinavians for dosimetry were rather problematical; not being spore-forming, they required frequent cultivation and calibration. Spores of B. pumilus were finally chosen and these successfully cultivated and used.

Comparison with physical dosimetry over a long period resulted in the inevitable conclusion that microbiological control was more expensive, more time-consuming and not more reliable than physicochemical control. Hence, it would only demonstrate the effect of radiation sterilization, rather than provide accurate evidence of its result [26].

3. DEVELOPMENT OF PRODUCTS AND PROCESSES

3.1. Disposable syringes

One of the main obstacles preventing a decision to establish a radiation-sterilization facility was the lack of know-how for the production of suitable products for the relatively small local market, and the difficulty in obtaining such know-how from reputable sources abroad at reasonable cost. It was therefore decided to develop locally the skills needed for the production of syringes (the most significant single item). A relatively small fund was jointly allocated to this project by SNRC, Isorad Ltd. and a local custom moulder of plastic products. A bench-scale production line was organized [27], permitting test production of ~ 75 000 disposable syringes (2- and 5-ml sizes). These were radiation sterilized at the SNRC irradiator in small batches.

The know-how thus achieved was officially recognized as an original contribution [28]. A time-and-motion study [29] was undertaken to plan a pilot-scale and large-scale plant for the production of disposable syringes. The know-how was licensed to the custom moulder, who plans large-scale production for 1975/76.

3.2. Disposable petri dishes

Within the framework of the above-mentioned project, know-how was also developed by the same group for the production of radiation-sterilized petri dishes, and a bench-scale assembly line allowed the test production of ~ 20 000 disposable petri dishes (60- and 90-mm diam.) [27]. This operation is now commercial, licensed by SNRC.

3.3. Sterile laboratory animal feed

Two major laboratory-animal breeding houses (at the Hebrew University in Jerusalem and the Weizmann Institute in Rehovot) showed considerable interest in radiation-sterilized rodent feeds. A detailed study at the Hebrew University laboratory [30] covering 13 months of test feeding to 15 groups of rats and 20 groups of mice, with normal rodent diets irradiated with 2.5-Mrad doses, indicated no adverse effects. The SPF colonies at this laboratory now receive the radiation-sterilized diet on a routine basis. The Weizmann Institute laboratory found that special diets, sterilized at doses of 4.5 Mrad gave good results with completely sterile animal colonies, and these are now routinely supplied with such diets.

3.4. Pasteurized farm animal feeds

A preliminary study [31] over a period of 70 d showed that radiation-pasteurized poultry feed gave no adverse effects to small poultry chicks. In a second study [32], one group of 350 chickens was fed for half a year a radiation-pasteurized (1.5 Mrad) feed, and compared with a second group fed non-pasteurized feed. It was found that the radiation-pasteurized feed was equivalent to the latter in nutrition value.

On the basis of the local feeding results and data obtained by the Canadian authorities [33-37], a petition was presented to the local Ministry of Health, and clearance of radiation-pasteurized poultry feed was granted in July 1973. A pilot-scale project (100-300 t/yr over a 3-year period) has been approved but is awaiting the completion of repairs of the Mobile Gamma Irradiator [38] on loan from the USAEC.

Another project undertaken as a result of the above study was the production of SPF eggs for veterinary vaccine production, from an SPF poultry flock fed radiation-sterilized poultry diets [39] for over 18 months. The results were very encouraging, as no mortality due to eventual disease was observed despite the fact that no vaccinations had been applied.

3.5. Sterile grafts

Preliminary experiments on the radiation sterilization of bone and nerve tissue intended for grafts have shown some promise. However, more detailed studies are necessary before the process is widely accepted.

3.6. Sterile soil

Several agricultural laboratories have repeatedly indicated their interest in radiation-sterilized soils for diverse research purposes. Samples have been irradiated successfully on many occasions [40], but no practical economically feasible process has yet been developed.

4. COMMERCIALIZATION ACTIVITIES

4.1. Industrial irradiation facility

The positive results of the economic feasibility study [21] encouraged a large international concern to form a joint enterprise with the local AEC (Sor-Van Radiation Ltd.). The major accomplishment of this firm was the erection of a service radiation-sterilization facility in 1971, based on an AECL J-6500 [41] irradiator. The facility started with a charge of 67000 Ci ^{60}Co in 1972 and was recently augmented to a level of 105000 Ci (i.e. 2200 m^3/yr). Although business has been slow at the beginning, considerable interest has been aroused. Recent R & D efforts (cf. section 6), as well as negotiations with local manufacturers based on former R & D projects (cf. section 3), should allow considerable expansion in the near future. (In addition, the company operates an EPS-550 electron beam accelerator system for modifications of plastic materials and electron beam curing).

4.2. Commercial production processes

Several local manufacturers of hospital supplies and devices have been using the services of the radiation-sterilization facility over the last two years. The following products have been adapted from another sterilization technique or especially designed for radiation sterilization (generally at doses of 2.5 Mrad): disposable surgical gloves, petri dishes, gauze pads, dressings, cotton swabs, dress packs (designed and manufactured by Sor-Van), blood-infusion sets, solution administration sets, mucus extractors and a few additional minor items, as well as fish meal, laboratory animal feeds and poultry feed. The total volume sterilized reached 1500 m^3/yr.

In addition to increased volumes of the above-mentioned items, significant volumes of disposable hypodermic syringes and needles are expected in 1975/76. This is based on renewed negotiations with a local manufacturer who has acquired the license for the local skills [28] and on new economic feasibility studies [42].

4.3. Physical control of the sterilization process

After each loading with ^{60}Co, the plant is recalibrated by means of ceric sulphate (accompanied by red perspex dosimeters [43b, 44]. The process is controlled electro-mechanically, i.e. records are kept of the time interval between intermittent box motions in the shuffle-dwell irradiator system, release of the source into the pool in cases of conveyor equipment malfunction, etc., as well as by two red perspex dosimeters attached to every 10th to 20th box [43c]. Because of the varying composition of the product mix in such a service facility, a detailed six-month survey is planned for the first half of 1975 to formulate an updated operating and control manual of procedure, which will ascertain the absence of errors at the higher throughput rates envisaged in the next few years.

The physical control will very likely be performed by the Dosimetry Group of SNRC on a part-time basis. This will probably be less expensive than employing a qualified quality control group at the industrial irradiation facility.

4.4. Microbiological control of the sterilization process [45]

Commercial strips impregnated with B. pumilus spore suspension are attached to the red perspex dosimeters and employed in the sterilization facility. The population of 10^5 spores per strip seems too low, since 2.5 Mrad should inactivate as many as 10^8 spores per strip (the number used in the initial development period, see section 2.3).

The count of initial contamination is the responsibility of the product manufacturer. We strongly favour more rigorous control of the initial contamination of the product. In a series of determinations of initial contamination of bench-scale manufactured syringes (see section 3.1), very few bacteria were found. Obviously the physical conditions in the plant play a major role in the amount of bacterial contamination of the product.

4.5. False positives and false negatives [46]

One of the main lessons emerging from the development of the microbiological control technique was the importance of a thoroughly trained bacteriological team. This became particularly apparent while handling bulky objects such as dressings. Although all work was conducted within a "laminar flow" bench, a high percentage [20-30] of false-positives was "recovered" from sterile bandages. The same findings were obtained by several additional reputable bacteriological laboratories who consistently reported a high number of false-positives. With additional practice, the know-how for proper handling of clumsy items was gradually achieved, resulting in an almost total absence of false-positives.

Another difficulty encountered was false negative results when attempting to determine the initial contamination of surgical gloves. It was established that the talcum powder, used as a lubricant, inhibited bacterial growth and very poor recovery was observed with intentionally contaminated gloves. For this reason, as well as for additional medical reasons, an absorbable lubricant is now employed for the gloves.

5. CODE OF PRACTICE

The possibility of developing a code of practice for radiation-sterilized products was considered in joint meetings of representatives of the Ministry of Health, SNRC, Sor-Van Ltd. and Tel-Aviv University Medical School. In these deliberations the difficulties inherent in striking the right balance between the desirable and the possible became apparent. Essentially the same reasons delayed the adoption of an international code of practice. The problem can be reformulated as follows: What is the value and cost of additional safety? Obviously any marginal increase in safety requires greater expenditures[2].

The local Ministry of Health has meanwhile formulated a code for good manufacturing practice (G.M.P.) [47] in the pharmaceutical industry. To complement this code with a special section on sterilization of devices by means of radiation or gas (ethylene oxide), a United States expert spent three months in 1974 (IAEA technical assistance programme) studying the local situation, and has formulated a manual of G.M.P. for sterile medical devices [48]. This is now under consideration by the Ministry of Health and the local industry. However, an established international code of practice is still awaited, particularly as our future plans centre on export markets.

6. RECENT RESEARCH AND DEVELOPMENT

6.1. Ready-to-use culture media

The Radiation Processing Section at SNRC and the Bacteriological Laboratories of the Ministry of Health Sheba Hospital have recently developed, on behalf of Isorad Ltd., a process for the production of radiation-sterilized culture plates and culture tubes of significant shelf life [49]. Several offers from local and United States firms to license the processes are now being considered.

[2] In this connection we would like to suggest a possible solution to the safety-versus-cost dilemma. For all practical purposes a sterilization dose of 2.5 Mrad is absolutely sufficient; there is generally no reason to exceed it. On the other hand, under certain conditions, even one non-pathogenic bacterium may become dangerous. Thus, in organ transplantations, when immuno-suppressive drugs are employed, and in some debilitating diseases, a much higher degree of safety is required. Therefore, it seems that two different irradiation doses should be administerd to syringes and similar devices according to their intended use (they should bear different colours to prevent confusion). In this way the majority would be sterilized with 2.5 Mrad, while the syringes intended for special uses would get as much as 5 Mrad.

6.2. Bioproducts (sera and related products)

Bioproducts, like sera, for laboratory use, were tested for their loss of potency and sterility after 2.5 Mrad gamma irradiation [50]. The products were found sterile and were not affected adversely. Work in progress on related products has reached the pilot scale and is being tested for product acceptance.

6.3. Surgical sutures

On behalf of a local pharmaceutical firm, a R & D effort, in progress at SNRC for over a year, has culminated in the development of a process of radiation-sterilized surgical sutures. The licensing of this process by the above firm is now being considered.

6.4. Sterile aqueous solutions

A basic study of the radiation sensitivity of several pharmaceutical aqueous solutions has been recently undertaken in conjunction with the Pharmacy Faculty of the Hebrew University. The results of this study will have serious implications for future programmes aimed at sterilizing pharmaceutical and physiological solutions as well as for the application of radiation sterilization to secondary and tertiary treatment of waste waters.

6.5. Antibiotics and other pharmaceutical products

Sporadic requests by several pharmaceutical firms have resulted in a series of preliminary attempts to sterilize antibiotics, ophthalmic solutions and ointments, and a few other pharmaceutical formulations.

6.6. Radiation sterilization of garbage components

Local household and agricultural garbage, as that of many other countries of similar climate, contains a large proportion (two-thirds by weight) of putrescible matter. A process has recently been developed for the separation of this matter (mainly vegetable and animal food remnants) from the other components of garbage, and for its processing to a valuable radiation-sterilized product [51]. A pilot plant is planned for 1975. The radiation sterilization will be applicable to other components, such as plastics and paper, intended for re-use.

7. FUTURE PROGRAMME

The main effort in the next two years will centre on pilot-plant and semi-commercial implementation of most R & D projects described, on the expansion of the sterilization volume at the commercial irradiation facility and on the completion of projects now in progress, mainly in the field of bioproducts and other radiation-sensitive pharmaceutical and non-pharmaceutical products.

8. FACTORS OBSTRUCTING ACCEPTANCE OF RADIATION STERILIZATION

Considering the existing local potential and know-how, there is relatively little exploitation and use of irradiation for sterilization purposes. There are various reasons for this, some of which are quite justifiable (see sections 1 and 4.5), but the psychological ones are by far the most interesting. The following examples, while not comprehensive, are quite representative:

(a) Initially, various medical care services were opposed to the introduction of disposable devices. It was feared that workers in charge of cleaning and re-sterilizing syringes and needles, and of sharpening the latter, would become redundant. This problem was eventually solved by transferring the workers to more sophisticated occupations.
(b) The idea of testing initial contaminations is always appealing on principle, but loses its attractiveness when translated into actual expenses.
(c) It is expensive and complicated to send pharmaceutical containers to a service sterilization plant, particularly since many plants have their own autoclaves.
(d) Many unversed officials and staff are unjustifiably apprehensive of possible radioactive contamination.
(e) Within the industry which produces sterile products there is an intellectual acceptance of the desirability of switching from sterilization by ethylene oxide to irradiation. However, the absence of pressure from the medical care services for the implementation of radiation sterilization allows the continuation of conventional procedures. This points to the absence of proper critical civic awareness concerning the quality and hygiene of medical products in this country. This is particularly deplorable since many products are still imported from abroad, which implies reliance on standards and controls of other countries.

9. CONCLUSIONS

The last decade has seen a considerable effort to introduce commercial radiation sterilization into Israel. Despite a highly developed and modern medical care system, this effort has been an uphill struggle. However, with the help of detailed and frequently updated economic feasibility studies, and very close co-operation with industry and medical care authorities, considerable progress has been achieved in the last few years. Recent development and promotional efforts should result in quicker implementation of additional products and processes aimed at doubling the radiation-sterilized products volume every two or three years for another decade.

REFERENCES

[1] FOA, E., LAPIDOT, M., Applications of Large Radiation Sources in Israel, Technologic and Economic Survey, IAEA (June 1965).
[2] LAPIDOT, M., Applications of Radiation to Sterilization of Medical Items, IAEC (1965) (in Hebrew).
[3] KAPILUTO, C., ALGAVE, A., LAPIDOT, M., DONAGI, A., CHAIMOW, I., Economic Feasibility of Radiation Sterilization of Hospital Supplies in Israel. I. Survey of hospital supplies requiring sterilization at Dr. Hillel Yaffe Governmental Hospital, Hadera, IAEC Special Rep., 1966 (in preparation).

[4] LAPIDOT, M., ibid., II. Preliminary estimates of the potential market of radiation disposable hospital supplies, IAEC Special Rep., 1967 (in preparation).
[5] ALGAVE, A., KAPILUTO, C., BEN MOSHE, H., LAPIDOT, M., ibid., III. Survey of hospital supplies requiring sterilization at Beilinson Kupat Cholim Hospital, IAEC Special Rep., 1967 (in preparation).
[6] ALGAVE, A. et al., ibid., VIII. Survey of hospital supplies requiring sterilization at Tel Hashomer Governmental Hospital, IAEC Special Rep., 1967 (in preparation).
[7] ALGAVE, A. et al., ibid., V. Survey of hospital supplies requiring sterilization at Meyer Kupat Cholim Hospital, IAEC Special Rep., 1967 (in preparation).
[8] ALGAVE, A. et al., ibid., XI. Survey of hospital supplies requiring sterilization at Hadassah Hospital, IAEC Special Rep., 1967 (in preparation).
[9] ALGAVE, A. et al., ibid., IV. Survey of hospital supplies requiring sterilization at Kaplan Kupat Cholim Hospital, IAEC Special Rep., 1967 (in preparation).
[10] ALGAVE, A. et al., ibid., IX. Survey of hospital supplies requiring sterilization at Psychiatric Governmental Hospital, IAEC Special Rep., 1967 (in preparation).
[11] ALGAVE, A. et al., ibid., VI. Survey of hospital supplies requiring sterilization at several Kupat Cholim dispensaries, IAEC Special Rep., 1967 (in preparation).
[12] ALGAVE, A. et al., ibid., XII. Survey of hospital supplies requiring sterilization at Central Microbiologic or Medical Laboratories, IAEC Special Rep., 1967 (in preparation).
[13] ALGAVE, A. et al., ibid., X. Survey of disposable and replaceable non-disposable hospital supplies requiring sterilization at Ministry of Health General Stores, IAEC Special Rep., 1967 (in preparation).
[14] ALGAVE, A. et al., ibid., VII. Survey of hospital supplies requiring sterilization at Kupat Cholim General Stores, IAEC Special Rep., 1967 (in preparation).
[15] COHEN, Y., BEN OR, Y., Economic feasibility of the production and sterilization of disposable hospital supplies, IAEC Special Rep., (Dec. 1967).
[16] COHEN, Y., BEN OR, Y., ibid., Appendices (Dec. 1967) (in Hebrew).
[17] LAPIDOT, M., Action Plan for ISORAD Ltd., A Plant for Disposable Hospital Supplies, IAEC Internal Rep. (14/2/67) (in Hebrew).
[18] JONAT, M., LAPIDOT, M., SHNEI DOR, K., Economic Feasibility of Plant dor Disposable Hypodermic Syringes and Needles, IAEC Internal Rep. (27/9/68) (in Hebrew).
[19] ERHARD, G., Economic Feasibility of Proposal for Manufacture of Disposable Syringes and Related Products, IAEC Internal Rep. (August, 1969).
[20] SHNEI DOR, K., LAPIDOT, M., Proposal for a Radiation Sterilization Facility for Medical Products, IAEC Internal Rep. (14/6/70).
[21] SHNEI DOR, K., LAPIDOT, M., Economic Feasibility of Radiation Sterilization Facility for Hospital Supplies, IAEC Internal Rep. (Jan. 1971).
[22] LAPIDOT, M., Specifications for Low Through-Put Irradiator for Sterilization of Disposable Hospital Supplies, IAEC (1970).
[23] LAPIDOT, M., HARAM, S., BARON, I., SATCHY, C., KAHAN, R.S., "A versatile cobalt-60 irradiation facility within a swimming pool reactor", Food Irradiation (Proc. Symp. Karlsruhe, 1971), IAEA, Vienna (1966) 755-66.
[24] ROSENBERG, K., The Ceric Sulfate Dosimeter, IAEC Internal Rep. (1970).
[25] ROSENBERG, K., Calibration of the Red Perspex Dosimeter, IAEC Internal Rep. (1971-72).
[26] MEDALIA, O., ARONSON, M., Use of B. pumilus Spores for Irradiation Dosimetry, IAEC Internal Rep., 1974 (in preparation).
[27] BEN MOSCHE, H., SHNEI DOR, K., LAPIDOT, M., Report on Bench-Scale Test Production of Radiation Sterilized Disposable Hypodermic Syringes and Petri Dishes, IAEC Internal Rep. (1971) (in Hebrew).
[28] LAPIDOT, M., BEN MOSCHE, H., SHNEI DOR, K., Improvements in the Production of Disposable Syringes Adapted to Sterilization by Gamma Radiation, Interministerial Inventions Committee Prize (1971).
[29] MOSCONA, A., Improvement of Systems in Pilot Scale Production of Disposable Syringes, IAEC Internal Rep. (1969) (in Hebrew).
[30] (a) ADLER, J., RODERIC, H., SALITERNIC, R., EISENBERG, E., LAPIDOT, M., Radiation Sterilized Laboratory Animal Feed Rations, IAEC Internal Rep. (1974) (in preparation).
(b) BRUCHIM, E., ADLER, J.H., EISENBERG, E., LAPIDOT, M., Use of Gamma Radiation to Eliminate Undesirable Microorganisms from Animal Feed Concentrates, Int. J. Radiat. Sterilization 1 4 (1974).
[31] BEN ABRAHAM, D., ADLER, J., ZIR, D., EISENBERG, E., LAPIDOT, M., Effect of Gamma Radiation on the Biological Value of Poultry Feed, IAEC Internal Rep. (1969).
[32] BEN ABRAHAM, D., ADLER, J., LAPIDOT, M., BRUCHIM, A., EISENBERG, E., Development of Poultry Feeds Sterilized by Ionizing Radiation, IAEC Internal Rep. (1970) (in Hebrew).

[33] Process for the Elimination of Salmonella from Animal Feeds by Gamma Radiation, II. Bio Research Laboratories Ltd. and Commercial Products, AECL, Canada (26 May, 1971).
[34] Ibid. Vol. III.
[35] Ibid. Vol. IV.
[36] Ibid. Vol. V.
[37] Letter of approval by C.C. Stevenson, Chief, Feed & Fertilizer Section, Plant Products Division, Canada Dept. of Agriculture (6 Dec., 1971).
[38] GUERRERO, F.P., STALLMAN, R.F., CHIERO, F., MAXIE, E.C., The Mobile Gamma Irradiator, U.S.A.E.C. UCD-34P80-5 Contract No. AT(11-1)-34 (1 Feb., 1966 - 30 Jan., 1967).
[39] ADLER, J.H., EISENBERG, E., BAR, D., ROSS, I., LAPIDOT, M., SPF Poultry Fed on Radiation Sterilized Diet, IAEC Special Rep., 1975 (in preparation).
[40] HOROWITZ, M., HULIN, N., Effects of Gamma Radiation on Soil and Diphenamid. Weed Sci. 19 (1973) 294-96.
[41] McKINNON, R.G., Design of Modern Medical Products Irradiators, Int. J. Radiat. Engng. 1 (1) (1971) 1-14.
[42] REDELMAN, A., et al., Economic Feasibility of Disposable Medical Products Facility, IAEC Internal Rep. (1973).
[43] (a) ROSENBERG, K., Calibration of J-6500 Sterilization Facility, IAEC Internal Rep. (1972).
(b) ROSENBERG, K., Recalibration of J-6500 Sterilization Facility, IAEC Internal Rep. (1974).
(c) ROSENBERG, K., RAHMIEL, B., Dosimetric Control at J-6500 Sterilization Facility for 1972-1973 period, IAEC Internal Rep. (1973).
[44] ROSENBERG, K., Calibration of J-6500 Irradiation Facility after Source Replenishment, IAEC Internal Rep. (October, 1974).
[45] ARONSON, M., MEDALIA, O., Proposal to Ministry of Health, IAEC Internal Rep. (1974).
[46] MEDALIA, O., ARONSON, M. Microbiological Control Experience in Industrial Radiation Sterilization, IAEC Internal Rep., 1974 (in preparation).
[47] Israel Ministry of Health - Code of Good Manufacturing Practice in the Pharmaceutical Industry (1973).
[48] STONEHILL, A.A., Report of Expert to Israel Government, IAEC Internal Rep. (March-June, 1974).
[49] EISENBERG, E., ALTMAN, G., BOBROWSKI, B., LAPIDOT, M., Ready-Made Sterilized Culture Media and Process of Preparation, Israel Patent Applied for 1974.
[50] EISENBERG, E., Private Communication.
[51] LAPIDOT, M., PADOVA, R., MOLCO, M., ROSS, I., LEIBOWITZ, Y., RAHMIEL, B., Assay of Radiation Sterilized Putrescible Matter from Garbage, IAEC Internal Rep. (1974) (in Hebrew).

DISCUSSION

Kikiben PATEL: Which method did you use for the pre-sterilization of microbial counts in the case of talcum powder?

M. ARONSON: The millipore-filtration method.

Kikiben PATEL: I should have thought that talcum powder would have blocked the millipore membrane.

M. ARONSON: We used the supernatant of the talcum powder suspension.

Kikiben PATEL: Yes, but if you use the supernatant for the pre-sterilization count, surely it will fail to show the presence of bacteria, since any bacteria will settle out together with the powder. I would suggest that in the case of talcum powder one would do better to use the shake-culture method on plate by applying serial dilution.

M. ARONSON: I can only speak from our limited experience. The gloves were cut, placed in a beaker containing a solution of saline and Tween, and rotated by a magnetic stirrer. The supernatant was first filtered (on a millipore filter), after which the talcum was added. The filter was placed on a petri dish, but no growth resulted. Recovery experiments with deliberately contaminated gloves were also negative. I agree that the shake-culture method might be better for talcum powder but we have not tried it.

J.O. DAWSON: As an example of the field of gnotobiotics, you might be interested to hear that at this time of year we are usually asked by a local children's hospital to irradiate Christmas stockings containing toys and coloured sweets for children undergoing treatment. This is one of the pleasanter aspects of irradiation!

PROPOSED PHILIPPINE RADIATION-STERILIZATION PLANT, AND A SURVEY OF MARKET POTENTIAL

Carmen C. SINGSON, L.D. IBE
Philippine Atomic Energy Commission,
Manila,
Philippines

Abstract

PROPOSED PHILIPPINE RADIATION-STERILIZATION PLANT, AND A SURVEY OF MARKET POTENTIAL.

The paper deals with a study to assess the market potential of radiation sterilization in the Philippines. A market survey conducted with the technical assistance of an IAEA expert from India shows that most of the pharmaceutical industries engaged in the manufacture of medical products unanimously agree that there is an urgent need for a centralized radiation-sterilization plant to meet the demands of sterilization of most of their products and packaging materials. The authorities of the government and some of the private hospitals surveyed are also very keen for the establishment of a sterilization facility since most modern medical products and devices are made of heat-sensitive thermoplastics which cannot be heat or steam sterilized. Availability of sterile products will help prevent cross-infection and ultimately help improve the public health standards of the population. The scope of the utilization of a radiation-sterilization facility in Diliman Quezon City is also discussed.

1. INTRODUCTION

Radiation sterilization using gamma radiation from radioisotope sources is now considered to be one of the most satisfactory methods of achieving sterility of medical supplies. More than fifty plants all over the world have been commissioned to sterilize medical supplies using gamma rays emitted from ^{60}Co sources since the first commercial radiosterilization plant was established in Australia in 1956. Aware of the successful commercial application of gamma radiation for medical product sterilization in advanced countries and also aware of many advantages of radiation-sterilization practices over conventional methods, a growing number of developing countries are also planning to set up radiation-sterilization plants.

The products are sterilized after packaging to permit their direct use without being contaminated with microorganisms. Usually known as the 'cold' sterilization technique, it is especially suitable for medical equipment made of plastics, which are heat sensitive and cannot be sterilized by conventional methods. Owing to the simplicity and safety of the plant operation, sterilization by gamma-ray irradiation has resulted in the wide use of single-use disposable items for the medical profession.

The Philippine Atomic Energy Commission (PAEC), created in 1958, is engaged in a programme for the application of radioisotopes and radiation in medicine, agriculture and industry. The acquisition and installation of a 20 000-Ci ^{60}Co source through the Technical Assistance Programme of the IAEA in March 1970 has initiated development work on the industrial applications of radiation, particularly in the sterilization of medical products.

With the Technical Assistance of an IAEA expert, Dr. V.K. Iya of India, and the author as the Philippine counterpart, a market survey was conducted from June to July 1972 to study the economic feasibility of setting up such a plant. At about the same time a similar plant was being constructed in Bombay, India, under the able supervision of Dr. Iya. It was very timely that we could avail ourselves of his expertise and experience gained in planning and constructing a radiation-sterilization plant.

The spectacular growth of the medical industry in the Philippines during the last 10 years has reached a considerable scientific magnitude. Most of those industries, located in the Greater Manila area, unanimously agree that there is an urgent need for a centralized radiation-sterilization plant to meet the demands of sterilization of most of their products and packaging materials.

We have approximately 750 government and private hospitals in the country with a total bed capacity of about 46 000. Numerous sterile medical supplies are required for daily use. Sterilization of these materials is usually carried out in autoclaves entailing considerable difficulties owing to frequent maintenance and breakdown problems. Another accepted practice of sterilization makes use of ethylene oxide gas. However, ethylene oxide is not manufactured in the country, and as yet very few manufacturers have installed gas-sterilization facilities. Hospital authorities agree that the establishment of a radiation-sterilization facility in the country is necessary since most of the modern medical products and devices are made of heat-sensitive thermoplastics which cannot be heat or steam sterilized. The availability at all times, of sterile cotton and gauze dressings, ready-to-use disposable plastic syringes, administration (intravenous) sets, gloves, catheters, kidney dialysis sets, heart valves and a host of other products, would not only be very convenient for our hospitals but would also help reduce cross-infection. The use of sterile sanitary pads also helps prevent vaginal infections. It has also been reported that the repeated use of syringes and needles is likely to cause serum hepatitis. An obvious solution to this problem is the use of disposable syringes and needles which can be efficiently and economically sterilized by radiation. Thus, the early establishment of a sterilization facility would not only be practical and useful in the improvement of public health standards in government and private hospitals but also to the population at large.

2. MARKET POTENTIAL

Drug and pharmaceutical companies engaged in the manufacture of absorbent cotton, sanitary pads, gauze dressings, talcum powder, dressing packs, sutures, disposable administration sets, pharmaceutical packaging materials, and also hospitals were visited in June to July 1972 to assess the need of a radiation-sterilization facility. A summary of the up-to-date market survey is shown in Table I.

The different medical products which were found to have a potential market for radiation sterilization are listed as follows:

2.1. Plastic administration sets

The raw material required for this purpose is polyvinyl chloride which is being manufactured in the Philippines. The tubings and other components

TABLE I. SUMMARY OF MARKET SURVEY

Product	Annual production	Estimated capacity requiring radiation sterilization (ft^3/yr)
1. Administration sets	2.5 million pieces	25 000
2. Disposable syringes & needles	10-20 million by 1977	15-30 000
3. Pharmaceutical packaging materials like polyethylene containers, eye-drop vials	10 million pieces	10 000
4. Absorbent cotton	700 t/yr	30 000
5. Sanitary pads	Over 600 t/yr	Not estimated, but potential market exists
6. Talcum powder	Production value of 200 million pieces annually	Not estimated, but potential market exists
7. Sutures	Production in 2 years likely – 10 million pieces	15 000

needed are also fabricated in the country. These sets are being assembled by a well-known pharmaceutical company, the annual production being about 2.5 million pieces with an estimated capacity of 25 000 ft^3. Formaldehyde and ethylene oxide gas are being used for sterilization but the manufacturer would be willing to switch over to radiation sterilization immediately a radiation facility became available. Another pharmaceutical firm which imports administration sets for distribution also intends to manufacture these sets in the country.

2.2. Disposable syringes and needles

It is estimated that about 2000 disposable syringes per day are being used by government hospitals and a similar number may be required by private hospitals.

There are at least two leading drug manufacturers in the country who have planned the production of disposable syringes and needles. The production target is 10 to 20 million syringes and needles per annum. Both are enthusiastic about the installation of a radiation plant.

2.3. Plastic containers and packaging materials

Several firms are engaged in the manufacture of polyethylene containers such as eye-drop vials, medicine droppers etc., and films for packaging. Since these cannot be sterilized by steam they are usually washed with water and rinsed with 70% alcohol. This method is time-consuming and unsatisfactory. Discussions with responsible staff of two companies show that they would be willing to utilize radiation sterilization.

2.4. Talcum powder

Talcum powder is at present manufactured by three leading firms. Raw talc of a sufficiently good quality is available at present. However, there are other more economic sources of talc which are otherwise not so satisfactory microbiologically. It is expected that radiation sterilization would be useful in upgrading and utilizing these new sources of talc and also in making available talcum powder free of any undesirable micro-organisms.

2.5. Absorbent cotton

There are two manufacturers of absorbent cotton, one of whom produces about 2 t/d. Hospitals purchase this absorbent cotton in big rolls and prepare gauze bandages, cotton balls, surgical packs etc. and steam sterilize them before use. Autoclaves often break down, which makes them unreliable devices for sterilization. The two companies mentioned earlier also manufacture sanitary pads for gynaecological and obstetric use. Discussions with the authorities of the hospitals visited reveal that many of these medical products used daily could qualify for radiation sterilization. The Hospital Association of the Philippines can help its members by providing them with irradiated and sterile medical products. It is expected that the surgical cotton, gauze, dressings and sanitary pads would alone provide a large potential market.

2.6. Sutures

Sutures are not manufactured in the country at present, but an internationally known firm plans to manufacture them should a radiation facility be installed. The production goal is 10 million per year, and it has a considerable export potential.

2.7. Other products

A chemical company intends to manufacture kidney dialysis sets in the Philippines. Others have also indicated their desire to manufacture various disposable medical products such as catheters, gloves, petri dishes etc. All these developments will depend to a considerable extent on a radiation-sterilization facility to ensure sterility of their products.

2.8. Volume of material

The volume of material to be treated initially has been estimated to be about 100 000 ft^3/yr. With the demonstration of this technique, the volume of administration sets, disposable syringes and needles, plastic containers, talcum powder, absorbent cotton and its formulations, sutures and other products requiring radiation sterilization, is expected to increase to about 300 000 ft^3 within 3 years after plant commissioning.

3. THE PROPOSED RADIATION-STERILIZATION PLANT

As observed already, the market survey indicated an urgent need for the installation of a radiation-sterilization facility. The PAEC has initiated a request with the United Nations Development Programme for a project on the setting up of a demonstration facility in Quezon City.

Inasmuch as this project will constitute the first large irradiation installation in the country, it is proposed that the PAEC, an agency under the National Science Development Board, administers the said project. A project team will be organized from adequately experienced personnel of the PAEC to meet the requirements.

To ensure adequate liaison and co-operation from the health authorities as well as from the industrial sector, a Project Advisory Committee composed of representatives of the PAEC, the Department of Health, Food and Drug Administration, the Pharmaceutical Industries and the Hospital Association of the Philippines would be formed. The proposed plant will be located in Diliman, Quezon City within the Philippine Atomic Energy Commission site, taking into account its accessibility to industries and, at the same time, ensuring the ready availability of technical personnel from PAEC. The proposed radiation sterilization plant will have an initial load of 100 000 Ci of ^{60}Co source with a built-in capacity to load about 300 000 Ci of ^{60}Co at a later date. It is intended to enable the practical demonstration of sterilizing initially about 100 000 ft^3/yr of various medical supplies to be stepped up finally to 300 000 ft^3/yr.

We are very hopeful that financial assistance of about $500 000 can be obtained from funds programmed under the UNDP. This amount will be used for the purchase of the ^{60}Co source and the necessary equipment including conveyor, source frame, control console etc. It will be necessary for PAEC to obtain a government contribution of about $1 million, which will be used to construct the building, and to purchase other equipment and laboratory facilities. Expenses for foreign experts and fellowships will be requested from the Technical Assistance Programme of IAEA, Colombo Plan, India-Philippine Collaboration Agreement and elsewhere.

A pre-project training programme was initiated and personnel from the existing PAEC staff were selected. Three scientific personnel have already completed training at the Bhabha Atomic Research Centre (BARC) on civil and mechanical engineering aspects. They were here during the construction of ISOMED, India's first radiation-sterilization facility in Bombay. The Philippine Government has deputed the principal author of this paper on an IAEA fellowship to BARC in the quality control of radiation-sterilized products. The BARC has excellent facilities for this training programme, which involves microbiological, physical and chemical tests of medical products and pharmaceuticals before and after irradiation. This training will be very helpful for setting up a quality control laboratory on return to the Philippines. The problems that we may face in the commissioning and operation of the plant in the Philippines could be similar to those that may have arisen in India. Hence, we believe that we can fruitfully avail ourselves of the expertise and experience of the BARC scientists at ISOMED.

4. OBJECTIVES

The aim of the project is to introduce to manufacturers of medical products and also to hospital centres, the technology and practice of radiation sterilization of medical products. Within the framework of the broad health policy, this project will help improve public health standards through the availability of sterilized single-use medical supplies.

Its immediate objectives would, therefore, be:

(a) To introduce a new technique for proper sterilization of locally manufactured medical products;
(b) To enable the practical demonstration of sterilizing approximately 100 000 ft^3 annually various disposable medical products and other items for clinical purposes;
(c) To provide the sterilization services which will improve the health standards in hospitals;
(d) To provide regional training opportunities for the use of radiation-sterilization technology;
(e) To conduct a research and development programme on new applications of radiation sterilization;
(f) To provide information on the economics and operation for future projects on a larger scale.

The project was favourably endorsed by authorities of the Department of Health, Hospital Association of the Philippines and the Philippine Drug Manufacturers' Association.

BENEFITS

The establishment of a radiation-sterilization facility will help in the development of a disposable plastic industry and can, therefore, help conserve our much-needed reserves and decrease our imports that amount to about $1 million annually.

It can also act as a centre for the irradiation of complicated medical devices such as kidney dialysis sets, plastic parts for lung machine, heart valves, etc.

One can confidently and optimistically say that the radiatiation-sterilized products would enjoy greater reliability in the quality of sterility and thus be safer for their final intended use. Owing to the enormous increase of electric power cost brought about by the energy crisis, it is expected that the setting up of a radiation-sterilization facility would serve to mitigate some of these problems and also provide the much-needed sterile medical products for the various hospitals and other health centres throughout the Philippines.

ACKNOWLEDGEMENT

The authors wish to express their grateful thanks to Mr. V.K. Iya and Mr. N.G.S. Gopal for their helpful discussions.

BIBLIOGRAPHY

Foreign Trade Statistics of the Philippines, R.P. Bureau of Census and Statistics (1972).

Foreign Trade Statistics of the Philippines, R.P. Bureau of Census and Statistics (1973).

Annual Report of the Bureau of Hospitals, Philippines Department of Health (1973).

DISCUSSION

G.O. PHILLIPS: I would have thought, generally speaking, that it was up to the manufacturers of the products to be irradiated to set up their own plant, rather than for the Government authorities to undertake the project.

Carmen SINGSON: Yes, perhaps so. But the manufacturers don't want to be involved in so much initial investment and would prefer to pay a fee to have their products irradiated. If the plant runs smoothly, however, the Government may turn it over to industry at a later stage.

V.K. IYA: Medical product manufacturers are generally not willing to make the heavy investments necessary for radiation-sterilization plants, so it is best for the Government authorities to initiate matters in this regard in the developing countries. In India, Johnson and Johnson were considering the question of putting up a plant some ten years back, yet today, when ISOMED is available, their utilization of this plant is very limited — not even all their sutures are being irradiated.

J.O. DAWSON: I'd like to mention that the Wantage plant in England was originally set up as a pilot plant to encourage industrial use, and it charged an economic price. It was not intended to be a permanent installation. But transport, even on 400 miles of good road, is difficult, not to mention the question of turn-round time.

R.A. VAUGHAN: I should like to support the comments of Mr. Iya and Mr. Dawson in suggesting that a small-batch plant would be useful in many cases. The Radiochemical Centre of the UK is at present actively engaged in collaboration with a well-known UK manufacturer, in developing designs for manually operated small-batch irradiators for use where a large plant cannot be justified. These should be particularly suitable for use in developing areas.

H.M. ROUSHDY: In connection with increased demands for the establishment of radiation-sterilization demonstration facilities in the developing countries, I wonder whether the establishment of such plants could not be based on a carefully considered policy to be evolved by the IAEA, with the assistance of UNDP. The idea would be to divide the developing countries in geographical sectors, and to aim at equitable distribution in establishing such units, i.e. in accordance with actual demands. Naturally, this policy should not affect any national programmes for the establishment of fully commercial plants financed by industry in different developing countries.

V.K. IYA: As a comment on Mr. Roushdy's point, I would point out that regional plants may not be feasible due to high transport costs. On the other hand, in smaller countries where the market is not very large, one could perhaps consider low-cost panoramic irradiators.

PROGRESOS DE LA RADIOESTERILIZACION EN LA REPUBLICA ARGENTINA

E.E. MARIANO, H.A. MUGLIAROLI
Comisión Nacional de Energía Atomica,
Buenos Aires, Argentina

Abstract–Resumen

PROGRESS OF RADIOSTERILIZATION IN THE ARGENTINE REPUBLIC.

The technological and economic advantages of radiation as a means of sterilization have led to a steady expansion of its use. With the significant increase in the radiosterilization of medical products, the health authorities have established standards for the use of radiation for that purpose. These standards will be incorporated in the next edition of the Argentine Pharmacopoeia. In general, Argentine legislation in this area is based on the IAEA's Code of Practice for Radiosterilization of Medical Products. The microbiology services of Argentina's National Atomic Energy Commission (CNEA) determine the number of viable microorganisms in each product and package the product before irradiation. The controls are carried out on random samples. At the same time, the chemistry laboratories carry out quality controls on new products which have received sterilizing doses of gamma radiation. The gas sterilization methods employed most widely in the industrial manufacture of thermolabile products involve the use of ethylene oxide or formaldehyde. The cost of this process is influenced considerably by the fact that Argentina has to import ethylene oxide. The low penetrating power of formaldehyde means that the package containing the product must still be open at the time of sterilization if an adequate safety margin is to be ensured; consequently, one needs sterile installations for the post-irradiation sealing of packages. The capital and operating costs of such installations are high. A comparison has been made of the costs of sterilizing 11 000 m^3 of a product with a density of 0.1 g/cm^3 and the following relative values obtained: ethylene oxide, 399.2; formaldehyde, 947.5; gamma radiation, 166.0. Studies of the effects of sterilizing doses on certain pharmaceutical products have demonstrated that the radiosterilization of some drugs is possible. The CNEA has designed an installation for the radiosterilization of medical products; the installation would have a 500 000-Ci cobalt-60 source and an annual capacity of some 11 000 m^3.

PROGRESOS DE LA RADIOESTERILIZACION EN LA REPUBLICA ARGENTINA.

Las ventajas tecnológicas y económicas que ofrece la radiación como método de esterilización, ha facilitado un sostenido incremento del uso de este método. Teniendo en cuenta el significativo aumento experimentado en la radioesterilización de productos médicos, las autoridades sanitarias han establecido normas para su uso. Estas normas serán incorporadas a la nueva edición de la Farmacopea Argentina. En general, la legislación argentina está basada en el Código de Práctica para Radioesterilización de Productos Médicos del OIEA. Los servicios de microbiología de la Comisión Nacional de Energía Atómica (CNEA) realizan la determinación del número de microorganismos viables, presentes en cada producto y su envase antes de ser irradiados. Estos controles se realizan en muestras tomadas al azar. Asimismo, los laboratorios de química realizan el control de calidad en nuevos productos sometidos a la dosis esterilizante de radiación gamma. Los métodos de esterilización por gas más frecuentemente usados en la producción industrial de productos termolábiles son los que utilizan óxido de etileno o formaldehído. La Argentina debe importar el óxido de etileno y este hecho influye significativamente en el costo del proceso. La escasa penetrabilidad del formaldehído obliga a tratar el producto en envase abierto para esterilizar con un adecuado margen de seguridad. Consecuentemente, se debe disponer de instalaciones estériles para el cerrado de los envases después de esterilizado el producto. Estas instalaciones y la operación en las mismas influye notablemente en los costos. Se han examinado los costos comparativos para la esterilización de 11 000 m^3 de productos de 0,1 g/cm^3 de densidad. Los resultados obtenidos, expresados en valores relativos por cada m^3, arrojan las siguientes cifras: óxido de etileno: 399,2; formaldehído: 947,5; irradiación gamma: 166,0. Se estudiaron los efectos de las dosis esterilizantes en ciertos productos de uso farmacéutico. Estos estudios demostrarían que la radioesterilización de algunas drogas resulta posible. La CNEA ha desarrollado y diseñado una instalación para radioesterilización de productos médicos. Esta instalación ha sido proyectada para trabajar con 500 000 Ci de ^{60}Co y puede tratar del orden de 11 000 m^3 por año.

INTRODUCCION

La Comisión Nacional de Energía Atómica (CNEA), mediante el uso de sus instalaciones de radiación gamma, realiza tareas de promoción en el campo de las aplicaciones de las radiaciones a nivel industrial.

Factores tecnológicos y económicos facilitaron un sostenido incremento en el uso de la radioesterilización. El significativo aumento que se produjo en el uso de esta tecnología y las perspectivas de sus proyecciones a corto plazo, creó la necesidad de establecer precisas normas de trabajo que permitieran usufructuar al máximo las ventajas de esta tecnología.

En general, para la elaboración de estas normas se tomaron como base las recomendaciones establecidas en el Código de Práctica preparado por el Organismo Internacional de Energía Atómica [1].

Este trabajo presenta una visión general de los avances logrados en el campo de la radioesterilización en la República Argentina.

DESARROLLO LOGRADO

La CNEA ha desarrollado y construido una planta de irradiación semi-industrial. Esta instalación ha sido prevista para operar con una actividad de 10^6 Ci de ^{60}Co. Es oportuno señalar que esta instalación no fue diseñada exclusivamente para realizar servicios de esterilización sino también para promover la utilización de las radiaciones ionizantes en otros procesos industriales.

La actividad de promoción se cumple con el apoyo de laboratorios experimentales que permite asistir a la industria en todos los problemas conexos con la radioesterilización efectuando el correspondiente control de calidad química y biológica de los productos tratados.

Los nuevos productos que se incorporan al nomenclátor de materiales esterilizados por radiación son previamente sometidos a ensayos físicos y químicos a fin de determinar su resistencia a las radiaciones, o verificar que las transformaciones que puedan sufrir no comprometen la salud de los usuarios o alteren las propiedades inherentes a su uso.

Servicios especializados realizan un riguroso control de los niveles de contaminación del producto a fin de determinar si los mismos encuadran dentro de los márgenes de seguridad adoptados.

La Comisión Permanente de Farmacopea Argentina ha aprobado la incorporación en su nueva edición de un capítulo referente a la radioesterilización, estableciendo las normas a que deberán ajustarse los procesos de radioesterilización de productos médicos descartables; a continuación se resumen algunos conceptos referentes a los criterios utilizados en la elección de la dosis:

«La dosis de esterilización estará determinada por el factor de inactivación adoptado. Como norma general el valor mínimo de dicho factor será de 10^5. Si un producto puede, por características propias, actuar como medio apropiado, obrar como vehículo o contener productos que facilitan el desarrollo microbiano, el factor mínimo será de 10^8.

«El factor de inactivación se determinará sobre cultivos de microorganismo más radiorresistente de la flora de contaminación habitual del producto, debiendo realizarse la experiencia en las condiciones ambientales más adversas a la acción inactivante de la radiación.

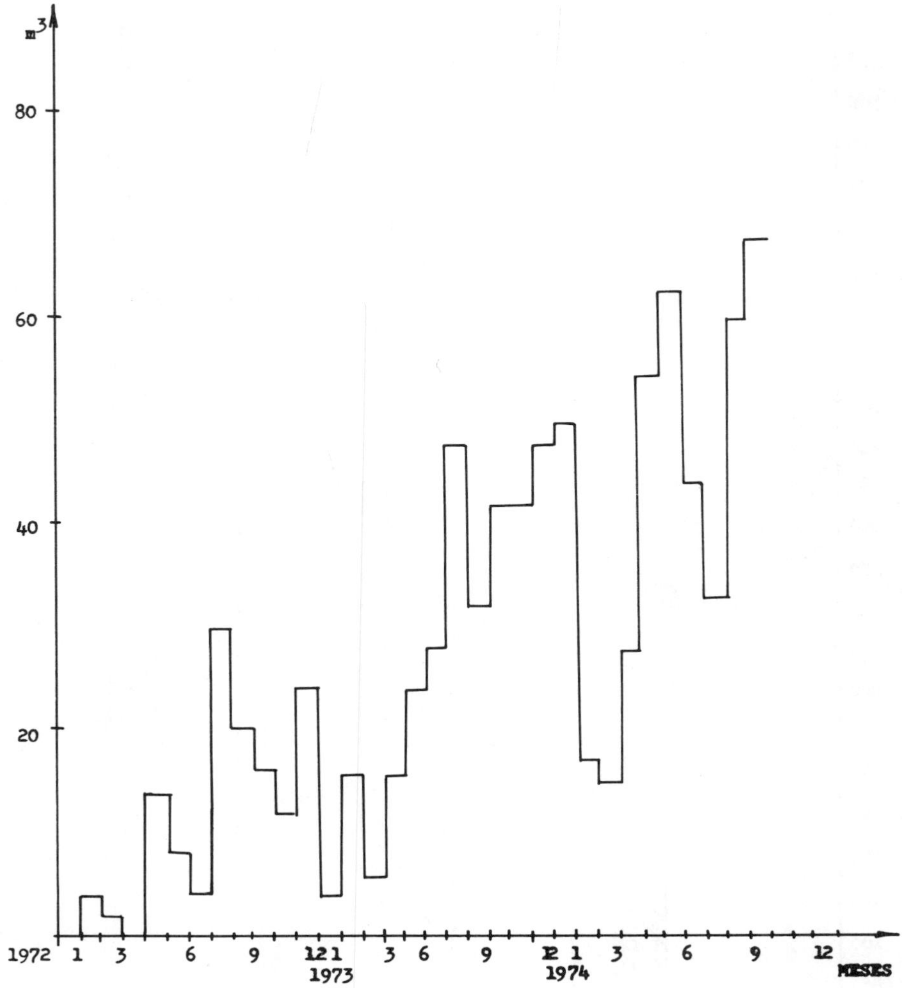

FIG.1. Volúmenes mensuales tratados.

« Cuando se esterilize con un factor de inactivación de 10^5 la contaminación inicial del producto no podrá superar a 10 microorganismos por unidad. Cuando dicho factor sea 10^8 la máxima contaminación inicial será de 50 microorganismos por unidad. Por cada orden de incremento de la contaminación inicial el factor de inactivación se incrementará en un orden.

« Si la contaminación inicial es superior a 5000 microorganismos por unidad, el producto no podrá ser irradiado.

« Los materiales con que se confeccionan los productos no deberán presentar, después de irradiados, cambios físicos, ni químicos que puedan originar sustancias tóxicas para el usuario.

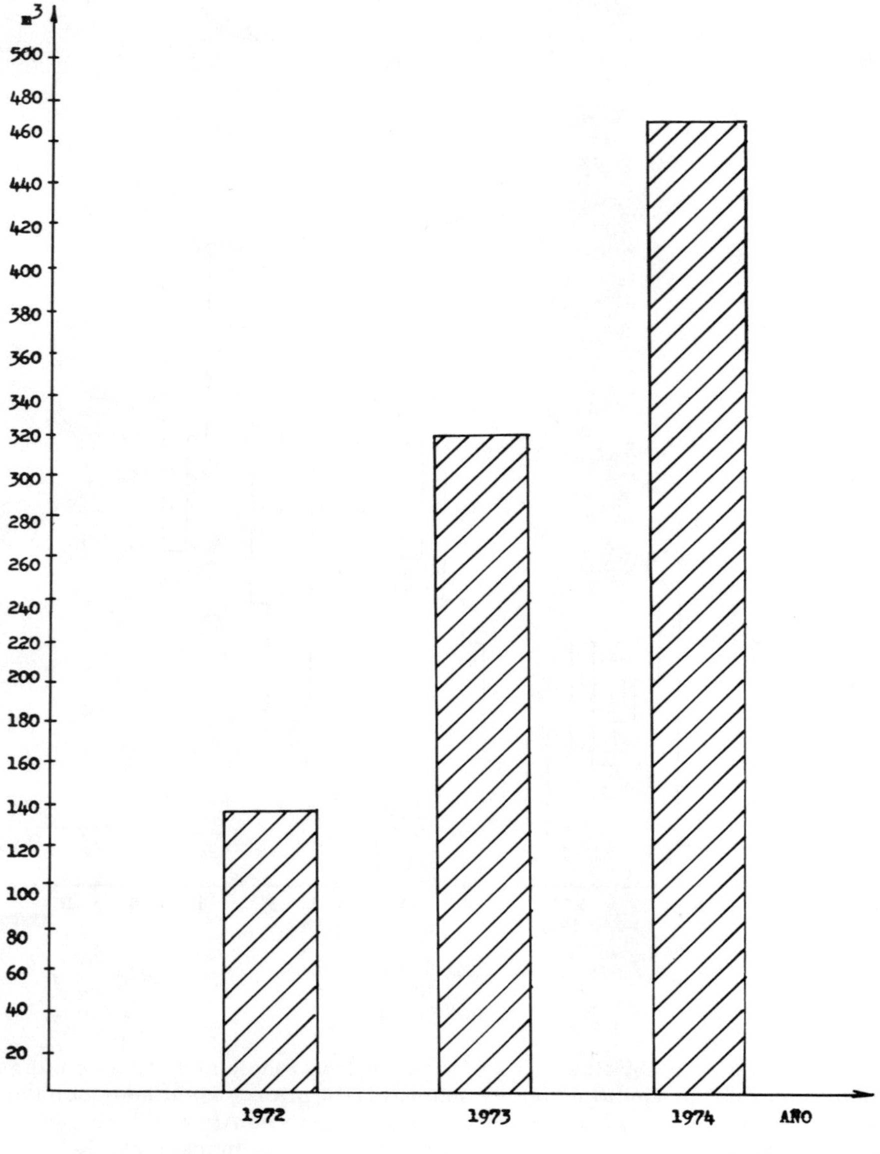

FIG. 2. Material tratado.

«En ningún caso la dosis de radiación empleada deberá ser inferior a 2,5 Mrads (o sea la que corresponda a una energía absorbida igual a $2{,}5 \cdot 10^8$ erg/g masa)».

Las ventajas ofrecidas por la radioesterilización han sido rápidamente captadas por la industria, lo que ha permitido una sostenida expansión de esta metodología en estos últimos años (figuras 1 y 2).

La lista de productos radioesterilizados en la actualidad incluye, entre otros, jeringas; equipos para transfusión de soluciones parenterales; vendajes; gasas; prótesis acetabulares para traumatología; acrílicos de uso ortopédico para cementaciones, fijaciones, reconstrucciones y prótesis; agujas hipodérmicas descartables; colectores urinarios pediátricos; apósitos para protección femenina; espirales anticonceptivas; oxigenadores; etc.

La radioesterilización ha permitido el desarrollo con éxito de algunos aspectos de la cirugía cardiovascular en ciertas anomalías congénitas, permitiendo la esterilización y conservación prolongada de homoinjertos.

RADIOESTERILIZACION DE FARMACOS

Los estudios realizados hasta el presente sobre diferentes fármacos indicarían la factibilidad del uso de la radioesterilización en algunos productos, preferentemente cuando los mismos son tratados en estado sólido. Entre los productos estudiados que parecían conservar sus propiedades terapéuticas o biológicas después de irradiados se destacan: gamma globulina liofilizada, sulfato de neomicina, sulfato de polimicina, cloruro de benzalconio, gramicidina, algunas fracciones de heparina.

Solamente se completaron los estudios de pancreatina irradiada y se han hecho las presentaciones pertinentes para obtener la correspondiente autorización de las autoridades sanitarias.

SEGURIDAD DEL PROCESO DE FABRICACION

A fin de ajustar los niveles de contaminación de los productos dentro de las normas establecidas para lograr la esterilidad del producto dentro del margen de seguridad deseado, se controlan las áreas de fabricación recomendando a las diferentes industrias las modificaciones que se consideran necesarias en sus instalaciones y en el proceso de fabricación de los productos.

De esta forma se han logrado mejoras significativas en los niveles de contaminación del producto final (figura 3).

La eficiencia microbiológica de la planta de irradiación se determina periódicamente mediante el uso de cepas de <u>B. sphericus</u> C_1A.

Las curvas de inactivación logradas en nuestra planta con estas cepas patrones muestran resultados coincidentes con los obtenidos por Christensen en los servicios de radioesterilización de Dinamarca [2, 3].

Es oportuno señalar que en ensayos realizados en nuestros servicios se ha podido comprobar que los microorganismos que contaminan habitualmente los productos son notablemente más sensibles a la acción esterilizante de las radiaciones que el <u>B. sphericus</u> C_1A.

Con el objeto de determinar si los niveles de contaminación de los productos a esterilizar encuadran dentro de las normas establecidas, se analizan muestras representativas por cada partida de producción.

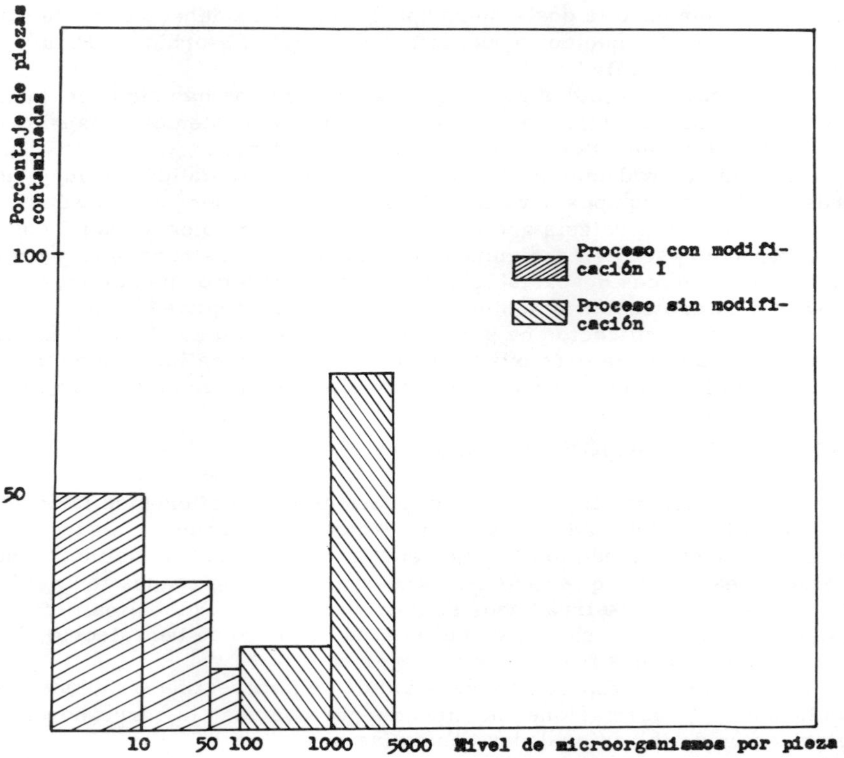

FIG.3. Mejoramiento de la calidad biológica de apósitos de uso femenino, mediante modificaciones del proceso de fabricación.

INGENIERIA DE RADIACION

De acuerdo con las cifras que surgen de una prospección sobre las necesidades a mediano término de la industria, la CNEA ha desarrollado un prototipo, factible de ser adoptado por la industria, con características que permiten la construcción total de la instalación por la industria nacional.

Para estas instalaciones se desarrolló y diseñó una máquina capaz de tratar anualmente 11 000 m^3 de material irradiado a una dosis de 3,2 Mrads, con una densidad de aproximadamente 0,1 g/cm^3.

La máquina diseñada está calculada para trabajar con una fuente de 500 000 Ci de ^{60}Co y permitirá tratar el volumen indicado en 8000 horas.

EVALUACION TECNICOECONOMICA

Las ventajas y problemas conexos con el uso de los métodos de esterilización que utilizan agentes químicos o radiaciones ionizantes han sido vastamente analizados en la literatura [4 - 7].

Por los tanto, sólo comentaremos aquellos detalles que por sus particularidades tienen incidencia en los costos del producto final, para las condiciones de operación en Argentina.

La esterilización por óxido de etileno y por vapores de formaldehído son los métodos más difundidos por la industria en la República Argentina; la particularidad que ofrece el formaldehído por su escaso poder de penetración obliga a considerarlo en forma especial, por su incidencia en el costo del producto final.

En efecto, para lograr márgenes de seguridad comparables a los obtenidos con óxido de etileno o radiaciones en la esterilización de productos biomédicos, es conveniente efectuar el tratamiento en envases abiertos. Consecuentemente, se debe disponer de áreas estériles y utilizar una escrupulosa técnica de cerrado del envase que evite la recontaminación del producto. Esta última etapa del proceso influye considerablemente en el costo del producto.

En base a las consideraciones apuntadas anteriormente se ha realizado un estudio comparativo para los tres métodos de esterilización mencionados: óxido de etileno, formaldehído y radiación gamma.

En el análisis se han tomado en consideración únicamente las etapas relacionadas con el proceso de esterilización, excepto en el caso del formaldehído en donde se incluye la amortización de las instalaciones especiales para el cerrado de envases en ambiente estéril del producto esterilizado.

El estudio se ha realizado para el tratamiento anual de 11 000 m^3 de material descartable de uso médico compuesto de jeringas, tubuladuras y agujas, cuya densidad es de 0,1 g/cm^3.

Para la esterilización por radiación del volumen del material mencionado, administrando una dosis de 3,2 Mrads, se requiere una batería de 9 autoclaves de capacidad real unitaria de 1,4 m^3.

Se ha considerado un período de 10 años para la amortización de las inversiones, con un interés anual de 1% considerado para moneda estable.

Costo del proceso de esterilización mediante radiación gamma

Se ha calculado para una instalación de iguales características del prototipo desarrollado por la División Fuentes Intensas de la Comisión Nacional de Energía Atómica.

Se consideró como parte amortizable de la instalación el 50% de la actividad de cobalto inicial, teniendo en cuenta que al final del período de amortización al remanente de cobalto, que sería prácticamente igual a la actividad inicial, se le asigna un valor de reventa del 50% debido a la disminución de su actividad específica.

La actividad a incorporar anualmente para cubrir la pérdida por decaimiento se la considera como combustible, en gastos fijos.

Gastos fijos:

a) Personal (operando 8000 h/año)
Incluye servicio de operación, dosimetría, control bacteriológico y mantenimiento — 60,00 $/m^3$ [1]

[1] Las cantidades están dadas en pesos argentinos: 10 pesos = 1 dólar de EE.UU.(enero de 1975).

b)	Varios Incluye consumo de ^{60}Co, servicios, repuestos, herramientas, seguros, etc.	50,00 \$/m^3	
c)	Amortización e intereses Incluye inmuebles, instalaciones y maquinarias	27,50 \$/m^3	
	Fuente de ^{60}Co amortizable	28,00 \$/m^3	
	Total	166,00 \$/m^3	

Costo del proceso mediante óxido de etileno

No siendo la República Argentina productora de óxido de etileno la industria debe importar este compuesto, recayendo sobre los gastos del producto final las costosas erogaciones que significa el transporte del óxido de etileno, el que por razones de seguridad es enviado mezclado con 80% de CO_2.

	Gastos variables:		
a)	Consumo de óxido de etileno + CO_2 incluyendo gastos de importación y flete	135,0 \$/m^3	
b)	Energía eléctrica (calefacción autoclave y bombas)	14,5 \$/m^3	
	Total de gastos variables		149,5 \$/m^3
	Gastos fijos:		
a)	Personal (operando 8000 h/año) Incluyendo servicio de operación, control bacteriológico y mantenimiento	48,00 \$/m^3	
b)	Varios Incluye repuestos y elementos para mantenimiento de inmuebles, instalaciones y equipos, seguros, etc.	4,90 \$/m^3	
c)	Amortización e intereses Incluye inmuebles, instalaciones generales y maquinarias, elementos y dispositivos auxiliares	146,80 \$/m^3	
	Total de gastos fijos		199,70 \$/m^3
	Total		349,20 \$/m^3

Costo del proceso mediante formaldehído

	Gastos variables:		
a)	Personal de área estéril de cerrado	803,40 \$/m^3	
b)	Consumo de formaldehído, servicio de electricidad	19,60 \$/m^3	
	Total de gastos variables		823,00 \$/m^3

	Gastos fijos:		
a)	Personal ídem óxido de etileno	48,00 $/m^3$	
b)	Varios Incluye repuestos y elementos para conservación de inmuebles, instalaciones y maquinarias, herramientas, seguros, etc.	3,80 $/m^3$	
c)	Amortización e intereses Incluye inmuebles, instalaciones generales, instalaciones especiales de ventilación, maquinarias, instrumentos, etc.	78,70 $/m^3$	
	Total gastos fijos		124,50 $/m^3$
	Total		947,50 $/m^3$

REFERENCIAS

[1] OIEA, «Recommended code of practice for radiosterilization of medical products», Radiosterilization of Medical Products (Actas Simp. Budapest, 1967), OIEA, Viena (1967) 423.

[2] CHRISTENSEN, E.A., Ugeskr. Laeger 131 (1969) 2123.

[3] CHRISTENSEN, E.A., «Radiation resistance of bacteria and the microbiological control of irradiated medical products», Sterilization and Preservation of Biological Tissues by Ionizing Radiation (Actas Panel Budapest, 1969), OIEA, Viena (1970) 1.

[4] ERNST, R.R., «Ethylene oxide gaseous sterilization for industrial application», Cap. 12, Industrial Sterilization (PHILLIPS, G.B., MILLER, W.S., Eds), Int. Symp. Amsterdam (1972) 181.

[5] EYMERY, R., "Design of radiation sterilization facilities", Cap. 11, Industrial Sterilization (PHILLIPS, G.B., MILLER, W.S., Eds), Int. Symp. Amsterdam (1972) 153.

[6] PHILLIPS, C.R., «Gaseous sterilization», Lectures on Sterilization (BREWER, J.H., Ed.), Int. Symp. Amsterdam (1972) 33.

[7] SPINER, D.R., HOFFMAN, R.K., Appl. Microbiol. 8 (1960) 152.

DISCUSSION

K.H. MORGANSTERN: You mentioned that if ethylene oxide were locally available in Argentina, it would be less expensive as a means of sterilization than gamma rays. However, the figures given in your paper do not seem to bear this out. In fact, if you take the cost of ethylene oxide as zero, the remaining operational costs appear to be greater than for gamma-ray sterilization. Would you please clarify this discrepancy?

E.E. MARIANO: On the basis of our study your observation is correct. But it is a fact that the difference in favour of radiation sterilization would be considerably reduced if ethylene oxide were produced in the country.

The difference in cost in favour of irradiation might not be competitive (or even be economically disadvantageous), as has been the case in the United State of America, to judge by the literature. You must remember that our calculation is theoretical and would have to be adjusted to fit in with actual conditions at a later stage.

V.K. IYA: I must congratulate you on a very comprehensive paper; I am very impressed by the excellent work done in Argentina in this field. I have two questions to ask you. First, do you propose to set up the plant in Buenos Aires, or in the provinces, and, second, in the course of your interesting comparison of the cost of radiation and ethylene oxide sterilization, have you compared the cost of irradiation with steam?

E.E. MARIANO: Thank you, Mr. Iya. The plant in question is to be set up in Buenos Aires.

In reply to your second question, we have not made comparative studies with steam, since steam sterilization is not a method normally used on an industrial scale in Argentina, or at least not for the heat-sensitive medical supplies.

A. KEALL: What was the frequency of isolation of radioresistant organisms? In what products were the organisms isolated, and what were the organisms?

E.E. MARIANO: I would not like to hazard any guesses as to the number, but I can say that the radioresistant organisms appeared very frequently, in fact, almost constantly. The product they were isolated in was industrial cotton. The radioresistant microorganism isolated was Aerobacter simplex, but the biochemical identification tests have not yet been completed.

K.S. AGGARWAL: Regarding the ^{60}Co source storage system, you mentioned that a wet storage facility was being used in the production plant. What has been your experience with this type of storage facility?

E.E. MARIANO: I personally regard this type of plant as easier to deal with in the event of an accident. We had an experience of this kind just this year. The source levelling mechanism went wrong and the source got stuck in the irradiation position. We flooded the irradiation chamber and sent in frogmen, who were able to repair the fault, and the plant was in working order again in about seven days. It should be stressed that the possibility of flooding the plant in the event of an accident had been provided for.

PROSPECTS FOR RADIATION STERILIZATION OF MEDICAL PRODUCTS IN EGYPT

H.M. ROUSHDY
National Centre for Radiation Technology,
Atomic Energy Authority, Cairo,
Egypt

Abstract

PROSPECTS FOR RADIATION STERILIZATION OF MEDICAL PRODUCTS IN EGYPT.

The pharmaceutical industry in Egypt is continually expanding its activity and each year marks new accomplishments and additions which enable the companies to apply the most modern scientific means in the production of pharmaceutical preparations and consequently to improve their market potentialities. The certainty of expansion and the possibility of increasing exports of sterilized medical products, particularly to Arab and African countries, indicate a need for a gamma-sterilization plant. This technology permits the introduction of the latest practices with regard to used disposables, thus greatly reducing the chances of cross-contamination which usually results in serious complications enhanced by local environmental conditions. This paper reviews the current state and future prospects for radiation sterilization of medical products and biological tissues in connection with other related industrial radiation processings. Moreover, the paper reviews the Egyptian scientific and technical experience with irradiation facilities and the parameters underlying the choice of Egypt's first industrial gamma and electron-beam irradiators designed for more than a single-purpose use, with hygienic measures taken to avoid biological contamination of sterilized medical packages throughout the sterilization process. In addition, the paper deals with the policy set up for establishing the Egyptian National Centre for Radiation Technology with a view to introducing irradiation techniques in the sterilization of medical products, and to improving the properties and increasing the competitiveness of Egyptian fabrics. Apart from medical sterilization, certain industrial processes have been mentioned to show how a multi-purpose irradiation facility may be utilized in a developing country to justify significantly the large investment required.

INTRODUCTION

Egypt occupies a central position between the major field of tropical production of basic material in Asia and Africa and that of industrial production in Europe. Realizing that the industrialization of the country is one of the cardinal pillars on which its economic independence rests, Egypt has encouraged all national industries and secured an atmosphere conducive to the investment of capital in industrial projects. An adaptation of custom duties to the requirements of local industries has helped to develop such industries to a level very beneficial to the Egyptian national economy. Special attention has recently been given to the establishment of free zones in order to encourage foreign investment.

Egyptian industry is being considerably encouraged to exploit all possible productive energies in the country. According to 1971 statistics, the value of industrial production amounted to 2258 million Egyptian pounds. Integrated industrial areas grouping several industrial projects have been progressively constructed; for example, the centre at Shoubra El-Kheima for industries of textiles, plastics, wood-working, varnishes and paints, electrical and telephone cables, jute, tyres, etc.; the centre of Mehalla El-Kubra where weaving and textile industries are famous for their high-quality products; the centre at Kafr El-Dawar dealing with modern textile industries,

TABLE I. STATISTICS (1973) ON LOCAL PRODUCTION OF PHARMACEUTICALS, CHEMICALS AND MEDICAL APPLIANCES
(Production values in 10^6 Egyptian Pounds)

Name of company	Year established	Investment	Production value	Export value	No. of workers	Major medical products
NILE	1962	5.839	8.141	0.406	2789	Pharm., antibiot., chem. extracts, solut. sutures
CID	1947	5.231	6.196	0.082	2520	Pharm., antibiot., chem. cosmot., antibilharz. drugs
KAHIRA	1962	3.425	6.552	0.040	1244	Pharm. cosmotics, gelatine caps.
MISR	1940	4.127	6.673	0.109	1749	Pharm., antibiotics
MEMPHIS	1940	2.881	3.752	0.118	1291	Pharm. antibiot. extracts, active compounds
ARAB	1963	2.220	3.046	0.036	883	Pharm., extracts, cosmot. anaesthet.
ALEXANDRIA	1962	2.458	3.640	0.025	565	Pharm. cosmot., adhesives, plasters, surgical gloves
EL NASR	1960	10.624	4.905	0.022	1744	Raw mat. for Pharm., laboratory chemicals
MED. PACKAGING	1965	2.644	2.551	0.037	2288	Plastic cont., metal, glass & paper packaging
		39.449	45.476	0.875	15073	

Companies in Joint Ventures: Pffizer/Misr - 1962
Hockest/Orient - 1962
Swiss Pharma - 1965

TABLE II. CAPACITY OF DRUG PRODUCTION AS PERCENTAGE OF TOTAL DRUG CONSUMPTION IN EGYPT
(Values in 10^6 Egyptian Pounds)

Year	Total consumption	Local prod. value	Percentage local production
1952/53	4.8	0.5	10.0
1960/61	14.9	4.3	28.0
1961/62	17.2	7.9	45.0
1962/63	22.5	12.1	53.0
1964/65	31.0	21.5	69.1
1965/66	34.0	26.0	76.5
1966/67	34.6	27.8	80.5
1968/69	37.7	32.6	86.5
1971/72	55.2	48.2	87.3
1973	62.7	55.0	87.6
1975 (estimated)		(66.2)	

particularly the production of synthetic and nylon fibres started in 1958; and the numerous industrial drug centres located in Cairo and Alexandria with an annual drug and pharmaceutical production amounting to more than 60 million Egyptian pounds. (See Tables I and II.)

In the health sector, the state has drawn up an overall comprehensive expansion plan aiming at efficient medical services for the whole population. The Egyptian General Organization for Pharmaceuticals, Chemicals and Medical Appliances has been set up, affiliated to the Ministry of Public Health to deal with production, research and development, import, distribution and the export of pharmaceuticals and medical products. The main objectives of the Organization are: substitution of foreign drugs by local production at moderate prices, production of different basic and specific drugs in accordance with local requirements, separation of active ingredients from Egyptian herbs and medicinal flora, creation of raw materials and new preparations, follow-up of drug research to encourage production imports of non-locally produced pharmaceuticals and medical products, distribution and commercialization of pharmaceuticals and medical supplies all over the country, and an increase in drug exports to comply with the needs of many Arab and foreign countries in Africa, Asia and Europe.

The pharmaceutical industry in Egypt continuously increases its activity. Vaccines and biological preparations are produced on a wide commercial scale by the General Organization for Vaccines and Biological Production, which is one of the largest and most modern centres for vaccine production in the Middle East. Naturally, the increase in production of medical products and biological preparations has been accompanied by a parallel increase in sterilization demands.

For example, cotton cellulose, because of its softness and ability to absorb moisture, is extensively used for the manufacture of bandages, gauze, swabs etc., while cotton fabrics are extensively used in hospitals for surgical drapes, dressings, underpads, napkins and many other purposes. In most of these, sterilization is essential and is routinely undertaken through autoclaving.

Owing to technical difficulties and a lack of appropriate facilities, some clinics in remote areas are not completely satisfied with their sterilization practice. The inadequacy of the autoclaving practice in such areas shows that such a conventional method of sterilization cannot, in many cases, satisfactorily fulfil requirements on purely technical grounds. The clinical use of items not properly sterilized, particularly under the climatic and environmental conditions such as prevail in most developing countries including Egypt, would undoubtedly increase the chances of bacterial contamination and cross-infection of diseases, leading to an enormous utilization of antibiotics.

On the other hand, surgical sutures are commercially produced in Egypt by the Nile Company for Pharmaceuticals and Chemical Industries at Cairo, covering less than 25% of actual local consumption. Measures are now being actively taken to increase production capacity. Because of the protein content of absorbable surgical sutures, commonly known as cat-gut, which constitutes a difficulty for heat treatment, the gas-sterilization method is used. However, the technique adopted is time-consuming and has limitations on increasing the production rate. Despite the rather limited gas-sterilization facilities available, a 60-h lapse in gas-sterilizing mixture under adjusted conditions has been found necessary for locally produced sutures to attain an acceptable degree of sterility. However, the use of ethylene oxide still raises many objections on medical grounds because of its secondar effect and related toxicity on the products to be sterilized.

The production of medical products, e.g. cotton wool, bandages, surgical sutures and gloves, infusion sets and pharmaceuticals, has already reached a scale that indicates a real need for an efficient sterilization technology to be developed in Egypt. The certainty of expansion and the possibility of increasing the export of sterilized medical products would justify the construction of a radiation-sterilization plant in Egypt, which would be the first plant of its kind in Africa. Radiation-sterilization techniques would permit the introduction of the latest practices with regard to used disposable medical products such as plastic syringes, thus greatly reducing the chances of cross-infection of hepatitis which causes measurable losses to the national economy in manpower and expenditure.

The project has been proposed by the national Atomic Energy Authority and has the full support of the Ministry of Public Health, which will be involved in its management. It has also attracted much interest from the medical products and pharmaceutical industries, who are willing to participate by supplying materials and by collaborating in the work, and by conducting quality control in their manufacturing operations. There is also keen interest in the plastic industry, which wants to introduce plastic packaging and disposable items such as syringes, which could not be introduced before owing to a limitation of appropriate sterilization techniques.

Internationally gained experience in radiation equipment design and development is now well advanced, and different types of irradiators have been operated to serve a variety of purposes in both research, pilot-plant programmes and application. Many experts in radiation technology believe that ^{60}Co gamma facilities and electron-beam accelerators are two major radiation sources complementary to each other. The selection of the source for a potential application is based on its own merits and the demand of the process.

RADIATION STERILIZATION OF MEDICAL PRODUCTS

The use of ionizing radiation for sterilization of medical products is now a well-established technology which is steadily replacing other sterilization techniques on the grounds of convenience, economy and freedom from cross-contamination. Recently, numerous new types of medical devices, many of which may only be sterilized by radiation, have been introduced into medical practice. In the past 15 years, pre-sterilized medical disposables have stimulated a drastic change in hospital design and planning, with less need for complex sterilization facilities. Furthermore, radiation sterilization allows a much wider choice of materials to be used for packaging than any other sterilization method.

The radiation-sterilization dose internationally accepted as adequate, namely 2.5 Mrad, has been described as more than sufficient to kill all contaminating bacteria. Two types of ionizing radiation are now in use for sterilizing medical products – gamma radiation and high-energy electrons. Gamma radiation is the more penetrating and may be used to process thick and dense articles. On the other hand, the penetrating power of electrons is dependent upon the energy at which they are accelerated and the density of the material itself.

Radiation sterilization has many advantages in the context of medical treatment, particularly in developing countries with denser population and less appropriate hygienic conditions. The product is sterilized within its sealed container and remains sterile as long as the latter is unbroken. The process offers the opportunity of sterilizing medical products within plastic packages, which is not always possible with steam sterilization, and it does not suffer from the disadvantages of sterilization by ethylene oxide gas. Since radiosterilized medical products are prepacked and hermetically sealed during manufacture, the nation-wide introduction of single-use disposable syringes, needles, infusion sets etc., to hospitals is safe. The extension of the radiosterilization technique to various hospital supplies, such as bone banks, pharmaceutical products, biological tissues and other items seems quite feasible. Such practice is a positive measure to safeguard public health against the cross-infection of hepatitis and other epidemics. The subject has attracted great attention and the IAEA has organized several international symposia and meetings of experts from this particular field [1-7].

Several commercial-scale gamma-sterilization plants exist throughout the world. In several cases, demonstration plants which may be later scaled up to full commercial production, simply by increasing the loading of ^{60}Co, are being installed in various developing countries with UNDP assistance, e.g. India, Republic of Korea, Hungary and Egypt. This is undoubtedly more economical than building a demonstration plant followed a few years later by a full production plant.

The introduction of the radiosterilization technique into Egypt will stimulate the industrial development of the manufacture of the single-use disposable items and exportable sterilized items for medical use; for example, absorbent cotton wool and other cotton products since Egypt is well known for high-quality cotton production.

OTHER INDUSTRIAL RADIATION PROCESSING

Radiation technology has contributed to the manufacturing industries for new product development. Another well-established area of irradiation technology is the use of electron-beam accelerators to modify the properties of plastics, fibres and textile materials. Cross-linking of plastic films and thin objects improves their physical properties, viz. higher mechanical strength, higher softening temperature, lower solubility in organic solvents etc., while their surface properties, viz. resistance to, or affinity for, water or organic materials, static charging, affinity for dyes etc., may be modified by the irradiation grafting of a suitable monomer on to the surface. Such grafting can also modify the surfaces of fibres and textiles, both natural and synthetic, confer soil-release and increase resistant properties to them, as well as improving their permanent press properties.

In applications such as surface coating, the modification of plastics and textiles which require large doses in a very short interval, ^{60}Co gamma irradiation is not adequate because of its limited dose-rate and its inefficiency in irradiating thin-sheet materials in a continuous manner. Thus, for such applications, electron-beam accelerators are used.

The Egyptian staple cotton production, despite its good reputation and outstanding qualities, now meets with a strong world competition which has necessitated the industrialization of a large part of the cotton crops in the form of yarn or fabric, thus widening the marketing scope. To maintain and, if possible, to increase cotton exports, its production must be extended to finished textile goods. It seems now necessary to make permanent press cotton fabrics to compete on the market with synthetic fibres. The problem of improving the poor crease-recovery and slow-drying characteristics of natural cotton through electron-beam processing, is of particular interest to the Egyptian economy since the development of drip-dry clothing, based on synthetic fibres, has taken over a part of cotton's share in the textile market.

Besides, there are flourishing rayon and nylon industries in Egypt, where elastic nylon fibres and helenca blends are routinely produced. In the next few years, with polyester production, blended fibres will be produced This has encouraged consideration of the current development of processes for crease-resistant fabrics for cotton synthetic blends.

Wooden bobbins and shuttles in textile mills becomes a substantial part of the manufacturing cost. These articles, made from irradiated wood plastics, have proved to be more durable and abrasion-resistant. The radiation treatment of cotton fabrics and wood plastics are two examples of upgrading various natural materials that may contribute to the national economy and would result in a measurable saving of hard currency.

EGYPTIAN SCIENTIFIC AND TECHNICAL EXPERIENCE WITH IRRADIATION FACILITIES

Gamma irradiators

Modest gamma irradiation facilities of few kilocuries, viz. ^{60}Co and ^{137}Cs teletherapy units, indoor gamma-cells, agricultural irradiation facilities "Agromat", Gamma-top, as well as a 2-MW research reactor, are in common use in Egypt since 1960. Up till now, the whole spectrum of activities in different disciplines of radiation research and pilot experimentation has been building up; e.g. radiotherapy, radiation biology, radio-parasitology, radio-agronomy, radiation chemistry, radiation protection, radiation sterilization of medical products, radiation preservation of food, seed disinfestation, pest control by the sterile male technique, radiation polymerization, radiography, radiation engineering etc. A considerable number of Egyptian experts in such fields are now available and a number of scientific publications and technical reports have been issued. Also review articles have been presented in various IAEA symposia and United Nations conferences; e.g., see Refs [8-10].

Accelerator machines

The first accelerator that was installed in Egypt in 1960 is a Van de Graaff machine of the electrostatic type with a peak energy of 2.5 MeV, and has been routinely used since then as a positive particle machine to accelerate protons, deutrons, alpha particles and helium-3. The machine is run and maintained by Egyptian experts and technicians. Continued heavy-particle beam research programmes carried out have not permitted the machine to be changed to accelerate electrons.

The accelerator research group of the Nuclear Research Centre at Inchass are working on different experiments besides the design and testing of ion and electron sources. To acquire good experience, the group intends to build up in stages, with IAEA and UNDP assistance, an 8-MeV electron linear accelerator so as to be able to study the technical problems at each stage of construction. Recently, the electron injector assembly has been constructed and tested. A number of scientific publications and technical reports have been issued by the group, a few in connection with the design, construction and industrial application of a wave guide electron linear accelerator [11-13].

The long-term objective of the accelerator group is to train and build up a school capable of undertaking research and development in accelerator technology. The trainees would effectively participate in the operation, maintenance and development of electron-beam accelerators to be installed at various institutes and industries under the National Development Programmes.

CHOICE OF EGYPT'S FIRST INDUSTRIAL IRRADIATION FACILITY

^{60}Co gamma irradiator

It is known that the large investment required for setting up an industrial irradiation plant makes it desirable, particularly in a developing country such as Egypt, to be used as a multi-purpose radiation-processing facility. Taking into account that the radiation source and the surrounding biological shield constitute the main capital investment in the irradiation plant, an extra conveyor has been added for use in research and development studies.

The plant's purpose is to practice the process of the gamma sterilization of medical products under local conditions, to train local staff in the techniques and to undertake research and development programmes for other applications. Knowing that the international recommendations for the manufacture of medical products emphasize that such products must be produced under good hygienic conditions, and that they must be physically separated from other products throughout the entire manufacturing process so as to eliminate any risk of biological contamination or introduction of toxic compounds [14-16], and since our irradiation facility is planned to deal with several different product categories, special precautions have been taken in order to meet the requirements of the regulatory authorities.

Our industrial ^{60}Co gamma irradiator is a JS 6500 manufactured by the Atomic Energy of Canada Limited, with a maximum capacity of 1 000 000 Ci of ^{60}Co and with an extra research channel. The plant, which will be put into operation in the outskirts of Cairo during 1975, is a self-contained unit with a ventilated concrete biological shield and a water pool for source storage. The plant is designed to sterilize by irradiation medical products contained in fibreboard boxes of 40.6 to 58.4 × 38.1 to 48.2 × 91.4 cm in size. The annual throughput is 9230 m^3 of sterilized medical products calculated at a product density of 0.25 g/cm^3 and a minimum dose level of 2.5 Mrad based on operating time not exceeding 800 h/yr. The product boxes are conveyed by pneumatic pushers and end-transfer devices completing eight passes by the vertical plaque source before being discharged outside the irradiation chamber.

To ensure the complete physical separation of medical packages from other industrial items, as well as not to interfere with the sterilization programme, such a separate conveyor system is provided for research and development work; thus, the sterilization line can operate on an uninterrupted, 24-h daily basis if required. The research and development area of the building is separate from the sterilization area, and access to the latter is strictly controlled to reduce chances of contamination, particularly after sterilization.

For demonstration purposes, the gamma-irradiation facility will initially be loaded with 400 000 Ci of ^{60}Co. To be able to keep pace with increasing requirement, the possibility of raising the capacity to 1 MCi has been maintained. It is believed that the throughput of a facility of this size can justify a better insight into the technical and engineering problems associated with the continuous operation of the radiation sterilization of medical products and should give a close estimate of operating cost, analogous to commercial scale costs. The availability of an extensive supply of different kinds of samples for various microbiological, pharmacological and materials-testing experiments is considered very important as the success of evaluation will depend upon the extent and correctness in quality control under the prevailing local conditions. When bearing in mind that the throughput of this facility of sterilized medical products will refer to many kinds of products, it is believed that this magnitude of throughput will permit a reliable statistical certainty to be attained for each kind of product in the evaluation programme.

According to our preliminary calculations, based on local data and international practice, we believe that the cost of irradiation should be lower than the cost of sterilization by alternative methods. It is expected that this project will make available materials of improved hygienic quality for medical uses at a sterilization cost of about US$ 10/m^3 on a commercial production scale.

However, it has been felt beneficial for a technical marketing and economic feasibility study to be carried out before setting up the irradiation facility. Recently, a feasibility study showing how much of the capacity is likely to be used, based on a manufacturing capacity for medical products for market requirements, has been proposed with the assistance of UNDP through UNIDO. It is also believed that a further view of the economic aspects of the radiosterilization of medical products in relation to other research and development programmes should aim at establishing a successful irradiation operation schedule.

Electron-beam accelerator

The appropriate choice of an electron-beam accelerator for such a project must be a compromise between desirability and cost. From investigations carried out, an electron-beam would be a suitable choice in the modification of cotton fabrics, treatment of plastic films for packaging purposes, and in the sterilization of cat-gut sutures and miscellaneous medical products, for which the volume might be too small for irradiation with a ^{60}Co source.

Although the bulk of the applications will be in thin-film materials, for which a voltage of 500 keV would be adequate, to be able to demonstrate the process on thicker articles, such as cable insulation and thicker plastic objects, a 1.5-MeV machine has been proposed in agreement with IAEA and UNDP Missions [17-18]. It is known that such a machine will not be sufficient for all technological purposes; however, it is believed that once the process is demonstrated and accepted by Egyptian industry, there will be a need for additional machines since industrial products cannot be brought to this accelerator machine to take their place on a continuous line, as is the case for textile, fibre and cable industries.

NATIONAL CENTRE FOR RADIATION TECHNOLOGY

In view of establishing the National Centre for Radiation Technology, the Egyptian Atomic Energy Authority has been establishing, for the last decade, the principles for creating an integrated self-sustained scientific school with competent scientific workers covering different disciplines of radiation research. Besides basic research, the groups have been intensively involved in developmental techniques underlying radiation technology in various fields; viz. the radiation sterilization of medical products and biological tissues, radiation preservation of food and food-products, seed disinfestation, modification of characteristics of textile fibres, plastics, wood and rubber material and fabrics etc. The work has been conducted in collaboration with many specialists from the medical, agricultural and industrial authorities.

Study leaves organized by the IAEA and UNDP for training Egyptian specialists in different countries as well as scholarships kindly offered by certain countries on bi-lateral agreements, viz. France, Denmark, Hungary, USSR and Italy, have contributed significantly to the development of the project's programmes and to building up the whole spectrum of its personnel.

In February 1972, a supreme National Council for Radiation Research and Technology was formulated by the Academy of Scientific Research and Technology, chaired by the President of the Atomic Energy Authority and drawing its specialists from the universities, research institutes and concerned ministries and industrial bodies. The council advises on planning policy in radiation technology to ensure co-ordination and to build up an infrastructure among the various organizations.

Objectives of the Centre

The Egyptian National Centre for Radiation Technology is a governmental institute established as a multi-disciplinary body whose activities are planned to serve the needs of the public and scientific community. The centre aims at identifying problems of national importance and contributing to their solution through irradiation processing.

The main task of the Centre will be to conduct feasibility experimentation on the radiation sterilization of medical products, and to undertake an industrially oriented programme on radiation processing, including materials evaluation, quality control and cost estimates before passing over to the wide commercial application. It is most important that the pilot experimentation is run on a reasonable economic scale in order to be able to convince industry to follow up this line which will certainly require large investments on its part.

Since the task requires co-ordination with the manufacturers as well as with health and industrial authorities, it has been agreed that the Centre should concentrate a major part of its available manpower and material support on the implementation of the programme as a nationwide project.

Long-range objectives

(1) To introduce the technique of gamma sterilization for medical products and pharmaceuticals.

(2) To introduce electron-beam treatment for plastics and textile industries with a view to improving and modifying the properties as well as to increasing the competitiveness and export potential of Egyptian products.

(3) To investigate the feasibility of using irradiation technology economically in other fields such as wood upgrading, rubber vulcanization, food pasteurization, etc.

(4) To advise concerned authorities on the application of radiation technology and to offer the technical support and experience for establishing new radiation facilities at various industrial and medical sites.

Immediate objectives

(1) To install and commission a ^{60}Co gamma sterilization plant and to operate it in conjunction with industrial personnel in order to demonstrate that gamma sterilization of medical products and biological tissues can be carried out effectively and economically with local labour under local conditions.

(2) To train industrial personnel in the technique of gamma sterilization and in the appropriate technological background.

(3) To carry out research and development on the use of gamma irradiation in other areas such as food pasteurization, rubber vulcanization and wood modification with the aim of investigating their technical and economic feasibility.

(4) To install, commission and operate a 1.5-MeV electron-beam accelerator in collaboration with industrial personnel to demonstrate under local conditions and with local labour such processes as cross-linking of plastic film, modification of surface properties of plastics, textiles and fibres, cross-linking of plastic insulation, and sterilization of special medical products such as cat-gut sutures.

(5) To suggest programmes to consider national demand, existing activities, and the possibility of the immediate application of results.

(6) To encourage oriented development programmes in using radiation technology to solve industrial problems.

(7) To evaluate, with the aid of industry, the long-term potential of radiation sterilization and irradiation processing for Egyptian industry, and to formulate future requirements in terms of plant and equipment.

Framework

The Centre will be organized as follows:

(a) Research and Development Division: This comprises Departments of Bacteriology and Virology, Parasitology and Radio-immunology, Cytology and Cytogenetics, and Radiation Biology.

(b) Pilot-Plant Division: This comprises Departments of Radiosterilization of Medical Products, Modification of Textile Fibres and Fabrics, Modification of Plastics, Rubber, Wood and Varnishes, Seed Disinfestation, Food Preservation and Pest Control.

(c) Services Division: This comprises Departments of Radiation Protection, Dosimetry, Materials Testing, and Micro-analysis.

(d) Ancillary Services: These include Workshops, Documentation, Animal House and Transport.

Major scientific and technical services performed by the Centre

Irradiation processing, sterility control testing, materials testing, free radical determination, amino-acid auto-analysis, electron microscopy, trace-elements determination, neutron-activation analysis, gamma spectroscopy, atomic absorption spectrophotometry, flame photometry, u.v./visible and infra-red spectrophotometry.

The Centre will have the following major equipment: indoor gamma cells, neutron generator, electron-spin resonance (ESR), nuclear magnetic resonance (NMR), atomic absorption, u.v./visible and infra-red spectrophotometers, auto-gamma spectrometer, liquid-scintillation system, planchet-counting system, multichannel analyser with Ge-Li detector, electron microscope, ultramicrotome, ultra-centrifuges, amino-acid auto-analyser, lamina flow sterile benches, autoclaves and incubators, freezing cabinets, pilot plastic syringe production line, Instron strength tester, vacuum pumps and impregnation units, paddlers and wash machines, wet-angle measurement apparatus, weatherometer, impulse heat sealers, and other necessary routine laboratory equipment.

Because of the importance and priority of the project, it has received full support from the Government. The Atomic Energy Authority, Academy of Scientific Research and Technology, and Ministry of Public Health have so far co-ordinated to implement the programmes. The active participation of manufacturers and the enthusiasm of the research workers from various institutes are the essential factors in having qualified personnel to work on the project.

Consideration of the UNDP inputs

The project has been given a high priority in the Egypt Country Programme for 1973/1977, and submitted to the UNDP Governing council.

The Egyptian Government has considered the procurement of the complete ^{60}Co gamma irradiator, research and development laboratories, buildings and manpower as its contribution to the project. The UNDP inputs have been concerned with the delivery, installation and commissioning of the electron-beam accelerator; has assisted in establishing laboratories for materials testing and evaluation, and industrial radiation processing. UNDP expert services have devoted their attention to getting the project industrially oriented and to advise on specific problems on a short-term consultancy basis. The assignment of fellowships has concentrated on training for the construction and operation of irradiation facilities and to acquire practical experience in medical product sterilization and industrial radiation processing.

CURRENT STATE OF RESEARCH ACTIVITIES PERFORMED IN CONNECTION WITH THE PROJECT

Research and development programmes carried out in medical and industrial fields utilizing available radiation facilities include adaptation research on the already established methods as applied to locally produced items and under local conditions.

In the medical field, besides adaptation studies on radiosterilization of suitable medical devices, the scientific research on the possible radiosterilization of other items, e.g. certain pharmaceuticals, is being carried out in parallel with international efforts on those lines.

On the other hand, in the industrial field experimental programmes of industrial radiation processing have been actively pursued in Egypt. Laboratory experimentation on grafting to textiles in relation to soil-release properties, and on related problems concerning graft polymerization, are being carried out. In both cases, novel systems have been devised which give confidence that new approaches to the problem are under way. Promising results have been obtained so far on radiation-induced grafting of vinyl monomers with cotton fabrics. Experimentation is also being conducted on the radiation vulcanization of rubber, grafting of acrylic acid to polyethylene to produce reverse osmosis membranes for water desalination purposes, and the use of wood plastics in textile bobbins and shuttles. Such research activities reflect the co-ordination of research efforts with the industrial development programmes and the effective collaboration established with the appropriate local industries.

Research programmes carried out in connection with radiation sterilization of medical products and biological tissues

(a) Sterilization technique

Believing that it is not possible to predict the effect of ionizing radiation at varying dose levels on the chemical and biological properties of particular medical products, each category has to be carefully studied. In this respect, the following topics are being dealt with: radiation sterilization of medical devices and containers, cat-gut sutures and certain solutions, effect of sterilizing-radiation level on certain drugs and pharmaceuticals, microbiological quality control aspects of medical products before and after radiosterilization, consideration of the IAEA Recommended Code of Practice for Radiosterilization of Medical Products, and possible induction of more radioresistant mutants from microorganisms surviving radiation exposure.

(b) Effect of radiation on packaging materials

The properties of the samples are determined before and after irradiation at a sterilization dose level of 2.5 Mrad compared with much higher dose levels to show the trends. Changes in physical characteristics including colour and odour, mechanical properties including tensile and elongation strengths, and chemical properties including solubility in organic solvents and moisture absorption, are being investigated. On this matter, the following programmes are being carried out: effect of gamma irradiation on packaging materials, viz. cotton, polythene, polypropylene, and glass; the effect of gamma irradiation upon the biological properties of polymeric materials; and physical, chemical and biological changes induced in irradiated materials and their quality control aspects.

(c) Dosimetry

As it is important to evaluate the radiation dose administered to medical products packages, two methods of control are being investigated; physical with red acrylic perspex, and bacteriological, with relatively radiation-resistant organisms. In this respect the following procedures are being investigated: dosimetry procedures for the radiation sterilization of medical and biomedical products, microbiological standard preparations used as biological monitors for control of sterilization efficiency of a radiation facility, and physical and chemical dosimetry of a radiation facility.

These research programmes are being carried out on a nationwide scale. The following research centres and industrial bodies are contributing to the accomplishment of various undertakings:

(a) The Nuclear Research Centre at Inchass; Departments of Radiobiology, Radiation Protection, Nuclear Chemistry, Nuclear Physics and Engineering; (b) the National Research Centre; Laboratories of High Polymers and Textile Research; (c) the Egyptian Organization for Vaccine and Biological Production; Sterility Control Division; (d) the General Organization for Pharmaceuticals, Chemicals and Medical Appliances; Production Plant for Surgical Sutures at the Nile Company for Pharmaceuticals and Chemical Industries, the Medical Packaging Company, the Alexandria Company for Pharmaceuticals and Chemical Industries, and the Drug Research and Control Centre; (e) the Egyptian Organization for Spinning and Weaving, Mehalla El-Kubra Company for Spinning and Weaving; (f) the Egyptian Organization for Plastic Industries, the Egyptian Company for Plastics and Electricity at Alexandria; (g) the Universities: Cairo University Faculty of Medicine; Department of Bacteriology, Cairo University Faculty of Pharmacy; Departments of Microbiology, and Pharmaceutical Chemistry, Al-Azhar University Faculty of Science; Department of Microbiology and Cairo University Faculty of Science; Departments of Chemistry and Biology.

It is intended that the programme of the National Centre for Radiation Technology pave the way for the application of radiation technology for various purposes, and the development of radiation research for the industrial production programme, and promotion of public health services. The Centre is planned to make progress in executing its programme in every one of its directions.

Nevertheless, we should bear in mind that while many long-range problems in the application of radiation technology are equally important when relating their significance to the national interest, it might disperse and diffuse the effort and resources and slow down the Centre's progress during its foundation period. To avoid such a risk, the concentration on definite specific problems related to national development has been maintained. However, the framework of the Centre may be considered later in a much broader sense so as to embody various disciplines which may be gradually added to the Centre when sufficient manpower and resources are available to support the desired expansion.

REFERENCES

[1] IAEA, Large Radiation Sources in Industry (Proc. Symp. Warsaw, 1959) 2 Vols, IAEA, Vienna (1960).

[2] IAEA, Industrial Uses of Large Radiation Sources (Proc. Symp. Salzburg, 1963) 2 Vols, IAEA, Vienna (1963).

[3] First Int. Conf. Radiation Sterilization of Medical Supplies, Risø (1964).

[4] IAEA, Radiosterilization of Medical Products, Pharmaceuticals and Bioproducts (Proc. Panel, Vienna, 1966), IAEA, Vienna (1967).

[5] Ibid., see Code of Practice.

[6] IAEA, Radiosterilization of Medical Products (Proc. Symp. Budapest, 1967), IAEA, Vienna (1967).

[7] IAEA, Sterilization and Preservation of Biological Tissues by Ionizing Radiation (Proc. Panel, Budapest, 1969), IAEA, Vienna (1970).

[8] HASHISH, S.E., ROUSHDY, H.M., ABDEL-WAHAB, M.F., HALLABA, E., MAHFOUZ, M., EL GARHY, M.A., 4th UN Int. Conf. peaceful Uses atom. Energy (Proc. UN Conf. Geneva, 1964) 13, IAEA, Vienna (1972) 325.
[9] KAMAL, T.H., ROUSHDY, H.M., EL KHOLI, A.F., WAKID, A.M., EL SAYED, S.A., ABO HEGAZI, A.M., 4th UN Int. Conf. peaceful Uses atom. Energy (Proc. UN Conf. Geneva, 1964) 12, IAEA, Vienna (1972) 361.
[10] HASHISH, S.E., in Peaceful Uses of Atomic Energy in Africa (Proc. Symp. Kinshasha, 1969), IAEA, Vienna (1970).
[11] ABDEL-AZIZ, M.E., MOONTAN, V., EL-NASHAR, A., RAGEH, M., GANEM, A., AMMAR, I., UARAEE/Internal Rep. 36 (1971).
[12] ABDEL-AZIZ, M.E., EL-NASHAR, A., MOONTAN, V., AMMAR, I., TAXEL, M., AREAEE/Internal Rep. 52 (1973).
[13] BAKR, M.H.S., AREAEE/Internal Rep. 68 (1973).
[14] See Ref.[5], p.423.
[15] IAEA, Recommended Code of Practice, Radiation Sterilization of Medical Products, IAEA-159, Unpubl. Doc. 1974.
[16] CHRISTENSEN, E.A., HOLM, N.W., 4th UN Int. Conf. peaceful Uses atom. Energy (Proc. UN Conf. Geneva, 1964) 2, IAEA, Vienna (1972) 345.
[17] YUAN, H.C., Report of UNDP Mission, Jan., 1973.
[18] AMPHLETT, C.B., Report of UNDP Project Appraisal Mission, May, 1974.

DISCUSSION

K.H. CHADWICK: In connection with a point which has come up on several occasions in the course of this symposium, I would be interested in knowing how accessible your gamma irradiation plant will be to the industries that are to provide the products to be irradiated.

H.M. ROUSHDY: The irradiation facility has been set up in the vicinity of the major pharmaceutical industries in Cairo and will be easily accessible to them. Other pharmaceutical industries are considering the establishment of a central storage facility near the plant, from where they can organize the distribution of sterilized medical products to different localities. The pharmaceutical industries will also participate in the management of the facility.

RADIOSTERILIZATION OF MEDICAL APPLIANCES USING A 5000-Ci COBALT-60 SOURCE

M.A.R. MOLLA
Health Physics Division,
Atomic Energy Centre, Dacca

M.A. LATIF, M. HAQUE
Institute of Public Health, Dacca,
Bangladesh

Abstract

RADIOSTERILIZATION OF MEDICAL APPLIANCES USING A 5000-Ci COBALT-60 SOURCE.

A study has been made on the radiosensitivty of the pathogenic microorganisms in order to develop a method for the radiosterilization of medical products and appliances in Bangladesh. The tests were carried out on the following microorganisms: Pseudomonas pyocyaneus, Staphylococcus aureus, Escherichia coli, Klebssiella pneumoniae and Bacillus subtilis. Each microorganism was irradiated with uniform doses ranging from 0.0313 to 2.5 Mrad at suitable dose intervals using a nominal 5000-Ci ^{60}Co source. A significant difference in the radiosensitivity was observed in different species of the microorganisms. For complete sterilization the doses varied between 0.25 to 1.75 Mrad depending on the microorganisms, the highest dose being required for the B. subtilis. To allow a safety factor a 2.5-Mrad dose has therefore been chosen for sterilization. This dose was then applied for sterilization of the intraveneous transfusion sets prepared in the Institute of Public Health, Dacca. Over 100 000 transfusion sets have so far been radiosterilized and successfully used in different hospitals. The radiation dosimetry of the ^{60}Co source was made by a Victoreen R-meter and verified by the Fricke chemical dosimetry method.

INTRODUCTION

In recent years, a significant progress has been made on the sterilization of medical products and appliances using ionizing radiation. The radiosterilization has been widely used in a number of countries on an industrial scale [1-3], because it has many advantages over the conventional methods of sterilization, namely ethylene oxide gas and autoclaving [4-5]. The main objective of the present work has been to develop a method for the radiosterilization of medical products and appliances so that it could be directly used to sterilize intravenous transfusion sets, and may also be used subsequently to sterilize other items when irradiation and other related facilities become available. To develop a radiosterilization method, it was necessary to study the radiosensitivity of some microorganisms of medical importance commonly found in the medical appliances. Although similar studies have been carried out elsewhere [1, 2, 7-11] we believe that it is important to study the radiosensitivity of the microorganisms under the existing tropical climate and humid environment of Bangladesh.

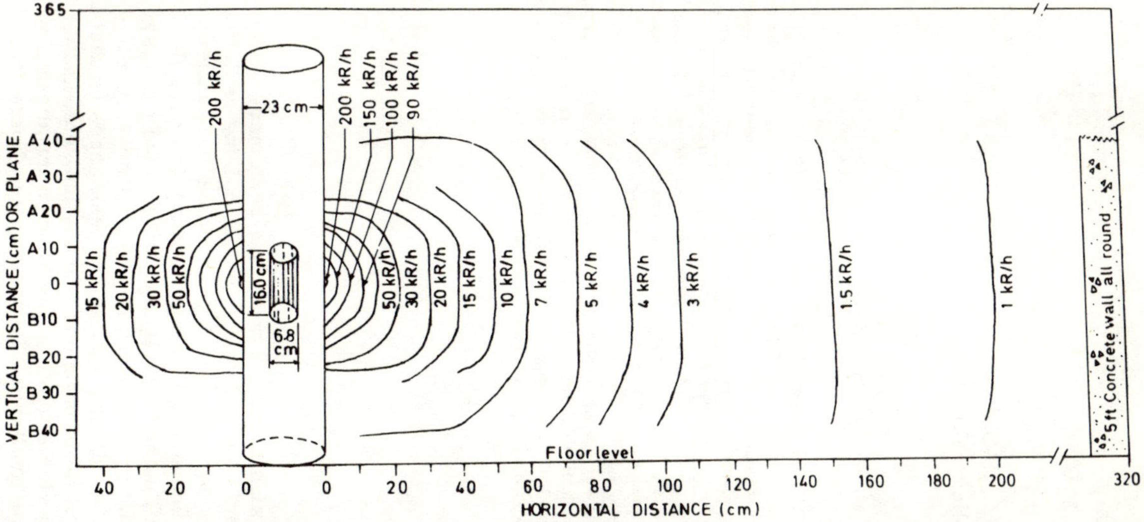

FIG. 1. Typical isodose curves for 5000-Ci ^{60}Co source (~3500 Ci at the time of the experiment) consisting of 5 independent needle sources circumferentially arranged in a ring. The symbols A10–A40 and B10–B40 signify horizontal planes at 10 to 40 cm above and below the '0' plane, respectively. The '0' plane represents the horizontal plane passing across the centre of the source while in irradiation position. (Details may be found in Ref. [12].)

IRRADIATION FACILITY

A nominal 5000-Ci ^{60}Co source (~3500 Ci at the time of this investigation) located at the Health Physics Division, Atomic Energy Centre, Dacca (AECD), has been used both for the radiosensitivity tests of the common microorganisms and the sterilization of the medical appliances. Although this irradiation facility is normally used for research and development activities of the Bangladesh Atomic Energy Commission (BAEC) and some outside organizations, the present radiosterilization programme utilizing this facility had to be undertaken on a priority basis because of the urgent need for the sterilized intravenous transfusion sets to be used for the Institute of Public Health (IPH), Dacca. The radiation dosimetry of this source was made by a Victoreen R-meter and verified by the Fricke chemical dosimetry method, following similar procedures described elsewhere [12]. To obtain isodose curves on both horizontal and vertical planes the radiation doses were measured at more than 1500 different positions within the source room at varying distances in eight different directions and at different vertical planes above and below the centre of the source. Typical isodose curves are shown in Fig.1. The percentage error in the dosimetry lies within ±6%.

IRRADIATION OF THE MICROORGANISMS

A variety of organisms may be found in medical products and appliances as contaminants [1,2,9]. The number of each species/strains of microorganisms which are likely to be present may possibly vary with the size and type of medical products and appliances and also with the natural and environmental conditions. Of the different contaminants, five microorganisms were selected for the present study. These are Pseudomonas pyocyanea, Staphylococcus aureus, Escherichia coli, Klebssiella pneumonia and Bacillus subtilis. These organisms were isolated locally at the IPH laboratory and the stock cultures were preserved in standard medium at 2-5°C. Irradiation of the microorganisms was carried out in nutrient broth medium in a 1-ml sample at normal temperature (about 22-26°C). For B. subtilis, fluid cultures containing a large abundance of spores were tested both in liquid form and after soaking the same in filter-paper strips. Samples of each microorganism were independently dispensed in sterilized polythene vials (about 2.5 cm in length by 0.25 cm in diam.) and sealed under air. For each species of organisms, at least 15 samples, each containing an equal-sized population, were prepared; two of these samples were used as control and the remainder was irradiated with uniform doses ranging between 0.0313 to 2.5 Mrad at suitable dose intervals (Table I) using a constant dose-rate of 200 krad/h.

STERILITY TEST

The complete absence of survivors after irradiation is normally taken as an indication that an effective dose has been applied. To obtain an effective dose for radiosterilization a survival test was carried out both for the irradiated and control samples. Both the irradiated and the control

TABLE I. THE PERCENTAGE SURVIVAL OF MICROORGANISMS FOLLOWING EXPOSURES TO DIFFERENT DOSES OF GAMMA RADIATION OBTAINED FROM CULTURE TESTS

Type of organisms	No. of organisms before irradiation (control) per ml	Special condition during irradiation	Percentage of organisms surviving irradiation (Unirradiated control samples = 100%)												
			Dose (Mrad)												
			0.0313	0.0625	0.125	0.25	0.50	0.75	1.00	1.25	1.50	1.75	2.00	2.25	2.50
Pseudomonas pyocyanea	7.0×10^8	Fluid	6.90	1.0×10^{-1}	2.3×10^{-5}	0	0	0	0	0	0	0	0	0	0
Staphylococcus aureus	7.9×10^8	Fluid	100	83.5	3.0×10^{-3}	0	0	0	0	0	0	0	0	0	0
Escherichia coli	4.3×10^8	Fluid	-	-	1.9	0	0	0	0	0	0	0	0	0	0
Klebssiella pneumonia	7.9×10^8	Fluid	-	100	5.1	2.2×10^{-3}	2.0×10^{-5}	0	0	0	0	0	0	0	0
Bacillus subtilis	6.1×10^8	Fluid	100	100	100	44.0	8.2	3.3×10^{-2}	1.5×10^{-3}	1.5×10^{-4}	1.2×10^{-5}	0	0	0	0
	7.0×10^7	Dried in paper strips	100	100	100	65.7	6.6×10^{-2}	4.3×10^{-4}	$1.4\ 10^{-5}$	0	0	0	0	0	0

All samples except B. subtilis in dried state were irradiated in nutrient broth media.
- indicates no experiment was carried out.
0 indicates 100% sterilization.

samples were plated directly and/or diluted to different factors up to 10^7 as and wherever necessary for plating in Bacto-Agar nutrient medium for culture. The formation of a colony by a living microorganism was taken as an index of its survival. The colonies were counted after 24 and 48 h of incubation at 37°C. In some cases, the incubation period was extended to 4 d in order to observe a probable change in the number of colonies with the incubation period. However, no significant change in the number of colonies was observed with an incubation period up to 4 d. The colonies were counted out at 48 and 96 h following incubation. Once again, there was no significant change in the number of colonies counted at 48 and 96 h following incubation. The percentage survival of each species of organisms was calculated with respect to the corresponding control sample, which was also cultured under conditions identical to those of the irradiated samples.

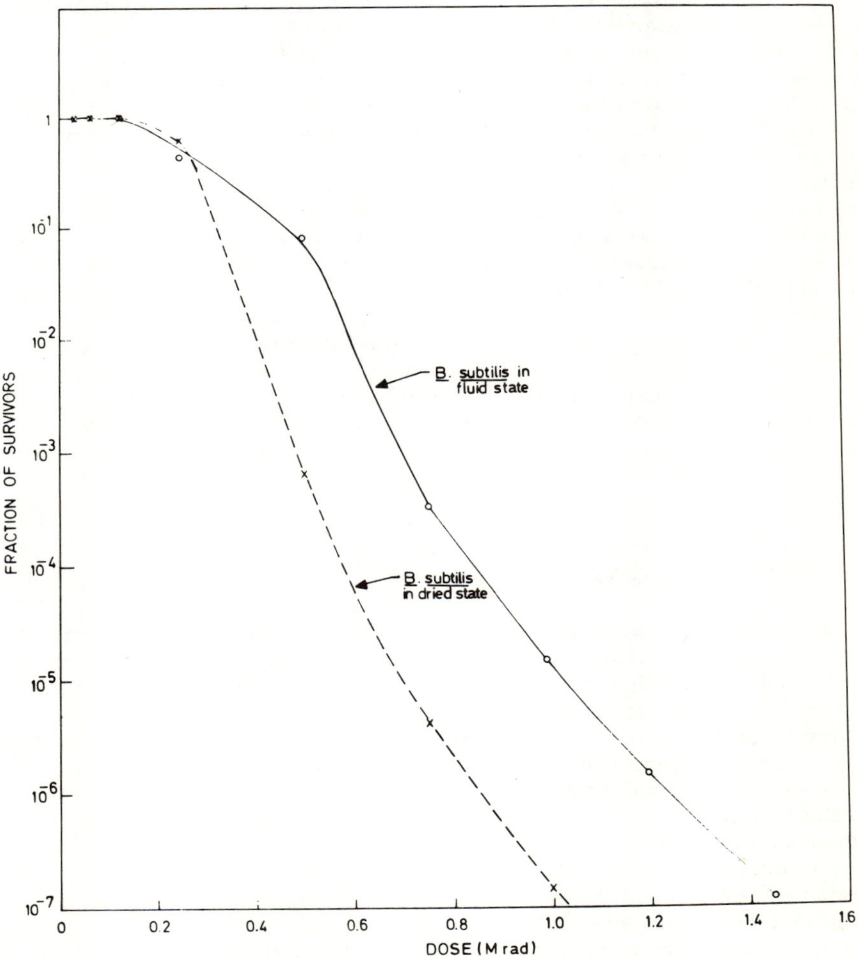

FIG. 2. Inactivation curves for B. subtilis.
- o - o -: fluid state and - x - x - x: dried state.

RESULTS

The data on survival studies carried out on the common microorganisms following exposures to different doses of gamma radiation are summarized in Table I. A significant variation may be observed in the radiosensitivity of the different microorganisms. Although a maximum dose of 0.25 Mrad is required for complete sterilization of Ps. pyocyanea, Staph. aureus and E. coli, there appears to be considerable difference in the percentage survival of these three species of organisms exposed to same dose of gamma radiation at doses below 0.25 Mrad. Of these three organisms, Ps. pyocyanea appears to be most radiosensitive. For Kleb. pneumonia, complete sterilization may be achieved by a dose not exceeding 0.75 Mrad. The mixture of spores and vegetative forms of B. subtilis, irradiated both in the form of fluid bacteria and dried state in paper strips, appears to be more radioresistant than other microorganisms studies in the present series. It is of interest to note that the B. subtilis culture in the dried form is relatively more radiosensitive than fluid bacteria of B. subtilis; the respective doses required for their complete inactivation are 1.25 and 1.75 Mrad. This difference in the radiosensitivity of the two forms of B. subtilis may be due to the difference in the media in which they have been irradiated; the nutrient broth might have possibly acted as a protective agent. The dose/survival curves of B. subtilis (dose versus log survival fraction) for both the dried and the fluid state would clearly exhibit a shoulder in each case at the beginning of the curve extending up to a dose of 0.125 Mrad, suggesting that the killing of these organisms requires multiple hits (Fig.2). Similar shoulder regions in the dose/survival curves for other organisms under the present investigation are indicated for Staph. aureus and Kleb. pneumonia and are likely to be obtained perhaps at a low dose range with varying magnitude for the others. Although the remainder of the dose-survival curve following the shoulder region roughly follows a single exponential function, each with different inactivation constants, there appears to have been a short tail at the end of each inactivation curve (Fig.2).

DISCUSSION AND CONCLUSION

Many factors may influence the inactivating effect of ionizing radiation on microorganisms [1,8,9,13-16]. The environmental and natural conditions under which the organisms are maintained before, during and after the irradiation play an important role on the radiosensitivity. The main factors involving radiosensitivity are the growth media, relative humidity, temperature, oxygen, nitrogen, protective and sensitizing agents etc., and some factors involving radiation, namely the dose-rate and dose fractionization. Even under identical experimental and environmental conditions the radionsensitivity is found to vary between two strains of the same microroganisms [5,7]. It is therefore desirable to perform radiosensitivity tests under those conditions which maximize the radiation resistance of the organisms in order to ensure an additional margin of safety. (Laboratory conditions normally increase the resistance of the organisms.)

Of the various common microorganisms tested for radiosensitivity in our laboratory, Ps. pyocyanea was the most radiosensitive whereas B. subtilis in the nutrient broth media proved to be most radioresistant. To achieve complete sterilization for various microorganisms, the doses varied between 0.25 and 1.75 Mrad depending on the species of the organism, the highest dose of 1.75 Mrad being required for B. subtilis in the liquid state. On the other hand, B. subtilis in the dried state required the second highest dose of 1.25 Mrad for complete inactivation. However, in general, our observed sterilization dose for each species of microorganisms is comparable with and consistent with the corresponding reported value [1,7-8,17,18] except for Ps. pyocyanea. For Ps. pyocyanea, the reported sterilization dose ranged from 0.05 to 0.1 Mrad [1] as compared with a dose of about 0.25 Mrad obtained in our experiment. This discrepancy, although not unexpected, because of reasons stated earlier, may not represent any serious problem since a dose much higher than 0.25 Mrad was used to sterilize the medical appliances. Although a 1.75-Mrad dose was found to inactivate completely the common microorganisms investigated, a higher dose of 2.5 Mrad was chosen to sterilize our medical appliances to allow a margin of safety and to account for the probable presence of some microorganisms of higher radiation resistance. The dose/survival curves produced under laboratory conditions for the most radioresistant bacterial spores recovered from dust yielded about one survivor in 10^7 organisms irradiated with a 2.5-Mrad dose [11]. Even the radioresistant Streptococcus faecium isolated from dust in a Danish laboratory and irradiated with 2.5 Mrad under laboratory conditions, which maximize its radioresistance, has shown about one survivor in 10^7 organisms [19]. Although the number and the type of different organisms may vary with the type and the size of the medical appliances and also with the environmental conditions, normally the total number of organisms found in the medical appliances as contaminants ranges from $1-10^3$ [1,5,20-22], about 99% of which is completely inactivated with a 2.0-Mrad dose or so [9]. Even on the assumption that the contaminants in the medical appliances comprise only high radiation-resistant microorganisms, an assumption, which may be unrealistic in practical situation, the inactivating factor is estimated to be 10^6 [11].

However, a comprehensive study has recently been made on the radiosensitivity of 36 non-sporogenic organisms, comprising 19 species of 11 generally including high resistant species of Streptococcus faecium and Micrococcus radiodurans [7]. The irradiation was carried out in phosphate buffer (pH 7.0) sealed in vacuum (150 mm Hg pressure) at a radiation temperature of -80°C for M. radiodurans and at different temperatures of 5, -30, -80, -140 and -196°C for 4 strains (FEC, Φ12, A21 and F6) of S. faecium. (Note that the experimental conditions would maximize the radiation resistance of the organisms). Although M. radiodurans survived a 1.8-Mrad dose (the maximum dose used in the experiment), all four strains of S. faecium were completely inactivated by a 1.8-Mrad dose. It has been stated by the same authors that M. radiodurans is neither a food spoilage organism, a public health hazard, nor a measure of food sanitation. Van Winkle, Jr. et al. [5] made a study of the radiosensitivity of three strains (A21, F6 and Φ12) of Streptococcus faecium in broth and aqueous media with a 2.5-Mrad dose. Complete sterilization was achieved for all the strains of S. faecium for 14 d incubation at 37°C except in three

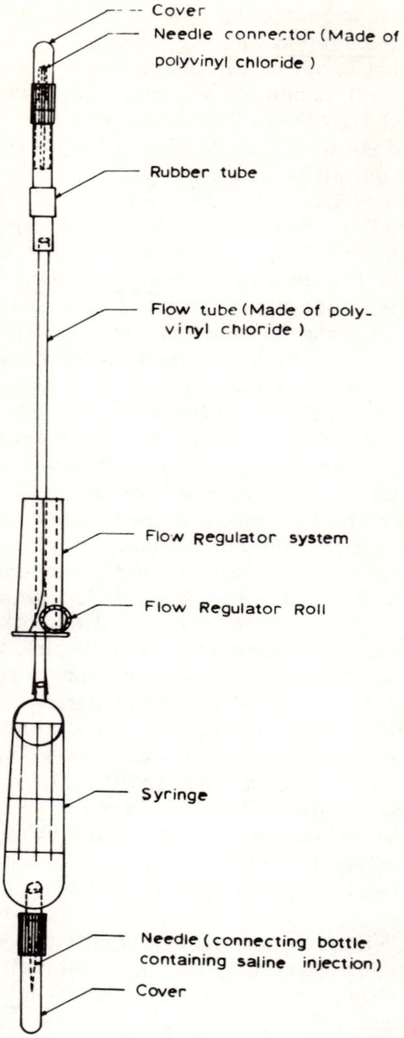

FIG. 3. Intravenous (I. V.) transfusion set.

tests out of 59 with the strain F6 in broth media, where growth was recorded. Probable factors responsible for the survival of strain F6 in nutrient media has been reported to be due to the presence of SH groups in the broth, which acted as a protective agent, and also the slow dose-rate from the ^{60}Co source permitting multiplication of the organisms during irradiation.

In view of the above discussions, the choice of a 2.5-Mrad dose for the sterilization of the intravenous (I.V.) transfusion sets may be considered as appropriate. The I.V. transfusion sets are used for intravenous administration of glucose solution (both aqueous and saline), saline and cholera fluid to the patients (Fig.3). More than 100 000 I.V. transfusion sets have so far been sterilized with a 2.5-Mrad dose and successfully used without any problem. During routine irradiation procedures of the I.V. transfusion sets, the B. subtilis in nutrient media has been used as a biological indicator. Although our preliminary studies did not show the presence of any high resistant organisms, namely the S. faecium and M. radiodurans in the transfusion sets, further work is in progress to isolate these organisms from our environment (if any) and to study their radiosensitivity.

ACKNOWLEDGEMENTS

The authors express their gratitude to Dr. K.A. Monsur, Director, Institute of Public Health, Dacca, for his continued interest in the project and valuable suggestions in the preparation of the manuscript. The assistance in the calibration of the source, irradiation of the test samples and the transfusion sets, by Mr. M. Rahman, Mr. F. Karim, Mr. Sahajahan, Mr. B. Rahman, and other scientific and technical personnel of the Health Physics Division, AECD, and of the IPH laboratory, is gratefully acknowledged. This project has been jointly financed by the Bangladesh Atomic Energy Commission and the Institute of Public Health, Dacca.

REFERENCES

[1] LEY, F.J., CRAWFORD, C.G., KELSEY, J.C., "Report from the United Kingdom", Radiosterilization of Medical Products, Pharmaceuticals and Bioproducts, Tech. Rep. Ser. No. 72, IAEA, Vienna (1967) 40-59.

[2] CHRISTENSEN, E., HOLM, M.W., JUUL, F., "Report from Denmark", Ibid., p. 60-74.

[3] ARTANDI, C., "Radiation sterilization of medical supplies and materials", Manual on Radiation Sterilization of Medical and Biological Materials, Tech. Rep. Ser. No. 149, IAEA, Vienna (1973) 169-86.

[4] ALLADINE, M.F., GIBBONS, J.R.P., "The use of tygon tubing sterilized by gamma radiation in heart-lung by-pass machines", Radiosterilization of Medical Products (Proc. Symp. Budapest, 1967), IAEA, Vienna (1967) 285-88.

[5] VAN WINKLE, W., Jr., BORICK, P.M., FOGARTY, M., "Destruction of radiation-resistant microorganisms on surgical sutures by ^{60}Co-irradiation under manufacturing conditions", Ibid., p. 169-80.

[6] JEFFERSON, S., CRAWFORD, C.G., "Development of industrial sterilization of medical products", Ibid., p. 361-65.

[7] ANELLIS, A., BERKOWITZ, D., KEMPER, D., Comparative resistance of nonsporogenic bacteria to low temperature gamma irradiation, Appl. Microbiol. 25 4 (1973) 517-23.

[8] LEY, F.J., "The effect of ionizing radiation on bacteria", Manual on Radiation Sterilization of Medical and Biological Materials, Tech. Rep. Ser. No. 149, IAEA, Vienna (1973) 37-63.

[9] CHRISTENSEN, E.A., "Hygienic requirements, sterility criteria, and quality and sterility control", Ibid., p. 131-52.

[10] CHRISTENSEN, E.A., "Radiosterilization of medical devices and supplies", Radiosterilization of Medical Products, (Proc. Symp. Budapest, 1967) IAEA, Vienna (1967) 265-85.

[11] COOk, A.M., BERRY, R.J., "Presterilization bacterial contamination on disposal hypodermic syringes: Necessary information for the rational choice of dose for radiation sterilization", Ibid., p. 295-308.

[12] MOLLA, M.A.R., ABDULLAH, S.A., KARIM, F., Calibration of a 500 Ci Co-60 source, Hlth Phys. 16 (1969) 135-43.
[13] BURT, M.M., LEY, F.J., Studies on the dose requirement for the radiation sterilization of medical equipment. I. Influence of suspending media, J. Appl. Bacteriol. 26 (1963) 484.
[14] BURT, M.M., LEY, F.J., II. A comparison between continuous and fractionated doses, Ibid., p.490.
[15] TALLENTIRE, A., "Aspects of radiation microbiology fundamentals to the sterilization process", Radiosterilization of Medical Products, Pharmaceuticals and Bioproducts, Tech. Rep. Ser. No. 72, IAEA, Vienna (1967) 1-22.
[16] TALLENTIRE, A., J. Appl. Bacteriol. 33 (1970) 141.
[17] WEBB, R.B., Radiat. Res. 18 (1963) 607-19.
[18] HOLLAENDER, A., STAPELTON, G.E., MARTIN, G.L., Nature 167 4238 (1951) 103.
[19] CHRISTENSEN, E.A., HOLM, M.W., JUUL, F., The basis of Danish choice for dose for radiation sterilization of disposal equipment, Risø Rep. No. 140, Danish Atomic Energy Commission (1966).
[20] VAN WINKLE, W., "Report from the United States of America", Radiosterilization of Medical Products, Pharmaceuticals and Bioproducts, Tech. Rep. Ser. No. 72, IAEA, Vienna (1967) 23-39.
[21] EMBORG, C., "Equipment for preservation of blood and blood transfusion", Manual on Radiation Sterilization of Medical and Biological Materials, Tech. Rep. Ser. No. 149, IAEA, Vienna (1973) 191-98.
[22] CHRISTENSEN, E.A., "Radiation resistance of bacteria and the microbiological control of irradiated medical products", Sterilization and Preservation of Biological Tissues by Ionizing Radiation (Proc. Panel, Budapest, 1969) IAEA, Vienna (1969) 1-13.

DISCUSSION

Nazly HILMY: How do you prepare your test pieces and/or material, and what are the humidity and temperature conditions?

M.A.R. MOLLA: The stock cultures, preserved in a nutrient broth medium at 2-5°C, were dispensed in sterilized polythene vials in Bacto-Agar nutrient medium at the ambient temperature and humidity, and then sealed in air and irradiated under these conditions.

Kikiben PATEL: When carrying out radiosensitivity tests in Bacillus subtilis, did you use vegetative forms or spore forms?

M.A.R. MOLLA: We used a mixture of spores and vegetative forms of B. subtilis, as we thought that our I.V. transfusion sets might be contaminated with both forms of B. subtilis.

Kikiben PATEL: In this event I think the inactivation curves for spores and vegetative forms will be different.

M.A.R. MOLLA: Yes, I quite agree that there may be a difference in the radiation sensitivity between the two forms of B. subtilis, and hence in the activation curves. But both the forms of B. subtilis together were completely inactivated at a dose of 1.75 Mrad, and therefore served our purpose.

Kikiben PATEL: Did you prepare an inactivation curve for Streptococcus faecium from test pieces obtained from Denmark (Dr. Christensen), or did you do the work yourselves?

M.A.R. MOLLA: We have not done any work using Streptococcus faecium. I merely referred to the data on S. faecium in the literature. Our preliminary findings did not show the presence of any S. faecium in our environment and for this reason this microorganism has not yet been a problem to us.

RADIATION-STERILIZATION PRACTICES IN KOREA

KANG-SOON RHEE
Division of Radiation Biology,
Korea Atomic Energy Research Institute,
Seoul, Korea

Abstract

RADIATION-STERILIZATION PRACTICES IN KOREA.

The need for cold sterilization to sterilize medical devices and supplies has been emphasized. It has recently become known that a Radiation Processing Demonstration Facility is one of the most effective tools for such sterilizing methods. Since January 1972 the Korea Atomic Energy Research Institute has planned building a Radiation Processing Demonstration Facility as a co-project of the Korea Atomic Energy Research Institute, the United Nations Development Programme and the International Atomic Energy Agency. It will be finished in June 1975. To ensure the effective use of this Facility after being set up, the Korea Atomic Energy Research Institute is now carrying out various sterilization experiments and training technologists.

INTRODUCTION

Heat sterilization methods have mainly been applied for sterilizing medical supplies produced in Korea, and the development of thermolabile medical supplies needing sterilization has been somewhat delayed. For a long time, many manufacturers of medical supplies have insisted on the need for cold sterilization.

Should cold sterilization be available, better quality and a variety of medical supplies can be expected.

Keeping in mind the peaceful uses of atomic energy, and the demand of many medical manufacturers, the Korea Atomic Energy Research Institute has proposed setting up the Radiation Processing Demonstration Facility with the assistance of the United Nations Development Programme (UNDP) and the International Atomic Energy Agency.

A total amount of $465 000 from the UNDP and the same contribution from the Korean government have been provided to construct the facility.

With technical assistance from the IAEA the Korea Atomic Energy Research Institute has been building a Radiation Processing Demonstration Facility (^{60}Co source of 100 000 Ci capacity) since January 1972 and it will be finished in June 1975.

When this Facility is in operation the radiation capacity will be a ^{60}Co source of 100 000 Ci. However, if more capacity is needed in the future the facility could be designed to increase its capacity.

At present, the Korea Atomic Energy Research Institute is discussing the problems of an increased production of high-quality medical supplies.

The Korea Atomic Energy Research Institute has also requested co-operation from the Ministry of Health and Social Affairs to deal with these problems. In addition, the authors are undertaking many kinds of experiments to assist in the effective operation of the Facility.

1. PROSPECTS OF RADIATION STERILIZATION

For the effective use of the Radiation Processing Demonstration Facility, which is expected to be completed by 1975, we have analysed the kinds of medical products, their production and the amounts of medical supplies to be imported and exported.

Figure 1 shows the annual increase of the total production of medical products from 1969 to 1973 [1], which in 1973 amounted to $436 000 000.

As shown in Figs 1 and 2, of the total medical products, those devices and supplies which can be sterilized by irradiation have also increased since 1969 when 1.8% ($2 630 000) of the total were medical devices and supplies which can be sterilized by irradiation, and this was increased to 2.5% ($11 010 000) in 1973.

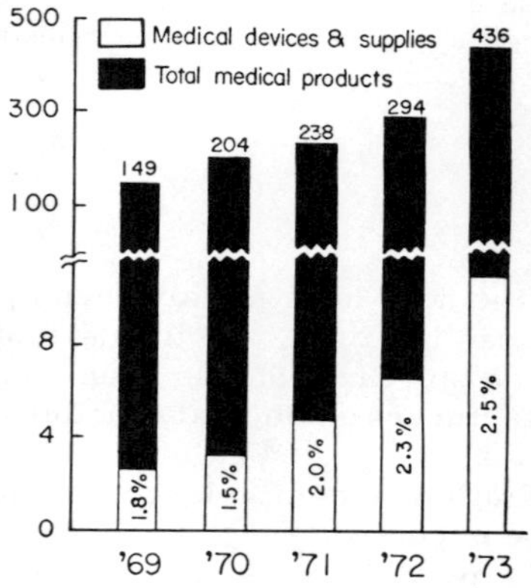

FIG.1. Amounts of total medical products and devices or supplies available for radiosterilization from 1969 to 1973 (in $\$10^6$).

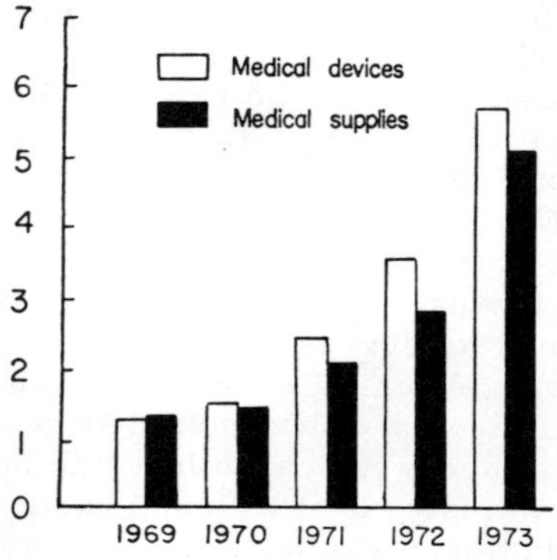

FIG.2. Annual growth-rates of medical devices and supplies available for radiosterilization (in $\$10^6$).

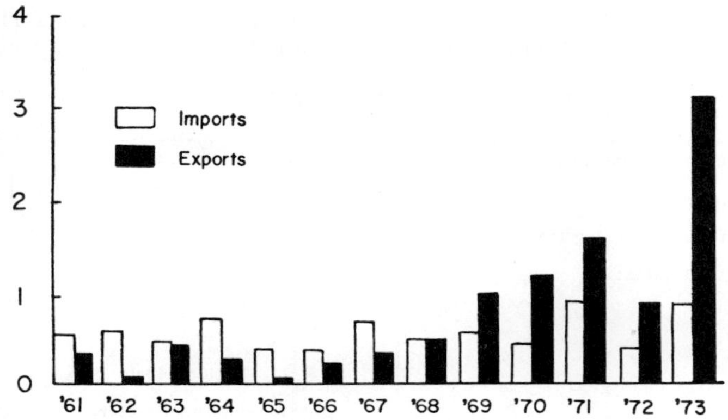

FIG.3. Comparison of the total amounts of imported and exported medical supplies in Korea from 1961 to 1973 (in $\$10^6$).

Figure 3 shows the results of analysis of the total amount of imported and exported medical supplies from 1961 to 1973 [2, 3], indicating that the import/export ratio was 1:0.62 ($600 000: $380 000) in 1961, and it was 1:2.74 ($1 100 000: $3 020 000) in 1973.

We can clearly see the annual export increase since 1961. Because of the sterilization problem we can export only those medical supplies which can be heat sterilized, which means that the exported thermolabile medical supplies must be sterilized on arrival in the country importing the supplies.

In general, medical science and pharmacy have progressed considerably but the production of medical supplies is only on a small commercial scale in the field of public health in Korea.

The author believes that many medical products which must be sterilized can be sterilized by irradiation, and that also many medical products need to be further developed.

As shown in Table I many medical supplies can be sterilized by irradiation now or will be so in the near future. Also, it is to be expected that there will be an increase in the future of medical devices and supplies which can be sterilized by irradiation.

Recently, Korea has greatly progressed in industrial development. If technical co-operation is possible between the technique or facility of medical supplies and industrial technology, better progress in good-quality medical supplies could be expected.

In view of this, the authors believe that the Radiation Processing Demonstration Facility could be the most useful instrument in sterilizing medical products, and we also could increase the export of medical supplies.

2. EXPERIMENTAL WORKS ON RADIATION STERILIZATION

The authors have carried out various experiments which mainly concerned applied field work, i.e., the training of technologists who will operate the Facility, gaining experience in radiosterilization techniques, and

TABLE I. MEDICAL PRODUCTS AVAILABLE NOW AND IN THE FUTURE FOR RADIATION STERILIZATION IN KOREA

Present	Future
1. Gauze and plain gauze	1. Sutures of all types (catgut, nylon)
2. Compress gauze	2. Prostheses
3. Cotton and cotton gauze	3. Plastic hypodermic syringes
4. Cotton applicators	4. Plastic and rubber catheters
5. Masks	5. Plastic gloves
6. Eye bands	6. Plastic scalpels
7. Bandages	7. Plastic petri dishes
8. Cast bandages	8. Cannulae
9. Sanitary pads	9. Urea bags
10. Dressing packs	10. Surgical blades
11. Swabs	11. Preparation razors
12. Maternity towels	12. Artificial organs
13. Syringes	13. Tongue depressers
14. Syringe needles	14. Plastic bags and containers for pharmaceutical preparations and pathological samples
15. Suture silk	15. A limited number of biological products are also to be sterilized
16. Intravenous injection sets	16. Biological tissues for grafting
17. Blood recipient sets	
18. Ringer sets	
19. Donor sets	
20. Scalp vein infusion sets	
21. Blood vacuum bottles	
22. Condoms	
23. Rubber gloves	
24. Talc powder	
25. Army dressing kits	

demonstrations of the effectiveness of radiosterilization by using the Facility for the medical manufacturers who are the most interested.

The effect of γ-rays on Salmonella typhi Ty 2 cells; the comparison of radiosensitivity of bacteria isolated from the different radiation exposure histories; the preparation of vaccine and antibiotics by γ-ray and radiosterilization of disposable plastic sets have been reported by several authors [4-8]. The radiation source used for irradiating samples was the 1000-Ci ^{60}Co panoramic irradiator.

It is known that the radiosensitivities of bacteria differ according to the environmental conditions during irradiation.

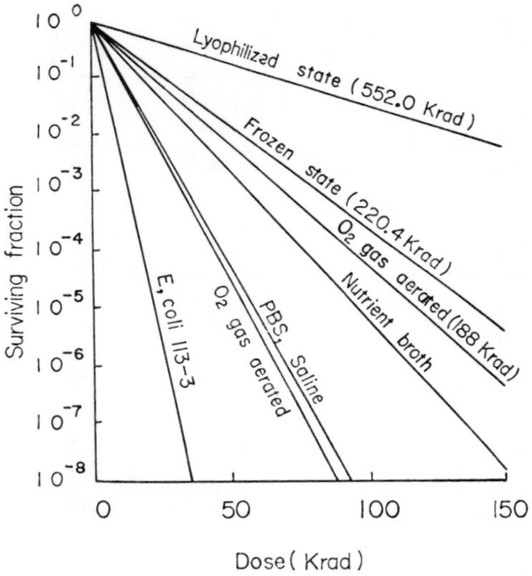

FIG.4. Radiosensitivities of Salmonella typhi Ty 2 according to different environmental conditions. Except for oxygen (16.2 μM O_2/litre) and oxygen (1.8 μM O_2/litre) in gas-aerated state, all the others were irradiated under atmospheric conditions.

To test the effects of temperature, degree of hydration and oxygen contents on radiation sensitivity, the Salmonella typhi Ty 2 cells were irradiated in physiological saline, in the lyophilized state or in the frozen state.

As shown in Fig.4, the experimental group suspended in physiological saline and phosphate buffer solutions has shown no difference in radiosensitivity to each other, but the experimental group suspended in nutrient broth have shown a little increase in radioresistance.

On the other hand, the group suspended in the frozen or lyophilized state has shown a relatively high radioresistance.

When two experimental groups were suspended in the same physiological saline, the group containing more oxygen (16.2 μM O_2/litre, has shown more radiosensitivity than the low oxygen group (1.8 μM O_2/litre).

Similar experimental results were also reported by Ley et al. [9]. Generally it is known that radiation damage to bacteria is higher when oxygen is present.

In the vegetative cells especially the radiosensitivities are three times higher in the presence of oxygen [10].

Howard-Flanders et al. [11] have also reported their experimental results of oxygen effects. They observed maximum radiosensitivity at a level of 4.0 μM O_2/litre; however, there is no increased radiosensitivity at more than 4.0 μM O_2/litre.

These and other results indicated that the maximum radiation effect can be obtained with an even smaller oxygen concentration

There are no different temperature effects when comparing the room temperature with the freezing-point on the radiation effects; however, the experimental groups in the frozen state showed considerable protection effects.

TABLE II. LETHAL DOSES OF MICROORGANISMS ACCORDING TO THE DIFFERENT CELL CONCENTRATIONS

Microorganisms	No. of cells per disc	Lethal dose (Mrad)
B. cereus	1×10^2	0.375
	1×10^5	0.901
	1×10^8	1.450
B. brevis	1×10^2	0.450
	1×10^5	1.150
	1×10^8	1.802
Cl. tetani, Harvard	1×10^2	0.625
	1×10^5	1.512
	1×10^8	2.425

TABLE III. EXPERIMENTAL RESULTS OF BIOLOGICAL AND PHYSIO-CHEMICAL TESTS ON THE DISPOSABLE PLASTIC SET

Type of test [16, 17]	Radiation doses (Mrad)					
	1.0	1.5	2.0	2.5[a]	4.5[b]	10.0[c]
Sterility test	+	+	+	+	+	+
Pyrogen test	+	+	+	+	NT	NT
Safety test	+	+	+	+	+	+
Toxicity injection test	+	+	+	+	+	+
Non-volatile residue	+	+	+	+	+	-
Residue on ignition	+	+	+	+	+	-
Heavy metal	+	+	+	+	+	+
Buffering capacity	+	+	+	+	+	-

Notes: + = Test passed; - = Test failed, NT = Not tested.
[a] Colour = Slight yellow; Appearance = No change, Extension rate = 0.98;
[b] Colour = Yellowish-brown; Appearance = No change, Extension rate = 0.92;
[c] Colour = Dark brown; Appearance = Rapid erosion on needle surface, Extension rate = 0.86.

Several reports indicated that bacteria had more resistance to radiation in the frozen state [12], in the dried state [13], or in the presence of some chemical compounds [14].

Our experimental results were also similar to these results and the radioresistance of Salmonella typhi suspended in nutrient broth could be attributed to the existence of protecting agents.

For the studies of radiation effects according to the different cell concentrations we have irradiated three kinds of bacteria which contained 1×10^2, 1×10^5 and 1×10^8 cells, respectively, and measured the lethal dosage. The results are given in Table II, which shows some evidence that the lethal dosages increase according to the increase of cell numbers.

Similar experimental results were also reported by Ley et al. [15]. Even with the same kinds of medical supplies, the dosage needed for sterilization of bacteria differs from manufacturer to manufacturer.

The author also tested the commercial value of disposable plastic sets after being radiation sterilized and the results are given in Table III, which shows that the whole experimental group tested at 4.5 Mrad showed positive results; however, when the dosage increased to 10.0 Mrad the non-volatile residue, the residue on ignition and the buffering capacity tests showed negative results.

In this experiment, where the disposable plastic set was irradiated at 10.0 Mrad, the attached needle and aluminium plate eroded within one month.

In addition, disposable plastic sets irradiated at 2.5 Mrad changed their colour from white to slight yellow, and the white colour was changed to a yellowish brown in the 4.5-Mrad irradiation group. At 10.0-Mrad sterilization, the disposable plastic sets cannot be used for commercial purposes.

REFERENCES

[1] MINISTRY OF HEALTH AND SOCIAL AFFAIRS, The Data of Productivities of Medical Products According to the Manufactures from 1969 to 1973 (1973).

[2] KOREA PHARMACEUTICAL TRADERS' ASSOCIATION, Korean pharmaceutical trade directory, 1961-1972 (1973).

[3] KOREA PHARMACEUTICAL TRADERS' ASSOCIATION, Total amounts of imports and exports of medical products in 1973 (1973).

[4] RHEE, K.S., MIN, B.H., CHANG, C.S., Kor. J. Microbiol. 11 (1973) 79.

[5] RHEE, K.S., MIN, B.H., KIM, K.S., Kor. J. Microbiol. 12 (1974) 69.

[6] RHEE, K.S., MIN, B.H., CHANG, C.S., J. Kor. Soc. Microbiol. 8 (1973) 37.

[7] RHEE, K.S., CHUN, K.J., KIM, K.S., to be published, 1974.

[8] RHEE, K.S., CHUNG, K.J., MIN, B.H., to be published, 1974.

[9] LEY, F.J., FREEMAN, B.M., HOBBS, B.C., Studies on the use of ionizing radiation for the elimination of salmonellae from various foods (to be published).

[10] THORNLEY, M., in Radiation Control of Salmonellae in Food and Feed Products, Tech. Rep. Ser. No.22, IAEA, Vienna (1963) 81.

[11] HOWARD-FLANDERS, P., ALPER, T., Radiat. Res. 7 (1957) 518.

[12] NICKERSON, J.T.R., CHARM, S.E., BROGLE, R.C., LOCKHART, E.E., PROCTOR, B.E., LINEWEAVER, H., Food Tech. Campaign 11 (1957) 159.

[13] BROGLE, R.C., NICKERSON, J.T.R., PROCTOR, B.E., PYBNE, A., CAMPBELL, C., CHARM, S., Food Res. 22 (1957) 572.

[14] ERDMAN, I.E., THATCHER, F.S., MacQUEEN, K.F., Can. J. Microbiol. 7 (1961) 207.

[15] LEY, F.J., CRAWFORD, C.G., KELSEY, J.C., in Radiosterilization of Medical Products, Pharmaceuticals and Bioproducts, Tech. Rep. Ser. No.72, IAEA, Vienna (1967).

[16] US Pharmacopoeia, 18th ed.

[17] National Formulary, 13th ed.

DISCUSSION

Nazly HILMY: What type of safety and toxicity tests did you carry out on your plastic material before you considered the findings conclusive?

KANG-SOON RHEE: We carried out monthly tests over a period of six months. The tests were based on the methods given in the 18th edition of the United States Pharmacopoeia and in the National Formulary (13th edition).

S.G. KODKANI: What was the type of the plastic used in your studies on plastic disposables?

KANG-SOON RHEE: We studied disposable plastic sets made of polyethylene.

REPORT OF WORKING GROUP ON THE REVISION OF THE IAEA RECOMMENDED CODE OF PRACTICE FOR RADIATION STERILIZATION OF MEDICAL PRODUCTS

(Session IX)

Chairman: J. FLEURETTE (France)

RECOMMENDATIONS FOR THE RADIATION STERILIZATION OF MEDICAL PRODUCTS

INTERNATIONAL ATOMIC ENERGY AGENCY
Vienna, Austria

Presented by R.N. Mukherjee

INTRODUCTION

The International Atomic Energy Agency (IAEA), in collaboration with WHO, prepared in 1967 a Recommended Code of Practice for the Radio-sterilization of Medical Products.[1] The primary objective of those recommendations was to serve as a guide in drawing up national standards and rules regulating radiation sterilization practices in Member States. In the light of new experience obtained from the operation of radiation sterilization facilities and practices in several Member States, the need for bringing up to date and revising the recommendations has been emphasized by many leading experts who have also given their comments and suggestions in this matter.

In response to the above-mentioned need the IAEA established a small working group to continue with the Revision of the Recommendations for Radiation Sterilization of Medical Products. The purpose of the small working group is to serve as the task force to continue and complete the revision of the Recommendations, taking into consideration the relevant scientific and technical developments and experts' comments. The first meeting of this working group was held in Vienna on 5 - 6 April 1974, in which the following experts participated: Dr. V.G. Khruschev (USSR), Dr. A. Bishop (UK), Dr. R. Chu (Canada), Dr. J.C. Kelsey (UK), Dr. V.K. Iya (India), Dr. E. Christensen (Denmark), Dr. N. Holm (Denmark), Dr. K.H. Peter (Federal Republic of Germany), Dr. R.N. Mukherjee (IAEA) and Dr. L. Sztanyik (IAEA) (Chairman). On the basis of the discussions and comments and at the request of the working group, Drs Bishop, Kelsey and Chu jointly helped in preparing the draft revised version which was subsequently circulated to all the members of the working group as well as to all the participants of the Agency Symposium on Ionizing Radiation for Sterilization of Medical Products and Biological Tissues, held in Bombay, 9 - 13 December 1974, the Proceedings of which appear in this volume.

One of the symposium sessions in Bombay was devoted to open discussion of the draft recommendations and to receiving further revisory comments for their inclusion, where essential. This document has resulted from those discussions. Two of the members of the working group, Dr. E. Christensen and Dr. N. Holm (both from Denmark), have expressed their disagreement with the current version of the document and the Recommendations therein, as they consider that many of the practices and standards incorporated in the current document may be inadequate in providing public health safety assurances as required by the Scandinavian countries.

---

[1] Radiosterilization of Medical Products (Proc. Symp. Budapest, 1967), IAEA, Vienna (1967) 423.

These Recommendations relate to the radiation sterilization of medical products[2], the precise definition of the term 'medical products' being the responsibility of national health authorities. The Recommendations are not intended to be construed as a working Code of Practice but only attempt to highlight the different areas which need consideration by the national health authorities of the Member States while developing the sterilization practices for the public health care items by ionizing radiation. As new technical information and experiences accumulate in the future, particularly from the implementation of such practices in the developing Member States in the tropics, the document may need further revising and up-dating. The Recommendations are intended for application primarily to single-use medical devices, although they may also be applicable (in part or entirely) to biological materials and drugs.

For ease of presentation the Recommendations for Radiation Sterilization of Medical Products will be discussed under the following separate broad headings:

(i) Manufacture
(ii) Microbiological control of the radiation facilities
(iii) Operation of radiation sterilization facilities

1. MANUFACTURE

The primary manufacturer, in order to exert the necessary microbiological control, must ensure that all steps in the preparation of articles to be sterilized by irradiation are carried out in accordance with an existing code of good manufacturing practice, e.g. WHO Technical Report Series No. 418, Annex 2 (1969), WHO 1211 Geneva 27, Switzerland. For the sake of convenience of the users and for completeness of the document, the major aspects of the Recommendations pertinent to the manufacturing practice and the production site are discussed in the following separate sections.

1.1. Premises

1.1.1. The medical products should be manufactured in areas designed to suit the operations involved and should minimize the risk of bacterial contamination of such products. The buildings should be constructed or modified to prevent the entrance and harbouring of vermin. Waste materials should be collected and sealed into suitable containers for removal to collection points outside the buildings. It should be disposed of at regular and frequent intervals. The extent of necessity and the standard should be determined for each manufacturing area, and would depend on the nature of the work and the degree of handling and exposure of the products concerned.

---

[2] Medical products may be defined as "drugs, materials and devices which are used for prophylactic, diagnostic and therapeutic purposes in surgical and clinical procedures and which are required to be sterile at the time of use and are stable to radiation. They would also include containers for pharmaceutical products".

1.1.2. Water used in the manufacturing process, or for any other relevant purpose, should be monitored microbiologically from time to time to prevent it becoming a serious source of contamination.

1.1.3. Storage conditions should be without physical or chemical effect on materials or components. Closed containers should be used for all storage.

1.2. Equipment

1.2.1. Equipment should be used in the production of single-use medical supplies to be designed to facilitate cleaning, and to prevent foreign matter, oil or lubricants coming into contact with the products or the components. Materials which may disseminate particles or fibres should not be used in clean rooms. Use of cardboards boxes for transport of raw materials, components, etc. should be avoided.

1.2.2. Written instructions for each piece of equipment should be made readily available to the person(s) responsible for cleaning the equipment. The necessary enforcement of those instructions should be the responsibility of the microbiologist in charge.

1.3. Hygiene

1.3.1. No person known to have skin lesions on exposed surfaces of the body, or to be suffering from a disease in communicable form or to be the carrier of such a disease should be employed on production processes. All operatives must be medically examined before employment.

1.3.2. All staff engaged in production should be supervised in respect of:
(a) systemic diseases
(b) their personal hygiene
The supervision should be the responsibility of a qualified person prefereably with a medical background.

1.3.3. Personnel should be encouraged to report to the supervisor the development of any symptoms. Maintenance staff should be adequately trained in hygiene and clean room routine.

1.4. Microbiological control

1.4.1. There should be a competent microbiologist with adequate laboratory facilities who should have responsibility for ensuring the production-site hygiene.

1.4.2. The laboratory should preferably be on the manufacturer's premises. The laboratory staff should be capable of carrying out routine tests with a low level of adventitious contamination.

1.4.3. The microbiologist should be responsible to the Board of Directors or the Quality Controller and should be independent of production at shop floor level.

1.5. The microbiologist should be responsible for:

1.5.1. Continuing monitoring and recording of environmental microbial contamination; isolating and identifying unusual contaminants and attempting to determine their sources.

1.5.2. Investigation of the level of microbial contamination of the product and elimination of the sources of contamination where possible.

1.5.3. Provision of written hygiene regulations and their implementation.

1.5.4. Provision of written cleaning schedules for all areas and equipment and their implementation.

1.5.5. Definition of plant (factory) performance, and routine monitoring of the performance (microbiological) of air conditioning and filtration equipment.

1.6. Packaging

The following are general requirements which may be modified or augmented by the requirements of individual specifications.

1.6.1. Each item should be sealed in a unit container.

1.6.2. The unit container should permit sterilization of the contents in situ.

1.6.3. The packaging of the unit container should be adequate to ensure maintenance of sterility after radiation sterilization and should provide adequate physical protection for the contents under normal conditions of handling, transit and storage.

1.6.4. It should be designed to ensure that the product can be presented for use in an aseptic manner, and that once opened it cannot be easily re-sealed.

1.6.5. A suitable number of unit containers should be packed into shelf containers for ease of storage and to assist maintenance of sterility.

1.7. Marking of containers

The following are general requirements which may be modified or augmented by the requirements of individual specification. Unit containers should include in their markings:

1.7.1. Description of contents (this description may include approved abbreviations if the container is too small for the full details)

1.7.2. The word STERILE in prominent lettering

1.7.3. The name or trade mark of the manufacturer

1.7.4. An identification reference to the batch and/or the date of manufacture

1.7.5. The words "Destroy after Single Use" or similar approved wording

1.7.6. Cautionary advice against hazard if applicable

1.7.7. Any necessary instructions for aseptically opening the pack

1.8. Shelf containers shall include in their markings:

1.8.1. Description of contents

1.8.2. The word STERILE

1.8.3. Name and address of manufacturer

1.8.4. The identification reference as in 1.7.4.

1.8.5. Date of sterilization (month and year), e.g. Sterilization Date: January 1975

1.8.6. A statement of storage requirements if applicable

2. MICROBIOLOGICAL CONTROL OF THE RADIATION FACILITY

2.1. Commissioning of radiation sterilization plants

The health authorities of some countries (e.g., the Scandinavian countries) require the use of standardized microbiological preparations as part of the commissioning process on the grounds that they measure what the process aim is designed to achieve, i.e. the killing of micro-organisms. In addition, they provide an independent check on physical and chemical dosimetric systems, as experienced from the instances where that monitoring has proved useful. They consider that the extra time and trouble taken, which they report as not being very great, is well justified by the provision of an additional and independent cross-check aimed at facilitating the public health safety assurances of the consumers.

In many other countries the health authorities do not require the use of standard microbiological preparations. They consider that once the sterilizing radiation dose has been established the object of the sterilization plant is to deliver the required dose accurately. The available physical measurement and chemical dosimetry is sufficient to ensure that this can be done effectively. They concentrate their efforts primarily on using reliable dosimetric systems by insisting on the employment of skilled staff and by the use of external reference standards for efficiency calibration of the irradiation facility. They consider that, for the use of microbiological preparations, a substantial expenditure of time and trouble is needed and that this should be better devoted to the improvement of the dosimetry system.

At present there appears to be little prospect of resolving this controversy and disagreement between the experts from the countries concerned. The situation is expected to be clarified with the passage of time and the accumulation of more operating experience. Meanwhile the health authorities of a Member State must decide this matter on the basis of relevant published work and such expert advice as they may seek.

2.2. Routine process control

As stated above (Section 2.1.) there is still controversy over the role of using the microbiological standard preparations in routine process control, although disagreement is less marked in this context than in that of commissioning. Most of those who advocate their use at commissioning do not generally recommend their routine use but would retain them as an additional technique for checking performance for use on special occasions and circumstances. Those who do not favour their use at the commissioning stage may fail to see any justification in using them thereafter. Only time and experience will resolve this situation.

2.3. Quality controls and assurance of sterility

When a correctly designed, commissioned and operated plant is used, the receipt by the product of the sterilizing dose is assured by dosimetry, and there should be no need for the routine use of microbiological preparations or for routine sterility tests. However, health authorities or purchasers may require either or both forms of test.

3. OPERATION OF RADIATION STERILIZATION FACILITIES

3.1. Suitability

The irradiation facility must be approved by the appropriate authority as suitable in its technical operations for the purpose intended. Of course, this situation should be resolved at the very initial stage of planning and designing the facility.

3.2. Safety

The installation must conform to the occupational safety and health requirements laid down by the country of operation. In the case of electron accelerators the energy of the radiation must be kept at a level precluding the chances of any induced radioactivity hazard to the consumer. The provision should be kept at many suitable points of the plant-operating schedule, so that it automatically shuts down in the event of any technical failure and/or malfunctioning.

3.3. Plant control

Recognized methods of plant control and dosimetry must be employed, and complete records of the data must be kept.

3.4. Note 1. Control of gamma-irradiation plants

(a) General

The dose delivered at a particular location in a product irradiation container[3] of a given size being conveyed through an irradiation field in a multipass mode depends on the source strength, the conveyor parameters and the densities and atomic numbers of the constituent materials undergoing irradiation.

The aim of plant control and dosimetry is to ensure that every item processed receives at least the required minimum dose and that excessive doses are not administered. These conditions should be verified on commissioning the plant and whenever the process parameters are altered. Control must be maintained during routine operation.

(b) Calibration on commissioning and after plant modification

For a given mode of conveyor operation and a given source and product geometry, the absorbed dose is a function of the conveyor parameters. These parameters should be monitored to ascertain that they are reproducible. There should be a positive indication that the source and product items are in the correct operating position during processing and that the conveyor is operating correctly. If the plant malfunctions, it must shut down automatically.

To achieve correct conveyor settings, the dose distribution in the production unit has to be evaluated for the particular mode of conveyor operation. The dose distribution may be determined by placing dosimeters (e.g., 25) in an actual production unit at locations representative of areas which might be expected to give the best indication of dose variability. Recognized dosimetry systems should be employed.

Once the dose distribution has been determined, conveyor settings should be fixed so as to ensure that at least the specified minimum dose is given to all the products. Provided it has been established that the conveyor can be set and controlled accurately, the dose distribution may be determined for doses other than those which it is intended to apply in production, and conveyor settings may consequently be calculated by scaling.

When the process (including materials and packaging) is altered, those parts of the foregoing procedures which are influenced by the alterations should be repeated.

(c) Choice of radiation sterilization dose

The choice of radiation dose will basically depend on one primary parameter — the degree of assurance of sterility required. From this will follow two other parameters, viz., the radiosensitivity of the contaminating microorganism and the level of initial presterilization contamination expected on the items to be sterilized. These parameters are discussed separately as follows:

---

[3] A package of standard volume fitting into the conveyor carrier.

(i) The degree of assurance of sterility required:

This is a matter for administrative decision but many health authorities require that no more than one organism should be expected to survive on one million sterilized items.

(ii) Radiosensitivity of the contaminating microorganisms:

This may be established experimentally as well as from the published data for the pathogenic and non-pathogenic microorganisms already investigated. The introduction of radiation sterilization practices in the tropical countries should be preceded by a survey of the commonly expected forms of contaminating microorganisms under the local environmental conditions. The radiation sensitivity of the uncommon and/or doubtful species should be established through suitable investigations.

(iii) Initial contamination level to be expected on items to be sterilized:

This must be established experimentally under the prevailing hygienic conditions of the manufacturing site. After the standard under local conditions has been assessed, its maintenance or deterioration must be detected in the course of routine survey and necessary measures should be immediately adopted.

Using these parameters and in the light of over twenty years' experience of commerical-scale radiation sterilization a number of health authorities have arrived at a recommended minimal dose for articles of low initial microbial contamination.

The pharmacopoeias of Belgium, Czechoslovakia, France, Italy, the United Kingdom (British Pharmaceutical Codex), United States of America and the Australian Code of Practice give this dose as 2.5 Mrad. The recommended minimum dose in the Scandinavian countries ranges from 3.5 to 4.5 Mrad depending upon the level of initial contamination.

Health authorities may wish to vary their recommended dose but it should be understood that in so doing they are accepting a variation in one or more of the process parameters. Such variation may alter the objective end-point of the irradiation process from sterilization to some lesser degree of microbial decontamination appropriate for some special purpose. Raising the irradiation dose to increase the degree of assurance of sterility carries the risk of the materials' degradation, which may in the case of some packaging materials subsequently reduce the assurance of sterility through package failure during handling, transport or storage. It is recommended that health authorities should require a single dose level for the sterilization of medical devices.

The operator of the radiation facility must bear the responsibility for delivering the required dose of radiation and for complying with the recommendations set out. He should satisfy himself that the manufacturer of the articles to be sterilized understands the principles embodied. To exert the necessary microbiological control the primary manufacturer must ensure that all steps in the preparation of articles to be sterilized by irradiation are carried out in accordance with an existing code of manufacturing practice.

(d) Routine process control

Once the plant has been calibrated and the required minimum absorbed dose in the production unit established as a function of conveyor settings, the acceptance or rejection of products is subject to the following control steps:

(i) Positive indication of the correct operational position of the source and items of product.
(ii) Continuous-line recording of conveyor speed and/or records of plant cycle times. The records should permit calculation of the proper correction factor to compensate for source decay since the time of calibration.
(iii) For inventory control, the production units should be labelled with a radiation indicator and precautions taken to prevent confusion between irradiated and non-irradiated production units.

Alternative process control methods (e.g., those based on dose measurements) may also be used.

3.4. Note 2. Control of electron irradiation plants

(a) General

The type of accelerator plant considered in this section has the following characteristics:

(i) A scanning system with an electron beam sweep which ensures a homogeneous surface dose distribution over the width of the product
(ii) A conveyor system which moves the product in a direction perpendicular to the beam and the scanning direction
(iii) A sweep velocity high enough to ensure good beam-spot overlap on products passing at the highest conveyor speed used in practice

With constant beam and scanning, the absorbed dose at a particular location in a product depends on the beam parameters (energy, average current and scan width), the conveyor speed and the thickness, density and atomic number of the materials placed between the beam exit window and the location in the product.

The aim of plant control and dosimetry is to ensure that every item processed receives at least the required minimum dose and that excessive doses are not administered. These conditions should be verified on commissioning of the plant and whenever the process parameters are altered. Control must be maintained during routine operation.

The beam and conveyor parameters should be monitored continuously during processing.

(b) Calibration on commissioning and after plant modification

For a given set of beam parameters and for a given geometry of the irradiation field, the absorbed dose is inversely proportional to the conveyor speed. The reproducibility of the conveyor speed setting and the reliability of the system during operation must be within the limits required to ensure that the correct dose is administered.

A system should be provided with a safety device that will act in the event of failure (e.g., automatic and immediate shutdown of the product conveyor).

The beam parameters may be standardized as follows:

Direct current measurements of the scanned beam can be made by arranging for the electrons to be absorbed in, for example, an aluminium absorber thick enough to absorb all the electrons entering it and by metering the resultant current flowing to earth. Correction must be made for electrons scattered back from the absorber.

Indirect current measurements may be carried out continuously by means of a non-intercept monitor. In the case of one simple technique, the scanner chamber of the accelerator is provided with secondary emission collectors mounted symmetrically around the beam to collect electrons scattered back from the accelerator exit window. It should be ensured that the system gives accurate readings which are not significantly influenced by other factors such as minor changes in beam focusing.

Measurement of the beam energy may be carried out by various methods, such as:

(i) Magnetic spectrometry
(ii) Threshold-detector methods
(iii) Film-stack dosimetry (or other range measurements)

One way of obtaining a relative measure of the electron energy is available when the accelerator is fitted with a bending magnet located before the scanning system. In this case, the current required to bend the beam through 90° (or any other selected angle) is a relative measure of the electron energy. If the process parameters are to be integrated into the plant control procedure, it should be ensured that the energy remains constant during processing. A system of secondary emission collectors such as the one referred to above may be used to control this parameter. If the energy varies, the beam – at a constant current in the bending magnet – will deviate from the selected angle and thereby cause a greater response from one of the collectors than from the other. This system can be made to trigger an alarm or to control an automatic energy-regulating circuit. Another system is based on the fact that the current necessary to scan the beam over a certain width is a function of the beam energy. This principle can be applied to a monitoring system by installing two bars of wires at each end of the scanner house (under the exit window) and making the scanned beam strike the inner, but not the outer, bars. The scan width may also be controlled by this method with an automatic feedback system. Whatever method is used, it should be ensured that the response is not influenced significantly by minor changes in other process parameters. Malfunctioning should give rise to automatic shutdown of the process equipment.

To obtain correct settings of the conveyor speed for a given set of beam parameters, the dose distribution in the production unit should be determined. Because of the absorption characteristic of high-energy electrons, the product in any production run should be of uniform weight and composition. The dose distribution may be determined by placing dosimeters (e.g., 25) at random throughout the production unit's volume. It will normally be found that the part of the product item furthest from the beam receives the smallest dose. Recognized film systems should be employed for dosimetry.

Once the dose distribution has been determined for a given set of beam-product parameters, conveyor settings should be fixed so that the specified minimum dose is delivered to all product locations.

Whenever the process is altered, those parts of the foregoing procedures which are influenced by the alteration should be repeated.

(c) Routine process control

Once the plant has been calibrated and the required minimum absorbed dose in the production unit established as a function of the conveyor speed, the acceptance or rejection of products should be subject to the following control steps:

(i) The employment of an interlock system between beam current monitor and conveyor;

(ii) Continuous-line recording of current, energy, scan width and conveyor speed. The electron current signal, as monitored by secondary emission collectors or some other non-intercept monitor, may be used to moderate the conveyor drive circuit, thereby ensuring a constant surface dose;

(iii) For inventory control, each tray or production unit should be labelled with a radiation indicator and precautions taken to prevent confusion between irradiated and non-irradiated products.

Alternative process control methods (e.g., those based on dose measurements) may also be used.

3.5. Protection in the event of mechanical failure

Suitable devices should be provided to ensure that a failure of the mechanical systems does not result in an incorrect dose.

3.6. Supervision of radiation facility

To ensure that excecution of the radiation sterilization procedure meets the foregoing requirements one person should be designated by the management and approved by the appropriate authority. The designated person should be of a senior status and well qualified by training and experience for this responsibility. He should have direct access both to the top management of his own organization and to the representative of the appropriate authority. In addition, he should be supported by a competent physicist (or himself be a physicist) with adequate laboratory facilities and staff for carrying out dosimetry and all process controls.

CHAIRMEN OF SESSIONS

Session I	B.B. GAITONDE	Haffkine Institute, Bombay, India
Session II	K.H. CHADWICK	EURATOM-ITAL, Netherlands
Session III	K.H. MORGANSTERN	Radiation Dynamics, United States of America
Session IV	V.G. KHRUSHCHEV	Institute of Biophysics of the USSR, USSR
Session V	V.K. IYA	ISOMED, Trombay, India
Session VI	G.O. PHILLIPS	University of Salford, United Kingdom
Session VII	Pamela A. WILLS	AAEC, Australia
Session VIII	J.O. DAWSON	Ethicon Ltd., Scotland, United Kingdom
Session IX	J. FLEURETTE	Laboratoire de bactériologie, Hôpital Cardiologique, Lyon, France

SECRETARIAT

Scientific Secretary	R.N. MUKHERJEE	Division of Life Sciences, IAEA
Administrative Secretary	Gertrude SEILER	Division of External Relations, IAEA
Editor	Monica KRIPPNER	Division of Publications, IAEA
Records Officer	J. RICHARDSON	Division of Languages, IAEA

LIST OF PARTICIPANTS

ARGENTINA

Mariano, E.E. — Comisión Nacional de Energía Atómica Argentina, Avenida Libertador 8250, Buenos Aires

AUSTRALIA

Wills, Pamela A. — Australian Atomic Energy Commission Research Establishment, Private Mail Bag, Sutherland 2232, N.S.W.

AUSTRIA

Flamm, H. — Hygiene Institut der Universität Wien, Kinderspitalgasse 15, 1095 Vienna

BANGLADESH

Molla, M.A.R. — Health Physics Division, Atomic Energy Centre, P.O. Box No. 164, Dacca

CANADA

Chu, R.D.H. — Atomic Energy of Canada Ltd., Commercial Products, P.O. Box 6300, Postal Station J, Ottawa K2A 3W3

CZECHOSLOVAKIA

Horáková, Vlasta — National Textile Research Institute, Centre for Research and Application of Ionizing Radiation, 664 71 Weverska Bityska, Brno

Klen, R. — Faculty Hospital, Regional Institute of Public Health, Hradec Králové

EGYPT

Mohamed, A.M. — Egyptian Embassy, New Delhi, India

Roushdy, H.M. — National Centre of Radiation Technology, Atomic Energy Establishment, 101 Kasr El Eini Street, 10th Floor, Cairo

FINLAND

Halonen, P.E.	Department of Virology, University of Turku, Kilinamyllynkatu 10, SF-20520 Turku 52

FRANCE

Eymery, R.	CEA, Centre d'études nucléaires de Grenoble, Centre de Tri, B.P. n°85, 38041 Grenoble-Cédex
Favre-Mercuret, R.V.L.	Commissariat à l'énergie atomique, Département des radioéléments, B.P. n°21, 91190 Gif-sur-Yvette
Fleurette, J.	Laboratoire de bactériologie, Hôpital cardiologique, Lyon Montchat, 69394 Lyon-Cédex 3
Laizier, J.	CEA, Centre d'études nucléaires de Saclay, B.P. n°2, 91190 Gif-sur-Yvette
Puig, J.R.	CEA, Centre d'études nucléaires de Saclay, B.P. n°2, 91190 Gif-sur-Yvette
Transy, Marie-José	Laboratoire de bactériologie, Hôpital cardiologique, Lyon-Montchat, 69394 Lyon-Cédex 3

GERMANY, FEDERAL REPUBLIC OF

Kistner, G.N.	Division Radiation Hygiene, Federal Health Office, Corrensplatz 1, 1 Berlin 33
Wallhäusser, K.H.	Hoechst AG, Pharma-Qualitäts-Kontrolle, 6230 Frankfurt 80

HOLY SEE

Fernandes, F.	Ciba-Geigy Research Centre, Goregaon East, Bombay 400 063, India

HUNGARY

Fehér, I.	Research Institute for Radiobiology and Radiohygiene F.J. Curie, B.P. 101, 1775 Budafok
Kerecsen, J.	MEDICOR Factory of Medical Supplies, P.O. Box 125, 4001 Debrecen
Stenger, V.	Institute of Isotopes of the Hungarian Academy of Sciences, Konkoly Thege u., Budapest XII

INDIA

Aggarwal, K.S.	Isotope Division, Bhabha Atomic Research Centre, Trombay, Bombay 400 085

Baichwal, R.S.	M/s. Raptakos Brett & Co. Ltd., 47 Dr. Annie Besant Road, Worli, Bombay 400 025
Balasubramanyan, R.	Central Drug Standard Control Organization, 23, Abubacker Mansion, Colaba Causeway, Bombay 400 005
Banerjee, L.K.	Office of the Director General, Armed Forces Medical Services, Ministry of Defence, New Delhi
Bhagath, R.K.	J.L. Morison, Son & Jones (India) Ltd., 9th Mile, Tumkur Road, Dasarhalli P.O., Bangalore 562 139
Bhatnagar, J.S.	Indian Schering Ltd., Sion-Trombay Road, Chembur, Bombay 400 071
Desai, M.W.	Bharat Laboratories, c/o Bharat Serum & Vaccines, A 371/373, 27th Road, Wagle Industrial Estate, Post Box No. 14, Thana 400 604 (Maharashtra)
Deshpande, R.G.	ISOMED, Bhabha Atomic Research Centre, Trombay, Bombay 400 085
Gaitonde, B.B.	Haffkine Institute, Parel, Bombay 400 012
Ghody, A.N.	India Surgical Implements, 101 Mittal Estate, Andheri-Kurla Road, Bombay 400 059
Gopal, N.G.S.	Isotope Division, Bhabha Atomic Research Centre, Trombay, Bombay 400 085
Gupta, B.L.	Division of Radiological Protection, Bhabha Atomic Research Centre, Trombay, Bombay 400 085
Iya, V.K.	Isotope Division, ISOMED, Bhabha Atomic Research Centre, Trombay, Bombay 400 085
Iyer, P.V.	McGaw Ravindra Laboratories, Amraiwadi Road, Ahmedabad 4
Iyer, S.N.	Organization of Pharmaceutical Producers of India, c/o Johnson & Johnson Ltd., Lal Bahadur Shastri Marg, Mulund, Bombay 400 080
Jussawala, D.J.	Tata Memorial Centre, Parel, Bombay 400 012
Kabra, S.P.	Burroughs Wellcome & Co. (India) Pvt. Ltd., 88-C, Old Prabhadevi Road, Bombay 400 025
Kamat, K.L.	Quality Control Dept., Glaxo Laboratories (India) Ltd., Dr. Annie Besant Road, Worli, Bombay 400 025
Kankonkar, S.R.	Department of Antitoxin and Sera, Haffkine Institute, Parel, Bombay 400 012
Kapadia, A.K.	Research and Development Dept., Indian Schering Ltd., Sion-Trombay Road, Deonar, Bombay 400 088

Kehr, K. D.	Kehr Surgical & Allied Products Pvt. Ltd., C-34, Panki Industrial Estate, P.O. Box No. 65, Kanpur 208 020 (U.P.)
Kodkani, S.G.	Johnson & Johnson Ltd., Lal Bhadur Shastri Marg, Mulund, Bombay 400 080
Krishnamurthy, K.	Isotope Division, Bhabha Atomic Research Centre, Trombay, Bombay 400 085
Kulkarni, R. D.	Department of Pharmacology, Grant Medical College, Bombay 400 008
Kumta, U. S.	Biochemistry and Food Technology Division, Bhabha Atomic Research Centre, Trombay, Bombay 400 085
Nagaratnam, A.	Institute of Nuclear Medicine and Allied Sciences, Probyn Road, Delhi 110 007
Narayan, V.M.	The Indian Drug Manufacturers Association, c/o Indo-German Alkaloids, Mahakali Road, Andheri (East), Bombay 400 093
Parekh, R. H.	Femme Pharma Ltd., 5, Krishna Udyog Bhavan, Off Nariman Road, Worli, Bombay 400 025
Patani, A.	The Indian Drug Manufacturers Association, c/o Indo-German Alkaloids, Mahakali Road, Andheri (East), Bombay 400 093
Patel, K.M. (Mrs.)	ISOMED, Bhabha Atomic Research Centre, Trombay, Bombay 400 085
Patel, V.K.	Elys Chemical Laboratories, Trans Thana Creek, Industrial Estate Area, Thana-Belapur Road, Thana 9 (Maharashtra)
Ramasarma, G.B.	The Indian Pharmaceutical Association, Kalina, Santacruz (East), Bombay 400 029
Rangnekar, M.K.	Food and Drug Administration, Grimanirman Bhavan, Bandra (East), Maharashtra, Bombay 400 051
Rao, M.V.	ISOMED, Bhabha Atomic Research Centre, Trombay, Bombay 400 085
Rathi, S.R.	Compax Engineering Ltd., 430 a/32, Senapati Bapat Road, Poona 411 016 (Maharashtra)
Rosha, R.	Hoechst Pharmaceuticals Ltd., P.O. Box No. 7755, Lal Bahadur Shastri Marg, Mulund, Bombay 400 080
Shah, V.R.	Medical Division, Bhabha Atomic Research Centre, Trombay, Bombay 400 085
Sundaram, K.	Bio-Medical Group, Bhabha Atomic Research Centre, Trombay, Bombay 400 085

Yellore, S.	Quality Control Laboratory, Hindustan Antibiotics Ltd., Pimpri, Poona 411 018

INDONESIA

Hilmy, Nazly	Pasar Jum'at Research Centre, P.O. Box 2, Kebayoran Lama, Jakarta Selatan

IRAN

Rouhanizadeh, N.	University Nuclear Center, P.O. Box 2989, Tehran

IRAQ

El-Hajji, Laika Ali-Mohammed	The State Company for Drug Industries, P.O. Box 271, Samarra - Baghdad
Ismail, B.A.K.	The State Company for Drug Industries, P.O. Box 271, Samarra - Baghdad

ISRAEL

Aronson, M.	Tel-Aviv University, Sackler School of Medicine, Ramat Aviv

KOREA

Rhee, K.S.	Radiation Biology Division, Korea Atomic Energy Research Institute, P.O. Box 7, Cheong Ryang-Ri, Seoul

NETHERLANDS

Armbrust, R.F.	Gammaster, Morselaan 3, Ede
Chadwick, K.H.	Association EURATOM-ITAL, P.O. Box 48, Keyenbergseweg 6, Wageningen (see also CEC)

PHILIPPINES

Singson, Carmen	Philippine Atomic Energy Commission, Diliman, Quezon City 3004

SPAIN

Arago Mitjans, J.M.	Laboratorio Arago, Salvador Mundi 11, Barcelona 17
Eguiluz Gordillo, J.M.	Nuclear Iberica S.A., Pintor Rosales, 36, Madrid 8

SRI LANKA

Amarasiri, W.A.
Atomic Energy Authority, National Science Bldg., 47/5 Maitland Place, Colombo 7

SUDAN

Abdel Razig, M.T.
Veterinary Research Administration, P.O. Box 293, Khartoum

UNION OF SOVIET SOCIALIST REPUBLICS

Balitskij, V.A.
Institute of Biophysics of the USSR, Ministry of Health, Moscow

Khrushchev, V.G.
Institute of Biophysics of the USSR, Ministry of Health, Moscow

Pavlov, E.P.
Institute of Biophysics of the USSR, Ministry of Health, Moscow

Semenenko, Eh.I.
All-Union Scientific Institute for Medical Polymers, Moscow

Terent'ev, B.M.
All-Union Research Institute for Radiation Techniques, Moscow

UNITED KINGDOM

Blackburn, R.
Department of Chemistry and Applied Chemistry, University of Salford, Salford M5 4WT, Lancs

Dawson, J.O.
Ethicon Limited, P.O. Box 408, Bankhead Ave., Edinburgh EH11 4HE, Scotland

Edwards, R.
H.S. Marsh, Nuclear Energy Ltd., 125, Southampton Street, Reading RG1 2RA, Berks

Keall, A.
Department of Health and Social Security, 14, Russell Square, London WC 1

Philipps, G.O.
Department of Chemistry and Applied Chemistry, University of Salford, Salford M5 4WT, Lancs

Tallentire, A.
Department of Pharmacy, University of Manchester, Manchester M13 9PL

Vaughan, R.A.
The Radiochemical Centre, Amersham, Bucks

UNITED STATES OF AMERICA

Gaughran, E.R.L.	Central Research Laboratories, Domestic Operating Company, Johnson & Johnson, 501 George Street, New Brunswick, N.J. 08903
Morganstern, K.H.	Radiation Dynamics, Inc., 1800 Shames Drive, Westbury (L.I.), N.Y. 11576
Razi, Mahpara S.	Harbor General Hospital, UCLA, Torrance, Cal.
Trauth, C.A.	Sandia Laboratories, P.O. Box 5800, Albuquerque, N. Mex. 87110

ORGANIZATIONS

COMMISSION OF THE EUROPEAN COMMUNITIES (CEC)

Chadwick, K.H.	(see also under Netherlands)

INTERNATIONAL ATOMIC ENERGY AGENCY (IAEA)

Slee, R. (IAEA Expert)	IAEA, P.O. Box 590, A-1011 Vienna, Austria
Yuan, H.C.	Division of Technical Assistance, IAEA, P.O. Box 590, A-1011, Vienna, Austria

AUTHOR INDEX

(including participants in discussions)

The numbers underlined indicate the first page of a paper by an author.
Other numbers denote the page number of a participant's first intervention in a discussion.
Literature references are not indexed.

TRANSLITERATION INDEX

Балицкий, В.А.	Balitskij, V.A.
Бочкарев, В.В.	Bochkarev, V.V.
Ванюшкин, Б.М.	Vanyushkin, B.M.
Глухих, В.А.	Glukhikh, V.A.
Горбунов, Ю.С.	Gorbunov, Yu.S.
Гринев, М.П.	Grinev, M.P.
Иванов, А.И.	Ivanov, A.I.
Ковальчук, В.М.	Koval'chuk, V.M.
Коньков, Н.Г.	Kon'kov, N.G.
Коняев, Г.А.	Konyaev, G.A.
Корженевский, Э.С.	Korzhenevskij, Eh.S.
Кривоносов, А.И.	Krivonosov, A.I.
Крылов, С.Ю.	Krylov, S.Yu.
Кудрявцев, А.А.	Kudryavtsev, A.A.
Левин, В.М.	Levin, V.M.
Мамиконян, С.В.	Mamikonyan, S.V.
Маркелов, М.А.	Markelov, M.A.
Михайлов, Л.М.	Mikhajlov, L.M.
Мунтян, В.И.	Muntyan, V.I.
Мышковский, В.И.	Myshkovskij, V.I.
Осипов, В.Б.	Osipov, V.B.
Павлов, Е.П.	Pavlov, E.P.
Пономаренко, С.И.	Ponomarenko, S.I.
Полунин, В.А.	Polunin, V.A.
Постригань, М.В.	Postrigan', M.V.
Потапова, З.М.	Potapova, Z.M.
Пикаев, А.К.	Pikaev, A.K.
Сарапкин, И.И.	Sarapkin, I.I.
Сафонов, Ю.И.	Safonov, Yu.I.
Седов, В.В.	Sedov, V.V.
Семененко, Э.И.	Semenenko, Eh.I.
Степанов, Г.Д.	Stepanov, G.D.
Терентьев, Б.М.	Terent'ev, B.M.
Тушов, Э.Г.	Tushov, Eh.G.
Харламов, В.Т.	Kharlamov, V.T.
Хрущев, В.Г.	Khrushchev, V.G.
Шубнякова, Л.П.	Shubnyakova, L.P.

INDEX OF PREPRINT SYMBOLS

The following conversion table is provided for the convenience of readers and to encourage the use of SI units.

FACTORS FOR CONVERTING UNITS TO SI SYSTEM EQUIVALENTS *

SI base units are the metre (m), kilogram (kg), second (s), ampere (A), kelvin (K), candela (cd) and mole (mol).
[For further information, see International Standards ISO 1000 (1973), and ISO 31/0 (1974) and its several parts]

Multiply			*by*	*to obtain*
Mass				
pound mass (avoirdupois)	1 lbm	=	4.536×10^{-1}	kg
ounce mass (avoirdupois)	1 ozm	=	2.835×10^{1}	g
ton (long) (= 2240 lbm)	1 ton	=	1.016×10^{3}	kg
ton (short) (= 2000 lbm)	1 short ton	=	9.072×10^{2}	kg
tonne (= metric ton)	1 t	=	1.00×10^{3}	kg
Length				
statute mile	1 mile	=	1.609×10^{0}	km
yard	1 yd	=	9.144×10^{-1}	m
foot	1 ft	=	3.048×10^{-1}	m
inch	1 in	=	2.54×10^{-2}	m
mil (= 10^{-3} in)	1 mil	=	2.54×10^{-2}	mm
Area				
hectare	1 ha	=	1.00×10^{4}	m^2
(statute mile)^2	1 mile^2	=	2.590×10^{0}	km^2
acre	1 acre	=	4.047×10^{3}	m^2
yard^2	1 yd^2	=	8.361×10^{-1}	m^2
foot^2	1 ft^2	=	9.290×10^{-2}	m^2
inch^2	1 in^2	=	6.452×10^{2}	mm^2
Volume				
yard^3	1 yd^3	=	7.646×10^{-1}	m^3
foot^3	1 ft^3	=	2.832×10^{-2}	m^3
inch^3	1 in^3	=	1.639×10^{4}	mm^3
gallon (Brit. or Imp.)	1 gal (Brit)	=	4.546×10^{-3}	m^3
gallon (US liquid)	1 gal (US)	=	3.785×10^{-3}	m^3
litre	1 l	=	1.00×10^{-3}	m^3
Force				
dyne	1 dyn	=	1.00×10^{-5}	N
kilogram force	1 kgf	=	9.807×10^{0}	N
poundal	1 pdl	=	1.383×10^{-1}	N
pound force (avoirdupois)	1 lbf	=	4.448×10^{0}	N
ounce force (avoirdupois)	1 ozf	=	2.780×10^{-1}	N
Power				
British thermal unit/second	1 Btu/s	=	1.054×10^{3}	W
calorie/second	1 cal/s	=	4.184×10^{0}	W
foot-pound force/second	1 ft·lbf/s	=	1.356×10^{0}	W
horsepower (electric)	1 hp	=	7.46×10^{2}	W
horsepower (metric) (= ps)	1 ps	=	7.355×10^{2}	W
horsepower (550 ft·lbf/s)	1 hp	=	7.457×10^{2}	W

* Factors are given exactly or to a maximum of 4 significant figures

Multiply		*by*	*to obtain*
Density			
pound mass/inch^3	1 lbm/in^3	= 2.768×10^4	kg/m^3
pound mass/foot^3	1 lbm/ft^3	= 1.602×10^1	kg/m^3
Energy			
British thermal unit	1 Btu	= 1.054×10^3	J
calorie	1 cal	= 4.184×10^0	J
electron-volt	1 eV	≃ 1.602×10^{-19}	J
erg	1 erg	= 1.00×10^{-7}	J
foot-pound force	1 ft·lbf	= 1.356×10^0	J
kilowatt-hour	1 kW·h	= 3.60×10^6	J
Pressure			
newtons/metre^2	1 N/m^2	= 1.00	Pa
atmosphere[a]	1 atm	= 1.013×10^5	Pa
bar	1 bar	= 1.00×10^5	Pa
centimetres of mercury (0°C)	1 cmHg	= 1.333×10^3	Pa
dyne/centimetre^2	1 dyn/cm^2	= 1.00×10^{-1}	Pa
feet of water (4°C)	1 ftH_2O	= 2.989×10^3	Pa
inches of mercury (0°C)	1 inHg	= 3.386×10^3	Pa
inches of water (4°C)	1 inH_2O	= 2.491×10^2	Pa
kilogram force/centimetre^2	1 kgf/cm^2	= 9.807×10^4	Pa
pound force/foot^2	1 lbf/ft^2	= 4.788×10^1	Pa
pound force/inch^2 (= psi)[b]	1 lbf/in^2	= 6.895×10^3	Pa
torr (0°C) (= mmHg)	1 torr	= 1.333×10^2	Pa
Velocity, acceleration			
inch/second	1 in/s	= 2.54×10^1	mm/s
foot/second (= fps)	1 ft/s	= 3.048×10^{-1}	m/s
foot/minute	1 ft/min	= 5.08×10^{-3}	m/s
mile/hour (= mph)	1 mile/h	= 4.470×10^{-1}	m/s
		1.609×10^0	km/h
knot	1 knot	= 1.852×10^0	km/h
free fall, standard (= g)		= 9.807×10^0	m/s^2
foot/second^2	1 ft/s^2	= 3.048×10^{-1}	m/s^2
Temperature, thermal conductivity, energy/area·time			
Fahrenheit, degrees −32	°F − 32	$\frac{5}{9}$	°C
Rankine	°R	$\frac{5}{9}$	K
1 Btu·in/ft^2·s·°F		= 5.189×10^2	W/m·K
1 Btu/ft·s·°F		= 6.226×10^1	W/m·K
1 cal/cm·s·°C		= 4.184×10^2	W/m·K
1 Btu/ft^2·s		= 1.135×10^4	W/m^2
1 cal/cm^2·min		= 6.973×10^2	W/m^2
Miscellaneous			
foot^3/second	1 ft^3/s	= 2.832×10^{-2}	m^3/s
foot^3/minute	1 ft^3/min	= 4.719×10^{-4}	m^3/s
rad	rad	= 1.00×10^{-2}	J/kg
roentgen	R	= 2.580×10^{-4}	C/kg
curie	Ci	= 3.70×10^{10}	disintegration/s

[a] atm abs: atmospheres absolute;
atm (g): atmospheres gauge.

[b] lbf/in^2 (g) (= psig): gauge pressure;
lbf/in^2 abs (= psia): absolute pressure.

HOW TO ORDER IAEA PUBLICATIONS

Exclusive sales agents for IAEA publications, to whom all orders and inquiries should be addressed, have been appointed in the following countries:

UNITED KINGDOM	Her Majesty's Stationery Office, P.O. Box 569, London SE 1 9NH
UNITED STATES OF AMERICA	UNIPUB, Inc., P.O. Box 433, Murray Hill Station, New York, N.Y. 10016

In the following countries IAEA publications may be purchased from the sales agents or booksellers listed or through your major local booksellers. Payment can be made in local currency or with UNESCO coupons.

ARGENTINA	Comisión Nacional de Energía Atómica, Avenida del Libertador 8250, Buenos Aires
AUSTRALIA	Hunter Publications, 58 A Gipps Street, Collingwood, Victoria 3066
BELGIUM	Service du Courrier de l'UNESCO, 112, Rue du Trône, B-1050 Brussels
CANADA	Information Canada, 171 Slater Street, Ottawa, Ont. K 1 A OS 9
C.S.S.R.	S.N.T.L., Spálená 51, CS-11000 Prague Alfa, Publishers, Hurbanovo námestie 6, CS-80000 Bratislava
FRANCE	Office International de Documentation et Librairie, 48, rue Gay-Lussac, F-75005 Paris
HUNGARY	Kultura, Hungarian Trading Company for Books and Newspapers, P.O. Box 149, H-1011 Budapest 62
INDIA	Oxford Book and Stationery Comp., 17, Park Street, Calcutta 16
ISRAEL	Heiliger and Co., 3, Nathan Strauss Str., Jerusalem
ITALY	Libreria Scientifica, Dott. de Biasio Lucio "aeiou", Via Meravigli 16, I-20123 Milan
JAPAN	Maruzen Company, Ltd., P.O.Box 5050, 100-31 Tokyo International
NETHERLANDS	Marinus Nijhoff N.V., Lange Voorhout 9-11, P.O. Box 269, The Hague
PAKISTAN	Mirza Book Agency, 65, The Mall, P.O.Box 729, Lahore-3
POLAND	Ars Polona, Centrala Handlu Zagranicznego, Krakowskie Przedmiescie 7, Warsaw
ROMANIA	Cartimex, 3-5 13 Decembrie Street, P.O.Box 134-135, Bucarest
SOUTH AFRICA	Van Schaik's Bookstore, P.O.Box 724, Pretoria Universitas Books (Pty) Ltd., P.O.Box 1557, Pretoria
SPAIN	Nautrónica, S.A., Pérez Ayuso 16, Madrid-2
SWEDEN	C.E. Fritzes Kungl. Hovbokhandel, Fredsgatan 2, S-10307 Stockholm
U.S.S.R.	Mezhdunarodnaya Kniga, Smolenskaya-Sennaya 32-34, Moscow G-200
YUGOSLAVIA	Jugoslovenska Knjiga, Terazije 27, YU-11000 Belgrade

---

Orders from countries where sales agents have not yet been appointed and requests for information should be addressed directly to:

Publishing Section,
International Atomic Energy Agency,
Kärntner Ring 11, P.O.Box 590, A-1011 Vienna, Austria